HANDBOOK OF STRATA-BOUND AND STRATIFORM ORE DEPOSITS

Volume 8
GENERAL STUDIES

HANDBOOK OF STRATA-BOUND AND STRATIFORM ORE DEPOSITS

Edited by
K.H. WOLF

PART I
1. Classifications and Historical Studies

2. Geochemical Studies

3. Supergene and Surficial Ore Deposits; Textures and Fabrics

4. Tectonics and Metamorphism
   Indexes Volumes 1—4

PART II
5. Regional Studies

6. Cu, Zn, Pb, and Ag Deposits

7. Au, U, Fe, Mn, Hg, Sb, W, and P Deposits
   Indexes Volumes 5—7

PART III
8. General Studies

9. Regional Studies and Specific Deposits

10. Bibliography and Ore Occurrence Data
    Indexes Volumes 8—10

ELSEVIER SCIENTIFIC PUBLISHING COMPANY
Amsterdam — Oxford — New York   1981

HANDBOOK OF STRATA-BOUND AND STRATIFORM ORE DEPOSITS

PART III

Edited by

K.H. WOLF

Volume 8

GENERAL STUDIES

ELSEVIER SCIENTIFIC PUBLISHING COMPANY
Amsterdam — Oxford — New York   1981

ELSEVIER SCIENTIFIC PUBLISHING COMPANY
Molenwerf 1, 1014 AG Amsterdam
P.O. Box 211, 1000 AE Amsterdam, The Netherlands

Distributors for the United States and Canada:

ELSEVIER/NORTH-HOLLAND INC.
52, Vanderbilt Avenue
New York, N.Y. 10017

Library of Congress Cataloging in Publication Data   (Revised)
Main entry under title:

Handbook of strata-bound and stratiform ore deposits.

    Errata slip inserted in v. 5.
    Includes bibliographies and indexes.
    CONTENTS: 1. Principles and general studies v. 1.
Classifications and historical studies. v. 2. Geochemical
studies v. 3. Supergene and surficial ore deposits:
textures and fabrics.--[etc.]--Pt. 3. v. 8. General
studies.
    1. Ore-deposits. 2. Geology. I. Wolf, Karl H.
QE390.H36   553   77-461887
ISBN 0-444-41401-0 (v. 1)

ISBN: 0-444-41823-7 (Vol. 8)

Printed in The Netherlands.

LIST OF CONTRIBUTORS TO THIS VOLUME

D.J. GROVES
Department of Geology, University of Western Australia, Nedlands, W.A., Australia

I.B. LAMBERT
Baas Becking Geobiological Laboratory, Canberra City, A.C.T., Australia

P. LAZNICKA
Department of Earth Sciences, The University of Manitoba, Winnipeg, Canada

K.H. WOLF *
Watts, Griffis and McOuat Ltd., Consultants; and Directorate General of Mineral Resources, Jeddah, Saudi Arabia

* Formerly: Laurentian University; Sudbury, Ontario, Canada

CONTENTS

VIII

Chapter 3. THE CONCEPT OF ORE TYPES – SUMMARY, SUGGESTIONS AND A PRAC-
TICAL TEST
by P. Laznicka

Chapter 4. PLATFORM VERSUS MOBILE BELT-TYPE STRATABOUND/STRATIFORM
ORE DEPOSITS: SUMMARY, COMPARISONS, MODELS
by P. Laznicka

X

PREFACE

"All science is a makeshift, a means to an end which is never attained".
Henry David Thoreau, in: *Autumn,* October 13, 1860.

"Science is rarely a matter of black-and-white reasoning or conclusive experimentation. Scientific truth (i.e., what most scientists believe to be true) emerges by the gradual accretion of new information and the open assessment of rival theories."
B. Dixon, in: *What is Science For?,* 1973, p. 241, Pelican/Penguin.

"Not so much to impart a detailed knowledge of a wide range of facts as to create a scientific spirit; a spirit of toleration and of co-operation, of intellectual adventure and intellectual honesty, which seeks ever to enlarge our knowledge of the external world and to found that knowledge, not on tradition or authority, but on a basis of ascertained fact."
A. Finlay.

"The factual burden of a science varies inversely with its degree of maturity. As a science advances, particular facts are comprehended within, and therefore in a sense annihilated by, general statements of steadily increasing explanatory power and compass — whereupon the facts no longer need to be known explicitly, i.e., spelled out and kept in mind."
P.B. Medawar, in: *The Art of the Soluble,* 1969, p. 128, Pelican.

"Science depends upon tolerance and a broad-minded readiness to listen to unorthodox ideas."
B. Dixon, in: *What is Science For?,* 1973, p. 59, Pelican-Penguin.

"Healthy tension between specialization and broad understanding is a necessary element in all education."
McMaster 1978-79. A short report to the community on the state of their university, McMaster University, Hamilton, Ontario, Canada.

"Paradoxically, the extent of our present ignorance is a telling reflection of progress. For a high level of prior understanding is, after all, essential to the formulation of the kinds of questions geologists are now asking and attempting to answer."
Preston Cloud, in: *The Veils of Gaia,* 1977, p. 390.

"Science belongs to no one country."
Louis Pasteur.

"Science and art belong to the whole world, and before them vanish the barriers of nationality."
J.W. Goethe, in: *A Conversation with a German Historian,* 1813.

"Science will fulfill what is demanded of it only on condition that it retains its freedom and independence in its work. In spite of all necessary limitations, liberty remains an essential characteristic of science, and spiritual and intellectual freedom is a necessity for scientists. Every true scientist desires to serve his nation. His work, however, cannot be complete if it is merely a service to his nation."
Ferdinand Sauerbruch, in: *Speech at Dresden,* Sept. 22, 1936.

"The main difference of modern scientific research from that of the Middle Ages, the secret of its immense success, lies in its cohesive character, in the fact that every useful experiment is published, every new discovery of relationships explained."
H.G. Wells, in: *New Worlds for Old.*

"Science, which is the organization of knowledge, must itself be organized. And this organization must be international; . . . ."

Francis Bacon, in: Will Durant, *The Story of Philosophy*, ch. III.

The presently issued Handbook (i.e., Volumes 8, 9, and 10) continues in the same philosophical spirit as the seven-volume compendium of 1976, in that specific topics related to the origin of strata-bound/stratiform mineralizations are critically, synthetically summarized. The individual chapters are evidently of a broader coverage than the normal journal articles, but shorter than average books.

During the preparation of the first seven volumes it was not possible to include reviews of numerous subject matters for a variety of reasons, e.g., the invited authors were unable to accept a writing assignment of that magnitude; certain schools-of-thought based on language and/or country had to be limited for reasons of space (however, in at least two instances no suitable authors could be found); and in rare cases the writers accepted the preparation of a review but had to withdraw at a critical time when no replacement was possible. On the other hand, spontaneous suggestions in response to the earlier volumes have resulted in the planning of new contributions for the present three volumes or for future issues. It is hoped that the readers will accept the editor's (K.H.W.) and the publisher's open invitation to submit suggestions, and possibly offer to write summaries on pertinent topics (with the understanding that this Handbook sequence is not a vehicle to have the same data and concepts simultaneously published in duplication elsewhere).

There is little to be gained by discussing in this preface the individual chapters — except for two.

(1) As exemplified by the contribution "The Nature and Origin of Archaean Strata-Bound Volcanic-Associated Nickel—Iron—Copper Sulphide Deposits" (Groves and Hudson, Vol. 9), in addition to low-temperature and low-pressure type of mineralizations, the Handbook series will also include as of this volume those strata-bound/stratiform deposits that originated under "higher" temperatures and pressures.

(2) On first and superficial consideration, one might feel that the two chapters by Förstner on "Recent Heavy-Metal Accumulation in Limnic Sediments" and "Trace Metals in Fresh Waters (with Particular Reference to Mine Effluents)" do not quite qualify to be included under the "leitmotiv" of the Handbook. However, several reasons can be provided in defense of these chapters' relevancy when they are placed in context with the other contributions. There is little difficulty in justifying incorporation of the chapter on metals in limnic sediments because, for example, (a) it supplements and expands the scope of the earlier review on "Freshwater Ferromanganese Deposits" by Callender and Bowser (Chapter 8, Vol. 7), (b) in order to understand the differences and similarities

between marine and freshwater mineral concentrations (both ancient and Recent), a geologist should have at his disposal information of both these types of deposits, and (c) in geochemical exploration the various natural and human contributions to the chemical budget of freshwater milieus have to be considered to recognize anomalies.

One variable that is particularly influential in Recent freshwater environments is pollution — a parameter which need not (as yet) be considered in the study of deep-sea (in contrast to shallow-water) mineral accumulations. In order to make the first chapter by Förstner more complete and, therefore, more meaningful and enhance its usefullness even further, the chapter with particular reference to mine effluents has been added. In order to extrapolate the geochemical data obtained from Recent lakes (for instance to explain the origin of ancient concentrations), one has to deduct the influence of pollution on the former — a formidable task to be resolved!

Researchers and practitioners (explorationists) should be familiar with some of the fundamentals of the "field of documentation" as it will assist them to comprehend the multi-type of publications available, some or all of which may be used within one organization. One of the most conspicuous examples of a failure in properly using "the documentation/communication method" was provided by the surprise-launching of the Russian Sputnik satellite in 1957. According to Ten Haken (1978), it seems that it had been described in several Russian journals, obtainable all over the world — but the information was *not noticed*. As to the earth-sciences, how many "launchings" and "completions" of the projects in one country have been missed by geologists in other sectors of the world — even among English-language nations? It is here that the "field of documentation" helps as "the method by which the scientific information which appears in various publications can be made easy to locate"; this information is the "vehicle by which scientists may pass on knowledge in their work or research and which has become great in both quantity and diversity, so that it may be difficult to know where to locate all this diverse knowledge in movement" (Ten Haken, 1978). Some of the basics of the "transfer of scientific information" are provided here, partly based on Ten Haken's summary. In particular since the present handbook series is part of the "data cycle", the discussions will illuminate the position of such topical reviews within the total information spectrum.

Under the section "the cycle of information", Ten Haken considered the four-phase cycle developed by Loosjes (1954, 1973) (depicted in Fig. 1): "Starting with the scientist's research (RE), the first result of his work, the manuscript (MS), is seen on the circle. The next step of processing is multiplication (MU) which results in a publication (PU). The publication is catalogued in the library (LI) and then called library material processed (LMP)". In addition to the large number of books stored and catalogued in and retrievable from libraries, a greater number of articles in periodicals, papers in proceedings, etc., are catalogued as volumes in the library, with the individual publications *not* being retrievable. By also cataloguing the separate contributions in the "documentation depart-

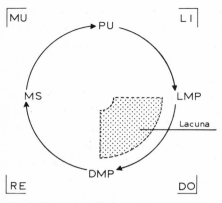

Two main parts are distinguished:
1. The *four phases of processing,*
   indicated outside the circle:
   RE = Research
   MU = Multiplication
   LI = Library
   DO = Documentation
2. The *four phases of the document,*
   indicated on the circle:
   MS = Manuscript
   PU = Publication
   LMP = Library material processed
   DMP = Documented material processed

Fig. 1. The cycle of information.

ment" (DO), either as part of the library (done from the point-of-view of collection) or separate from it (in the documentation department from the point-of-view of subject), "we get documented material processed (DMP). This material is available to the scientific researcher as a basis for further research, possibly resulting in a new publication, of which the first step will be the manuscript. And the cycle will be repeated." "On the circle, between LMP and DMP, we see the significant indication *lacuna.* Here lies the great obstacle for the control of the literature. For the presentation of different kinds of publications (books, reports, articles in periodicals, papers contributed to congresses) is so overwhelming that it is hardly possible to cover all these publications in a documentation system. And it is not without reason that the meaningful terms 'literature explosion' and 'literature avalanche' are used, whereas the desperate attempt to make this mass of literature accessible is indicated as 'operation deluge'." I go as far as saying that should an educational or industrial institution have accumulated within a scientific discipline so much data or publications that its available staff cannot properly "digest" it, an "information and literature constipation" is the consequence. Since a "moratorium on the collecting of data" is an impractical, unacceptable proposal, the only solutions rest on one or several expensive pills: first of all an increase in *competent* man power is needed of which a certain proportion is to be *assigned to constructive–creative reviews.*

As an extension of the above, Ten Haken discussed the "bibliographic control", which is "the term generally used to describe the whole process of documentation and retrieval as applied to literature; when used on a worldwide scale this is called 'universal bibliographic control' .... Making use of the cycle of information let us see which aids are at our disposal in various stages of the information process, to achieve bibliographic control (Fig. 2)." "There is an essential difference between the activities of library and those of the documentation service, perhaps not always realised at first by the user of catalogued information. A feature of the library is that it builds up and controls its own collection, whereas a documentation service tries to deal with all relevant literature

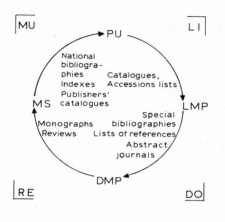

<table>
</table>

In the *four sectors of the information cycle* we distinguish as producers of bibliographic aids:

1. *Multiplication:* National bibliographies
   (= publishers) Cumulative indexes on periodicals
         Publisher's catalogues

2. *Libraries:* Catalogues
         Accessions lists

3. *Documentation:* Special bibliographies
         Abstract journals
         Lists of references

4. *Research:* Monographs
         Reviews
         (Both of them normally well provided with references)

Fig. 2. The bibliographic aids produced in the various stages of the information process.

wherever it may be. A basic rule to be observed is that a *catalogue* refers to a certain collection, and a *bibliography* to literature in general, often covering a certain field of science."

"Summarising and comparing the nature of the bibliographic aids in the four sectors of the information cycle we see the following:

MU: Registration of the literature produced

LI: Registration of a collection

DO: Registration of the literature in a certain field

RE: Review of the literature of a specific part of a certain field."

From the above considerations, it is rather obvious to deduce where the present Handbook series on strata-bound/stratiform ore deposits fits into the cycle of information, and what the "bibliographic control" effects are of such reviews or monographs.

That the synthesis of knowledge in both the arts and sciences has played an important role in past centuries has been pointed out by Bowder (1979) in an article entitled "Synthesis = Progress Early Arab Style". To draw from Bowder's analysis: "Since the Quran specifically encouraged learning, wherever the Arabs came into contact with the old, established cultures they made an effort to study them and in many cases adopted what they learnt to conform with the principles of their religion. . . . The main method used to acquire a knowledge of . . . was through the translation of texts and treatises. This translation movement reached its peak in the year 830 . . ." when a translation school and research academy was established, so that by the "end of the 9th century the scholars of Islamic civilization had available to them most of the existing theories in the so-called important fields of learning." Through history there were several such centers of learning which not only summarized or synthesized the existing knowledge of various cultures, but made their own significant, fundamental contributions, e.g., Baghdad, Sicily, Spain (Toledo).

The above deliberations should offer a lesson to our present technical—scientific age, in that similar synthesis/assimilation, plus translation schools in various countries and in diverse institutions, could be a tremendous benefit to organize knowledge, disseminate information, eliminate unnecessary duplications, and formulate new problems.

To allow everyone (at least those eager enough) to catch up with the dispersed pertinent concepts, the most ideal solution (and that's all it is, namely ideal, theoretical and impractical) would be to call a moratorium on research — that is to agree on a geological covenant to suspend research for a "shorter" period of time until the older data has been digested. (Actually, in very specific geological sub-disciplines such a moratorium on publishing has been seriously proposed — but to no avail!). Consequently, to enable an intellectual inventory, periodic synopses, syntheses, reviews, overviews or summaries (whatever one might like to call them) are necessary — in some fields of endeavour even paramount to further progress. Having stated this several times in print, I have had readers relay to me that they wholeheartedly agree, some went as far as assuring that they "violently" agree!

It should be reiterated (see introduction of Vol. 1 of the Handbook) that any of the modern techniques of manipulating data will not replace the summarizing and meaningful synthesis of information guided by the human mind. Even computers, which are of great help in storage—retrieval methods, have their limitations in analyzing and synthesizing data and cannot (as yet anyway, and may (hopefully?) never be able to) totally replace the unique discriminatory judgement ability of the human brain. Electronic computers depend on input, and this input-data is selected by the human operator, so that in the final analysis the output is still at present the "electronically supplied *effect* of the human selective or preferential *cause*". To use a worn-out phrase: "Garbage in, garbage out". As this may be too derogatory and offending to some, one could replace this statement by my own concoction: "Half-digested information in, half-digested information out".

The chapters that have been and will be issued in this Handbook cover the data obtained from the whole spectrum of "pure" (= "curiosity-oriented") and "applied" (= "mission-oriented") investigations (see Dixon, 1973, p. 103). Actually, these two artificial end-members are more based on the theory of philosophy than related to practice; no rigid compartmentalization is possible. It is true that one may begin in research with satisfying one's curiosity, but as soon as the results show "practical" implications, a general, often sudden, increase in the demand for more information may induce "mission-oriented" work. It is equally true that some geologists have to perform their professional duties at an unsatisfactory breakneck rate and consequently have no time for the "luxury" of "intellectual contemplation" – but too frequently the absence of an opportunity for such "economically luxurious" curiosity-oriented work has had to be paid for dearly in the end, both in terms of shoddy and incomplete data, which in many instances led to wrong expensive geological decisions.

Related to this, Professor Tom Cottrell (see Dixon, 1973, p. 29) has levelled strong criticism against much professional science (in this case referring to chemists) on two accounts: first, that the average scientist is reluctant to master and apply scientific theory, thus resorting to the purest empiricism; second, that some use "an insufficiently critical and inquiring attitude to the research problem itself. Much effort is spent in investigating problems that do not require solution, and which could have been shown by a careful preliminary analysis not to require solution." In both instances, an interest in the "field of documentation" would assist in removing the above-cited dilemmas.

On numerous occasions the present writer (K.H.W.) has reiterated the need for synthetic research, reviews and comparative studies, and one may rightfully enquire what specific proposals he has to enhance such theoretical, but practical, research. That there actually exists such a need has been recognized by more progressive and enlightened quarters of the scientific-technological community. Those companies, for example, that can financially support it, have one or more specialists preferentially concentrating on "data reviewing", i.e., collecting both descriptive information and interpretive concepts (hypotheses, theories, etc.), which are then intellectually digested and compared. Contrasting and comparing older with newer ideas frequently results in the formulation of new concepts. Just to propose two, most likely very fruitful, comparative studies to be undertaken, I have already recommended to some colleagues: (1) Use Brown's (1948) book as a basis to compare the modern metallurgical with the hydrothermal ore-forming theories, and (2) employ Butler's (1935, 1967) approach, as outlined in his pamphlet "Some Facts About Ore Deposits", to up-date the general geological–geochemical–geophysical exploration principles on a continual basis.

(3) To broaden the scope of coverage and to increase the pace of comparatively summarizing geological information, I would suggest that all major companies, institutions, universities and governmental surveys establish positions for "scientific reviewers" or "geological reviewers" in at least one specific field. The coordination of these efforts, would prevent duplications. Some companies and other "institutions" have already undertaken this formally, and have advertised vacancies to fill such positions. Certain individual researchers have taken on the job as part of their contribution to science, which has provided us with the *Benchmark* books, *Encyclopedia of Earth Sciences* by Prof. Fairbridge, and the present *Handbook* series on ore deposits. Many others have done this by writing reference and text books (which certainly constitute a noteworthy contribution in comparison to the possibly better-selling first-year texts that re-digest more-or-less the same fundamental information), but many of these publications *lack the comparative approach*.

(4) Another particular proposal is required in regard to the personnel to fill the position of "scientific reviewer", because such an individual cannot be created by merely giving him/her a few courses in writing, creative editing, and scientific subjects. As in the case of technical editors (especially the "creative" variety), the geological scientific reviewer has to have qualities and experiences (beyond the introductory and beyond the superficial) that are attained only after many years of being either "practically" and/or "academically" occupied in, preferably several, fields of geology (a good combination, for example, would be if this interests cover sedimentology/stratigraphy/environments; petrology and geochemistry of sedimentary, igneous and metamorphic rocks; and ore petrology). Whatever the combination of experiences may be needed to satisfy specific job-requirements, particularly pertinent is that the reviewer has a strong up-to-date theoretical knowledge. Many of the personal qualities are either "inherited" or "developed"; namely having a persistent curiosity, a mind capable of synthesizing, be able to reason both deductively and inductively, and have a flair for writing and organizing data, to mention only a few. As I have pointed out the discussion period of the ELSE/EDITERA (European Life and Earth Science Editors) Fourth General Assembly meeting, April 1979, in Budapest, retired geologists from companies, universities, and surveys with the necessary qualifications could be drawn into this new "corps of technical reviewers and editors". An untapped source is offered by the many retired, highly qualified scientists which together constitute a tremendous — often unused — pool of knowledge. If only one would encourage and financially support them to summarize their experiences during the evening of their life-long professional activity! To repeat: financial support should come forth for such, often very hard work, which could consist of two modes of approach:

(1) The individual writer would comprehensively review his *own* data and ideas, and contrast them with those of others (i.e., use the comparative method). He may choose to do the compilation with a close associate as co-author (-collaborator) to share the burden of literature research, for example.

(2) The individual reviewer may wish to select a broader scope of coverage in comprehensively summarizing a discipline or sub-discipline, and invite numerous contributors to review defined topics. He will then act as an editor on behalf of the authors, and preferably will also make specific contributions in the form of writing his own chapters and ascertaining that all major data and concepts have been treated.

Properly approaching this "community editorial research", the elder scientist could ascertain that his life-time experience and knowledge find a suitable place in the literature. A very gratifying occupation for those who wish to be intellectually occupied in the spirit of "Men love to wonder, and that is the seed of our science" (Ralph Waldo Emerson, in: *Society and Solitude — Works and Days,* 1870).

REFERENCES

Bowder, G., 1979. Synthesis — progress early Arab style. *Middle East Mag.,* Jan. 1979: 95–96.
Brown, J.S., 1948. *Ore Genesis — A Metallurgical Interpretation: An Alternative to the Hydrothermal*

Theory. The Hopewell Press, Hopewell, N. J., 204 pp.

Butler, G.M., 1935, 1967. Some facts about ore deposits. *Ariz. Bur. Mines Bull.,* 139: 99 pp.

Dixon, B., 1973. *What is Science For?* Penguin/Pelican Books, Harmondsworth, Middlesex, England, 284 pp.

Loosjes, Th.P., 1954 (1973). *On Documentation of Scientific Literature*. Butterworths, London.

Ten Haken, J.H., 1978. On scientific information and documentation. *Int. Inst. Aerial Surv. Earth-Sci. J.,* 1978, (4): 680–700.

KARL H. WOLF (Jeddah)

Chapter 1

TERMINOLOGIES, STRUCTURING AND CLASSIFICATIONS IN ORE AND HOST-ROCK PETROLOGY

KARL H. WOLF

INTRODUCTION

"Science is knowledge of the truth of proportions, and how things are called."
    Thomas Hobbes (in: *Human Nature,* 1651)

"Gathering facts, without formulated reason for doing so and a pretty good idea as to what the facts may mean, is a sterile occupation and has not been the method of any important scientific advance. Indeed, facts are elusive, and you usually have to know what you are looking for before you can find an important one."
    G.G. Simpson

"Becoming acquainted with the literature is a considerable part of the lifetime work of any progressive geologist, who must, as Newton said of himself, stand on giants' shoulders in order to see farther. The rediscovery of already published facts is a tragic waste of time and energy. The library habit is utterly essential to sound scientific research ... "
    R.M. Pearl (in: *Guide to Geologic Literature,* 1951)

    The current accumulation of voluminous geological data has resulted in the modification of earlier methodologies and hypotheses, or has led to the proposal of new approaches, and our efforts continue in the theoretical as well as applied multi-disciplinary earth sciences. The older investigative techniques and simpler scientific interpretive and extrapolative philosophies have given way to more precise, intricate, often highly quantitative methodologies and more complex, profound, interrelated-comparative theoretical evaluations. In the past, the ore petrologist's descriptions and classifications were relatively "simple" (judged by hindsight); they have gradually evolved with the methodologies and concepts (although at times with an unfortunate time-lag), so that a need will arise, now and in the future, for hitherto unknown detailed descriptions and classifications.

    The ever-broadening scope of each of the numerous geological subdisciplines, that has been the consequence of the refinement and introduction of new techniques and concepts, is accompanied by an ever-increasing demand on each researcher (or group of investigators) to concomitantly expand their range of observations, hypotheses, reconstructions, and comparisons. That is to say, as more methods and interpretative concepts become available, more indirect and direct pressure is exerted to consider these new approaches – to fail to do so (either knowingly or unknowingly) means to risk one's study being euphemistically called "preliminary", "reconnaissance", or less complimentary even "incomplete". To mention just one example: in the past ore petrology was often considered a "hard-rock" subject (admittedly, there were always those who never tolerated intellectual boundaries); in contrast, current thinking compels a *multi*-disciplinary approach spanning the *total spectrum* of the igneous–sedimentary–metamorphic field. This situation is, no doubt, welcome and to be desired from the overall point of view of the earth sciences, but it does create a general problem to the individual as his experience(s) may be restricted; and *relative* to the total volume of accessible information, his knowledge seems to recede, although his *actual* knowledge may increase. There is no easy escape from this – it is one problem to recognize that our studies (including the nomenclatures and classifications to which this chapter is devoted) must transgress any former artificial boundaries between the scientific disciplines; it is quite another enigma to solve the psychological stresses imposed by the increasing demands to widen the scope or coverage of our investigations. Methodologies and hypotheses are the basic tools of the scientific method in geology, so that some comments are in order.

It was Wegemann (1963) who said that " ... commonly the notions, concepts and hypotheses control the selection of facts recorded by the observers. They are nets retaining some features as useful, letting pass others as of no immediate interest. The history shows that a conceptual development in one sector is generally followed by a harvest of observations, since many geologists can only see what they are asked to record by their conceptual outfit." Important to remember in this context is that no plausible concept or hypothesis can be formulated prior to obtaining unambiguous observations (descriptive data), which presupposes the existence of a generally agreed-upon terminology (see sections below), classifications and basic tenets ("axioms" and "paradigms", if you wish).

The fundamental importance of proper descriptions was emphasized by Hubaux (1973) by pointing out that " ... of the next ten mineral deposits of economic significance which will be developed within the coming years, at least six or seven are already mentioned in the files of some mining company. The geologists of the mineral companies to whom the author was able to pay a visit agree on this prediction." The "mentioning in the files" may be obvious (e.g., a mineralized locality or a gossan), but the reference may be subtle and obscurely worded, as exemplified by such terms as "minor anomaly", "favorable host-rock environment", or "hydrothermal alteration". It is quite clear that vital data (not "vital" at the time they were obtained, but becoming so during the development of subsequent geological principles) can be easily obscured inadvertently by incomplete and/or wrong or ambiguous descriptions and classifications. The question arises here: How can one properly utilize data by computer technology in interregional comparisons, for example, if the primary information is unreliable and unstandardized? A more coordinated effort is required. Hubaux continued: "The storage of geological data is in a less advanced stage in the mineral industry than in the oil industry." (To repeat: this fact directly depends on nomenclatures and classifications.) "A mining geologist has said to me: 'We are using less than 5% of the potential information contained in our archives and yet completely new ore deposits are increasingly harder to find; the majority of ore deposits which will be developed and exploited within the next years are contained somewhere in these archives."

The multiplicity of terminologies and classificatory schemes that has been offered to the geologist can be overwhelming for those who wish to keep abreast, as already mentioned, and a summary is in order before additional developments may make it increasingly difficult to obtain a meaningful overview. Future supplementary synopses are thereby simplified. The present chapter purposes to offer a review covering numerous sub-disciplines of ore petrology (cf. also Gabelman's chapter in Vol. 1 of this Handbook on eleven tabular classifications of strata-bound ores), in the vein of Carl L. Hubbs' succint observation: "Science's only hope of escaping a Tower of Babel calamity is the preparation from time to time of works which summarize and which popularize the endless series of disconnected technical contributions" (in: *Copeira,* 1935). Periodic compilations are as "useful" as "fundamental" research — each is making its own kind of contribution, and the "original" studies are often guided and kept in perspective by comprehensive reviews. Any controversy about this moot point is unnecessary (cf. the "Introduction" to Vol. 1 of this Handbook, especially pp.7–8); both modes of research have their peculiar challenges, demands, hazards and rewards. As A.A. Weinberg opines: "As I see it, at least part of the conflict amounts to a philosophic judgement whether science is the search for new knowledge or the organizer of existing knowledge" (in: *Reflections on Big Science,* Pergamon Press, 1967).

It has been said that the importance of a subject is often, in reality, no more than a reflection of the quality of mind brought to bear on it — in the present case, if the study of terminologies and classifications is regarded as academically trivial and unworthy, the fault does not lie in the nature of this subject, but in the nature of study or mode of approach utilized. Our understanding has not grown alongside our knowledge — we have to prevent that the more facts we know, the more people seem to fail to come to a real understanding of the mass of unrelated data (and sometimes we have the problem to decide between what is too complex in detail and too simple in essence) — systems analysis, modelizations (based in the first phase on terminologies and classifications) will be indispensable in fulfilling the task by furnishing a relatively simple framework of reference and explanation.

Under the individual headings of the present chapter, the numerous interrelated topics will be treated only briefly; not so much by discussions which are beyond the scope of the contribution, but instead by concisely summarizing the subject matter in the form of tables, diagrams and conceptual models. Lengthy descriptions and discussions must also be avoided because of space limitation — moreover, they are not required in the present instance, for much of the data is self-explanatory to the experienced geologist;

one table or diagram can be equivalent to one or several pages of verbal (verbose) description.

IMPORTANCE OF TERMINOLOGIES AND CLASSIFICATIONS

"If you wish to converse with me, define your terms."
  Voltaire (François Marie Arouet)

"In the fields of observation chance favors only the minds which are prepared."
  Louis Pasteur

"The history of science consists not so much in the progressive accumulation of facts as in the progressive clarification of problems. What makes a natural scientist is not only knowledge of facts about nature but his ability to ask questions about nature; first, to ask questions at all, instead of merely waiting to see what turns up; and secondly, to ask intelligent, answerable questions, as distinct from questions that would be answerable only if he had access to facts which are hidden from him."
  R.C. Collingwood

The scientific nature of geology demands unambiguous terminologies and classifications as well as conceptual models (cf. Chapter 2, Vol. 1 of this Handbook), for both descriptive and interpretive schemes are of prime importance in lending economy and versatility to our thinking and communication. In the processes of grouping, structuring and classifying, we must find some empirical rules, so that even though we draw arbitrary lines — like in pigeon-hole schemes — meaningful grouping is the first step toward clarification in the manipulation of data. To put it differently, science is any body of knowledge arranged or classified in such a way that the phenomena can be understood — natural laws are not man-made, they are only discovered by inquisitive man and to do so he has to describe nature and sort out (= group, classify) the complex observations (or structure the imbroglio of data). Precision and a high degree of uniformity (and reproducibility, therefore) is important, as differences of opinions may be the consequence of a sometimes scandalously imprecise vocabulary.

Ideas cannot exist without verbalization, and the two are interdependent: diminish the vocabulary, and one reduces ideas; increase the word power, and one increases ideas — one simultaneously also eliminates muddled thinking; the more words at one's command, the more precisely and expressively one can describe even the subtlest thing, concrete or abstract — conversely, not to know the right word, not to use it, may mean not being able to convey ideas, because words are the tools of thought and the instrument by which we transmit ideas or grasp those of others. As Lewis (1974) stated: "Words are the symbols of knowledge and understanding." (See also the Introduction to Dutch's 1962, or later, edition of Roget's Thesaurus.) These generalizations are equally applicable to non-scientific and scientific communications — in geology one speaks of nomenclatures or terminologies and classifications — they are part of our "recipe knowledge". Therefore, to be interested in philology, etymology and semantics is not to be viewed as an ostentatious or delatory preoccupation.

In any written communication, the first step is to be aware of the meaning of the terms used — if this is not the case, ambiguity will arise among the parties involved, often without either party being aware of it. A clarification (definition) of the terms is required (e.g., Beveridge, 1957). On many occasions, there is still careless and subjective utilization of the available nomenclature, which is partly the consequence of the separation of practice from theory. Frequently, too frequently one might emphasize, scientists are not interested in semantics (e.g., Ullman, 1951; Hayakama, 1962), to the detriment of their own precision in communications. Beveridge (p.198) quoted R.A. Daly, Augustus Locke and T.B. Nolan, who respectively stated that it has been " . . . said that discussions generally end where they should have begun — with the definition of terms."; "Let's first define our terms."; and "The profession . . . needs to use words accurately if we are to ask people to believe we think clearly and accurately." Accordingly, "We are faced with the need of building our science on a better and more logical base with the clarification and improvement of terminology as a first step . . . " and " . . . there appears to be a need for a new, complete and carefully thought out glossary of terms for mining geology . . . "

The above requirements, to be sure, are not new and have resulted in the preparation of several, periodically up-dated and expanded, scientific dictionaries (e.g., Gary et al., 1974) that became the tutelary guides of geological writing. The same applies to classification schemes and conceptual models. A point to be made here is that differences of opinion in regard to the use of many terms persist, so whenever such terms are applied they must be defined; and with the ever-increasing complexity in geology we have an unavoidable mandate for even more discriminatory terminologies, and especially for classifications plus models.

The availability of good glossaries does not imply total inflexibility, because "neither common usage nor glossaries should be considered absolutely authoritative; the former is undisciplined and transient with overemphasis on fashionable conceptual jargon, the latter in some cases appears to be too rigid, with over-emphasis on original meanings." Scientific "authority must be based on philosophical sensibility, scientific utility, and wherever possible on philological purity" (Rickard and Scheibner, 1976). ". . . on the one hand lexigraphers and glossaries are needed in order to prevent a language from degenerating into babel, while on the other hand conceding that standardization of terms should not be so rigid as to stultify growth and change, that is not ambivalence. The two processes, standardization on the one hand and growth and change on the other, are not incompatible if taken in the proper sequence".

Nevertheless, especially at a particular point in time, standardization is required to some degree. To provide one example; in discussing the several tenets in designing a metallogenic mapping program, Bowman and Stevens (1978) stressed the importance of *objective* maps (as descriptive or non-interpretive as possible) in contrast to *subjective* maps. The former can be utilized by *workers other than the compiler* for subsequent multi-purpose genetic-interpretive extrapolations, whereas the subjective maps represent *only the compiler's* interpretations; thus the latter are less useful (sometimes short-lived) as genetic ideas continually evolve. In some cases, one particular subjective map was prepared by *several* geologists, each with a different professional background, with different descriptive preferences, and varying genetic tastes. In short, for objective maps, *standardized* nomenclatures and classifications are paramount. In the preparation of subjective maps, *one mode of approach for one particular map* is a necessity, although the same investigator may choose a separate set of criteria and principles in another study.

As late as 1974, Jancovič pointed out that both *geophysical* and *geochemical* methods have been developed to a high level for prospecting of metalliferous deposits, but that the theoretical *geological* principles useful in the search for mineralizations have not been developed at the same rate and to the same level. The geological methodology covering the relationships between ore genesis and relevant parameters used in the exploration for specific ore types, as well as in general metallogenic studies, needs refinement. There exists a " . . . need for systematical development of geological methodology in the search for ore deposits." (op. cit., p.348). "The present geologic methodology applied in the search for ore deposits is more qualitative than quantitative, more general than specific, more regional than applicable for local prospecting targets." (op. cit., p.354.) Although Jancovič's publication is merely an introductory note to a complex field, he made it quite clear (even if indirectly) that classificatory schemes, models and analogues are of fundamental importance in the development and/or refinement of concepts/hypotheses/ theories, and offered an example of a tentative tabular model for the formation of hydrothermal ores (op. cit., his table 1, p.350).

Roscoe (1965) can be cited as another researcher who strongly felt the need for a precise and meaningful terminology. He has been engaged in a metallogenic study of part of the Canadian Precambrian shield and has used the comparative approach on some occasions. Roscoe opined that especially in comparative work " . . . the decisions must be made regarding family relationships of many deposits for which we have relatively little data. The *selection of an appropriate procedure for grouping, or classifying, deposits* has been found to be *a more crucial consideration than was realized.* In order to show mineral deposits on a map, to describe deposits, or to evaluate possibilities of mineral discoveries in a given area, it is necessary to *decide on some type of classification or order of presentation.* Mineral deposits may be classified in many ways depending on the emphasis allocated . . . The type of classification used and the priority with which relevant factors are treated in arriving at a classification has a direct bearing on the type of information sought in the field, in the laboratory and in the literature." By scanning classifications, they give us a sense of direction, provide an orientation, sort out and select, and even structure similar members from different ones, so that a degree of order is being put into an, at least apparent, haphazardly arranged system.

Such deliberations lead us directly to the application and restrictions of the *Scientific Method.* This method dictates to every researcher and explorer a certain sequence of steps he has to follow in his daily professional duties in order to assure maximum success (cf. Wolf, 1973 and Chapter 2, Vol. 1 of this Handbook). As discussed in some sections below, both descriptive and interpretive techniques are employed in each one of these investigative steps in field and laboratory work and correspondingly several types of terminologies and classifications may have to be employed.

The importance of clearly *delineating* first of all the (more-or-less) *precise nature and extent* of a geological investigation, followed by *choosing the most relevant and pertinent mode of communication* from a *myriad of schemes available,* cannot be emphasized too strongly. These two steps are frequently ignored as an individual or a team of geologists may believe this to be unnecessary, thus relying purely on *past experience in the application of favourite methods and preferentially selected terminologies/classifications to be applied to any and all geological settings.*

A "hierarchy of terminologies and classifications" (as one might suitably call it) has been offered — a "hierarchy" only for those who can see beyond the, at present, seemingly haphazard and uncoordinated (or "unsorted" or "unstructured") availability of the myriad of approaches. "At present" is emphasized, because it is hoped that in the near future geologists will have reached a concensus on guidelies in (1) the application of the more plausible and practical classifications, (2) outlining, at the same time and where practical, the conditions under which the classifications are useful (e.g. study methods used), and (3) employing hierarchically structured multiple terminologies and classifications as offered in Tables IV–VI and VIII. The need for some fundamental uniformity among various earth sciences to ensure reliability in making comparisons is beyond question, e.g., in computer studies where one geologist collects the data of several hundred or thousand descriptions from numerous sources. As clarified above, the uniformity need not eliminate all flexibility, and it certainly should never prevent the incorporation of new ideas.

DESCRIPTIVE OR GENETIC CLASSIFICATIONS? – THAT IS THE QUESTION!

"Learning, experimenting, observing, try not to stay on the surface of facts. Do not become the archivists of facts. Try to penetrate to the secret of their occurrence, persistently search for the laws which govern them."
I. Pavlov

The above headline is only an eye-catching query and in reality does not convey the true communication or semantical/etymological situation in geology (all related to or traceable back to logic *per se*), for *there is no question* about the requirements for *both* descriptive and genetic terminologies and classifications! As will be demonstrated here, as both approaches are required, the only problem that remains is related as to *when, where,* and *how* to employ these two basically different schemes. To be more pragmatic and fac-

8

TABLE I

Sequence of investigations in geology (simplified as complex transitions/gradations and overlaps are not shown) (K.H. Wolf, 1973, unpublished)

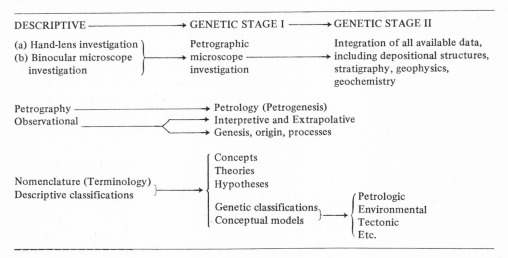

DESCRIPTIVE ──────────→	GENETIC STAGE I ───────→	GENETIC STAGE II
(a) Hand-lens investigation (b) Binocular microscope investigation	Petrographic microscope investigation	Integration of all available data, including depositional structures, stratigraphy, geophysics, geochemistry

Petrography ────────────→ Petrology (Petrogenesis)
Observational _____→ Interpretive and Extrapolative
                          ↘→ Genesis, origin, processes

Nomenclature (Terminology)
Descriptive classifications →
{ Concepts
  Theories
  Hypotheses

  Genetic classifications
  Conceptual models } → { Petrologic
                          Environmental
                          Tectonic
                          Etc.

tual, these two schemes *should* preferably be separate from each other — one descriptive and the other genetic — but in reality many nomenclatures and classifications do not fully meet with this wishful requirement, because many descriptive terms are "genetically flavoured" (see below) and numerous schemes consist of a hybrid agglomeration of both descriptive and interpretive names.

A fundamental axiom in geological work, and related to semantics, is that the separation of purely observational data from interpretations and extrapolations is a necessity. As Table I indicates, this is done by subdividing the operations of a geologist into petrography (i.e., based on naming, describing and classifying the observational objects) which is commonly succeeded by petrologic studies (i.e., involving genetic reconstructions or interpretations guided by concepts, hypotheses and theories of the processes envisaged to have been operative). Noteworthy are the fundamental requirements of descriptive nomenclatures on *all scales* (see also below) in geological enquiries, i.e., for thin sections and hand specimens, as well as for investigations of the igneous, sedimentary and metamorphic rocks, and in tectonic environmental reconstructions.

Many researchers (individually) and companies or institutions (collectively) are using standard "petrographic/lithologic forms" and logs to record field and laboratory observations — many variations have been proposed of which comparative specifics cannot be discussed here. In detailed reports, more use could be made of these forms instead of verbose descriptions — a table may be preferable. When the information of several specimens, or the averaged-out data of members of particular lithologic units, are arranged in juxtaposition, one can immediately pick out the differences and similarities —

TABLE II

Handspecimen petrography — descriptive summary sheet/comparisons (Example only, modifications to be made according to requirements)

Type of specimen or locality or formation, unit / Parameters	Member X			Member Y			Member Z		
	arkose	gray-wacke	tuff	rhyo-lite flow arkose	ignim-brite tuff	volcan-iclast	gneiss	schist arkose	meta-basalt flow

Minerals

Textures

Size

  Maximum
  Minimum
  Average
Sorting
Matrix
Cement
Ratios of Gr./M/C
Colour [1]
Rounding
Sphericity

Maturity [2]

Mineralogical
Textural

Fabric

Structures [3]

Alterations

Primary
Secondary
Paragenesis

"Ore" mineralization

Weathering

Rock name/lithology [4]

NOTE: 1. This type of form needs to be modified for each study; e.g., if the "arkose" unit of member X consists of several lithologies (= sub-types), then *one separate* comparative form may be required to describe these varieties.

2. Not all parameters (= variables) along the *X*-axis are applicable to all rock types, and particularly in the study of complex stratigraphic sequences separate forms may be required for each of the sedimentary, pyroclastic, volcanic-flow, and plutonic sequences for summary and comparisons.

---

[1] Use of colour-chart numbers and/or names is recommended.
[2] After Folk (1974).
[3] See separate check lists.
[4] List the type of classifications utilized for each lithology.

thus enhancing *comparative* work (still often neglected). These latter tables could be termed "Summary and Comparative Petrographic Tables" as exemplified in Tables II and III.

TABLE III

Thin-section petrography — descriptive summary/comparative form (Example only, to be modified according to requirements)

T.S. nos., type of sample, localities, formation, etc. / Parameters	Sample No. 1	Sample No. 2	Sample No. 3
Colour			
Pleochroism			
Index of refraction			
+or − than balsam			
compared to others			
relief			
birefringence			
Shape			
Habit			
Crystal structure			
Cleavage			
Twinning			
Elongation			
Extinction			
Uniaxial/biaxial			
2V			
Dispersion			
Orientation			
Textures			
Fabrics			
Structures			
Intra-mineral			
Inter-mineral			
Associated minerals (includ. opaques)			
Alterations			
Paragenesis			
Specimen's field occurrence			
Field rock association			
Method of identification (references)			
Mineral name			
Composition			
References re origin			

TABLE IV

List of "modes of investigations" (I—VI and 1—23) during various stages of geological studies as used in Table VI (After K.H. Wolf, 1975, unpublished)

General increase in scale, complexity and completeness of investigation

I.	(1) Definition	V. (1) Qualitative
	Problem	(2) Quantitative
	(a) qualified	VI. (1) Composition
	(= selective)	(2) Texture
	(b) unqualified	(3) Fabric
	(= non-selective)	(4) Paragenesis No. 1 (micro- to
	Parameters	megascopic)
	(2) Identification	(a) descriptive
	(3) Description	(b) genetic
	(4) Classification:	(5) *Comparison No. 1*
	A. Descriptive	Structures
	B. Selective analoques, models	(6) —microscopic
	(5) Interpretation	(7) —megascopic
	C. Genetic	(8) —macroscopic
	D. Selective analoques, models	(9) Occurrence
	(6) Classifications of No. (5)	(10) Origin No. 1
	E. Genetic	(11) *Comparison No. 2*
	(7) Extrapolation	Stratigraphy
	F. Descriptive	(12) —form
	G. Genetic	(13) —arrangement
	(8) Comparisons	(14) —chronology
	H. Descriptive	(15) Geographic distribution
	I. Genetic	(16) Paragenesis No. 2 (macro-
II.	(1) Descriptive in scope	scopic/regional)
	(2) Genetic—interpretive in scope	(17) Correlation No. 1
III.	(1) Field work	(18) *Comparison No. 3*
	(2) Laboratory work	(19) Environments (interpretive!)
IV.	Handspecimens	(20) Evolution/History (interpretive!)
	(1) Hand lens	(21) Correlation No. 2 (regional to
	(2) Binocular	inter-/intracontinental)
	(3) Thin- and polished sections	(22) *Comparison No. 4*
	(4) Chemical studies	(23) etc.
	(5) X-ray	
	(6) Mathematical techniques	

A generalized, simplified outline of the steps taken during geologic investigations, coupled with indications of the nature of the studies, is given in Tables IV and VI. The Roman and Arabic numbers in both tables correspond to each other; the horizontal lines in the latter table depicts the particular approach applicable.

The need for both descriptive and genetic schemes to facilitate research and exploration (cf. Wolf, 1973, for example) can be well illustrated by examining the Scientific Method. By returning to its underlying principles of our "thinking patterns" and logic, we may eventually be able to agree on procedures, supported by the appropriate terminology, that will allow us to *move with relative ease from one scale or type of study to*

TABLE V

Descriptions (definitions) of fields of geological investigations listed in Table VI

1. *Mineralogy*. The study of minerals: their formation and occurrence, their properties and composition, and their classification. "Mineralogy" comprises both descriptive and genetic (interpretive) work; however, "*minerogenesis*" could be used for the latter mode of study (genetic). The investigation of "ore" minerals and the field of "ore microscopy" (= *mineragraphy*) is also included in mineralogy.

2. *Petrography*. That branch of geology dealing with the description and systematic classification of rocks in the field and in the laboratory. (Petrography is more restricted in scope than petrology.)

3. *Petrology*. That branch of geology dealing with the origin, occurrence, structure, and history of rocks. Petrology is broader in scope than petrography as it genetically interprets the petrographic descriptive observations. *Petrogenesis* is a branch of petrology which deals with the origin and formation of rocks; it is the interpretive—extrapolative part of study after all the petrographical and fundamental petrologic data has been accumulated.

4. *Lithology*. The description and classification of rocks, in both the field and laboratory, on the basis of such characteristics as colour, structures, mineralogical composition, and grain size (and other textures).

5. *Lithogenesis*. The origin, formation and historical development of rocks — the science of the formation of rocks. The scope of lithogenesis is broader than lithology, as it is based on the latter's descriptive data.

6. *Stratigraphy*. The branch of geology that deals with the definition, description and classification of major and minor natural divisions of rocks (sedimentary, volcanic and layered plutonic and metamorphic rocks) investigated in outcrop and in the subsurface, in the field and in the laboratory, and their historical reconstruction (i.e., study of form, arrangement, geographic distribution, chronologic succession, and correlation plus mutual relationships, as well as interpreting their origin, occurrence, thickness variations, lithology, composition, age, evolution, environments of deposition). Stratigraphy encompasses all the previously listed geologic disciplines (from mineralogy to lithogenesis), but also uses the methods of others such as geophysics and geochemistry.

Increase in Scope of Investigations

another. As the situation stands at the moment, as a consequence of training and habit, as well as the result of an absence of a logically comprehensive system, a change from one stage of investigation to another is accompanied by a change from one set of terminologies and classifications to a different set or sets — no wonder that conflicts arise in geologic opinions which, when carefully analyzed, rest too often on an inconsistency in communication.

The basic phases imposed by the Scientific Method are outlined in Fig.1 and Table VII. After recognizing and identifying a problem, the first stage involves the collection of data which, in turn, is based on describing the observations made, followed by attempts to organize the information into a meaningful pattern or patterns. In the first instance, *descriptive* terminologies and classifications are required; but as soon as data are sorted, organized, structured and classified, *other types* of nomenclatures and classifications may come into play, some of which could be either "genetically flavoured" or are purely inter-

13

TABLE VI

Diagrammatic scheme summarizing the "modes of investigations" (I–VI and 1–23) during various stages of geological studies (cf. Tables IV and V) (After K.H. Wolf, 1975–1977, unpublished)

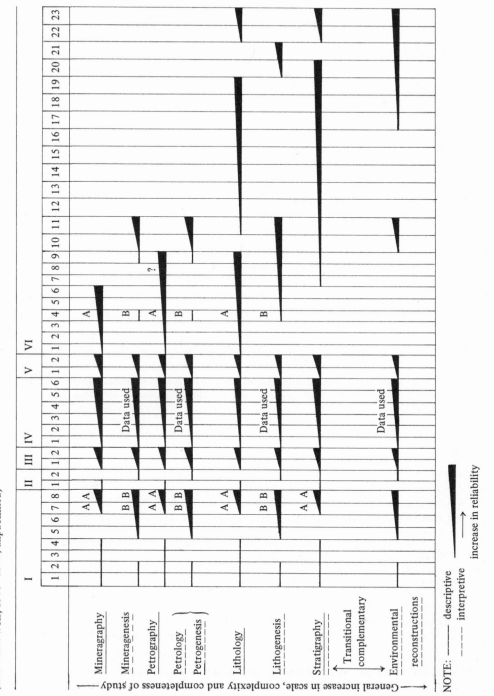

NOTE: ——— descriptive
      - - - - interpretive
      ——→ increase in reliability

14

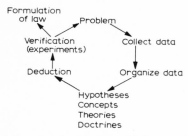

Fig.1. Circular diagram outlining steps of scientific method. (After Wolf, 1973.)

pretive/genetic. The last phases of the investigation are a prerequisite to put order and meaning into the data, and to set them into a useful natural context or framework in which the individual observations become interrelated — thus multiple hypotheses are applied and/or formulated concurrently.

In a *very simplified* form, the semantic discussions, agreements or prescribed rules, and solutions follow the sequence: terminologies (definitions) → structuring (sorting, ordering) → classification(s) → analyses → conclusions → concepts, hypotheses. Only "structuring" may require a brief explanation here. As employed in the present semantic context, structuring (in both descriptive and interpretive work) means to sort observational data, parameters or variables and arrange them with the aim of building a complete whole or a supporting framework that brings out the presence or absence of interrelationships or some order or disorder. (It is particularly here that mathematical—statistical techniques find their application, none of which will be treated in this chapter.) "Sorting—

TABLE VII

Scientific Method (After Krumbein and Graybill, 1965; cf. also Wolf, 1973, 1976b)

---

(1) Formulate the *question*; define the *problem*; define *purpose* [*].
(2) Plan an *experiment* (or set of experiments) to answer question; establish conditions; plan the field of inquiry [†].
(3) State the *exact steps* involved (the experimental procedure) in obtaining the raw data[†].
(4) *Classify* and *summarize* the observations (raw data) [*†].
(5) *Analyze* the data [†].
(6) Draw *conclusions* (inferences) from the analysis [*†].
(7) Erect a *hypothesis* on the basis of the inferences [*†].
(8) *Predict* new inferences from the hypothesis [*†].
(9) *Test* the inferences by further experimentation [†].
(10) *Accept or reject* the hypothesis on the basis of the analysis of the data in step 9 and its agreement or otherwise with prediction [*†].

---

[*] Stage at which models are newly formulated and used.
[†] Models that have already been formulated are used at this stage.

ordering" involves the first steps of "classifying", but this is often done on a trial-and-error (empirical) basis, because a formal classification scheme is not used (either as a consequence of being *absent* or by being deliberately ignored), or because the scheme utilized was of a transitional variety. Once a certain degree of generality and wide applicability is observed, the "structured arrangement" or "structured order" can be considered as a formal classification for further testing, preferably on a world-wide scale.

In as far as one particular problem can be studied by *different* scientific disciplines, *varying types* of descriptive and interpretive nomenclatures are, or can be, the result. However, within *one* method of approach as part of *one* particular discipline, the terminology should be agreed upon and reasonably consistent (with the qualifications expressed elsewhere in this chapter). Due to variations in the mode of "sorting—ordering" one may achieve differently structured data, which has as a consequence varying descriptive and genetic classifications — all of which may be valid until modifications are required.

In the earth sciences, tenets, paradigms, concepts, principles, doctrines, hypotheses, and theories are part of our scientific communicational repertoire, so that especially here one needs agreed-upon multiple — but integrable — schemes. These schemes, as logic demands, must be founded on the earlier descriptions. The differences between observable "facts" and hypotheses are, however, often not recognized, resulting in various "schools of thought". To draw an anology, as Thornton (1961, pp.2—3) put it: " . . . as there is in law, a clear distinction between evidence and verdict; and that this distinction is being obscured or ignored if it is suggested, as it sometimes seems to be suggested, that we can argue equally well in each direction — from evidence to verdict, and from verdict to evidence."

As Table VIII demonstrates, during investigations from the time of outlining a problem to the final phases of deduction — verification — formulation of a law (most day-by-day studies end at stages 5 or 6), and depending on the purpose and scope of any particular study or the stage attained, either descriptive or genetic or both approaches may be required. It is up to the investigator(s) in *each case* to make a logical decision on that aspect *prior* to commencing each phase of the research or exploration. In Table VIII, the vertical bars indicate when the various approaches find their maximal application — note that certain ones are exclusive of others, whereas some overlap. Generally, there is a progression from phases 1 to 7 and from the descriptive to the interpretive — with transitional possibilities and/or simultaneous use of certain phases.

Jaques (1965; see also Brown, 1974, pp.79—84) presented the idea that the nature of the relationship between people and their work is reflected by different (at least five) "levels of abstraction", these levels (or "strata") being directly correlatable with scientific methodologies and abstract concepts employed in the earth sciences. To understand at which strata one uses which terminologies, classifications and models, a brief summary of Jaques' hierarchy of abstraction may be welcome here.

Stratum 1. Perceptual—concrete. The person's mind deals with work restricted to what the eyes can observe (visual observation—concrete matter).

16

TABLE VIII

Utilization of terminologies, structuring and classifications in the geological sciences (After K.H. Wolf, 1976, unpublished)

Steps in Scientific Method	Semantics (Linguistics)					
	Descriptive phase			Interpretive/genetic phase		
	T	S	C	T	S	C

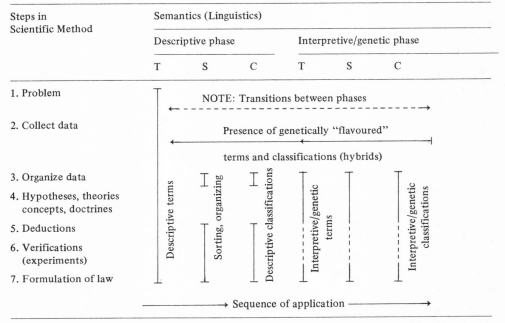

NOTE: T = terminologies/nomenclatures, S = structuring (sorting, ordering), C = classifications.

Stratum 2. Imaginal—concrete. The mind handles not only what the person visually recognizes, but also works at the image-making level, i.e., can compare it with previously seen concrete items (mental picture—concrete matter).

Stratum 3. Conceptual—concrete. Not being able to cope with an expanding volume of data at either the concrete or image levels, the mind thinks in the form of classes of things and concepts (concept-forming — concrete thinking — concrete modeling).

Stratum 4. Abstract modeling. This type of thinking (versus concrete-model-thinking; simple descriptive vs. complex interpretive/genetic/extrapolative-model-thinking) goes beyond the confinement of the concrete and is able to formulate complete (often complex) models of the present, past and future situations.

Stratum 5 and above. Theory and strategy construction. The mind is capable of recognizing and integrating interactions of models and systems of thought as a first step in constructing hypotheses and theories to interpret the past, current and future and to formulate strategies.

Some explanatory notes:

(1) During the first level of abstraction (or stratum), the object of the task or the material to be examined must be *physically* present to be observed or perceived, i.e., it must not be totally abstract but concrete in form. Descriptions are made with or without simple classification.

(2) At the second level, a physical or visual contact with the object to be studied is not required so long as the investigator has had sufficient contact in the past to have retained an image in his mind as a model, i.e., comparisons are being made based on past experience.

(3) The next-higher level, the stratum "conceptual—concrete", operates with extrapolations into the unknown/uncertain in both space and time by formulating and relying upon models. New techniques and models must be developed to permit progress — here too classifying and conceptual models plus comparative methods find their place in creative thinking. This level, as pointed out by Jacques (1965) is still tied to the concrete and to models in physical terms.

(4) The "abstract modeling" or "conceptual-modeling", level is the stratum where the investigator is engaged (and has to have the capacity to do so) in detaching himself from the concrete physical object with which he had past experience and direct association; this is creative thinking by dropping the restrictive concrete frame of reference. Thus by beginning afresh in one's deliberations, one widens the mental horizons and may come to a "break-through" or an "invention". In the geological sciences, this level of abstraction has been exemplified when data were pulled together, related to each other in a new, hitherto unfamiliar, manner, resulting in a sometimes initially called "hair-brained" concept. Here too, on a new untried stratum, both classificatory and comparative studies find their deserved use. As Brown (1974) stated: "The jump from the third stratum to the fourth requires the ability to start thinking from the basic data, instead of starting off from what others have said or written, or from the individual's own past experience."

(5) The mental operations in the "theory-constructing" stratum are related to strategy formulation, because a number of tested models of complex situations are applied — apparently conflicting data, for example, are sorted out and rationally explained and structured. (See also Park, 1963, and Gomez, 1978.)

Some correlation between Jacques' (1965) and Bloom's (1956) approaches with the Scientific Method has been attempted in Table IX.

Another mode of examining the work performed by the geologist has been discussed by Kosygin (1970), who divided the natural systems studied into four groups: (1) static, (2) dynamic, (3) retrospective-historical, and (4) retrospective-genetic. Without entering discussions of the details, Table X lists the respective elements, types of relationships, principles and time-relationships involved — all of which demand the proper application of the Scientific Method in the steps mentioned above and the utilization of both descriptive and genetic terminologies and classifications.

As dictated by the *Exploration and Research Methodologies and Philosophy* (cf.

TABLE IX

Comparison of phases of investigative method (= "scientific method") (K.H. Wolf, 1978, unpublished)

Stratum	1	2	3	4	5
Jaques (1965), (cf. Brown, 1974)	Restricted evidence (visible)	Increase in quantity of data; use of "image level"	Use of concepts; building of models (i.e., not possible to cope with increasing data even in terms of images of concrete objects)	Think of relationships, structures, techniques and processes which have not previously existed; creation of new framework	Correlation of one system to others; compare data, explain and/or integrate contradictory data; offer new strategies
Bloom (1956)	Remember useful information	Apply given information to a familiar situation	Apply given information to an unfamiliar situation	Start without given information and construct a method or methods for tackling an unfamiliar situation	
"Scientific method"	Mainly descriptive of the concrete	Use of sorting techniques, classification of data; some interpretations; preliminary or simple comparisons	Increase in sophistication and complexity in descriptive, interpretive, comparative and ⟶ ⟶ ⟶ conceptual methods, terminologies, classifications, and models/analogues (including "systems analysis") ⟶ ⟶ ⟶		

TABLE X

Types of systems of investigations, their elements, relationships, principles, time considered, and examples (After Kosygin, 1970, table 1, p. 77)

Types of systems	Elements	Types of relationships	Principles	Time	Examples
Static	Geologic bodies	Spatial	Specialization, correlation	Fixed	Geologic maps and sections of various types and contents, graphs and functions with fixed time
Dynamic	Conditions and processes	Spatial-temporal	Those of physics, mechanics and chemistry	Physical	Functions and graphs with time as one of the variables
Retrospective (historical)	Reconstructed events	Successions in time	Successive formation of geologic bodies	Logical	Systems of chronostratigraphic subdivisions
Retrospective (genetic)	Reconstructed processes	Causal	Actualism	Logical	Genetic diagrams: facies maps and sections

chapter in supplemental volume), one of the first attempts in geological work is to scan the literature for data and concepts related to one's field of activity (i.e., synthesize available information). In order to accomplish this efficiently, one first has to identify and classify the entities one is concerned with properly. As common-sensical as this sounds, it is not so easily put into practice (cf. Roscoe, 1965; and earlier section), and the investigation of ore deposits may be a good example. A fair amount of haphazardly or unorganized data is available which, upon careful scrutiny and by employing the trial-and-error method initially, finally succumb to our attempts of finding the correct nidge in a classification; or a few common variables are identified that permit the formulization of a new scheme. From then on — and note, only after an appropriate scheme has been chosen — it is possible to proceed meaningfully with the assemblage of further data, either from additional published sources and/or from field observations.

It is the thesis of much of the present discussion that a clear differentiation between descriptive and genetic terms and classifications is to be maintained — this is particularly so in the preliminary phases of a study, when one may still be "groping" by trial-and-error until one has established a clear investigative path. However, in the *intermediate* and *later stages* of the study (sometimes even during the *very early stages* without having difficulties), a correlation between the available descriptive and genetic schemes may be advantageous, so that classifications combining both descriptive and interpretive terms and concepts are useful as a transitional and/or correlative scheme. The final member of this spectrum would then be a purely genetic/interpretive classification.

Although it is comparatively easy to outline the ideal theoretical requirements, in practice a certain compromise may have to be attained, as numerous so-called descriptive terms are in reality "genetically flavoured". Hence, as soon as we see a descriptive term, we "jump to conclusions" as to genesis. Take the examples of granite, basalt, bauxite, glauconite, oolitic limestone, graded-bedded greywacke, . . . All of these have been defined purely descriptively — but do they not arouse or induce a genetic mental image in most of us?

One last word on the above philosophical–philological aspects; the relevance of maintaining a strict separation of descriptive versus interpretive information is reflected in another rule, namely, that geologic reports must make a clear distinction between these two semantic and classificatory styles of communication. Unless this is done, a certain state of confusion will prevail.

Let us briefly demonstrate the geological applications of the above deliberations. Among several types of classifications, Roscoe (1965) discussed genetic ones and emphasized that they "must not be deprecated as academic. Exhaustive tests of hypotheses of origin provide us with the best possible insurance that we have learned the things we need to know about an ore deposit in order to exploit it efficiently and search most efficiently for others. Genetic terms, moreover, provide a most succint way of conveying a great deal of information, albeit inferentially, about an ore deposit . . . We should seek agreement and consistency in the way we use descriptive and genetic terms, but a universal system for using such terms in a classification does not seem necessary or desirable. Requirements vary . . . it may be convenient to think of some deposits primarily in terms of their genesis, others in terms of their lithological associations, age or stratigraphic position, and others in terms of a distinctive mineralogical association or a characteristic structural configuration." The needs for both descriptive and genetic classifications are unequivocal, therefore.

So-called ore "types" have played a significant role in the delineation and characterization of specific varieties of mineralizations (cf. Laznicka's Chapter 3, this volume). The "type"-classification schemes are either purely descriptive or genetic, or of a hybrid nature. Routhier (1969b, p.213) appears to be one of the opponents of divisions according to "types" and maintained that "we should depart from known 'types' in ore genesis and assume that any element can be enriched during various stages of the geological and geochemical cycle." The present writer contends, on the other hand, that the use of "types" is advantageous, but only under restricted investigative conditions, when one is aware (1) of the basis upon which the types have been established, (2) that there are certain restrictions involved in employing them in comparative work, and (3) that the modelization or conceptualization of the various types of ores is at present incomplete and open-ended (i.e., future comprehensive developments and modifications should be anticipated). Pigeon-holing information has always been dangerous, but as long as one is aware of the limitations and of all transitional or gradational possibilities (cf. Chapter 4 by Gilmour, Vol. 1) between subtypes, a properly established classification of ore types can be a proper guide, as demonstrated by Laznicka's Chapter 3 (this volume).

Hypotheses of ore genesis evolve (cf. three chapters by King, Ridge and Vokes in earlier volumes of this Handbook), and so do terminologies and classifications as the following demonstrates. As Petrascheck (1968) pointed out, the classical theory of ore genesis related to magmatic evolution became questionable, particularly during the past twenty years, when it was realized that there are at least three types of granites: (1) granite originating as a result of magmatic differentiation from an "Ur-magma" (relatively rare occurrence); (2) granite formed by melting of a complex rock section of the earth's crust, when the latter sank during mountain building or when a heat front rose towards the surface (many palingenic granites in fold mountains); and (3) granite originating as a consequence of "soaking" and penetration by solution of available quartz-feldspathic rocks (metasomatic granitization). From then on, the various hypotheses on the mobilization of metals and the metal-carrying solutions (hydrothermal, pseudo-hydrothermal, i.e., the latter non-magmatic, recycled groundwater, different types of connate fluids, sea water, continental surface waters, types of exhalative solutions, etc.) were either revived and modernized through recent information, or were newly developed. These, of course, then resulted in refined and new nomenclatures (petrographic and petrologic) and new genetic classifications as based on ore-creating processes and milieus of formation (as reflected, for example, by host-rock characteristics); it also led to various new ideas on the stages of ore genesis within the geotectonic evolution of magmatism and volcanism together with the development of specific sedimentary-pyroclastic piles. These geotectonic classifications are very similar to those proposed by H. Stille, more recently expanded by A. Bilibin, W.I. Smirnov, and A. Cissarz, for example. A few similar schemes are presented in the section on metallogeny.

It may seem a bit far-fetched to mention here a study of ore genesis on the moon by Borchert (1969). On second thoughts, however, it will become clear that the more we are able to delineate the numerous ore-forming mechanisms on earth where one set of conditions prevails, the more successful we can predict the types of mineralizations on other planets where other physical and chemical milieus occur. Borchert theorized that magmatic, igneous-hydrothermal and volcanic deposits might be expected on the moon, but that "sedimentary" and "exhalative-sedimentary" (or dependent on surface waters) will be absent. Interesting from this point of view are the discussions by Wilson et al. (1974), who compared the geologic cycles of the earth, moon and mars, and those by O'Keefe (1978) on the tectite problem. The question is: to what extent can we utilize the established terminologies/classifications to study other planets? Are new, unique sets of criteria required?

As will be further developed in the subsequent sections, classifications depend upon *specific purposes* so that many different schemes can be prepared — they frequently incorporate details which permit a further division into *subgroups,* in contradistinction other schemes may reflect only *major* characteristics. In organizing geological objects, one has to remember (and although we realize this, it slips one's mind when discussing individual classifications), that "there are qualities in wholes that are not discoverable in the parts" (Titus, 1953, p.110). If one analyzes an object into its individual elements or simple constituents, these units are no more real than the object or event *en toto* — we merely shifted the scale of operation! Although there is a trend to analyze objects into their constituent parts in the fundamental sciences, especially in geology this trend has always been supplemented by synthesizing the units into larger and larger groupings (often with spectacular results as demonstrated by ERTS photography) (cf. Chapter 2 on modeling, Vol. 1 of this Handbook). Consequently, the so-called "fallacy of reduction" (according to which simple, smaller units of an object have a reality or properties not possessed by the larger and more complex object) has never been an obstacle in geological thinking and only the absence of certain *methods* in the past has imposed restrictions on us.

SOME GENETIC CONSIDERATIONS

"Man cannot make principles; he can only discover them."
    Thomas Paine (in: *The Age of Reason,* 1794)

"Science is knowledge, not of things, but of their relations."
    Lucien Poincaré (in: *Science and Hypothesis*)

"Geology is a particularly alluring field for premature attempts at the explanation of imperfectly understood data."
    R.H. Dana

It has been concluded in the above sections that genetic nomenclatures and classifications are employed in all phases of ore petrology, so that it seems advisable to comment relevantly on the origin of metalliferous deposits, augmenting the discussions offered in the chapters of this multi-volume compendium. This brief bird's-eye view will demonstrate the genetic spectrum of geological complexities one has to deal with; consequently, a carefully agreed-upon terminology must be used in each type or phase of investigation of the ore-forming processes. The following aspects are considered: (1) the variables controlling ore-forming mechanisms; (2) the importance of scale and size in geologic studies; and (3) the transitional nature of mineralization processes, environments and products.

(1) The paragenetic sequence depicting the phases in the origin of mineralizations from a whole gamut of high-to-low pressure and temperature fluids, including hybrid solutions, is shown in Table XI. In Table XII, on the other hand, a number of regional, local and chemical factors are listed to exemplify the variations in scale involved. The modes of investigations range, therefore, from large to small scale and from general to particular, as indicated below the tabulated information. Although no simple correlation is possible, studies varying from general to particular are frequently accompanied or paralleled by an application of "classical" methods and concepts, supported by "modern" ones in exacting examinations. With the change in scale, of course, one also has to expect a progressive increase in the employment of experimental, laboratory and general theoretical approaches with an emphasis on obtaining more precise data and probing deeper into the natural phenomena.

Only very few attempts have been made to synthesize all factors controlling the origin of ore deposits into one scheme — comprehensive lists cannot even be found in the literature, let alone attempts to illustrate their genetic relationships. An exception is the table by Ovchinnikov et al. (1964), listing five major primary sources of metals, and their respective "mobilization and transportation factors" (total of 28) plus "extractor and carrier media" (totalling 19) — this list being applicable to the endogenic ores only. More all-inclusive "stock-taking lists" of variables of both exo- and endogenic ores may offer several dozen parameters. What researcher is willing to undertake this immediately needed task?

TABLE XI

Controls on the origin of sedimentary ores (After Beales and Jackson, 1968)

Source → Concentration → Release → Transportation → Route → Precipitant → Precipitation

---

TABLE XII

Regional, local, physico-chemical, biochemical and miscellaneous parameters controlling the origin of ores (with mode of studies — very generalized and simplified — indicated at the bottom)

Regional factors	Local factors	Physico-chemical/ biochemical factors	Others
Orogenic vs. epeirogenic (cratonic)	Source rocks	Eh	Time
Geosyncline vs. geanticline	Reservoir rocks	pH	Evolution of earth's
Eu- vs. miogeosyncline	Fracture, etc. for passage	Bacteria	physical and chemical make-up
Sedimentation vs. metamorphism vs. volcanism vs. plutonism	Hydrodynamic system	Composition of fluids	
	Fluid chem- istry	Reactive host- rocks	
Continental drift (?)		Etc. (for more	
Plate tectonics (?)		complete details see below)	
Large-scale ——————————————→ Small-scale			
General ————————————————→ Particular			
"Classical geology" ———————————————————→ "Modern geology"			
Field approaches ————————————————————→ Experimental, labora- tory, and theoretical approaches			

(2) For reasons that will become clear below, special emphasis is laid in this treatment upon the variations in scale and size which a geologist has to deal with, although his particular specialty may confine him to either large-scale (e.g., mapping and field geology) or small-scale work (e.g., petrology), whereas others cover the whole spectrum in persuing their duties.

The "scope of structural investigations" in geology has been divided by Kirchmayer (1961, cf. his table 1) according to the scale (or size-ranges of the objects under scrutiny) into the sub-, meso- and supermicroscopic; sub-, meso- and supermacroscopic; and megascopic spectrum. This variation in scale is accompanied by a change in methodology and type plus amount of data obtained (see also the papers by Adler, 1970a, b). The information by these two authors, conceptually tabulated, is directly applicable to the techniques utilized by ore petrologists.

Table XIII presents five orders of scale, listing the corresponding types of studies undertaken and the commonly used terms describing them. Particularly this table should make it clear that the terminologies developed for each of the subdisciplines (i.e., 1st to 5th order) must be carefully applied to ascertain a *meaningful shift* from one scale to another (i.e., to *supplement* the data of one order by that of any other order). It seems that, in general, structural geologists have succeeded in doing this as they advance from

TABLE XIII

Range of scale and types of geological studies: an example (cf. Table XVI) (K.H. Wolf, 1974–75, unpublished)

Order of deformation	Type of study	Geologic terminology used for particular study
1st order	Continental and world-wide	Regional tectonics
2nd order	Specific areas of regions	Structural geology (macro-structural geology)
3rd order	Specific outcrop and hand-specimens studies	Mega-structural geology
4th order	Structures in thin sections	Petrofabrics
5th order	Structures of individual minerals and down to atomic size	Rock mechanics (X-ray studies)

micro- through macro- to mega-structural investigations. Much remains to be done in developing working techniques that would induce and permit similar shifts in other geological disciplines, which at present still use a "trial-and-error", "pragmatic", or even haphazard approach in correlating results from studies on varying scales.

Tables XIV and XV present two more cases to illustrate the significance of not only the scale *per se,* but the *continuum* or *transition* of the scale and size to be taken into account in establishing a generally acceptable nomenclature and classification

TABLE XIV

Influence of tectonism on a regional and local scale (K.H. Wolf, 1974–75, unpublished)

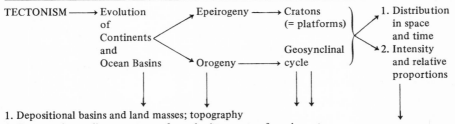

1. Depositional basins and land masses; topography
2. Topography → climate → type of weathering → rate of erosion, etc.
3. Lithologic types of sediments, volcanics and plutonics, and ores
4. Facies distribution and stratigraphy (and related traps)
5. Hydrodynamic system of whole region (subsurface and surface) controlling origin of petroleum and mineral-forming fluids
6. Deformational structures
7. Diagenesis → metamorphism → anatexis → remobilization of fluids and solids

TABLE XV

Range in scale in studies of ore genesis: a Canadian example (K.H. Wolf, 1974–75, unpublished)

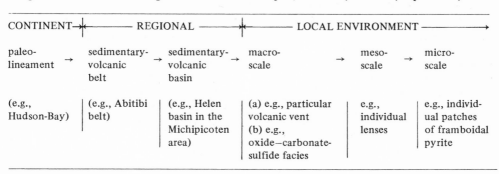

CONTINENT →	←——— REGIONAL ———→		←——— LOCAL ENVIRONMENT ———→			
paleo-lineament →	sedimentary-volcanic belt →	sedimentary-volcanic basin →	macro-scale	→	meso-scale →	micro-scale
(e.g., Hudson-Bay)	(e.g., Abitibi belt)	(e.g., Helen basin in the Michipicoten area)	(a) e.g., particular volcanic vent (b) e.g., oxide–carbonate-sulfide facies		e.g., individual lenses	e.g., individual patches of framboidal pyrite

schemes. Table XV is based on the geology of the Canadian Precambrian shield of Ontario. In the final analysis, the most useful divisions or classifications of geological units are those based on unit values (or a range of values with a maximum or minimum limit), although these values are not always precise and absolute; they are flexible and approximate in most instances. Below are three examples: Baumann and Tischendorf (1976) classified the minerogenic (metallogenic) units according to size (in km^2) and form, ranging from individual ore bodies to, what they called, "planetary"-sized belts and provinces (Table XVI). (Cf. also Laznicka's Chapters 3 and 4 in this volume.)

The second and third size-models were offered by Carey (1962), who pointed out that the spectrum from the field or rock deformation, structural geology and geotectonics ranges through sixteen orders of magnitude; in contrast the range from global geology to astrophysics encompasses merely three orders (Fig.2). As time has a similar magnitude, Carey prepared his well-known "size–time" model for geotectonic phenomena (Fig.3), suggesting that errors are possible if empirical hypotheses or concepts, developed about the properties and behaviour of matter related to *one size–time range,* are then *extrapolated beyond this range.* These deliberations have found their applications in structural geology/tectonism; however, a more direct relationship to ore genesis can be visualized by referring to the controversies of the influence of thixotropy and diffusion (cf. Wolf, 1976c) as possible ore-concentration mechanisms.

Numerous indirect time- and space-influences have been proposed in ore and rock petrology, about which only general speculations are offered as quantitative data are still lacking: basin developments and ore distribution (e.g., separation of ore types in a geosyncline, such as mineralizations associated with flysch and ophiolite versus those in molasse deposits); magmatic differentiation with contemporaneous or subsequent hydrothermal pulsatory releases; metamorphism and expulsion of metals from rocks with subsequent diffusion; maturation of organic matter accompanied by release of metals in sedimentary basins, followed by migration; accumulation of stress leading to periodic shear-

TABLE XVI

Classification of minerogenetic (metallogenetic) units according to size and shape (After Baumann and Tischendorf, 1976, table 3.2, p. 172) (cf. Table XIII)

Size of minerogenetic unit		Shape of minerogenetic unit		Remarks
Surface area in km^2	size classification	± linear	± isometric	
$n \cdot 10^6$	planetary	} belt	province	regional minerogenetic units
$n \cdot 10^5$	subplanetary			
$n \cdot 10^4$	regional I	zone	} subprovince	
$n \cdot 10^3$	regional II	subzone		
$n \cdot 10^2$	local I	ore (= mineralization) region/area or district		local minerogenetic unit with concrete mineralizations
$n \cdot 10^1$	local II	ore (= mineralization) local (environs)		
$n \cdot 10^0$	local III	ore (= deposit)		
$n \cdot 10^{-1}$ to $n \cdot 10^{-2}$	local IV	ore body *per se*		

ing/veining and pulsatory filling by ore minerals. One complex interrelationship, where among other variables both time and space must be considered, is the following genetic sequence: duration of (subsurface) magmatism → duration and amount of hydrothermal activity → duration of (surface) volcanism → composition and viscosity of lava → force of expulsion (quiet vs. explosive) → extent of spread and thickness of flow and pyroclastics → properties and proportions of flow and pyroclastics → amount and composition of associated ores (e.g., Precambrian mineralization associated with "mill-stones"; cf. Chapter 5 by Sangster and Scott in Vol. 6).

The practical usefulness of the serious contemplation of the *size of ore deposits* was dramatized by a unique study. Folinsbee (1977) undertook an interesting statistical examination by plotting the size (in tons) versus the "rank order" of ore deposits using Zipf's Law (Zipf, 1949), arranged in sequence of decreasing size. This law is based on the limiting case of the Pareto distribution (the harmonic series $1, \frac{1}{2}, \frac{1}{3}$ ..., $1/n$, which tends to zero but does not converge). In other words, in any numerical sequence, the biggest item equals rank 1 and is twice as big as number 2 and three times that of number 3, etc. (Although Zipf used this law in human affairs studies, it has been found to apply also to natural phenomena — see Folinsbee, 1977, for examples other than those discussed.)

The practical utilization of Zipf's Law is exemplified by Rowlands and Sampey's (1976) study of the Zambian ore belt reserves (see also Chapter 6 by Fleischer et al. in Vol. 6) based upon the ore-body size distribution (fitted to Zipf's line), concluding that a number of open unoccupied spaces along the line suggest as yet undiscovered mineralization.

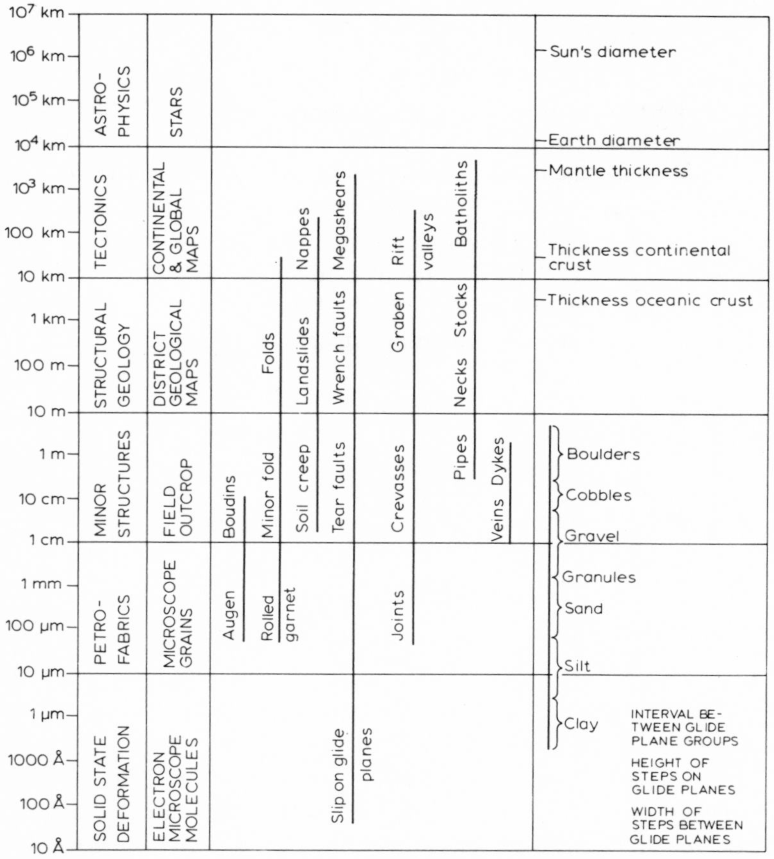

Fig.2. The size of geotectonic phenomena

Left-hand scale (top to bottom): 10⁷ km, 10⁶ km, 10⁵ km, 10⁴ km, 10³ km, 100 km, 10 km, 1 km, 100 m, 10 m, 1 m, 10 cm, 1 cm, 1 mm, 100 μm, 10 μm, 1 μm, 1000 Å, 100 Å, 10 Å.

Disciplines (columns, top to bottom along scale): ASTRO-PHYSICS — STARS; TECTONICS — CONTINENTAL & GLOBAL MAPS; STRUCTURAL GEOLOGY — DISTRICT GEOLOGICAL MAPS; MINOR STRUCTURES — FIELD OUTCROP; PETRO-FABRICS — MICROSCOPE GRAINS; SOLID STATE DEFORMATION — ELECTRON MICROSCOPE MOLECULES.

Reference sizes (right margin): Sun's diameter; Earth diameter; Mantle thickness; Thickness continental crust; Thickness oceanic crust.

Phenomena labels: Megashears, Nappes, Rift valleys, Batholiths, Folds, Landslides, Wrench faults, Graben, Necks, Stocks, Boudins, Minor fold, Soil creep, Tear faults, Crevasses, Pipes, Veins, Dykes, Boulders, Cobbles, Gravel, Augen, Rolled garnet, Joints, Granules, Sand, Silt, Clay, Slip on glide planes.

Right-hand notes: INTERVAL BETWEEN GLIDE PLANE GROUPS; HEIGHT OF STEPS ON GLIDE PLANES; WIDTH OF STEPS BETWEEN GLIDE PLANES.

Fig.2. The size of geotectonic phenomena. (After Carey, 1962; courtesy of Geological Society of India.)

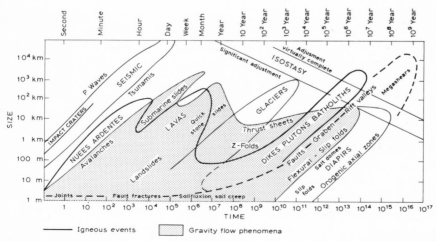

Fig.3. A size–time model for geotectonic phenomena. (After Carey, 1962; courtesy of Geological Society of India.)

Folinsbee (1977) applied Zipf's Law to the Australian uranium ore (Fig.4a), Yukon Pb—Zn(Ag) (Fig.4b), Northwest Territory base-metal (Fig.4c), Canadian North-of-60° base metal (Fig.4d), Irish Pb—Zn (Fig.4e), and Pacific Cordillera placer-Au (Fig.4f) deposits. For example, the Navan deposits of Ireland ranks of No. 1 in Fig.4e on the Zipf line, whereas the Abbeytown prospect is the smallest with several others in between. According to Folinsbee (p.905), "the Zipf prediction is that only about one-third of the potential ore has been discovered, . . . " To the present writer it seems that the real challenge in the use of Zipf's Law — and its ultimate utility — lies in determining a positive correlation between *size* of the ore deposits within one district and the *spatial* and *time*

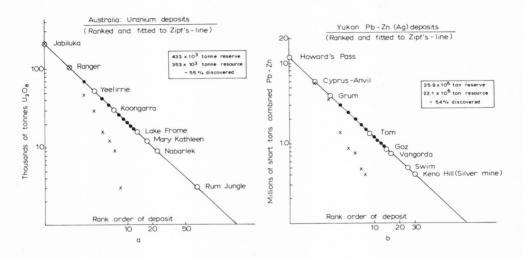

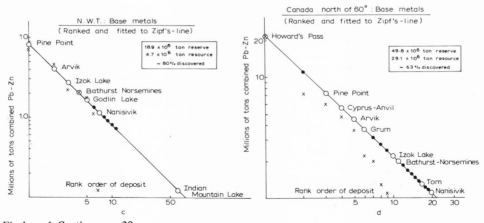

Fig.4, a—d. Caption on p.29.

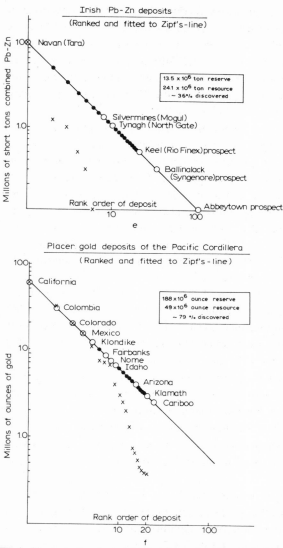

Fig.4. Six (a to f) selected types of ore deposits plotted on a "Tonnage" vs. "Rank Order" diagram illustrating Zipf's Law. (After Folinsbee, 1977; courtesy of Geological Society of America.)

distributions. For example, after establishing the largest and smallest more-or-less simultaneously formed deposits in a mining area, one should have an idea of the intermediate-sized ore bodies theoretically expected by Zipf's Law. As one also knows the localities of the largest and smallest deposits, can one assume that some of the intermediate-sized mineralizations are to be found *between or near these two end-members*? Does one first have to establish the locality of the source of the ore-forming fluids and thus possibly also their direction of movement to find the answer to such questions? Are such investigations confined to ore deposits that have originated penecontemporaneously, i.e., within a certain time-span? Or could the investigations apply to ores that may have been formed by pulsating or periodically supplied solutions, as long as the source remained the same and the directions of fluid movements were in general identical? From these subsidiary queries, a host of related auxiliary questions could be posed!

(3) One of the disadvantages of our geological terminologies and the "older-fashioned" classifications is their inability to incorporate *intermediate* cases or situations between two or more end-members, thus not emphasizing the *transitional* or *gradational* characteristics of all natural phenomena and their products (but see Chapter 4 by Gilmour in Vol. 1). To go beyond the familiar restrictive rules of nomenclatures, one must have the ability to escape from the rigidity of words and classifications; sometimes by the use of certain "mental crutches". Especially the flow-chart and pictorial-diagrammatic variety of conceptual models (Chapter 2 by Wolf, Vol. 1) (as well as certain numerical schemes not treated here) have remedied the shortcomings inherent in the pigeon-hole classifications. Although it takes only a moment of reflection to recognize numerous gradational genetic possibilities in the formation of ores, it appears to be only recently (with exceptions) that *several* ore-creating mechanisms were *used in combination* to find the more plausible explanation. As an example, consider Figs.5 and 6 and Tables XVII and XVIII, which illustrate the transitions between (1) various types of solutions, (2) varieties of sedimentary-volcanic deposits, (3) types of magmatic-volcanic ores according to depth of formation, and (4) sedimentary to hydrothermal or syngenetic to epigenetic ore types. Fig.8, then shows diagrammatically the changes from syngenesis through diagenesis (and epigenesis) to burial or incipient metamorphism and finally to surface weathering. It is the lack of consideration of these *gradations* that has led to misunderstandings — reliable petrologic, geochemical as well as regional geologic data, however, have been made available during recent years, which require logical integration with the classical information and concepts.

Fig.5 depicts seven ore-forming processes, each being described by one term or expression. It is mainly the diagram itself (i.e., the *pictorial* presentation of complexities) that compels the reader to envisage the numerous possibilities of mixing of the various solutions to give rise to hybrid fluids (i.e., $1 + 4$; $3 + 7$; $3 + 7 + 6$; $5 + 2$; $4 + 3 + 2$; $1 + 2 + 5$; $1 + 5$; $1 + 2$; $2 + 5$; etc.) Fig.6, on the other hand, lists seven major geologic settings where strata-bound ores can originate — again with all transitional cases indicated by arrows. Such *gradational processes,* or *gradational environments* under which the mechanisms operate, result in a consequent sequence of *transitional products,* as outlined in Table XVIII.

Table XVIII depicts the continuum between syngenetic surface origin of ores on the one hand, and igneous-hydrothermal epigenetic subsurface deposition, on the other. To illustrate the transition, four cases have been selected (I to IV, aligned horizontally)

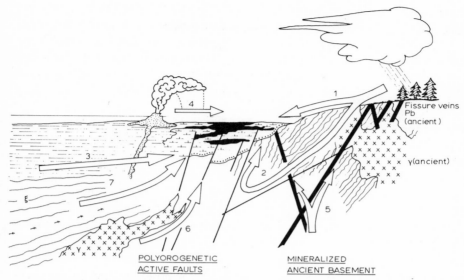

Fig.5. Diverse processes contributing contemporaneously and/or successively to the origin of stratiform ore deposits in a transgressive marine milieu after an orogenic episode. (After Laffite, 1967; courtesy of *Econ. Geol.*) *1* = sedimentary (chemical or biochemical); origin of the metal: leaching; *2* = circulating hydrologic-hydratogenic; origin of the metal: leaching; *3* = connate waters; *4* = volcanic–sedimentary; *5* = rejuvenation of ancient veins and hydrothermal migration; *6* = postmagmatic-telethermal; *7* = postmetamorphic migrations. ξ = compacting, permeable sediments, γ = intrusions, in some cases supply heat driving/hydrologic cell (hydrothermal); *f* = fluid.

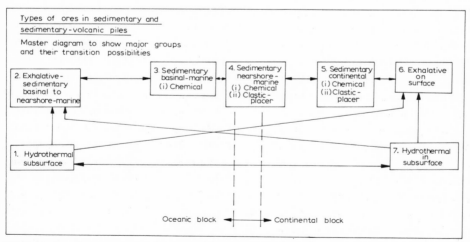

Fig.6. Mega-environments in which corresponding major genetic groups of ore deposits can form, and the transitional/gradational types depicted by arrows. (After Wolf, 1973.)

TABLE XVII

Transitions between main types of ores formed by igneous activity * (K.H. Wolf, 1974—75, unpublished)

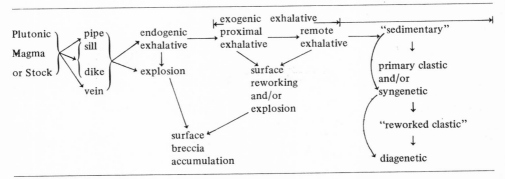

* Modified from the published literature.

and five variables (1 to 5 in vertical succession). No. I comprises the ores of purely sedimentary-syngenetic origin, meaning that both the locality of precipitation (Nos. 1, 2 and 4) as well as the source is of the sedimentary milieus. Mineralizations of type No. II, on the other hand, already show the first influence of hydrothermalism — namely volcanic exhalations bringing ions and/or anions for ore-mineral precipitation into the surface waters. All other variables are the same as in column I. Types I and II are strictly stratabound and no "epigenetic" features are present, unless produced by later remobilization. The type III deposits may exhibit features that are both syngenetic and epigenetic, exemplified by cases where the ores were precipitated from volcanic exhalations along the earth's surface as well as within earlier-formed sediments and possibly volcanics (either prior to, during or after their cementation/consolidation). The type IV ores constitute the other end-member by being purely epigenetic, hypabyssal to exhalative (mainly of *subsurface* origin) and exhibiting textural and structural features indicating later-stage mineralization. Cross-cutting relationships are the norm and any strata-bound material was precipitated by solutions moving within bedded permeable units. In summary, as Smirnov (1972) has pointed out for example, the three mechanisms of sedimentary-exhalative, subvolcanic exhalative-hydrothermal, and plutonic-hydrothermal deposits are genetically related, are "converging" towards each other, and are often associated in both space and time. Each one is represented respectively by (1) the Precambrian banded siliceous iron ores, Rammelsberg—Meggen and Lahn—Dill ores (cf. the three respective chapters in the other volumes of this Handbook); (2) the black ores of the Kuroko deposits; and (3) the porphyry copper and molybdenum ores. Smirnov listed the characteristics of each of these three types, which need further investigation.

As to the association of mineralization with host-rock types (and consequently reflecting modes of origin), several transitional cases can be envisaged here too, possibly

TABLE XVIII

Scheme indicating transitional/gradational genetic possibilities (K.H. Wolf, 1974–75, unpublished)

SEDIMENTARY ←——————— all transitions ———————→ HYDROTHERMAL
                                                                      (IGNEOUS)

Syngenetic   ←——————— many combinations ———————→ Epigenetic

I	II	III	IV
(1) *Type of deposit*: syngenetic-sedimentary	ditto (as I)	combination of syngenetic and epigenetic (as I and II plus epigenetic)	epigenetic only
(2) *Location*: along water–sediment interface	as I	as I and II plus in underlying rocks	confined to subsurface rocks
(3) *Metal source*: from sea water and/or land source	from volcanic exhalations into surface waters with possible spreading out prior to precipitation	volcanic-exhalative	volcanic-exhalative to hypabyssal
(4) *Precipitation*: purely physico-chemical and/or bacterially	as I	as I and II	mainly physico-chemically but bacterial processes also possible
(5) *Form*: strata-bound	as I	strata-bound and cross-cutting stratiform, plus possibly veins and disseminations	as III, with strata-bound deposits as a result of open-space fillings and replacements by fluids and gases moving parallel to beds

←——————— General decrease in degree of volcanism involved, closer to earth surface, and decrease in amount of volcanics in ———————→
host-rock assemblages

by utilizing Fig.7 (Amstutz and Bubenicek, 1967). The two groups, depicted in the figure on the left- and right-hand sides, respectively, were only offered separately for the sake of convenience in the preparation of the diagram. Certainly, the authors did *not* mean to convey the impression that one is dealing with two distinctly non-gradational classes of mineralizations.

    Stanton (1972) reviewed four theories of mineralization, namely (1) ore magma differentiation, injection and related processes; (2) lateral secretion; (3) igneous-hydrothermal emanations; and (4) marine sulfate reduction. He concluded that in particular

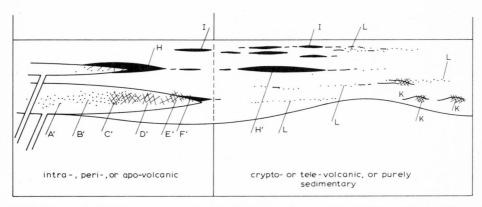

Fig.7. Schematic drawing of types of mineral deposits. To the left ($A'-H'$) are types of deposits clearly associated with volcanic rock; on the right ($H'-L$) are those which may or may not be connected with volcanic exhalative activity. K is largely contained in late-diagenetic compaction fractures in and near organic reefs. Types $A'-F'$ merely refer to common types of distribution patterns in lavas, whereas $H-L$ are located within sediments, with which they were formed. (After Amstutz and Bubenicek, 1967.)

combination (3) plus (4) offers the most likely process in the origin of many strata-bound ores. Another process has found wide acceptance since. It is a variety of No. 2 above and based partly on ore genesis related to the source-bed concept. In this process, concentrated connate and compaction fluids released during the evolution of sedimentary basins can migrate to suitable sites to precipitate ores (see Wolf, 1976c, for a review). Much progress has been made in ore petrology that has led to revisions in the genetic concepts since Stanton (1972) stated: "It seems that the time has come to look critically at the whole structure of current ore genesis theory, and to examine in an objective and logical way the primary facts and assumptions on which this structure has been built."

Another future problem to be given its deserved attention is related to setting the boundaries between syngenesis, diagenesis, burial- and higher-grade metamorphisms, and surface weathering as a result of exposure of the rocks (Fig.8). The question is not purely academic as seen by the confused terminology sometimes used by both "soft-" and "hard"-rock petrologists in describing complex paragenetic relationships. All the different alteration stages to which a pile of rock is exposed (i.e., ranging from syngenesis to metamorphism to weathering) are of practical concern to ore petrologists and gitologists, inasmuch as in each stage depicted in Fig.8 either ore-forming solutions can be released from the pile; metals can be leached from the clays, organics and pyroclastics; hypogene or supergene fluids can move within the beds; or remobilization can take place. To convey our ideas unambiguously, our terminology has to be appropriately meaningful — the first step would be to ascertain when one uses "syngenetic" and "epigenetic" as to whether these and similar terms are referring to (1) the *source* of the metals, (2) the *origin* of the

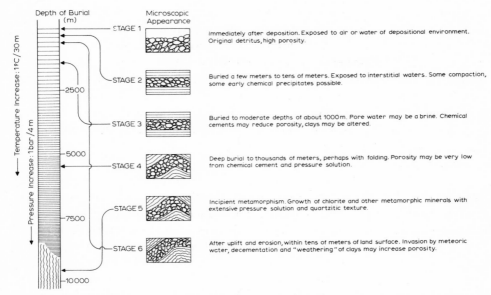

Depth of Burial (m)
Microscopic Appearance

STAGE 1 — Immediately after deposition. Exposed to air or water of depositional environment. Original detritus, high porosity.

STAGE 2 — Buried a few meters to tens of meters. Exposed to interstitial waters. Some compaction, some early chemical precipitates possible.

STAGE 3 — Buried to moderate depths of about 1000 m. Pore water may be a brine. Chemical cements may reduce porosity, clays may be altered.

STAGE 4 — Deep burial to thousands of meters, perhaps with folding. Porosity may be very low from chemical cement and pressure solution.

STAGE 5 — Incipient metamorphism. Growth of chlorite and other metamorphic minerals with extensive pressure solution and quartzitic texture.

STAGE 6 — After uplift and erosion, within tens of meters of land surface. Invasion by meteoric water, decementation and "weathering" of clays may increase porosity.

Fig.8. The stages of diagenesis in relation to depth of burial and increase of pressure and temperature. (After Pettijohn et al., 1972; courtesy of Springer, New York.)

host-rock, (3) the *processes* of transportation and precipitation, or (4) to the *fabric* characteristics of the mineralization.

VARIABLES USED IN CLASSIFICATIONS OF ORES

"Science is knowledge gained by systematic observation, experimenting, and reasoning."
Louis Pasteur

"Science is the knowledge of consequences, and dependence of one fact upon another."
Thomas Hobbes (in: *Leviathan,* 1651)

The parameters utilized in specific classifications vary in accordance with the requirements of the investigators, and it is an illuminating experience to examine the schemes by Bateman (1950), Sullivan (1957), Bates (1960), Schneiderhöhn (1962), Pereira (1963), Cissarz (1964, 1965), Amstutz (1965), Gross (1965), Snyder (1967), Bilibin (1968), Dimroth (1968 and Chapter 5, Vol. 7 of this Handbook), Korolev and Shekhtman (1968), Bostick (1970), Little (1970), Nicolini (1970), Gilmour (1971 and Chapter 4, Vol. 1 of this Handbook), Ruznicka (1971), Pélissonnier and Michel (1972), Botvinkina (1973), and Wolf (1973, 1976b). The following variables have been used by

the above investigators in their classification schemes: process and mode of metal transportation, distance of ore precipitation from original source, types of sources, type of
magmatic source, level of intrusion, environment of precipitation, mode of precipitation,
depth of ore-precipitation below the earth's surface, pressure, temperature, tectonic setting, types of host-rock, chemical/mineralogical facies and zonation, time of ore deposition related to host-rock, major mineralogy, types of metallic ions, minor and/or trace
element content, isotope composition, texture and fabric of the ore *per se*, structure (e.g.,
shape, form ore–host-rock relationships), and stratigraphic relationships, presence or
absence of alterations, presence or absence of remobilization, degree of metamorphism,
and transitional possibilities between types of ores.

Highly specialized classifications are needed for detailed work, and Bostick's (1970)
scheme (cf. also below) for computer statistical examinations of uranium-bearing sandstones was based on the following variables: average dip of the host-beds, original environment of deposition of the host-rocks, types of host-rocks, sandstone/shale ratio, colour of
rocks, grain-size distribution, sorting, grain composition, presence of organic matter, uranium content, presence or absence of pyroclastics, and type of under- and overlying beds.
Using these parameters in his classification, Bostick was able to delineate the most favourable conditions for uranium accumulation, i.e., information then used for exploration.

For additional variables utilized in classifications, the reader is referred to Chapter 3
by Gabelman in Vol. 1 of this Handbook, e.g., the degree of host-rock chemical reactivity
with the ore-forming fluids.

The numerous classifications presented in the subsequent sections demonstrate that
only a limited number of the above-listed variables can be considered in any one tabular
scheme (sometimes only two or three, at other times up to five or six).

Certain factors should be listed that cause difficulties and confusion in the application of terminologies as well as classification schemes. These are: (1) non-separation of
descriptive and genetic terms, and the indiscriminate use of "genetically flavoured"
expressions; (2) indiscriminate use of synonyms and little-accepted local terms; (3) poor
choice of prefixes and suffixes in terminologies; (4) only a restricted number of variables
can be incorporated in any one scheme; (5) gradational/transitional cases in processes,
variables and products may be ignored, thus hybrid possibilities are not considered; (6)
oversimplification; (7) ignoring of evolutionary changes in geologic time and space
(important for some types of mega-classifications); (8) non-adherence to logical application of certain terminologies and classifications when changing from one scale of study
to another, or from particular problems to more generalized situations (and vice versa);
(9) use of insufficient or wrong data "forced" to suit the purpose, e.g., lack of plausible
evidence, evidence preferentially/selectively choosen, degree of reliability ignored; (10)
personal judgements and preferences of the investigator misapplied; and (11) adherence
to "types" that do not stand the test of new information and concepts.

Let us consider one example that demonstrates the "type-categorizing" of ore
deposits accepted for some time and which needs a careful re-examination as a con-

sequence of new data. This is particularly required as these "types" have frequently acted as an intellectual straight-jacket, diminishing proper communication and even cancelling progress. Amstutz (1972) maintained that it is more appropriate to use the expression "Mississippi Valley—Bleiberg—Silesia" ore for those Pb—Zn and related metal accumulations that have been referred to collectively as "Mississippi Valley-type". His reasons for this nomenclatural expansion were: (1) the strata-bound Pb—Zn—Ba—F—(Cu—Co—Ni) deposits with sedimentary diagenetic features are not restricted to continental platforms, but also occur in geosynclinal belts; (2) they are neither confined to one geological period nor to one continent; (3) in a certain milieu, they grade into the Kupferschiefer- or red-bed type or massive sulfide-type of deposit; (4) they grade from those ores without any volcanic-exhalative affiliations to those with a positive affiliation; and (5) they occur in undisturbed to highly metamorphosed and folded geologic settings. To put order into the categorization of the above ores, Amstutz proposed a simple, but systematic, set of criteria useful for both exploration and theoretical work. Without providing specifics here, it is of interest to point out that these criteria are based on observations (*not* interpretations) on the regional, local/outcrop, handspecimen and microscopic scales.

In establishing his hypothesis of a diagenetic-sedimentary origin of the Pb—Zn ores in carbonate host-rocks, Amstutz used certain prerequisites (pp.209—210): "(1) interpretations should not be made on evidence from one scale only; (2) none of the congruencies are *a priori* more important than others; (3) chemical or compositional evidence also works with congruencies (histograms, phase diagrams, etc.); (4) the compositional (chemical, physicochemical) and the geometric (textural) evidence should be used 'at par', i.e., none should be overrated (or underrated); and (5) an interpretation or theory should be considered to be a working hypothesis, because our interpretations are all subjective, i.e., full of cultural input, which means historical and geographical relativity; and we need the next younger generation or colleagues with a different cultural background to tear us loose from our own idiosyncracies, our own subconscious ties to stiff dogmas."

In conclusion of this section, it may be quite appropriate to follow Joralemon's (1975) light-hearted — yet factual and realistic — treatment of defining and classifying "ore" (i.e., economically viable deposits). He offered nine categories of mineralization based on *individuals best qualified to discover them*: (1) geologist's ore, (2) geophysicist's ore; (3) metallurgist's ore, (4) engineer's ore, (5) miner's ore, (6) assayer's ore, (7) lawyer's ore, (8) politician's ore, and (9) prospector's (bar-room) ore. At first sight one may have the inclination to consider Joralemon's scheme with "tongue-in cheek" as though he has taken a condescending attitude; however, after appreciating his discussions it becomes clear that each one of the above-mentioned individuals is engaged in professional activities that have either discovered ore deposits or was responsible in changing known uneconomic mineralizations into "ore" through influencing the political—sociological—economic milieus.

BASIC TERMINOLOGIES IN ORE AND HOST-ROCK PETROLOGY

"The aim of all science is to cover the greatest number of empirical facts by logical deduction from the smallest number of hypotheses or axioms."
    Albert Einstein (in: *Life, 9,* 1950)

Adhering to the main leitmotiv of this multi-volume handbook, first consideration should be given to the clarification of stratiform, strata-bound and related terms. "Stratiform" refers to the form or shape of a layer, bed, or stratum of either sedimentary or igneous origin; consisting of roughly parallel bands or sheets, such as a "stratiform intrusion", without reference to the relationship between the ore and the surrounding host-rocks. "Strata-bound" is used for a mineral deposit confined to a single stratigraphic unit and is parallel to the enveloping rocks. This term should be clearly discriminated from "stratiform" whenever convenient. "Stratofabric" refers to the arrangement of constituents in any stratified body ranging from the dimensions of a thin section to those of a sedimentary basin. "Strata-bound" is equivalent to "conformable" and "concordant", therefore, in contrast to the disconcordant, transgressing or cross-cutting relationships. ("Pene-concordant" has been used for those that are *approximately* conformable.) Strata-bound deposits can be considered also as stratiform depending on the shape of the unit or its individual parts. The scale or size-relationships are a determining factor too: micro-lenses appearing disconcordant in a thin-section, may be part of a concordant (i.e., conformable = strata-bound) unit as seen along an outcrop. Also, as can be deduced from the above definitions, "stratiform" bodies are either concordant or disconcordant within *bedded* host-rocks [1].

The above semantical problems have been discussed and summarized by Nicolini (1970) and depicted in several figures and tables. Although these terms are based on geometric structural and stratigraphic relationships, certain genetic implications (or more precisely specific restrictions) are involved, so that both "stratiform" and "strata-bound" can be, according to some researchers, *"genetically flavoured"* names. Strata-bound ores (whether strictly stratiform or not on the *microscopic scale*) are either of syngenetic, syngenetic—diagenetic, diagenetic, or epigenetic origins; the epigenetic mineralizations being the consequence of solutions moving parallel to the bedding, possibly controlled by permeability. On the other hand, the well-developed discordant stratiform units can hardly be thought of as syngenetic in origin, but can range from diagenetic to epigenetic because the cross-cutting relations require a pre-existing host (either unconsolidated as in the case of early-diagenetic mineralizations or lithified as exemplified by post-cementation ores).

---

[1] Note that the presence or absence of this "concordancy relationship" can be established only when the host/country rock is bedded, i.e., is structurally heterogeneous; in a homogeneous (="massive") rock this is not possible.

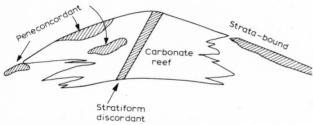

Fig.9. Diagram of carbonate biohermal reef with three styles of ore occurrences. (After Nicolini, 1970; courtesy of Gauthier-Villar, Paris.)

Strata-bound deposits can be viewed also as being either exogenous or endogenous types (or a hybrid of both in specific cases), i.e., the former originate *at or near the surface* of the earth (e.g., along the sediment—water interface or just below it), whereas the latter form *within* a stratigraphic unit (i.e., somewhat more remote from the surface or deep underground). The problem here too is that transitions and gradations exist between the two. This stands to reason, because a whole spectrum is evident in transitional processes and products from diagenesis to epigenesis, consequenting the transitional/gradational nature of the exogenous and endogenous deposits. (Here also, as in the case of supergene, hypogene and related terms, one should be unambiguous in clarifying whether "endogenous" and "exogenous" refer to *source of material,* types of *transportation processes,* or mode and locality of *concentration/precipitation mechanisms.*) At the two ends of the spectrum are the clearly definable exogenic and endogenic ores, but the nomenclatural problems may arise with the deposits that exhibit characteristics of both; and the grouping of the ore into one or the other class may depend on how willing the investigator is in shifting the boundary between exogenesis and endogenesis one way or the other along the spectrum. It should be understood, that in such deliberations we are comparing merely *primary* exogenous and *primary* endogenous mineralizations. Another variety of transitions—gradations between these two is that produced by *secondary* processes, namely when for example a *primary syngenetic exogenous* deposit is remobilized to give rise to a *secondary epigenetic endogenous* accumulation (and at the same time may change from a concordant strata-bound to a discordant stratiform deposit!).

Figs.9–13 and Tables XIX and XX summarize the various styles of relationships based on concordancy; Table XIX includes some comments on the connection between syngenesis and epigenesis and the types of concordance to be expected.

Mineralizations in sedimentary and volcanic rocks have been described from different genetic deliberations using "time", "location of source", "direction of fluid movements", "type of solution" or "mode of accumulation" and "influence of volcanicity" as parameters, summarized in Table XXI. Some definitions of key-words are offered in Table XXII.

Discriminations should be made between "ores in sediments" and "sedimentary

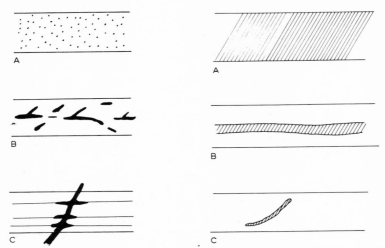

Fig.10. Mineralizations related to host-bed. (After Nicolini, 1970; courtesy of Gauthier-Villar, Paris.) *A* = stratabound-disseminated; *B* = stratabound-fissure filling; *C* = stratiform-discordant (versus other types of discordant bodies).

Fig.11. Types of concordant occurrences. (After Nicolini, 1970; courtesy of Gauthier-Villar, Paris.) *A* = mineralization occupies whole host-bed; *B* = only a certain part of the host-bed is mineralized parallel to bedding; *C* = mineralization is discordant within the host-bed.

ores", inasmuch as the former expression merely indicates that the mineralization is localized within sediments *without* suggesting any mode of origin. In contrast, "sedimentary ores" has genetic implications by indicating that the mineralization is the product of low-temperature and low-pressure sedimentary processes along the earth's surface. "Ores in sediments" requires further clarification because numerous transitional types have been incorporated in this class. The "syngenetic" and "diagenetic" ores, which formed more-or-less at the same time as their host-rocks, are of course the most obvious varieties of the "sedimentary ores". However, many "epigenetic" types of mineral accumulations have been wrongly included in this group merely because the primary features of the sedimentary rock (e.g., bedding, permeability zones, units with certain textures) controlled the *localization* of the secondary mineralization (see, for example, the diverse views on the origin of the Pb—Zn ores of the Mississippi Valley-type; Wolf, 1976c). As the textures—fabrics—structures of the ores clearly support a late-diagenetic to epigenetic mechanism of metal-concentration, the deposits are to be considered as "ores in sedimentary host-rocks". (Such discussions are not irrelevant or pedantic, as related "muddled thinking" has led to many communicational confusions.) Among the "ores in sediments" are included some varieties of telethermal—epithermal—hydrothermal (=igneous hydrothermal) origin. Therefore, among this group of mineralization one has to recognize transitional varieties into those that were formed by processes related to sedimentary basin-

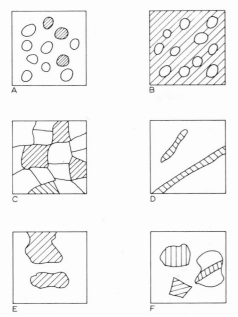

Fig.12. Six relationships between mineralization and the host-rock components. (After Nicolini, 1970; courtesy of Gauthier-Villar, Paris.)

A = specific detrital/clastic grains are replaced; B = ore mineral(s) constitutes cement (=intra-granular ore mineral-cement); C = coarsely crystalline rock of which some crystals constitute the ore mineral(s) - various possible origins; D = fissure fillings; E = irregular replacement of host-bed; F = detrital/clastic grains of ore minerals or with patches of mineralization ("reworked" ore).

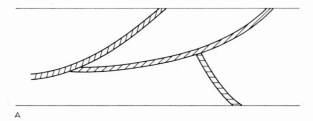

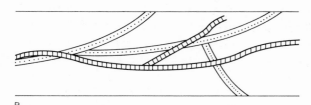

Fig.13. Relationships between mineralization and internal depositional structures within the host-bed. (After Nicolini, 1970; courtesy of Gauthier-Villar, Paris.)

A = minerals parallel to cross-laminae, e.g., Rhodesian copper belt; B = mineralization cutting across internal structures.

TABLE XIX

Genesis and concordance of ore deposits (After Nicolini, 1970, table XXVII; courtesy of Gauthier-Villar, Paris) (Freely translated from the French)

	Stratiform *per se*	Pene-concordant	Aggregates	Fractures	Veins
Relationships with the constituent parts of the geological environment	The ore is present as one of the constituents of the rock: i.e. as ore cemented by barren material (in the economic sense) or whose elements have the same appearance as the country rock, e.g. phosphorite, oolitic iron, detrital minerals.	Ore generally forms cement, or has the same aspects as the constituents of the rock.	Ore generally has the same aspects as the constituents of the rock.	Ore is discordant to the structure of the wall rock, but is confined to a specific host rock.	Ore is discordant to the structure of the wall rock and is *not* confined to a specific host-rock.
Concordance	Generally concordant with the properties and morphology of the surrounding rock.	Concordance is variable with the internal structures and the morphology of the surrounding rock.	Generally non-concordant to the internal structures, but the morphology of the concentration and the surrounding rock are sometimes analogous.	Discordant with both the internal structures and the surrounding rock.	Discordant to the structures.
Genesis and relationships between ore and country rock					

Genesis and relationships between ore and country rock:

Syngenesis ←————— Diagenesis ————→ Epigenesis Endo- ------ Exudation ------→

       Early | Late   Epigenesis Exo- ------→

Ore cemented ————→

←———— Ore has the same characteristics as the constituents of the country rock ————→

Ore is cementing the constituents of the host-rock ————→

←———— Ore is filling fractures ————→

←———— Ore in the form of "pockets"

TABLE XX

Hierarchy of the morphologic types of ore concentrations and their relationships to host rocks (After Nicolini, 1970, table XXX, p. 626; courtesy of Gauthier-Villar, Paris) (Freely translated from the French)

	Relationships with the constituents of the geologic environment	Degree of concordance
Stratiform *per se*	The ore is generally appearing as one of the constituents of the rock, cemented by the barren material (in the economic sense and not in the geochemical sense), e.g., phosphates, iron oolites, heavy-mineral detritus; or whose constituents have the same appearance as those of the wall rock, e.g., sulfides of Cu, Pb, Zn, etc.	Generally concordant in various degrees with the wall-rock constituents and/or the internal structures, and the morphology of the surrounding rock.
Pene-concordant	The ore generally forms the cement of the wall rock, or has the same appearance as the components of the wall rock.	Concordance is variable with the wall-rock constituents and/or the internal structures and the morphology of the surrounding rock. In general, there is better concordance when the ore is a sulfide (Cu, Pb, Zn) than when it is an oxide (K, V).
Massive	Ore has generally the same appearance as the constituents of the wall rock.	Generally non-concordance with the internal structures, but the morphology of the concentration is often determined by the general morphology of the surrounding rock.
Fractures	Ore is independent of the wall rock constituents, but is confined to a specific host-rock.	Ore deposit discordant with the internal structures and the surrounding rock.
Veins	Ore is independent of the wall rock constituents and can transect more than one type of wall rock.	Ore deposit discordant with the internal structures and the surrounding rock.

development, e.g., ore produced by connate solutions (i.e., *non*-igneous fluids) [1].

Although it may appear difficult in *practice* (and at the present time is often impossible) to determine the precise origin of many of the transitional—gradational and hybrid ore types, our *early hypotheses* and *simpler* genetic *models* have not given due consideration to them. Our *terminologies* and *classifications,* consequently had to be *too simplistic* also. As logic dictates (based on philological reasons), if our concepts and

---

[1] Connate fluids in basins as young as Cenozoic in age may be sufficiently hot to constitute one variety of "non-igneous hydrothermal" solutions (cf. review in Wolf, 1976c, figs.3-258A and 3-258B).

TABLE XXI

Subdivisions of sedimentary ores (and related terminology) (K.H. Wolf, 1977, unpublished)

Time

1. Syngenetic
2. Diagenetic   } genuine sedimentary
3. Diplogenetic   } a. sedimentary
4. Epigenetic   } b. non-sedimentary

Location of source

1. Endogene *-syngenetic
2. Exogene-syngenetic   } "sedimentary"
3. Endogene-epigenetic
4. Exogene-epigenetic
                         } a. "sedimentary"
                            b. non-sedimentary

* Cf. "lithogene"
of Lovering (1963)
"stratafugic" of
Beales and
Jackson (1968)

Direction of fluid movement

1. Hypogene-syngenetic   } a. sedimentary
2. Hypogene-diagenetic   } b. exhalative
3. Hypogene-epigenetic   } c. hydrothermal
4. Supergene-syngenetic } a. sedimentary
5. Supergene-diagenetic } b. indirectly exhalative
6. Supergene-epigenetic } c. indirectly hydrothermal

Source and mode of origin

A. "Sedimentary"

1. Direct precipitation from surface water (e.g., lake or ocean)
2. Direct precipitation from connate water plus compaction fluids
3. Direct precipitation from circulating meteoric fluids plus compaction fluids
4. Direct precipitation from heated fluids
5. Direct precipitation from descending groundwater
6. Weathering-residual
7. Placer

B. "Sedimentary—vulcanic"

8. Exhalative-sedimentary
9. Hydrothermal, epithermal, telethermal

---

nomenclatures ignore transitional and hybrid processes and products, then during routine work we are not compelled to consider them — our observations are incomplete and the communication oversimplified. To close this argumentative cycle: if we do not have the *theoretical* conceptual and semantic tools, we are not likely to search for specific criteria to recognize the above-mentioned hybrid mineral occurrences in *practice*. Only more recently have attempts been made to eliminate these dilemmas.

Amstutz has been "preaching the gospel" on the importance of terminology for

TABLE XXII

Definitions of terms

Allo-: a prefix derived from allochthonous and indicating that the material has been transported before accumulation.

Allochthonous: a term used to designate sedimentary and pyroclastic constituents which did *not* originate *in situ*; they were derived from outside of or within the area of depositional site and underwent transportation before final accumulation.

Allogenic: a term meaning "generated elsewhere", and applied to those constituents that came into existence outside of, and previously to, the rock of which they now constitute a part.

Authigenic: meaning "generated on the spot". Applied to those constituents that came into existence with or after the formation of the rock of which they constitute a part and were formed *in situ*.

Diagenesis: it includes all physicochemical, biochemical and physical processes modifying epiclastic, pyroclastic, and various types of chemical deposits between deposition and lithification, or cementation, at low temperatures and pressures characteristic of surface and near-surface environments. In general, diagenesis is divisible into pre-, syn-, and post-cementation or lithification processes. Diagenesis takes an intermediate position between syngenesis and epigenesis, the former grading into diagenesis by syndiagenesis, and the latter grading into metamorphism. Under unusual conditions, however, diagenesis may grade directly into metamorphism, as in the Red Sea today where hot hydrothermal solutions pass into unconsolidated sediments (see epigenesis).

Diplogenesis: processes that are in part syngenetic—diagenetic and in part epigenetic (Lovering, 1963).

Endogenic: referring to components derived from within the sedimentary formation.

Epigenetic: of later origin than the rocks among which they occur. (One should note that "epigenesis" has been used for both sedimentary and igneous deposits. In the former case, epigenesis was used as the stage following diagenesis and prior to metamorphism.)

Exogenic: referring to components derived from outside, i.e., from either above or below, the rock unit, so that exogenic can be either hypogene or supergene.

Hypogenic: a term applied to material that is derived from within the earth interior in contrast to supergenic components (see supergenic).

Primary: characteristic of or existing in a rock at the time of its formation. This definition is too all-inclusive and vague in detailed studies and its use should be discouraged. It can be used unambiguously as a very general colloquial term in connection with genetic discussions only if the context leaves absolutely no doubt (see secondary).

Secondary: a general term applied to rocks, minerals, elements, structures, etc., formed after the rock in which they are found. This term, like "primary", is too all-inclusive and ambiguous in detailed studies and should be used only as a very general colloquial term when misinterpretation is absolutely impossible (see primary).

Supergenic: a term applied to those processes and products caused by material derived from descending fluids and gases (see hypogenic).

Syngenesis: the processes by which rock components are formed simultaneously and penecontemporaneously. (Note that "syngenesis" has been used for both sedimentary and igneous processes. In the former, syngenesis grades into diagenesis.)

---

several years in a number of publications (e.g., Amstutz, 1959a, b, 1972; Amstutz and Bubenicek, 1967; Zimmermann and Amstutz, 1964; see also Chilingar et al., 1967; Wolf, 1976c; Lovering, 1963) and some of the ideas are presented below, supplemented by editorial comments. Amstutz (1959a) discussed criteria for and against syngenetic and epigenetic zoning as based on time, space or shape, and physicochemical conditions of ore precipitation or composition, supported by logical criteria. By using iso-chemical, iso-geometric and synchronous features of both minerals and rocks of a number of well-known ore districts (all of them described in the present Handbook), he concluded that in as far as a gradational transition between syngenesis, diagenesis and epigenesis exists, there should also be a gradational change in textures, fabrics and structures. The latter have been discussed (see Amstutz, 1959a, pp.97—102; fig.1 from Niggli, 1952) from the purely geometric view point by using the concept of "concruency" and "inconcruency" (=disconcruency) in the investigations of ores in sedimentary, volcanic and plutonic host lithologies. Amstutz (1959b) then outlined convincingly the various "patterns of thoughts", or combinations of syngenetic and epigenetic processes, by deliberating the origin of at least twenty "types" of mineral deposits. He combined the time and space factors in classifying ore-forming processes and their products: syn- and epigenesis indicating relative time, and endo- and exogeneous (or could the suffix "-ic" be appropriate also?) or super- and hypogene denoting space or direction of supply of the mineralizing fluids (Tables XXIII and XXIV). This approach was then used to explain the formation of specific ores, namely the Mississippi Valley Pb—Zn and the Arkansas barite deposits, as diagrammatized in Figs.14 and 15.

Amstutz (1959b, p.36) warned of duplication in terminologies by demonstrating that "syngenesis" may be equivalent to "volcanic—exhalative" in some situations and that, in turn, equals "hypogene-hydrothermal". Also, "transported weathering product" may be described as "syngenetic-supergene". He maintained also that "descending", "ascending", "autochthonous", "allochthonous", "authigenic" and "allogenic" have been given various meanings by ore petrologists. The distinction between epi- and syngenesis is especially difficult in plutonic-hypabyssal-volcanic ores, e.g., porphyry copper,

TABLE XXIII

Basic patterns of genetic interpretations of rocks and ores (After Amstutz and Bubenicek, 1967)

Time	Space
Syngenetic formation	*Endogenous formation*
= contemporaneous with the enclosing rock	= origin same as, or from within the host rock
Epigenetic formation	*Exogenous formation*
= formation later than that of the host rock	= ore matter originates from without the host rock
Possible combinations: syn-endo syn-exo epi-endo epi-exo	

TABLE XXIV

Time–space relationships of mineralizations of syngenetic or epigenetic (I or II) origin and of endogenous or exogenous (A or B) formations (After Amstutz, 1964, fig. 1) (cf. Table XXIII)

Time Δt — Δs Space [1]	I Syngenetic formation $\Delta t = 0$	II Epigenetic formation $\Delta t = x$
A Endogenous formation $\Delta s = 0$	Possibility Ia (or A1)	Possibility IIa (or A2)
B Exogenous formation $\Delta s = n$	Possibility Ib (or B1)	Possibility IIb (or B2)

molybdenum and iron deposits. The relationships between host rocks and mineralizations have the following possibilities:

Host-Rock	Ore
syn	syn
syn	epi
epi	syn
epi	epi

More recently, Amstutz (1972) again felt the need to maintain that the above terms must be more carefully used — implying, therefore, that even at the present time the confusion persists among some researchers. Although he has spoken of "diagenesis" in his works, it remained to be encorporated into his general scheme (cf. Wolf, 1967, p.181, in Chilingar et al., 1967):

diagenesis—endogenous
diagenesis—exogenous
    (1) diagenesis-hypogene
    (2) diagenesis-supergene
    (3) diagenesis-stratifugic

Table XXV summarizes all the possible combinations, and it should be noted that "stratifugic" refers to the source *within* a specific rock; for additional comments see above. It must be pointed out once more, that what is "diagenesis" to some, is already "epigenesis" to others; however, the three-fold time division into syn-, dia- and epigenesis is well entrenched and remains practical as long as the terms are unambiguously defined.

Inasmuch as the source of the ore-creating constituents has been proposed to be of prime importance in some terminologies and classifications, it is refreshing to read about

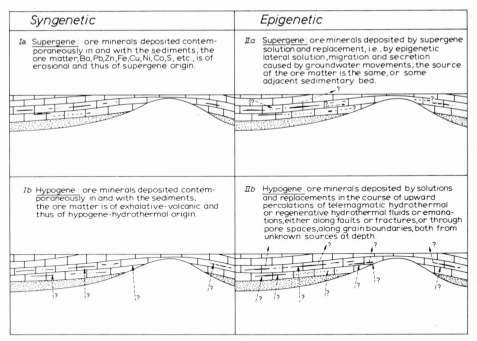

Syngenetic	Epigenetic
Ia <u>Supergene</u>: ore minerals deposited contemporaneously in and with the sediments; the ore matter, Ba, Pb, Zn, Fe, Cu, Ni, Co, S, etc., is of erosional and thus of supergene origin.	*IIa* <u>Supergene</u>: ore minerals deposited by supergene solution and replacement, i.e., by epigenetic lateral solution, migration and secretion caused by groundwater movements; the source of the ore matter is the same, or some adjacent sedimentary bed.
Ib <u>Hypogene</u>: ore minerals deposited contemporaneously in and with the sediments; the ore matter is of exhalative-volcanic and thus of hypogene-hydrothermal origin.	*IIb* <u>Hypogene</u>: ore minerals deposited by solutions and replacements in the course of upward percolations of telemagmatic hydrothermal or regenerative hydrothermal fluids or emanations, either along faults or fractures, or through pore spaces, along grain boundaries, both from unknown sources at depth.

Fig.14. The four basic theories on the genesis of the Mississippi Valley-type ore deposits. (Diagenetic fabrics point to a syngenetic mode of formation, according to Amstutz and Bubenicek, 1967.)

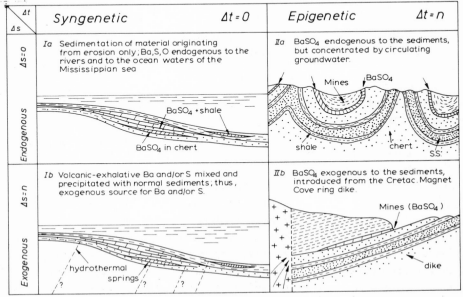

Fig.15. The basic theories on the genesis of the Arkansas barite belt (*S.S.* = sandstone). (Zimmermann and Amstutz, 1964.)

TABLE XXV

Basic patterns of genetic interpretations, considering time, space, source directions and possible combinations thereof (K.H. Wolf, 1975, unpublished)

	Time	Space	Source direction	
All transitions and combinations possible	*Syngenetic* = contemporaneous with enclosing rock	*Endogenous* = origin same as, or from within the rock with ± movement	*Stratifugic* = intraformational, ± from stratigraphic equivalents, ± horizontal	**All transitions and combinations possible**
	Diagenetic = all physicochemical, biochemical and biological plus physical processes modifying deposits between accumulation and lithification near the surface of depositional environment	*Exogenous* ⟶ = ore material from outside the locality of precipitation	⟶ *Supergene* = source from above *Hypogene* = source from below	
	Epigenetic = formation later than host rock			

Possible combination

syn-endo	
syn-exo	
dia-endo	→ syn-, dia-, epi- → stratifugic / supergene / hypogene
dia-exo	
epi-endo	
epi-exo	

All transitions and combinations possible

(Note: using any of the three terms that indicate "source direction", the prefix exo- is superfluous. The collective term exo- is only necessary in cases where the source direction is not known.)

NOTE: (1) It is very important to distinguish between the sources of (a) the fluids, (b) the cations, e.g., Cu, Fe, U, Pb, Zn, etc., and (c) the anions, e.g., sulfur, all of which may or may not have the same or may have different sources and origins. Cations may be physicochemically mobilized, whereas sulfur biochemically. From a strict point-of-view, it may be that many ores are diplogenetic.

(2) The above scheme is mainly based on Amstutz (various publications) and Wolf (1967, in Chilingar et al., 1967).

a more logical approach (at least at this stage of knowledge). Amstutz (1972, pp.213–214; supported by Maucher, pp.214–215) pointed out that the *source* of the material of the ore minerals is *not* as important in the *descriptive* classification schemes — one reason being its highly conjectural and interpretive nature — in contrast to *directly observable* or directly determinable geometric, mineralogical and geochemical criteria of both the host rock and ore. Only in genetic/interpretive classifications should one resort to genetic nomenclatures that implicitly indicate, for example, the location and/or direction of the sources of the anions and cations of the ore and gangue minerals.

Lovering's (1963) contribution should be mentioned here as he too discussed the semantical disagreements over "epigenesis" and "syngenesis". He clarified that some deposits are in part syngenetic and in part epigenetic, if one considers the *source* of the constituents; in such instances he would speak of "diplogenic" sources and processes of mobilization/transportation. For example, syngenetic sulfur can combine to form pyrite which becomes later partly replaced by copper, the latter being epigenetic (e.g., White Pine, Michigan). In another case of diplogenesis, syngenetic cations united with epigenetic anions; in other instances, epigenetic cations combined with syngenetic sulfur.

The three above-listed terms chiefly refer to *time* relationships between ore and host lithologies — only in a subsidiary sense are they *space* terms. They do not, in general, imply processes of formation or type of source, although they may be used in this way (Snyder, 1967) (Table XXVI), as long as a *meaningful definition* is furnished.

"Lateral secretion" is now too vaque, hence Lovering's (1963) proposal to use "lithogene" for ore-forming components derived from a rock (within the same basin?) through remobilization, transportation and reprecipitation, is preferable. The source can be earlier-produced syngenetic, diplogenetic and/or epigenetic components, and according to distance between source and loci of precipitation, one may distinguish between "local lithogene", "remote lithogene", "intra-host-rock lithogene" and "extra-host-rock lithogene". Further deliberations: "lateral secretion" and "lithogene" are to be distinguished from "stratifugic". Admittedly, the latter is of "lateral secretion" or "lithogene" origin,

TABLE XXVI

Syngenetic, diagenetic and epigenetic theories of origin (After Snyder, 1967, table 6; courtesy of *Econ. Geol.*)

		Mode of origin
Type of deposit	Syngenetic	Detrital
		Metals in solution in seawater
		Metals derived from volcanic exhalations
	Diagenetic	Recrystallization of materials *in situ*
		Minor migration, metals from host
		Minor migration, metals from volcanic exhalations
	Epigenetic	Cold meteoric water
		Hydrothermal
		Heated meteoric water
		Metals from known igneous source
		Metals from unknown igneous source
		Connate water

but the source is *within* the same unit in which the mineralization occurred, i.e., *intra*formational. In contradistinction, "lateral secretion" or "lithogene" are less precise in this regard. The consequent semantical question arising is: are all three terms theoretically and practically recommendable and acceptable?

A few supplementary comments are necessary on *gradational–transitional* processes and products in regard to the enigma of syngenesis versus epigenesis. Table XXVII presents two end-members, namely, syngenetic ores in clay-rich sediments (=mudstone, shale, slate) (cf. also Wolf, 1976c, table 5-VIII, and this Handbook, Vol. 1, Chapter 2, table II, p.65, for the shale–pyroclastic spectrum) and epigenetic mineralization within carbonate host-rocks. The shale–limestone lithologies are merely two possible combinations, and numerous others could have been chosen. Four envisaged genetic situations (Nos. 1 and 4) have been superimposed on the diagram depicting the "degree of syngenicity" involved, ranging from solely syngenetic (No. 1) through two cases of diplogenesis (Nos. 2 and 3) to purely epigenetic features (No. 4).

In order to avoid confusion in communications, when using the expressions "syngenesis" and "epigenesis" (and others), a clear distinction has to be made between (a) original or primary and (b) subsequent or secondary processes and products, as well as (c) between the source of the components and their final place of precipitation, because a secondarily formed epigenetic deposit may have been the result of remobilization (during

TABLE XXVII

Scheme depicting syngenetic–diplogenetic–epigenetic transitions (an example based on clayey to calcareous host-rock types) (K.H. Wolf, 1975, unpublished)

Syngenetic	*Diplogenetic*		*Epigenetic*
= contemporaneous ⟵ ⟶ with host-rock	= of mixed syngenetic ⟵ and epigenetic origins	⟶	= formed later than host-rock
deposits within clay-rich *source* ⟵ lithologies		⟶	deposits within calcareous/dolomitic *host*-rocks
e.g., Kupferschiefer Mount Isa			e.g., some Mississippi Valley-type deposits
(1)	(2)	(3)	(4)
no remobilization ⟶	various degrees of remobilization within shale source rock ⟶	remobilization leading to precipitation of some material in limestone host after transportation, whereas rest of constituents remained behind in clayey source lithologies ⟶	remobilization resulted in transportation and removal of material from the source to the host lithologies (i.e., no material left behind in source

one or more stages) of an earlier, primary syngenetic precipitate. Depending on various geologic circumstances, some features of the syngenetic constituents may be left for study or may have been completely obliterated. Thus, one also has to clarify as to whether one speaks of the *processes* or the *products* when applying terms like "syngenetic", "epigenetic" and "diplogenetic", although certain investigators do have certain preferences and thus restrict the use of these expressions.

Having recognized the confusion that can arise from employing simple terminologies and classifications to represent complex natural situations, appropriate modifications have to be introduced. The pigeon-hole scheme will remain a useful guide, but is misleading when applied too strictly. What one might call "transitional-type classifications" are probably more indicative of natural occurrences, as illustrated by Gilmour in Chapter 4, Vol. 1 in this Handbook, and Wolf (1976c) (cf. also Gabelman's Chapter 3, Vol. 1 in this Handbook). The application of two or three end-members, and then finding all possible intermediate gradational cases between them, is only in its infancy, so that one can expect numerous such schemes to be worked out in the near future (possibly by systems-analysis techniques). Many physical, chemical and geological parameters lend themselves as end-members — one set of these could be left constant, whereas another can then be systematically changed — thus offering a large number of combinations in presenting differences and similarities (thus forcing a comparison) between types of mineralizations. This comparative work, it is predicted, will demonstrate that parameters at present lumped together into one class, when used as independent variables, will be instrumental in the recognition of hitherto unknown subgroups and groups; see Amstutz (1972) for example. Gradational characteristics based on gross composition; minor and trace element composition; isotope make-up; textural, fabric and structural features; host-rock variations; original and/or secondary environmental settings; tectonic milieu; and others, may eventually all be employed in classifying specific ore types.

Nomenclatures and classifications, as clarified earlier, are tools of communication because a term represents an idea, sometimes a whole hypothesis; consequently, a particular term sets one's mind in motion along a certain line of thinking. Some "intellectual coaxing" by the force of one expression can redirect or rechannel one's trend of thought to other possible interpretations. The coining of a new word, where needed to denote a new situation, process, product or abstract case, is one means to achieve taking stock of all alternatives; unless one prefers to conveniently ignore the newly coined expression! Example: Wolf (1967, in Chilingar et al., 1967, table III, p.228) has expanded the term "micrite matrix" by using the prefixes "pseudo-", "ortho-", "allo-" and "auto-" (see also Dickinson, 1970). Originally, it was not realized that "matrix" is a collective name, representing several genetic types of matrices of limestones, sandstones and pyroclastics — the introduction of the prefixes *will demand a consideration* of all other types or sub-types thus established. Ignore these prefixes (i.e., remove them from our petrologic nomenclature) and the demand has gone. (For studies of matrices in iron ores, see Chapter 5 by Dimroth, Vol. 7 in this Handbook.)

Figs.16 and 17 demonstrate that syngenetic—epigenetic transitions are to be expected in the same ore district as a consequence of the mineralizations being genetically related in space and/or time (Smirnov, 1972). In the former, the geologic ages of various igneous rocks are listed: the volcanic—sedimentary mineralizations are associated with volcanic and subvolcanic processes and rocks in the Devonian (D_2, lower D_3), but these magmatic mechanisms did not result in ore concentrations during the upper Devonian (upper D_3) and Carboniferous (C_2, C_3). The plutonic-hydrothermal (i.e., epigenetic) deposits are the consequence of strictly plutonic activity during the C_2 and P_1 geologic periods — the magmatism was not accompanied by subvolcanic and volcanic processes. Other associations of syngenetic and epigenetic ore bodies are diagrammatized in several other figures. In all such occurrences it is rewarding to (1) list all comparative data (i.e., similarities and differences) of the syngenetic and epigenetic deposits, and (2) establish the various gradational/ transitional products between them.

As many contributions to the present Handbook refer to plutonic—volcanic and metamorphic processes, some relevant introductory terminologies and classifications are provided prior to the sections on the specific subject matters. Schneiderhöhn and Borchert (1956) discussed the zonal classification of ore deposits (Fig.19, see below) as proposed by several researchers and found them wanting in several respects, and they then proposed a scheme based on (1) temperature of formation, (2) depth of ore-forming magmatic—volcanic intrusions, (3) depth of mineral precipitation, and (4) pressure — each zone being properly named. The chief subdivisions are abyssal, "deep"-plutonic, "high"-plutonic, and sub-volcanic (volcanic *per se* being along the surface), whereas the "facies" terminology begins with eclogite to basalt and granite, and above these from pegmatite to telethermal deposits. The pressure values were calculated from the rock overburden with a density of $D = 2.75$. The above well-established approach is useful in subdividing host-rocks of mineralizations, as well as ores and alteration products. More recent information, one should realize, supports the proposal that the connotations tele-, epi- and meso-thermal ores, for example, refer to *several possible types* of deposit-forming conditions and mechanisms, and are *not* restricted to genuinely igneous- or magma-related processes. For example, deep intrastratal connate or compaction fluids over 100°C within sedimentary basins would be "telethermal" or "epithermal" solutions that can form both strata-bound and vein-like deposits (cf. Wolf, 1976c for review) — thus being similar to igneous-hydrothermal mineralizations of the upper zones formed under identical pressure and temperature ranges. The same arguments can be extrapolated to deposits that originated at greater depth to include truly magmatic granites, palingenetic and other "metamorphic granites". Any terminology formerly in voque for igneous derivatives can now be employed for a gamut of deposits produced by other processes recognized during the past few years.

The first of three diagrams offered below consists of temperature—depth curves (ranging from 10.0 m/1°C to 130m/1°C geothermal lines) extending to a depth of about 100 km of the earth's crust (Schneiderhöhn, 1958, p.37), upon which the granite iwth 4%

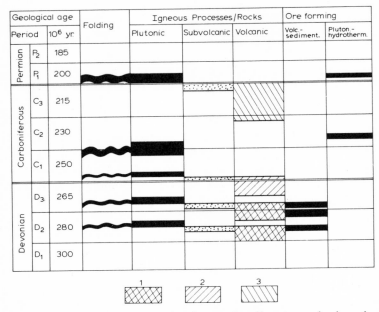

Geological age		Folding	Igneous Processes/Rocks			Ore forming	
Period	10⁶ yr.		Plutonic	Subvolcanic	Volcanic	Volc.-sediment.	Pluton.-hydrotherm.
Permian P₂	185						
Permian P₁	200						
Carboniferous C₃	215						
Carboniferous C₂	230						
Carboniferous C₁	250						
Devonian D₃	265						
Devonian D₂	280						
Devonian D₁	300						

Fig.16. Relation between syngenetic volcanogenic-sedimentary and epigenetic plutonogenic-hydrothermal pyrite-type ore mineralization at Rudny Altai. Legend: *1* = basaltic−rhyolitic rocks; *2* = andesitic−dacitic rocks; *3* = mixed shallow-water rocks. (Smirnov, 1972.)

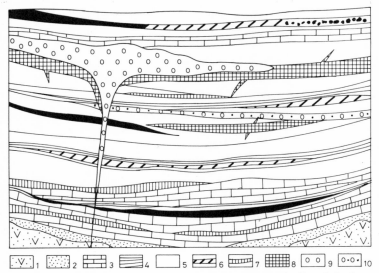

Fig.17. Relationships among syngenetic and epigenetic orebodies of the Atasu-type deposits (Smirnov, 1972). Legend: *1* = volcanics; *2* = aleurolite-sandstones; *3* = limestones and dolomites; *4* = cherts; *5* = argillaceous-carbonate rocks; *6* = iron, iron−manganese and manganese ores; *7* = sedimented zinc and zinc−lead ores; *8* = metasomatic lead−zinc−barite ores; *9* = barite metasomatites; *10* = barite−sulphidic ores.

and 7% H_2O-contents, gabbro and dunite melt-temperatures, and the volcanic (hypabyssal), sub-volcanic, plutonic, sub-Pacific, and the sub-Atlantic geothermal zones, were plotted (Fig.18). Extrapolating from such basic diagrams, more complicated models have been developed, as the subsequently presented two diagrams indicate (Figs.19 and 20).

Borchert (1957, 1962) offered a scheme (Fig.20) showing the convergence between plutonic and hypabyssal processes and deposits, on the one hand, and metamorphic ones, on the other. Similar to Fig.19 by Schneiderhöhn and Borchert (1956), this diagram is (1) a useful visual support for the reader to accompany the discussions of various types of igneous and metamorphic facies considered in specific chapters in this Handbook; and (2) Figs.19 and 20, when applied in ore petrology, will force a researcher to consider (a) the *transitional* characteristics of the mineralization and (b) the submission of early-formed strata-bound ores to numerous metamorphic alterations up to conditions that result in

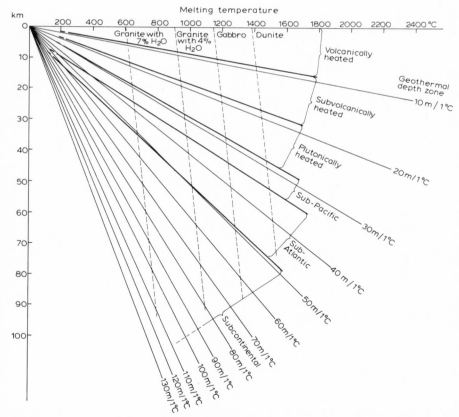

Fig.18. Temperature diagram for differently heated parts of the upper 100 km of the earth's crust. (After Schneiderhöhn, 1958; courtesy of Gustav Fischer, Stuttgart.)

remelting. In Fig.20, the following variables are given: pressure (*P*), depth (*D*), and numerous temperature/depth (*T/D*) gradients ranging from 10°/km (cratonic and Precambrian shields) to 90°/km (steepest gradient in orogenic areas), *P/T* zones of metamorphism as related to sialic-palingenic and juvenile tholeiitic-basaltic magmatism (including regional, contact, low-P/high-T, and migmatite-forming metamorphism), lines depicting 2% and 9% H_2O in igneous rocks, position of Conrad- and Mohorovičić-discontinuities, more common locality of feldspathisation, boundaries outlining conditions of migmatisation and melting of solid rocks to form magma, and the line depicting the *P/T* relations which markedly changes from 10,000 atm to 50,000–100,000 atm with a concomitant

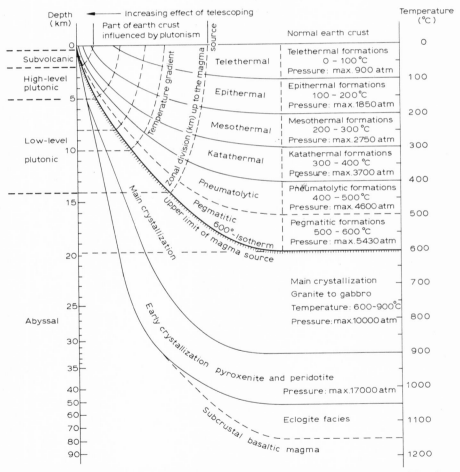

Fig.19. Schematic model of the zoning of the earth's crust. (After Schneiderhöhn and Borchert, 1956; courtesy of *Neues Jahrb. Mineral.*)

rate of crystallization from 1 mm/hour to 1 mm/second. The above-mentioned H$_2$O-content is important because the melting of granitic material begins under high pressures at an H$_2$O-content of 9% already at 600°C; or at 4000 atm the granitic and slaty rocks, for example, will commence to melt at 660°C under H$_2$O-vapour pressure; when the water contains 4% (by weight) HF the melting starts at 590°C. Important to realize is that such

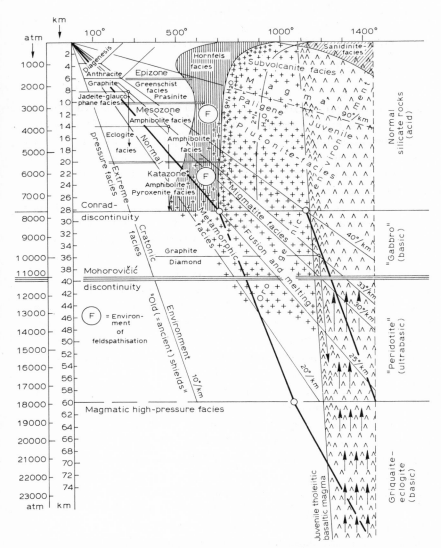

Fig.20. Pressure—temperature environment of metamorphism (mainly after A. Schüller, 1961) and their relationships to sialic—palingenetic and to juvenile tholeiitic—basaltic magma. (Mainly after Borchert and Tröger, 1950; cf. Borchert, 1957, 1962; courtesy of Glückauf, Essen.)

diagrams, although very useful, are idealized and simplified. For instance, old Precambrian shields do not exhibit clearly demarcated Conrad-discontinuities, and the Mohorovičić-discontinuity may be completely absent. If "discontinuities" did exist, they cannot be clearly equated with those in younger geosynclinal localities and conditions.

COMPREHENSIVE HOST-ROCK AND ORE CLASSIFICATIONS

"The growth of science consists in a continual analysis of facts of rough and general observation into groups of facts more precise and minute."
Walter Pater (In: *Coleridge,* 1865)

Mineralizations within sedimentary and volcanic rocks have to be placed into context with all other types of ores to ascertain a fuller genetic comprehension, thus also explaining their possible associations in space and time. Stratiform deposits, as a distinct group, have found a niche in most omnibus classifications, although they have often been treated in varying ways. The following schemes by the respective investigators are presented to offer a cross-section from which one can select the most suitable type for one's purposes, or as a challenge, an attempt can be made to combine two or more into one classification with broader coverage: (1) diagrammatic schemes by Taupitz (1954), Borchert (1950, unpublished), Skinner and Barton (1973); (2) tabular schemes by Sullivan (1957), Bateman (1950), Cissarz (1951, 1964, 1965), Pereira (1963), Clark and Lewis (1964), Pélissonnier (1962; see Nicolini, 1970) and Ridge (1972); and (3) verbal classification by Stanton (1972). (The reader is referred to the schematic classification of ore-genesis theories by Amstutz (1964) in Chapter 2, Vol. 1, fig.35, p.62). No claim is made that the genetic implications of the examples of specific ore districts cited in the classifications are at present acceptable in all details, for reinterpretations have occurred in several cases. Complementary to the above, a number of other schemes are available from the published literature, e.g., Schneiderhöhn (1941) and Beyer (1971).

The classifications provided below, in addition to conveying a comparative perspective between stratiform and remaining classes of deposits, illustrate that most of the classifications are oversimplified and convey a bias and simplistic situation. Consequently, the subsequent parts of this chapter will increasingly concentrate on schemes that selectively treat specific ore varieties, many of which are from multi-component stratiform/stratabound mineralizations.

Figs.21, 22 and 23a, b present a "bird's-eye view"-like diagrammatic classification of all major ore types. One may well ask about the purposes of preparing such "pictorial" or "cartoon"-like schemes, when in the past tabular arrangement of the data has been thought to be efficient. But are tables really and truly representative? The answer has already been partially provided above: first, it helps to have a visual (i.e., pictorial or flow-chart) support in any discussion, whether pedagogic or otherwise, and secondly,

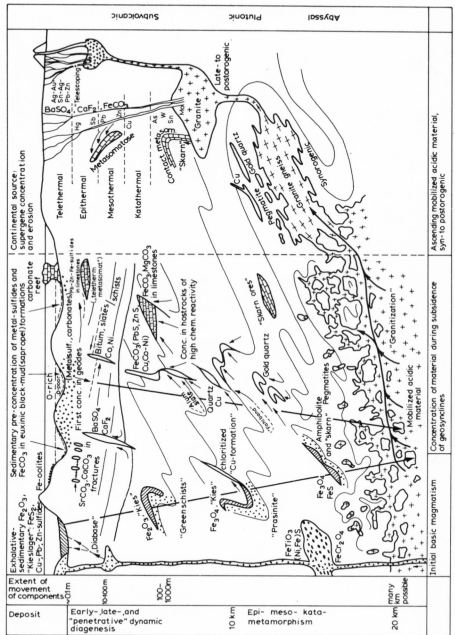

Fig.21. Rock and geochemical cycle, tectonism and the origin of ore deposits. (After Taupitz, 1954; courtesy of Glückauf, Essen.)

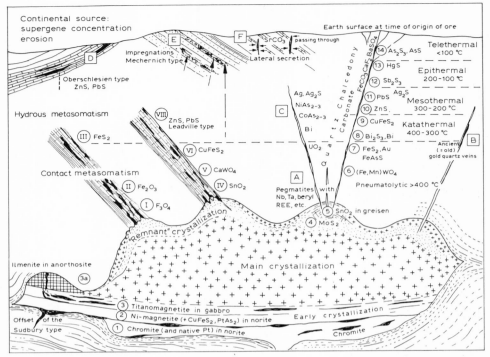

Fig.22. Genetic types of ore deposits. (After Borchert, 1950, unpublished; 1978; courtesy of Glück-auf, Essen.)

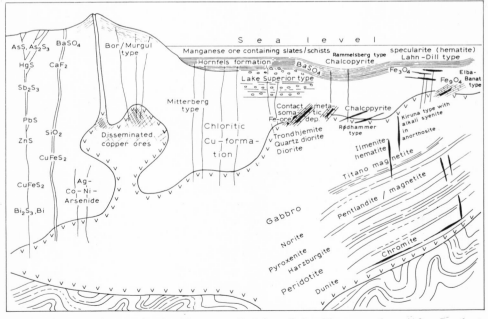

Fig.23a. Ore deposits associated with geosynclinal juvenile-basaltic magmatism. (After Borchert, 1960b, 1961, 1978; courtesy of Glückauf, Essen.)

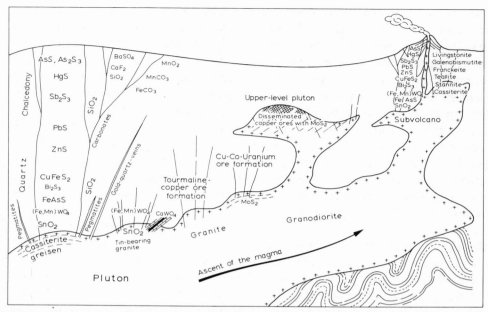

Fig.23b. Ore deposits associated with sialic-palingenetic magmatism. (After Borchert, 1960b, unpublished; 1978; courtesy of Glückauf, Essen.)

and more importantly, diagrams do not have the artificial boundaries of pigeon-hole schemes and, therefore, all transitional and gradational processes and deposits can be properly placed in an over-all "classification". More complicated and all-inclusive models will be prepared in the future, as has already been attempted by employing the flow-chart technique by Wolf (1973 and 1976b,c).

Figs.21 and 22 complement each other in so far as in the former drawing the geochemical cycle, tectonism and stage within the development of a geosyncline, depth of formation and association with plutonism/volcanism have been emphasized. Fig.22 is an expansion of the right-hand part of Fig.21 with some additions to list several major genetic ore types (see A—F, I—VII, and 1—14). Fig.23a (see also Fig.33 in the next section and compare both with Fig.23b) is a schematic simplification of a basalt mega-intrusion which invaded along a weak zone between an older basement and a younger pile of geosynclinal sediments; the amount of magma being in the order of tens of thousands of cubic kilometers. The early-magmatic crystallization formed chromite associated with peridotites, Ni-magnetite with norite mainly, Ti-containing iron ore with gabbros and anorthosites. The genesis of the Ni-magnetite is somewhat debatable as it has also been interpreted to have originated by pneumatolitic—hydrothermal activity. The Kiruna-type ores are associated with special basaltic magmas and if they occur intra-crustally in combination with iron oxide and alkali-silicate enriched remnant magma portions in contact with reactable carbonate host-rocks, then the contact-pneumatolitic ores of the Elba-and/or Banat-type can be produced. (However, see Parák, 1975, and Frietsch, 1978, for new genetic interpretations of the Kiruna deposits.) More intense differentiation during the early geosynclinal stage may give rise to a high degree of concentration of hydrothermal fluids which, in turn, may escape to the surface into the sea as chloride exhalative solutions to form, for instance, the Lahn—Dill Type iron ores (cf. Chapter 6 by Quade, Vol. 7 in this Handbook). The latter's genetic relationship with the Kiruna-type mineralization has been clarified by several researchers, inasmuch as the Lahn—Dill ores are frequently associated with diabases and keratophyric lavas as well as with acidic tuffs and ignimbrites; all the result of fractional crystallization and differentiation of a deep-seated magma. Reference to similar magmatic differentiation mechanisms in the subsurface relate the origin of the surface-formed Rammelsberg—Meggen type accumulations (see Chapters 9 and 10 in Vol. 9) to the intra-crustally produced Rødhammer-type deposits; the latter often being connected to the differential material of a trondhjemite magma. The high barite content of the Meggen and Rammelsberg mineralizations, which also has been reported from the Lahn—Dill ores, is considered to be characteristic of a transitional ore type into the manganese—slate—chert lithologic association. The chalcopyrite-containing massive sulfide deposit of submarine environments is an additional gradational variety. The intra-crustal chloritic copper formations find their surface or near-surface expressions as native copper deposits of the Great Lake-type.

In many instances, the group of disseminated copper (=porphyry copper) in the upper-plutonic zone is associated with intermediate tonacidic differentiates which, in turn, are derivatives of an original basaltic magma. The sub-volcanic boron-type deposits, associated with kaolinized and silicified andesitic rocks, are also present. Under these geologic settings, the processes of "hydrothermal replacement" play an important function in contrast to the mechanisms that produced the Rammelsberg ores.

One further aspect needs examination: during the past few years Amstutz (1974) has been the main exponent of the hypothesis (i.e., as yet in a state of flux and unproven) that spilitization is not restricted to secondary near-surface metasomatic processes (in a marine environment, for example), but that spilites can also be "primary" products of basaltic magma differentiation. By extrapolation, and believed to be supported by observational data, Amstutz advocated that the mineralization associated with spilites are also differentiation products. He reasoned that ore-rich spilites and quartz-keratophyres occur in all gradations and are closely associated in both space and time with basaltic and ultrabasic deep-seated rocks; hence the importance of the processes of fractional crystallization and differentiation in the origin of basic to acidic ore-rich end-products is believed to be established. With the addition of this hypothesis to solve the enigma of spilitization, two schools of thought exist, namely the so-called "secondary metasomatic" and the "primary differentiation" advocates.

The spilitization problem exemplifies the fact that certain *descriptive* terminologies and classifications of ores can remain valid, although the underlying *genetic* ideas may develop continually. If a classification is founded on host-rocks, then the spilite-associated ore remains part of the (descriptive) spilite-hosted mineral group, even though the *genetic* concepts of the spilites may be modified. This situation remains viable as long as "spilite" is a descriptive lithologic term. The only terminological and classificatory change to take place is when either (1) new data lead to the recognition of two or more *descriptive* types of spilites and one can establish a correlation between different ores with these new lithologic varieties, or (2) when *several genetic* spilite types are formulated, each with a specific genetic mode of mineralization. In both cases (1) and (2), any earlier classification that simply lists "mineralization in spilites" as one variety, must then be expanded to accommodate the newly recognized ore varieties. This classificatory expansion can take

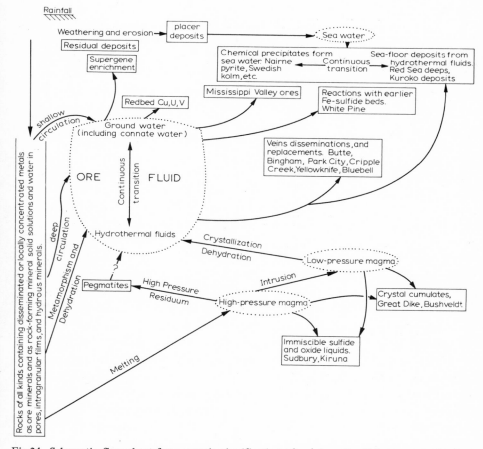

Fig.24. Schematic flow sheet for a genetic classification of ore deposits. This is a geologic cross-section. Rounded areas denote movement of material; boxes denote deposits; arrows show genetic relationships. (After Skinner and Barton, 1973; courtesy of *Annu. Rev. Earth Planet. Sci.*)

place either within the same scheme or a new classification could be proposed.

Skinner and Barton (1973) presented a schematic flow-sheet-like genetic classification of mineralizations (Fig.24) that resembles the rock or geochemical cycle (Fig.25) to be found in many textbooks. Inasmuch as ore concentrations are considered as naturally formed rocks, the comparison between the rock cycle and the ore-forming mechanisms is indeed appropriate — all processes resulting in the origin of "ordinary" lithologies can also form economical mineral concentrations, as illustrated in Fig.25A and B. Amstutz (1964) also offered a diagrammatic scheme of ore-genesis theories, which he divided into

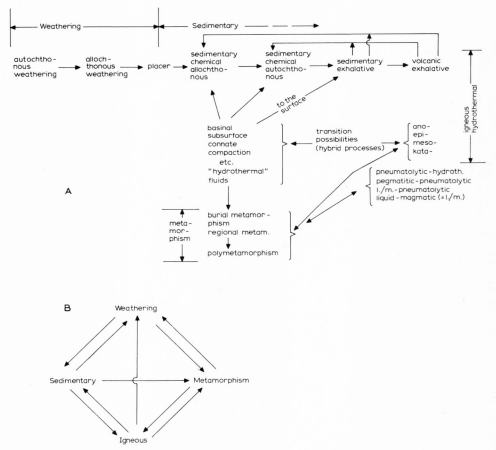

Fig.25. Relationship between geological mega-processes and the origin of ore deposits (A) (transitional/gradational situations depicted by arrows), and their identity to the "normal" rock/geochemical cycle in the earth's crust.

two parts (see fig.35, Chapter 2, Vol. 1) to depict the "conventional" and the "new" patterns of thought. Fig.26 gives an additional scheme.

The six tabular classifications (Tables XXVIII—XXXIII) and the list by Stanton (1972) (Table XXXIV) need little explanation.

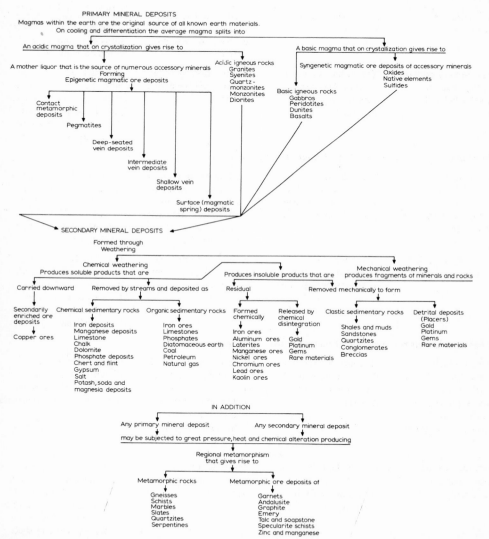

Fig.26. A classification of mineral deposits. (After Clark and Lewis, 1964; courtesy of Wiley, New York, N.Y.)

TABLE XXVIII

Classification of metalliferous provinces and deposits (After Sullivan, 1957; courtesy of *Can. Min. Metall. Bull.*)

Rock environment	Examples	Possible origin	Remarks
		Ores derived from sediments	
Syngenetic sedimentary ores			
Beach pebble conglomerate	Rand (Au); Blind River (U)	Gravity- and wave-concentrated black sands rejuvenated by metamorphism	Dimensions are those of sedimentation. Concentrations of economic grade comparatively rare.
Deltaic and closed basin (euxenic) sediments. Shore-line beds. Reef facies	White Pine, Michigan; Cu belt, Rhodesia— Katanga (mostly); Mansfield	Precipitation by reduction of $CuSO_4$-bearing solutions	Deposits of great importance though economic grade rare. May in future include deposits of Ni, Zn, U. Mo(?), as well as Cu, Co known at present.
Sedimentary ores re-concentrated by circulating meteoric waters			
Continental and lacustrine beds. Alternating oxidizing and reducing conditions of sedimentation	Red-bed Cu ores of U.S.A., Bolivia, Nova Scotia—New Brunswick; Colorado Plateau U ores.	Gradations from purely syngenetic deposits to ores which were re-concentrated by circulating waters both in ancient and in recent times.	Deposits generally small and erratic. May be superficial. However, this type grades into White Pine type which has red-bed affiliations.
Banded Fe formation	Lake Superior; Labrador; most Fe ores of Lake Superior type	Re-concentration by meteoric waters from Fe-rich source rocks	It is not clear in some cases that granitization has not played a part. Where this is the case the deposits should not be superficial.
Metal-accumulations re-concentrated thermally from sediments			
Dolomitized limestone, at times of algal reef origin, associated with shore lines and evaporites. Gentle folding, low-grade metamorphism	Missouri Pb; Tri-State Pb—Zn; Pine Point Pb—Zn; Zn in reefs of Alberta	Primary (biochemical ?) accumulation in limestone: re-concentration by low-temperature metamorphism over comparatively short distance	Commonly of large dimensions, perhaps reflecting comparatively low degree of re-concentration.
Limestone intruded by granite or converted to paragneiss	Pb—Zn deposits common throughout the world. Some Cu in skarn zones	Re-concentrated from limestone during granitization	Provinces co-extensive with particular lime-rich formations. Rum Jungle (Australia) U, Cu, and Pb—Zn province co-extensive with algal reef and shelf sediments.

(continued)

TABLE XXVIII (continued)

Rock environment	Examples	Possible origin	Remarks
Euxenic shales intruded by granite	Kimberley (B.C.) Pb–Zn ores associated with pyritic sediments.	Re-concentrated from sediments during granitization	Sedimentary facies important in prospecting.
Carbon-rich shales and allied sediments granitized	Lake Athabaska (Canada), Rum Jungle (Australia) (in part) U provinces	Syngenetic accumulations in sediments, re-concentrated during granitization	Paragneiss may be developed (Athabaska) or homogeneous granite (Rum Jungle).

Ores derived from tuffs

Tuffs and tuffaceous sediments converted in places to porphyry, granite, or paragneiss	Pb–Zn–Cu province, West Coast, Tasmania; New Brunswick Pb–Zn–Cu province; Swedish Pb–Zn ores	Accumulation of metals in tuffs concentrated during granitization	Extensive geochemical work shows widespread Pb–Zn–Cu in tuffaceous sediments, with ore concentrations associated with porphyritized tuffs. Porphyries and ores are related to structural foci.

Ores derived from basalts

Ores concentrated during cooling of basalt

Basalt, little altered; folding gentle	Coppermine province, N.W.T., Canada. Native Cu, bornite in cracks, vesicles, etc. Examples numerous throughout world	Limited concentration during cooling.	Ores of limited continuity though rich in places.

Ores derived during metamorphism of basalt

Basalts considerably folded and altered, but no granite	Michigan native Cu in flowtops and sediments intercalated with lavas	Re-mobilized from basaltic source-rock during folding and metamorphism	Deposits extensive.

Ores concentrated during granitization of basalt

Basalts and other lavas intruded by granite	Nicola volcanic province, British Columbia. Cu province of Chile, Peru?, Kennecott, Alaska?	Cu concentrated from volcanics during granitization	Many major provinces of this type. Very deep erosion does not favour preservation of these ores and many deposits in mountain terrains.

Ores due to granitization of chloritized carbonated lavas (greenstones)

Chloritized carbonated greenstone belts. Lavas	Au provinces of Ontario, Quebec; Yellowknife	Cu appears to be expelled from lavas during	Province is co-extensive with granitized greenstone

(continued)

TABLE XXVIII (continued)

Rock environment	Examples	Possible origin	Remarks
commonly submarine. Albite porphyry dykes, granite porphyries, and granites.	(Canada); West Australia	development of "greenstones". Au is concentrated ahead of front of granitization and "porphyritization"	belt. Au not associated with granite of same age and type intruding other source rocks.

Ores derived from gabbro—diabase

Ores concentrated during cooling of gabbro—diabase

Diabase dykes or sills intruding wide variety of rocks. Ores along walls	Co—Ag ores, Cobalt, Ontario; Co ores, Cloncurry, Queensland	Sulfides expelled from consolidating magma	This type of province not as productive as sedimentary ores of Rhodesia—Katanga.
Diabase dykes. Larger gabbroic masses in neighbourhood. Structurally prepared rock	Cu ores of Noranda district, Quebec; Cloncurry district, Queensland; home of Bullion mine, Australia	The comparatively thin dykes associated with these deposits in time and place may be channels for movement of ores, probably derived from larger cooling gabbroic masses	Ores commonly massive sulfides, pyrite—chalcopyrite—sphalerite. This type of province is of relatively secondary importance though major deposits are found. Dykes are guides in prospecting.

Ores concentrated by granitization of gabbro—diabase

Gabbro intruded by granite and allied rocks	Sudbury, Ontario Cu—Ni ores?; Chibougamau, Quebec?	Cu—Ni derived from norite, concentrated during replacement of norite by granite?	There are post-norite granites at Sudbury, and unfolded, unaltered gabbros are not normally productive of ore. However, Sudbury and similar areas have volcanic affiliations — pebble dykes, breccia pipes, tuffs, etc.
As above	Bird River, Manitoba, Cu—Ni ores	As above	Ore associated with contact of gabbro and granite. The granite is the younger rock.
Dykes and sills of diabase followed by localized porphyries and granites	Ray? Globe?, San Manuel?, U.S.A.	Cu in granite and porphyry derived from digestion of diabase	Monzonite probably hybrid rock. In some cases diabase almost totally digested; in others large remnants remain in monzonite. Not all porphyry coppers are of this type. Basalt or Cu-rich sediments may also supply Cu.

(*continued*)

TABLE XXVIII (continued)

Ores derived from ultrabasic igneous rocks

Ores concentrated during cooling of ultrabasic rocks

Rock environment	Examples	Possible origin	Remarks
Peridotite dykes and masses	New Caledonia chromite ores; Chromite ores generally; Bushveld complex Pt deposits; osmiridium deposits of Tasmania	Separation from silicates during cooling	Basic rocks not altered.

Ores due to alteration of ultrabasic plutonic rocks

Rock environment	Examples	Possible origin	Remarks
Serpentinized peridotite	Asbestos deposits, Quebec	Constituents of peridotite recrystallized and hydrated during metamorphism and granitization	Virtually all important asbestos deposits associated with altered peridotites.
Peridotite intruded by granitic dykes.	Asbestos deposits of north-west Australia	As above	In some cases, asbestos confined to peridotite near walls of granitic dykes.
Peridotite altered or replaced during granitization	Mystery Lake (Canada) Ni deposits; Lake Renzy (Canada) Ni deposits	Ni sulphides concentrated during re-heating, serpentinization, and granitization of peridotite?	Mystery Lake ore confined to serpentinized zones with paragneiss walls or occurs in paragneiss itself. Lake Renzy has a peridotite–granite association.

MAGMATIC-VOLCANIC AND RELATED TYPES OF ORE AND HOST-ROCK CLASSIFICATIONS

"The development of an observational science like geology depends upon the personal experience of individual workers. I suggest that with certain reservations the best geologist is he who has seen the most rocks."
H.H. Read

The genesis of many (in the past researchers claimed that "most") strata-bound/stratiform mineralizations cannot be dissociated from magmatism, so that the whole "plutonic-subvolcanic/hypabyssal-igneous hydrothermal exhalative-volcanic-derived syngenetic-sedimentary" spectrum has to be given attention in disscussions on ore genesis. This of course applies to deposits where a direct link with igneous or volcanic rocks can be demonstrated as well as to those with an absence of a visual evidence of volcanism. In

70

TABLE XXIX

Proposed classification (After Bateman, 1950, p. 363; courtesy of Wiley, New York, N.Y.)

Process	Deposits	Examples
1. Magmatic concentration	I. Early magmatic:	
	A. Disseminated crystallization	Diamond pipes
	B. Segregation	Chromite deposits
	C. Injection	Kiruna magnetite?
	II. Late magmatic:	
	A. Residual liquid segregation	Taberg magnetite
	B. Residual liquid injection	Adirondack magnetite, pegmatites
	C. Immiscible liquid segregation	Insizwa sulfides
	D. Immiscible liquid injection	Vlackfontein, S. Africa
2. Sublimation	Sublimates	Sulfur
3. Contact metasomatism	Contact metasomatic: Iron, copper, gold, etc.	Cornwall magnetite Morenci (old), etc.
4. Hydrothermal processes		
A. Cavity filling	Cavity filling (open space deposits):	
	A. Fissure veins	Prachuca, Mexico
	B. Shear-zone deposits	Otago, New Zealand
	C. Stockworks	Quartz Hill, Colo.
	D. Ladder veins	Morning Star, Australia
	E. Saddle-reefs	Bendigo, Australia
	F. Tension-crack fillings (pitches and flats).	Wisconsin Pb and Zn
	G. Breccia fillings	
	a. Volcanic	Bassick pipe, Colo.
	b. Tectonic	Mascot, Tenn., Zn
	c. Collapse	Bisbee, Ariz.
	H. Solution-cavity fillings	
	a. Caves and channels	Wisconsin–Illinois Pb and Zn
	b. Gash veins	Upper Mississippi Valley Pb and Zn
	I. Pore-space fillings	"Red bed" copper
	J. Vesicular fillings	Lake Superior copper
B. Replacement	Replacement:	
	A. Massive	Bisbee copper
	B. Lode fissure	Kirkland Lake gold
	C. Disseminated	"Porphyry" coppers
5. Sedimentation (exclusive of evaporation)	Sedimentary: Iron, manganese, phosphate, etc.	Clinton iron ores
6. Evaporation	Evaporites:	
	A. Marine	Gypsum, salt, potash
	B. Lake	Sodium carbonate, borates
	C. Groundwater	Chile nitrates

(continued)

TABLE XXIX (continued)

Process	Deposits	Examples
7. Residual and mechanical concentration		
A. Residual concentration	Residual deposits:	
	Iron, manganese, bauxite, etc.	Lake Superior iron ores, Gold Coast manganese, Arkansas bauxite
B. Mechanical concentration	Placers:	
	A. Stream	California placers
	B. Beach	Nome, Alaska, gold
	C. Eluvial	Dutch East Indies tin
	D. Eolian	Australian gold
8. Surficial oxidation and supergene enrichment	Oxidized, supergene sulfide	Chuquicamata, Chile Ray, Ariz., copper
9. Metamorphism	A. Metamorphosed deposits	Rammelsberg, Germany
	B. Metamorphic deposits	Graphite, asbestos, talc, soapstone, sillimanite group, garnet

some cases, this relationship is a simple one-to-one correlation as, for example, when exhalative metal-bearing solutions plus volcanic debris and/or flows formed accumulations in a marine milieu; the latter constitute the host-rock, whereas the former produced the mineralization. These syngenetic ore deposits can then be penecontemporaneously intra-environmentally (note: often wrongly termed "intraformationally") reworked by currents, waves, and/or slumping to form an ore-breccia horizon (e.g., one sub-type of the Kuroko ores). In instances where a deep-seated magma supplied the hydrothermal fluids (which precipitated ores along the conduits and upon reaching the earth's surface deposited exhalative-sedimentary ores) *without* any volcanic debris, the nexus between mineralization and magma can be suggested only by inference or conjecture (i.e., the interpretations are purely "hypothetical"). The fact remains, however, that igneous processes *in toto* are accepted to be one of the fundamental causes of stratiform ore accumulations — as clarified in numerous interrelated sections in this contribution as well as in several chapters in other volumes. (See also Korolev and Shekhtman, 1968, for post-magmatic mineralizations.)

The following topics are treated below:

(1) Pressure—temperature and temperature—composition diagrams of phases of magmatic differentiation.

(2) Mineralogic—lithologic stages (A to L and corresponding temperatures) of magmatic differentiation.

(3) Factors influencing chemistry of granitoid rocks.

TABLE XXX

Classification of ore deposits (After Cissarz, 1964; courtesy of Schweizerbart, Stuttgart)

Mode of formation	Location of formation	Formation from epeirogenically mobilized simatic magmas	Formation from orogenically mobilized sialic magmas	
Magmatic		*final* Exhalative deposits / Thermal springs	Exhalative deposits / Thermal springs	
Sedimentary	Continents (surface to groundwater)			
	Shelf areas			
	Ocean basins			
Magmatic-Sedimentary		Magmatic sedimentary sulfide deposits / Magmatic sedimentary Mn deposits / Magmatic sedimentary Fe deposits		
			Descending—ascending mixed type	
Magmatic	Small depth	*initial* Hydroth. Hg—Sb deposits / Hydroth. Pb—Zn deposits / Hydroth. Cu deposits / Hydroth. Ag—Co—Ni—As deposits / Pegmat.-pneumatolytic deposits / Pressure-injected ores / Crystallization-differentiation Fe—Ti / Liquidimmiscibility segregation Ni—Cu—Pt / Crystallization-differentiation Cr	Regeneration-type deposits / *subsequent* Subvolc.-hydroth. Hg—Sb deposits / Subvolc.-hydroth. Pb—Zn deposits / Subvolc.-hydroth. Cu—As deposits / Subvolc.-hydroth. Au—Ag deposits	
	Medium depth		*late orogenic* Plutonic-hydroth. Ba—F—Mg—Fe deposits / Plutonic-hydroth. Hg—Sb deposits / Plutonic-hydroth. Pb—Zn deposits / Plutonic-hydroth. FeCO$_3$ deposits ± Cu / Cu and Mo impregnations in silicate rocks / Plut.-hydroth. Au—Cu deposits / Plut.-hydroth. Co—Ni—Bi—U deposits / Pneumatolyt. Sn—Wo—Mo deposits / Contact-pneumatolytic deposits / Pegmatites	
	Great depth		*maximal orogenic* Pegmatitic injections	

Origin of weathering products during times of relative tectonic quiescence	Sedimentary formations in association with epeirogenic movements	Transformation as a result of pressure and temperature increases
Oxidation and cementation zones Kaolin deposits Weathering-type deposits Fe−Mn−P Weathering-type deposits Al−Ni−Mg Precipitations in inland waters Descending-type deposits Fe−V−U−Sr, etc. Arid concentration deposits Cu−U, etc. Terrestrial salt deposits	Fluviatile placer deposits Peat deposits	
	Marine placers $FeTiO_3$, monazite, etc. Clastic (breccia/conglomerate) Fe deposits Oolitic Fe−Mn deposits Marine phosphate deposits Marine salt deposits Sulfide impregnations in sediments Bituminous rocks	

anchimetamorphic

Salt transformation
Coal maturation
Mobilization of bitumen

Hydrothermally recrystallized deposits

metamorphic

epizone — Contact-metamorphosed deposits

mesozone — Metamorphic sulfide deposits
Metamorphic Fe deposits
Metamorphic Mn deposits
Metamorphic graphite deposits

katazone — Complex polymetamorphic deposits
Pegmatitic injections

Palingenic melting

TABLE XXXI

Genetic/environmental classification of ore deposits (After Pereira, 1963)

	A	B	C	D	E
Class and genesis	In volcanic pipe or near site of pipe	Volcanic associations	Sedimentary associations	Deformed (remobilised)	Rejuvenated
Setting	(1) Cordilleran volcanoes (2) Alkali Volcanoes in continental rifts (3) Basic vulcanicity	Island arc Volcanoes and Geosynclines	Shore lines and basins	Tectonic and metamorphic (Modified types of A, B or C)	Granitisation and magmatic intrusions
Type	(1) Porphyry Copper (2) Carbonatites etc. (3) Basic volcanoes and lava flows	Massive Sulphides (1) Pyritic (2) Banded with Pb/Zn/Cu (3) Unbanded with Pyrrhotite and Chalcopyrite (4) Ironstones and Manganese deposits	(1) Conglomerates (2) Cu belt types (3) Limestones with Pb/Zn (4) Mixed types — Mt. Isa Pb/Zn-	Originally A, B or C — (Remobilized — (a) Mt. Isa Cu	Originally A, B, C or D converted to vein type deposits.
Examples	(a) Toquepala, Peru (b) Palabora, S. Africa (c) Cripple Creek, Colorado	(a) Rio Tinto, Spain (b) Rosebery Tasmania (c) New Brunswick, Canada (d) Rammelsberg, Germany	(a) Witwatersrand and Blind River (b) Copper Belt, N. Rhodesia (c) Zellidja, Marocco Tynagh, Ireland (d) Algoma Iron Ores	(b) Mt. Lyell Tasmania (c) Broken Hill Australia (d) Mary Kathleen Australia (e) Singbhum Cu/U, India (f) Ore Knob, N. Carolina (g) Zawar, India	(a) Karagwe Tin deposits. Tanganyika (b) Bolivian Tin deposits Innumerable examples of vein deposits of Cu/Pb/Zn/Sn/W/U etc. e.g. Gatooma Goldfield, S. Rhodesia
Remarks			TRANSITIONS		

For the sake of simplicity, deposits such as Kimberlite diamond pipes, sedimentary deposits redistributed by circulating meteoric waters,

TABLE XXXII

Classification of ore deposits based on sources, modes of mobilization, transportation, precipitation, and environment (After Pélissonnier, 1962, in: Nicolini, 1970, p. 620; courtesy of Gauthier-Villar, Paris)

Origin	Processes of mobilization	Mobilization agent	Transportation		Distance between source and deposit (growing from left to right)	
			Mode of transportation	Cause	Weak (large components)	Significant (small constituents)
Mechanical erosion		Climate and geomorphology, rivers	Solid particles Colloidal suspensions	Mass potential gradient		
Chemical erosion { Surficial; Subsurface { Meteoric, Hydrothermal		Vadose waters			Remobilization of eluvium	Transport of the fluvial or vadose waters
					Lateral secretion	Telethermal phenomenon
Volcanism		Connate waters Juvenile waters Fumaroles	Solutions and possibly gas	Pressure gradient	Proximate exhalatives	Remote exhalative, waning
Hypovolcanism	Hydrothermal	Connate waters Juvenile waters Fumaroles			Hypobatholitic / Endobatholitic / Embatholitic	Epibatholitic / Acrobatholitic / Cryptobatholitic
	Pneumatolytic	Connate waters Juvenile waters				
	Pegmatitic	Idem, or input of thermal energy in a zone of tectonic tension				
	Orthomagmatic	Magma or heat input (1)		Gradient of the chemical concentration and thermal gradient (3)		
General metamorphism { Epi-, Meso-, Kata-		Heat input in an environment of tectonic tension without pressure	Diffusion of microsolutions or of atoms (2)		Lit-par-lit produced by metamorphic differentiation	Chemical fronts

(continued)

TABLE XXXII (continued)

Type of deposit	Channels	Constrictions	Environment of deposition through increasing depth (from top to bottom) and indication of mode of concentration		
			LACUSTRINE OR MARINE	**TERRESTRIAL**	
	Fluvial or submarine	Terrestrial valleys, submarine valleys	Lacustrine	Fluviatile — Continental zone temperate: Meteoric alteration, Pedogenesis (eluvial concretions) / Continental zone desert: Eolian deposits, Saline crusts / Continental zone tropical: Laterites, Pedogenesis (eluvial concretions)	
			Lagunal	Detrital	
	Fracture systems, permeable massifs	Domes or depressions opened depending on direction of circulation. Preferential fractures, intersections, or fractures	Marine { Littoral, Neritic, Bathyal, Abyssal }	**SUBSURFACE** — Descending epicrustal: Oxidation zone / Cementation zone { Residual concentration; Precipitation from solutions in cavities, Specific metasomatism; Remobilization of pre-existing ores }	
		. . .			
	Massif generated through diffuse penetration controlled by tectonism or lithology	Centers of metamorphism	Note - In all cases in this column the mode of concentration could be: – detrital – chemical – biochemical except in abyssal environment where it is chemical	Ascending epicrustal / Hypocrustal / Hypabyssal / Abyssal — with or without influence of mineral zoning: filling of cavities { in large-scale (fractures), in small-scale (fractures) }; metasomatic "substitution" (simultaneous filling and metasomatism)	

TABLE XXXIII

Modified Lindgren Classification of ore deposits [*] (After Ridge, 1972, p. 676; courtesy of Geological Society of America)

Type	Conditions of formation		
	temperature (°C)	pressure (atm)	depth (ft)
I. Deposits mechanically concentrated (plus normal mechanical sediments of economic value)	– – – – – Surface conditions – – – – – – –		
A. Residual placers			
B. Eluvial placers			
C. Alluvial placers			
II. Deposits chemically concentrated	Differ within wide limits		
A. In quiet waters			
1. By interaction of solutions (sedimentation)	0–70	low	0–600
a. Inorganic reactions			
b. Organic reactions			
2. By evaporation of solvents (evaporation)	0–70	low	0–600
3. By introduction of fluid igneous emanations and water-rich fluids	0–80	low	0–600
4. By diagenesis	0–70	low	0–600
B. In rocks			
1. By rock decay and weathering (residual deposits) (overlap with I-A)	0–100	low	shallow
2. By ground-water circulation (supergene processes)	0–100	low–moderate	shallow–medium
C. In rocks by dynamic and metamorphism	up to 500	high–very high	great
D. In rocks by hydrothermal solutions			
1. With slow decrease in heat and pressure			
a. Telethermal	50–150	low–moderate (40–240)	shallow (500–3000)
b. Leptothermal	125–250	moderate (240–800)	medium (3000–10,000)
c. Mesothermal	200–350	moderate–high (400–1600)	medium (5000–20,000)
d. Hypothermal			
(1) In non-calcareous rocks (Lindgren's hypothermal)	300–600	high–very high (800–4000)	great (10,000–50,000)
(2) In calcareous rocks (e.g., tactite, skarns)	300–600	high–very high (800–4000)	great (10,000–50,000)
2. With rapid loss of heat and pressure			
a. Epithermal	50–200	low–moderate (40–240)	shallow–medium (500–3000)
b. Kryptothermal	150–350	low-moderate (40-280)	shallow-medium (500–3500)
c. Xenothermal	300–500	low-moderate (80–700+)	shallow-medium (1000–4000)

Note: items B.1, B.2 and C share the bracketed phrase "with or without introduction of material foreign to rock affected".

(continued)

TABLE XXXIII (continued)

Type	Conditions of formation		
	temperature (°C)	pressure (atm)	depth (ft)
E. In rocks by gaseous igneous emanations	100–600	low	shallow
F. In magmas by differentiation or in adjacent country rocks by injection			
1. Early separation–Early solidification	500–1500	very high (1200)	great (15,000+)
a. Disseminations			
b. Crystal segregations			
c. Crystal segregations, plus injections as crystal mush			
2. Early separation–Late solidification	500–1500	very high (1200+)	great (15,000+)
a. Early immiscible sulfide melt accumulation			
b. Early immiscible sulfide melt accumulation, plus later fluid injection			
3. Late separation–Late solidification, with or without fluid injection			
a. Silicate pegmatites			
Usually gradational { i. Simple	575±	high–very high (800–4000+)	great (10,000–50,000)
ii. Complex	200–550	high–very high (800–4000+)	great (10,000–50,000)
iii. Barren Quartz	100–300	high–very high (800–4000+)	great (10,000–50,000)
b. Immiscible melts, metal-oxygen rich, metal-phosphorus rich	500–1500	very high (1200+)	great (15,000+)
c. Immiscible (carbonate-rich)	500–1500	low–very high (0–4000+)	shallow–great (0–50,000+)
4. Late formation-Deuteric alteration	Less than 575	moderate–very high (400–4000+)	medium–great (5000–50,000+)

* Terms not used by Lindgren were proposed by Graton, Buddington, Bateman, and Ridge.
+ Initially appreciably higher than the lithostatic pressure would produce.

(4) Two possible process–response models for granite plutons.

(5) Changes of Eh- and pH-values during magma differentiation.

(6) Scheme of late magmatic-postmagmatic metasomatism (e.g., tin greisenization).

(7) Geological and geochemical features for assessing tin-bearing granitic rocks.

(8) Classification of volcanic exhalations.

(9) Schematic bird's-eye view of the classification and nomenclature of magmatic deposits.

TABLE XXXIV

Major groups of ores based on rock association, environment, origin and composition (After Stanton, 1972)

Ores in igneous rocks, No. I: in mafic and ultramafic rocks

1. Ultramafic–mafic–Cr/Ni platinoid association
2. Mafic–ultramafic–Fe/Ni/Cu sulfide/platinoid association

Ores in igneous rocks, No. II: in felsic rocks

1. Carbonitite association
2. Anorthosite–Fe/Ti oxide association
3. Quartz monzonite–granodiorite–Cu/Mo sulfide association

Stratiform sulfides of marine and marine-volcanic association

1. Spatially related to both sedimentary and volcanic features
2. Spatially related to volcanism but not to any specific features of sedimentation
3. Spatially related to specific zones or environments of sedimentation but which show no apparent association with volcanism
4. Spatially related to neither volcanism nor to specific environments of sedimentation

Iron concentrations of sedimentary affiliation

1. Bog iron deposits
2. "Ironstones"
3. Banded iron formations

Manganese concentrations of sedimentary affiliation

1. Modern marine and non-marine deposits
2. Mn of the orthoquartzitic–glauconite–clay association
3. Mn of the limestone–dolomite association
4. Mn of the volcanic association

Additional strata-bound ores of sedimentary affiliation

1. Limestone–Pb/Zn association
2. Sandstone–U/V/Cu association
3. Conglomerate–Au/U/pyrite association

Ores of metamorphic affiliation

1. Ores of regional-metamorphic origin
   (a) produced
   (b) modified
2. Ores of contact-metamorphic origin
3. Ores of metasomatic origin
4. Ores of dislocation-metamorphic influences.

Seven major groups of ores based on environment and origin

1. Greenstone-sedimentary belts of the early Precambrian.
2. Ultrabasic rocks found along the arcs of the Pacific.
3. Porphyry-type ore of the North and South American Cordilleras and parts of Asia Minor, Poland and Russia.
4. Sediments of the late Precambrian age which include the copper deposits of Zambia, the Witwatersrand gold deposit.
5. Younger (post-Precambrian) sediments containing ores, e.g., Pb–Zn ores in carbonates.
6. Breccia pipes of South Africa and the carbonitites.
7. Deep-sea nodules.

(10) Distribution of chemical elements in magmatic to hydrothermal ores.

(11) Normal sequence of ore genesis in the "pegmatitic—pneumatolytic—intrusive hydrothermal" range.

(12) Diagram of deposits associated with geosynclinal juvenile-basaltic magmatism.

(13) Diagram of deposits associated with sialic-palingenic magmatism.

(14) Schematic division of time- and space-distribution of siamic and sialic magmatism.

(15) Diagram depicting telescoping (distance-factor) effect of ores.

(16) Influence of distance from an ophiolitic magma source on ore genesis.

(17) General model of influence of distance on composition and structural control on endogenic deposits.

(18) Genetic sequence of magmatogenic ores.

(19) Natural classification of magmatogenic ores.

(20) Host-rock variation exemplified by Cu-porphyry ores.

(21) Tabular scheme of strata-bound mineralization associated with volcanism.

(22) Diagram relating igneous rocks to mineralizations.

(23) Tabular classification of magmatic-to-exhalative ore deposits.

The normal (=idealized) scheme of magmatic differentiation (after Bowen, Goldschmidt, Niggli, Schneiderhöhn, Borchert, among others) is presented in a flow-chart-like model (Fig.27), which is divided into three major mineralogic-petrographic/lithologic phases (=early-crystallization, main-crystallization and late-crystallization phases) with one transitional stage (=pegmatitic—pneumatolytic). The minerals, rocks and ores formed during these respective igneous phases are depicted by basalt through diorite and granite to hydrothermal deposits. The supplementary temperature—concentration $(T-X)$ and pressure—temperature $(P-T)$ curves (Fig.28) illustrate the physicochemical, physical and compositional changes during differentiation, as visualized by Niggli. Fersman (e.g., 1952) has offered a subdivision of the igneous phases in Fig.29 (see also Schneiderhöhn, 1961), and related them to the evolution of magmas (cf. also numerous related diagrams in the next section).

The early concepts on the development of magmas were by necessity simple, but with an increase in data the models have become refined, and thus will approach the natural complex geological settings. For example, Baumann and Tischendorf (1976) provided charts depicting the most important influential factors in determining the composition of granitoid rocks (Fig.30), and Whitten (1966) considered two possible process—response models for a granite pluton (Table XXXV). Borchert (1964, unpublished; 1978) outlined the pH—Eh changes from the early- to late-crystallization phases of a magma (Fig.31), whereas Baumann and Tischendorf (1976; see Smirnov, 1976, fig.96, p.200 for a temperature—pH diagram) offered the pH-values of the solutions that caused late-magmatic—post-magmatic metasomatic alterations (in this case forming tin-bearing greisens). Six stages are recognized during which certain chemical elements are either precipitated or mobilized or removed in solution (Table XXXVI).

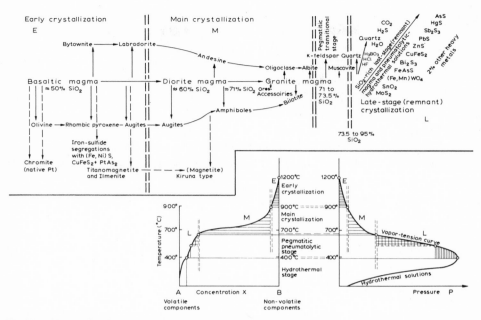

Fig.27. "Normal" scheme of magmatic differentiation. (After V.M. Goldschmidt, 1954; Borchert, 1950, unpublished; 1978; courtesy of Glückauf, Essen.)

Fig.28. Temperature–composition $(T–X)$ and pressure–temperature $(P–T)$ diagrams. (After P. Niggli, 1937; Borchert, 1950, 1978; courtesy of Glückauf, Essen.)

The exploration potential of the detailed knowledge of the origin and "indicator-characteristics" of metal-bearing host-rocks is exemplified by the work of Tischendorf (1968) (see also Rösler and Lange, 1972, table 174, p.377) on the tin-bearing granitoid rocks. [Compare with Štemprok (1978a,b), Rundkvist (1978) and Taylor (1978, for example).] The comparative Table XXXVII summarizes the geological phenomena (totalling 14) of the "perspective" versus the "non-perspective" (i.e., tin-rich and tin-poor, respectively) lithologies. There is no doubt that in the future we will see many "comparative classifications of indicator-characteristics" emerging from research laboratories which are useful in the search for all metals and industrial minerals.

Although the mega-divisions of the magmatic-volcanic products into the plutonic– volcanic types (with various intermediate phases, all transitional and gradational) may be convenient, it is clear that for the study of the hypabyssal and exhalative ores a clear understanding of the whole gamut of exhalations is paramount. In interpretations all possibilities are to be considered, but how often does one see in the literature discussions that commence with a list of "genetic possibilities" of which the unlikely ore-forming parameters are one-by-one dismissed by a process of elimination until the most plausible variables are isolated to form part of the provisionary hypothesis? In attempts to unravel

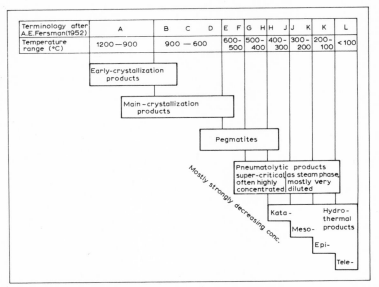

Terminology after A.E.Fersman(1952)	A	B	C	D	E	F	G	H	H	J	J	K	K	L
Temperature range (°C)	1200—900	900 — 600			600-500	500-400	400-300	300-200		200-100		< 100		

Fig.29. Magmatic phases of mineralization. (Mainly after Schneiderhöhn, 1961; courtesy of Gustav Fischer, Stuttgart.)

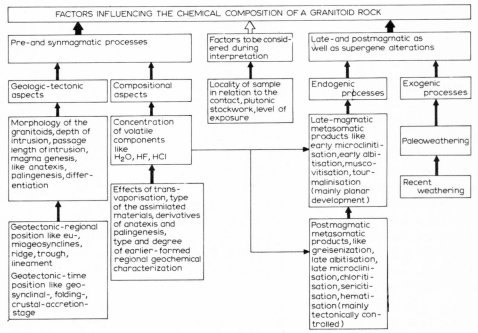

Fig.30. The most important factors influencing the composition of a granitoid magma and rock. (After Baumann and Tischendorf, 1976.)

TABLE XXXV

Two possible process—response models (with feedback loops) for a granite pluton (After Whitten, 1966)

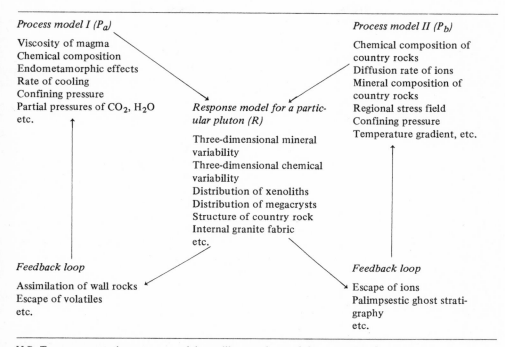

Process model I (Pₐ)

Viscosity of magma
Chemical composition
Endometamorphic effects
Rate of cooling
Confining pressure
Partial pressures of CO_2, H_2O
etc.

Response model for a particular pluton (R)

Three-dimensional mineral variability
Three-dimensional chemical variability
Distribution of xenoliths
Distribution of megacrysts
Structure of country rock
Internal granite fabric
etc.

Process model II (P_b)

Chemical composition of country rocks
Diffusion rate of ions
Mineral composition of country rocks
Regional stress field
Confining pressure
Temperature gradient, etc.

Feedback loop

Assimilation of wall rocks
Escape of volatiles
etc.

Feedback loop

Escape of ions
Palimpsestic ghost stratigraphy
etc.

N.B. Two representative process models are illustrated out of the *n* possible models.
P_a = magma intrusion
P_b = ionic migration.

the origin of volcanic-exhalative mineralizations, for instance, Table XXXVIII by Naboko (1959) (see also Rösler and Lange, 1972, table 105, p.262) can serve as a guide. This "classification of volcanic exhalations" lists two major modes of exhalations, namely, from magmas and lavas near or at the surface and from those located at great depth. Based on "point of separation of the exhalations", the temperature, and the chemical compositions, these exhalations show a wide variation of characteristics. Investigations like those of Naboko (1959) and Ovchinnikov et al. (1964) (see brief comment in the section "Some Genetic Considerations") will eventually provide us with a complete understanding of the multiple sources of ions and anions that constitute ores.

A schematic overview of the types and terminology of magmatic ores is given in Fig.32 (Schneiderhöhn, 1961), whereas the distribution of elements concentrated or precipitated during the five main igneous phases, and the "normal" or idealized sequence during the "pegmatitic—pneumatolytic—intrusive hydrothermal" ore formation are

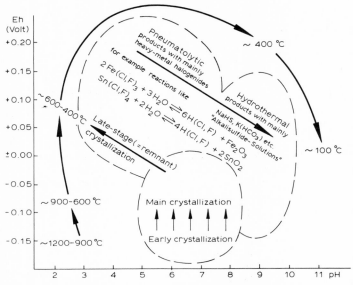

Fig.31. Development of the Eh- and pH-values and the temperature from the early-crystallization to the hydrothermal phases. (After Borchert, 1964, unpublished; 1978; courtesy of Glückauf, Essen.)

depicted, respectively, in Table XXXIX and Fig.33.

In three cartoon-like conceptual models, Borchert (1957, 1959, 1960b, 1961) illustrated the genetic relationship between mineralizations and (a) geosynclinal juvenile-

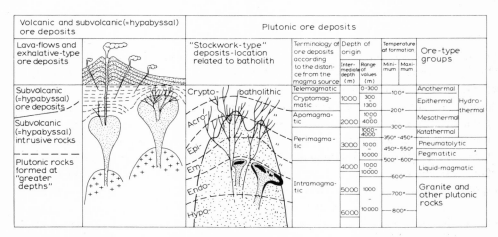

Fig.32. Schematic overview of the subdivision and terminology of the magmatic deposits. (After Schneiderhöhn, 1961; courtesy of Gustav Fischer, Stuttgart.)

TABLE XXXVI

Scheme of the late-magmatic metasomatic processes of the origin of tin-bearing greisen formations (After Beus and Soboler, 1964, in: Bauman and Tischendorf, 1976)

Stage	Precipitation of	Mobilization and/or migration of	Development of the pH of the solution
0 Late-stage solution (initial stage)	–	K, Rb, Si(Na)(Al)	$7 \to 9$ because of enrichment of K, Rb and Si
1 Early microclinization (Early muscovitization)	K, Rb	Na, Si(Li)(Al)(F)	$9 \to 8$ because of K−Rb precipitation and enrichment of Na and Si
2 Early albitization	Na	(Si)(F)(Li)(Al)(K)	$8 \to 6$ because of Na precipitation and further enrichment of Si and F
3 Silicification	Si	F, Li, Al, K, Na	$6 \to 4$ because of Si precipitation and strong enrichment of F
4 Topazation, protolithionitization	F, Li, Si, K, Al	Na	$4 \to 6$ because of F precipitation and enrichment of Na
5 Late albitization	Na	K	$6 \to 8$ because of Na precipitation
6 Late microclinization	K	–	$8 \to 9$ because of K precipitation

basaltic magmatism (Fig.23a) and (b) sialic-palingenic magmatism (Fig.23b), referred to earlier.

Fig.33 summarizes the time and space relationships between magmatisms which have been divided into four phases (I to IV) according to Stille (1939, in particular 1950), for example. Stage I represents the "*initial siamic magmatism*" of the orthogeosyncline and is of an ophiolitic nature. The derivation of the material is "juvenile-hypogene" in character with the magma moving upwards from great depth along zones of weaknesses. The magmatism is predominantly submarine-extrusive to form the so-called "volcanic associations" with or without sedimentary deposits. Compositionally, the material is basaltic, with numerous differential derivatives from the basalt magma. The ores to be expected are connected to both the early crystallization of the siamic magma as well as to the fractional differentiation of the original basalt. Here, quartz-keratophyres and weilburgites are the "antipoles" of peridotites and, correspondingly, the Lahn−Dill (iron ore) and Ergani−Leksdal−Meggen (Cu- and ZnS ores) types are the antipoles of the chro-

TABLE XXXVII

Geological and geochemical features for assessing the tin-bearing state of granitic rocks (After Tischendorf, 1968)

Geologic phenomenon	Perspective formation	Non-perspective formation
Geologic-tectonic position of the granitoid rock	palingenetic granites with intrusive contacts; post-tectonic, post-orogenetic-subsequent; multiphase, intrusive bodies of complicated structures	migmatites, granitisation phenomena, syntectonic, synorogenetic; frequently single-phase, monotonous intrusive bodies
Relative age in orogen	belonging to younger intrusive complexes	belonging to older intrusive complexes
Granite morphology	distinctly ragged, steep stocks and crests, protuberances, granite tongues, flat to medium steep dips of the contact	from insignificantly rugged to non-rugged, smooth course and relatively steep dip of the contact
Type of granitoid rock	biotite-granite, partly bearing muscovite, alaskite-granite, adamellite, tourmaline-granite; acid — ultraacid	granodiorite, granite, partly bearing hornblende, intermediately acid
Feldspar	Or > Pl; Ab-rich plagioclase	Pl > Or; plagioclase poor in Ab
Dark mica	protolithionite, Li-bearing siderophyllite, Li-bearing lepidomelane	Fe-biotite, Mg-biotite
Accessories	cassiterite, topaz, tourmaline (fluorite)	titanite, allanite, apatite
Chemistry of the granitoid rock	relatively rich in Si, K, Na, F and Li; K > Na	relatively rich in Ca, Mg, Mn, Ti
Sn concentration in the granitoid rock (ppm)	15—50	3—10
Sn concentration in dark mica (ppm)	(100) 150—500	10—100(150)
Trace element distribution in dark mica	rich in Sn, Ga, F, Li	rich in Pb, Ni, Cu
Trend of tin distribution in the rock	concentration in the most recent granite phases of the complex	partly concentration in the older granite phases of the complex
Autometasomatism	albitisation, microclinisation, muscovitisation, sericitisation	insignificant or none
Dispersion of trace elements, especially of Sn, Li, F	relatively great (more than double that of the non-perspective formation)	relatively small

TABLE XXXVIII

Classification of volcanic exhalations (From Naboko, 1959)

Character of exhalations	Point of separation	Temperature ($°C$)	Chemical composition
(I) Exhalations from surface magma	(1) from the crater at the instant of eruption		little data available; assumed are H_2O, H_2, CO, H_2S, SO_2, HCl, HF
	(2) from lava masses above the volcanic vent	varies	(a) fumaroles of the haloid stage: H_2O, H_2, CO, HCl, HF, SO_2, halides and sulphates of Na, K, Fe, Cu (b) fumaroles of the sulphidic stage: H_2O, CO, H_2, SO_2, H_2S, sulphates of Na, K, Ca, etc. (c) fumaroles of the carbonic stage: H_2O, CO_2, traces of H_2S (d) fumaroles of the water vapor stage: water vapors
	(3) from lava masses detached from the volcanic vent	1,200–700	(a) fumaroles from liquid lava: H_2O, H_2, CO, HCl, HF, S, SO_2, H_2S, halides and sulphates of Na, K, Fe, oxides of Si, Cr, Fe
		500–300	(b) fumaroles from solidified lava; salt fumaroles: NaCl, KCl, $FeCl_3$, NH_4Cl, Na_2SO_4, K_2SO_4, $CaSO_4$, etc. ammonium chloride fumaroles: NH_4Cl, $FeCl_3$ water vapor fumaroles
(II) Exhalations from magma located at great depth	calderas, craters, volcanic slopes	above the critical temperature	(a) fumaroles of the haloid stage (b) fumaroles of the sulphidic stage (c) fumaroles of the carbonic stage
		near the critical temperature	(a) solfataras: water vapors, CO_2, SO_2, H_2S, H_2SO_4 (b) mofettes: water vapor, CO_2, H_2S (c) jets of water vapor: water vapor
		under the critical temperature	hydrosolfataras: water vapor, CO_2, CH_4, H_2S

mite or titaniferous iron mineralization. Such relationships were not easy to recognize because of the absence of exposure that would unequivocally demonstrate these genetic connections — *either* the ultrabasics with the chromite *or* the volcanic differentiates with the Lahn–Dill ores are exposed, *but seldom both* (i.e., note the influence of "Level of Erosion"). However, Borchert (1957, 1960b) has described cases where this relationship

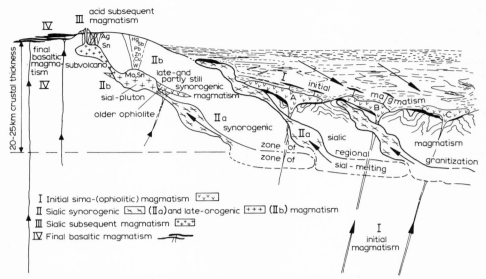

Fig.33. Schematic model of the sequence of plutonism—volcanism in space and time from the initial basaltic to the sialic magmatism, and its relation to the initial, synorogenic, late-synorogenic, subsequent and final magmatic stages, as proposed by Stille (1939, 1950). (Courtesy of Glückauf, Essen.)

between intrusive and extrusive deposits can be observed. According to Amstutz (1959b, 1974), "primary" spilitization has been neglected as an ore-forming mechanism inasmuch as it has been found that ore-containing spilites and quartz-keratophyres occur in all transitions with basic and ultra-basic plutonic rocks, as already pointed out in the preceding section.

Stage II is characterized by *"synorogenic and late-orogenic sialic magmatism"* which is part of the orogenic phases during the geosynclinal development. The material is "lithogenic-palingenic" or even "vadose" in origin, which was produced by the remelting of older rocks that became exposed to higher temperature and pressure conditions at depth. They are "intrusions" constituting the "plutonic association", composed mainly of granodiorites and granites (not to be confused with those formed by the consolidation of genuine magmatic differentiation). Here, too, one should note that subsurface magmatism has its surface or near-surface volcanic—subvolcanic equivalents, possibly with their own types of ore associations. Above the zone of sialic remelting lies a zone of regional granitization and pegmatization, above which, in turn, is the zone of synorogenic sial-intrusion within the older slates. Contact-metamorphism is to be expected. Borchert (1960b) has described at length the various ore types of this stage, many of which are the consequence of remelting—remobilization—reprecipitation of older mineralizations in sedimentary rocks. The ores are partly listed in the upper left-hand part of Fig.33. For tectonic reasons, submarine volcanic-exhalative mineralization is not extensive during this

TABLE XXXIX

Geochemical distribution of elements in specific genetic lithologic and ore-type groups (After Schneiderhöhn, 1962, table 8, p. 29; courtesy of Gustav Fischer, Stuttgart)

Vertical column of the chemical periodic system	Liquid-magmatic deposits	Main crystallization of the eruptive rocks	Pegmatites	Pneumatolytic deposits	Hydrothermal deposits
Ia		H, Na, K, Rb	H, Na, K, Cs, Li	H, Li	H
IIa		Mg, Ca, Sr, Ba	Be	Be	Mg, Ca, Sr, Ba
IIIa		Al	B, Al, Sc, Y, La	B, Sc	
IVa	Ti	Si, Ti, Zr	Si, Ti, Zr	Si	Si
Va	V		Nb, Ta	Nb, Ta	
VIa	Cr		Mo, W, U	Mo, W	W, U
VIIa		Mn	Mn		Mn
VIIIa	Fe, Co, Ni Ru, Rh, Pd Os, Ir, Pt	Fe		Fe, Co, Ni	Fe, Co, Ni, Pd
Ib	Cu			Cu, Au	Cu, Ag, Au
IIb				Zn	Zn, Cd, Hg
IIIb					Ga, Zn, Tl
IVb	C		C, Sn	C, Sn	C, Ge, Sn, Pb
Vb		P	P	P, As, Bi	As, Sb, Bi
VIb	O, S	O	O	O, S	O, S, Se, Te
VIIb			F, Cl	F, Cl	F, Cl

stage. The types of plutonic and subvolcanic ores formed depend on the material undergoing remelting.

Stage III depicts the *"subsequent magmatism"* of the post-orogenic phases and is associated with the first quasi-cratonic environment. The igneous rocks are the product of "lithogene-palingenic" mechanisms, but mainly extrusive activity takes place, with the "volcanic associations" predominated by quartz porphyry rocks. The synorogenic and subsequent magmatic phases combined comprise a sialic intermezzo in between a dominating siamic, i.e., initial and final magmatism.

The *"final magmatism"* (Stage IV) is part of the, by that time, fully developed cratonic phase of the geosynclinal-craton evolutionary trend. The material appears to be "juvenile basaltic effusives" which ascended along deep-reaching fault or fracture zones to form the "volcanic association".

The above information may incorrectly suggest that specific ore deposits or districts consist of minerals that originated under a narrow range of P(=depth)$-T$ conditions. On

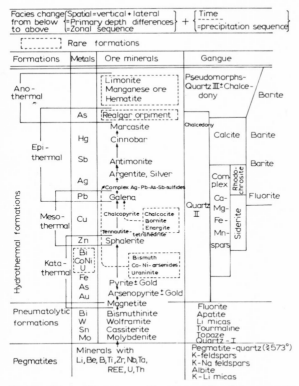

Facies change from below to above	Spatial=vertical + lateral {=Primary depth differences =Zonal sequence	+	Time =precipitation sequence

[- - - - - -] Rare formations

Formations	Metals	Ore minerals	Gangue	
Ano-thermal		Limonite Manganese ore Hematite	Pseudomorphs-Quartz III ± Chalce-dony	Barite
	As	Realgar orpiment	Chalcedony	
Epi-thermal	Hg	Marcasite Cinnobar	Calcite	Barite
	Sb	Antimonite		Barite
		Argentite, Silver	Com-plex	
	Ag	Complex Ag-Pb-As-Sb-sulfides	Rhodo-chrosite	
	Pb	Galena	Ca-	Fluorite
Meso-thermal	Cu	Chalcopyrite Chalcocite Bornite Enargite Tennantite-tetrahedrite	Quartz II Mg- Fe- Mn- Siderite	
	Zn	Sphalerite	spars	
Kata-thermal	Bi CoNi U Fe	Bismuth Co-Ni-arsenides Uraninite		
	As Au	Pyrite± Gold Arsenopyrite± Gold Magnetite		
Pneumatolytic formations	Bi W Sn Mo	Bismuthinite Wolframite Cassiterite Molybdenite	Fluorite Apatite Li micas Tourmaline Topaze Quartz - I	
Pegmatites		Minerals with Li, Be, B, Ti, Zr, Nb, Ta, REE, U, Th	Pegmatite-quartz(≥573°) K-feldspars K-Na feldspars Albite K-Li micas	

Fig.34. "Normal" sequence ("paragenesis") of the ore mineralizations in the pegmatitic—pneumato-lytic—intrusive hydrothermal environment. (After Schneiderhöhn, 1962; courtesy of Gustav Fischer, Stuttgart.)

Note: In the original figure "Fahlerz" (=fahl ore) was used instead of "tennantite—tetrahedrite. Fahlore is any grey-coloured ore mineral consisting essentially of sulfantimonides or sulfarsenides of copper; specifically tetrahedrite and tennantite (synonym: fahlerz).

the contrary, especially when genetically related ore districts or metallogenic provinces were examined in detail, the data demonstrated a "telescoping" of the ore and gangue mineral-precipitation, as summarized in Figs.34 and 35. Turning to a more specific example, Borchert (1957) proposed a direct connection between the type of ore deposits and the depth below the surface from which the igneous solutions were derived. Some discussions seem advisable in the following paragraph.

The scheme in Fig.35 offers an interesting division of volcanic-exhalative deposits, based on the distance of the ophiolitic magma from the earth's surface. This figure is not a comprehensive coverage of all possible related ore types, it is only a preliminary subdivision which may stimulate future investigations that will assist in establishing a spectrum of ores ranging from one extreme to another in regards to depth of igneous source. Fig.36 offers three major mineralization types formed along the "ocean water—sedi-

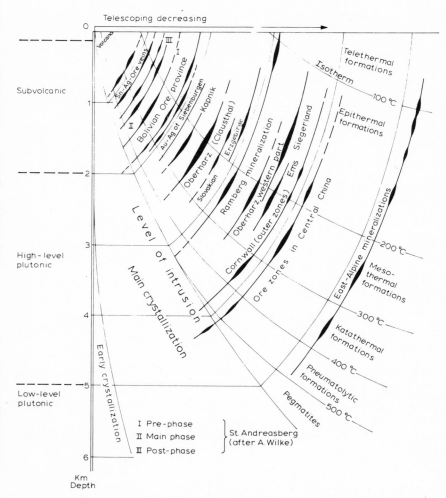

Fig.35. Telescoping effect as illustrated by a number of ore districts, regions, or provinces. (After Borchert, 1955, unpublished; 1978; courtesy of Glückauf, Essen.)

ments ± volcanics" interface, with the shallowest intrusion giving rise predominantly to lava flows, tuffs and pyroclastics. According to Borchert (1957, pp.561–562), the subsurface depth of the ophiolitic (basaltic) magma as well as the primary content of H_2O, HCl, HF, H_2S and CO_2 is important in determining whether oxidic, sulfidic or carbonate ores are formed. The most unstable tectonic setting, with a complex facies as a consequence, is characteristic of the Lahn–Dill ore type (cf. Chapter 6, Vol. 7 in this Handbook), whereas the radiolarite–manganese ores formed under tectonically stable situations (cf. Chapter 9, Vol. 7). In the latter case, the stable conditions may have been one

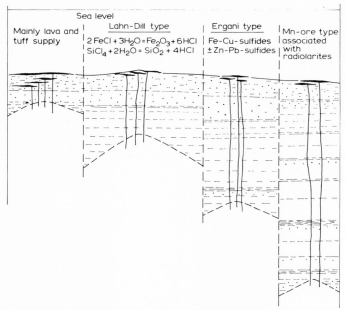

Fig.36. Scheme of ore deposits of the initial-magmatic type and their relation to the distance from the "roof" of the ophiolitic magma to the sea floor. (After Borchert, 1957; courtesy of Glückauf, Essen.)

explanation for the proposal of a few researchers that both the Mn and the SiO_2 were derived as weathering products from a continental source. The solutions forming the Lahn—Dill iron ores were chloridic, acidic volcanic exhalations in contrast to the Ergani-type ores (as well as the Rammelsberg and Meggen deposits) that were precipitated from alkali-sulfide solutions (cf. chapters in Vol. 9), i.e., the pH and Eh variables are important (see above).

In combination with the pH—Eh and other chemical properties of the fluids, the *composition of the rocks* through which these solutions pass are of importance in controlling the final chemical make-up of the fluids and gases that emerge at the earth's surface. For example, many minerals may react with the solutions to create alteration zones and pipes or replacement haloes. In particular, the feldspars in volcanics and pyroclastics as well as the calcite of limestones is very reactive, so that the type and degree of chemical reactions along the conduits of the ascending solutions is another variable that will determine the composition of the exhalative material.

In addition to the distance from the magmatic source[1], numerous other factors must be given attention before a well-founded classification — as attempted in Fig.36 —

---

[1] The zonal division of igneous hydrothermal deposits is partly based on distance from the source (see Fig. 34). Theoretically, it seems that a well-developed zonation cannot form if the magmatic intrusion approached the earth's surface too closely.

can be established that ranges from one end-member to another with numerous inter-mediaries in between. One of the more influential parameters is the *depth of the ocean water column*. Ridge (1973) discussed this problem at some length and concluded that "solutions above 220°–230°C cannot reach a sea floor on which the depth of water is 180 m or less in the liquid state, but will boil at some appreciable depth beneath the sea floor ... ore fluids probably never get close to the sea floor of shallow seas at temperatures high enough to permit boiling. Ore fluids significantly above 230°C can reach the sea floor in the fluid state only if the depth of the sea in the area in question is well above 180 m, for example; solutions at temperatures of ±300°C reach the sea floor in the liquid state only if the depth of sea water is slightly over 915 m. Ore fluids reaching the sea floor at such depths may be the parents of such fluids as the Red and Salton Sea brines." (Cf. Chapter 4, Vol. 4.)

The above information should serve as a warning against indiscriminantly employ-ing *zonal* classifications of ores based on two variables only (i.e., Figs.19 and 20).

Implicit in several of the classification schemes and models considered above is another "distance-factor" (additional to the "depth-below-earth's surface" parameter), namely, the distance from the magmatic source to the site of ore precipitation, men-tioned in both Fig.32 and Table XL, for example. Table XL (Baumann and Tischendorf, 1976) summarizes the six magma-related mineralizations. Let us now consider some spe-cific classification schemes.

An idealized "sequence of formation of magmatogenic ores" is given in Table XLI, which ranges from the liquid-magmatic to the epithermal types and according to the envi-ronment from the intra-magmatic to the subaerial produced mineralizations. Superim-posed on this tabular model are the simatic-juvenile and sialic-palingenic derivatives.

Amstutz' (1965) classification of magmatic mineral deposits (Table XLII) is sub-divided into four vertical columns according to the type of magmas (i.e., ultramafic to granitic) and into two major horizontal rows, each of the latter being split into plutonic and subvolcanic–volcanic sections. From our present leitmotiv of stratiform ores asso-ciated with volcanics (with or without sedimentary rocks), it can be seen that both intra- and extra-magmatic (the former syngenetic and the latter either syn- or epigenetic) pro-cesses give rise to stratiform and strata-bound types of ores. (Note: see above on spilite-associated mineralizations.)

A small digression from the general theme may be in order here to point out that:

(1) A particular magma may have supplied the material of *both* the ore and host-rock components, but where the ore-creating solutions moved into much older country rock this is not the case.

(2) Mineralization of the same metal (e.g., copper), or even of the same or related genetic type (e.g., copper porphyries) are not confined to the same host-rock and/or are associated by a variety of lithologies, as shown in Fig.37 (Smirnov, 1977).

Routhier and Delfour (1974) comprehensively summarized the massive sulfides associated with acid volcanism and presented a simplified classification of strata-bound

TABLE XL

Sequence of formation of the magmatic ore deposits (After Baumann and Tischendorf, 1976)

| Genetic types (physico-chem.) | → Endogenic dynamic (magm. diff.) | | | | | |
Environment of origin (type of structure)	liquid-magmatic	pegmatitic	super-critical to katathermal	hydrothermal katathermal	mesothermal	epithermal
Subaerial						As, S
Submarine — volcanic				Fe–Cu	Fe–Mn	Ag?
Submarine — subvolcanic				Au, Cu–Fe	Fe–Mn Zn–Pb, Ba–F	Ag, Sb–Hg
Intracustal — metasom.		Si–Al, Li, Be B, SE Pt, Ti	W–Mo–Sn Fe, Pt	Cu–Fe–As	Zn–Pb Fe–Mn, Ba–F	Sb–Hg
Intracustal — impregn.		Si–Al, Li, Be, B, SE	Sn–W–Mo, Bi, F	Cu–Fe–As Cu–Mo	Zn–Pb	Sb–Hg
Intracustal — void-filling	C, (Cr) Ni–Fe–Cu	Si–Al, Li, Be, B SE Pt, Ti, P, Ca–Mg	Sn–W–Mo, Bi Ti–Zr–Nb–Ta	Au Cu–Fe–As Cu–Mo	Zn–Pb U–Fe Fe–Mn, Ba–F	As–Sb–Ag Bi–Co–Ni (W)–Sb–Hg
Intramagm. — inject.	Ni–Cu–Co Ti–Fe, (Cr–Pt)					
Intramagm. — segreg.	Ni–Fe–Cu–Co–Pd Cr–Pt, Ti–Fe					

Magma type: ☐ simatic-juvenile ▦ sialic-palengenic.

TABLE XLI

The significance of the compositional and structural controls as a function of distance "source-to-precipitation locality" on type of endogenetic ore deposits (After Baumann and Tischendorf, 1976)

Type of deposit	Spatial arrangement of metal concentration in relation to magmatite	Distance between metal source and location of concentration	Temperature of formation of metal concentration	Syngenetic and/or epigenetic components of mineralization	Relative importance of the parameters controlling the mineralization at the location of concentration
Cr−Pt,Ti−Fe; Ni−Cu−Co	intramagmatic	small	very high	syngenetic	predominantly compositional control
Sn, W, Li; Be, Nb, Ta	perimagmatic	intermediate	high	syn- to epigenetic	compositional and structural controls
Au, Cu; U, Pb−Zn−Ag	apomagmatic	great	intermediate	epigenetic	predominantly structure control
Bi−Co−Ni; Ba, F; Sb, Hg	krypto- to telemagmatic	very great	low	epigenetic	

ores (Table XLIII). (Their comparison of the Precambrian Canadian sulfide ores and the Tertiary Japanese mineralization of the Kuroko-type is particularly recommended, but cannot be duplicated here; see also Chapters 5 and 12, Vol. 6.) The scheme divides the volcanism-associated ores into three groups, i.e., (1) those in direct contact with volcanic products (=true volcano-sedimentary types), (2) mineralizations associated with volcanic products from a distant source which makes the volcanic influence more difficult to prove, and (3) ore deposits where the volcanic material is separated from the mineralization by chronologic discontinuities. Group No. 1 is split into two models: (1) an ophiolitic ("spilite-keratophyre") model and variations with predominantly basic lava and (2) a volcanic model from more-or-less differentiated, and occasionally recurrent, calc-alkaline rock to abundant rhyolite.

A drawing by Bateman (1950) (Fig.38) pictures the relationship between igneous rocks and ore deposits, subdividing the latter into those produced from magmatic concentrations, emanation, and surface weathering of igneous rocks. (A classification of mineralizations related to acid intrusive magmatism has been offered by Rundkvist, 1978.)

To conclude this section, Schneiderhöhn's (1962) classification is quoted in Table XLIV.

TABLE XLII

Classification of magmatic mineral deposits (After Amstutz, 1965)

Location \ Magmas	Magmas	Ultramafic	Basaltic–gabbroic	Dioritic to quartzdiorite	Granitic to granitic-aplitic
Intramagmatic mineral deposits (syngenetic) The ore fluids do not leave the parent magma	plutonic	layered and disseminated oxide, native, and sulfide deposits in peridotites, etc. (almost no deuteric alteration)	disseminations and veins (little deuteric alteration) transition to spilites	disseminations and veins (medium deuteric alteration, beginning propylitiz.); transition to spilites and keratophyres	pegmatites; dissemination and networks of veinlets (strong propylitic alterations; e.g.: some porphyry coppers)
	subvolcanic–volcanic	rare	spilitic mineral deposits: sulfides, oxides (Fe, Mn, etc.), native metals	spilitic–keratophyric mineral deposits; (Fe, Mn, etc.) oxides, native metals, sulfides (transition to propylitic alterations)	mineral deposits associated with strong propylitic alterations ("extrusive porphyry copper deposits")
Extramagmatic mineral deposits (syn- or epigenetic).	plutonic	"filter-pressed-away" injections with or without marginal replacements	veins and marginal injections and/or replacements	veins and marginal injections and/or replacements	"classic" hydrothermal vein and replacement deposits
The ore fluids leave magma	subvolcanic volcanic	rare	subvolcanic veins and/or replacement deposits in adjacent rocks, and exhalative–sedimentary deposits (often in connection with spilites, keratophyres, and/or ophiolites; examples: many massive and disseminated sulfide deposits in orogenic belts; also the Michigan copper deposits)		subvolcanic veins and replacements, or exhalative–sedimentary deposits

From basic to acidic rocks an increasing tendency of the volatile fractions to be separated from, and to be in disequilibrium with the host rocks; increase of viscosity and increase of intrusive (batholitic) formation, but decrease of extrusive formation. The following parameters could be used individually as third dimensions to this binary classification: T, P, degree of accumulation of volatiles, distribution of volatiles, etc. The two main principles with regard to which this classification departs from previous classifications are: (1) All magmas produce their normal share of ore fluids. There is no missing link anymore in the basaltic–gabbroic and dioritic area. This gap is filled largely by the spilite–keratophyre group of rocks. (2) These ore fluids may or may not leave the parent magma. It is all too often assumed that they always leave.

TABLE XLIII

Classification of strata-bound mineralization associated with volcanism (simplified) (After Routhier and Delfour, 1974)

GROUP 1

MINERALIZATION IN DIRECT CONTACT WITH VOLCANIC PRODUCTS IN TUFF AND/OR CHEMICALLY DERIVED SEDIMENT
True Volcano-Sedimentary Deposits

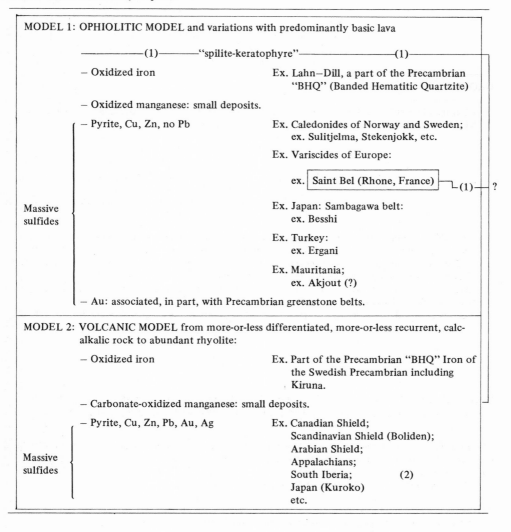

MODEL 1: OPHIOLITIC MODEL and variations with predominantly basic lava

————————(1)————"spilite-keratophyre"————————————(1)—

— Oxidized iron   Ex. Lahn—Dill, a part of the Precambrian "BHQ" (Banded Hematitic Quartzite)

— Oxidized manganese: small deposits.

— Pyrite, Cu, Zn, no Pb   Ex. Caledonides of Norway and Sweden; ex. Sulitjelma, Stekenjokk, etc.

Ex. Variscides of Europe:

ex. Saint Bel (Rhone, France) ⎯(1)⎯ ?

Ex. Japan: Sambagawa belt: ex. Besshi

Massive sulfides   Ex. Turkey: ex. Ergani

Ex. Mauritania; ex. Akjout (?)

— Au: associated, in part, with Precambrian greenstone belts.

MODEL 2: VOLCANIC MODEL from more-or-less differentiated, more-or-less recurrent, calc-alkalic rock to abundant rhyolite:

— Oxidized iron   Ex. Part of the Precambrian "BHQ" Iron of the Swedish Precambrian including Kiruna.

— Carbonate-oxidized manganese: small deposits.

— Pyrite, Cu, Zn, Pb, Au, Ag   Ex. Canadian Shield;
Scandinavian Shield (Boliden);
Arabian Shield;
Massive sulfides   Appalachians;
South Iberia;   (2)
Japan (Kuroko)
etc.

(*continued*)

TABLE XLIII (continued)

GROUP 2

MINERALIZATION WHOSE RELATIONSHIPS WITH VOLCANISM ARE MORE DISTANT AND MORE DOUBTFUL AND CANNOT BE USED AS A GUIDE; the mineralized bodies are outside the volcanites in beds where the volcanic distribution is slight or doubtful.

Massive sulfides	Ex. Rammelsberg, Meggen (Germany); Mt. Isa (Au, Pb, Australia); Cobar (Cu, Australia); Kilembe (Cu, Uganda); Ducktown (Tennessee, U.S.A.); etc.

GROUP 3

MINERALIZATION SEPARATED FROM VOLCANISM BY A CHRONOLOGIC DISCONTINUITY (exundation, continental weathering and leaching or late hydrothermalism).

Ex. Mn of Tiouine, etc. (Morocco); Cu of Boleo (Mexico); Cu of Lake Superior (in part); Co of Bou Azzer (Morocco); Au derived from Precambrian greenstones.

(1) The arrow joining the two models of Group 1 indicates the transitions that exist between them when considered on a world-wide scale; for example between the pure ophiolitic model with no acid lava (e.g. Alps) and assemblages rich in acid lava (keratophyre) such as in the Caledonides. The question mark signifies the doubts relative to the individuality of the keratophyre in relation to rhyolite (see text). This difficulty led to the doubtless extremely arbitrary exclusion of Saint Bel from Model 2 of Group 1.

(2) Take heed! Prudence is necessary when attempting to insert the Iberian deposits into this model in the same manner as the Canadian and Japanese models.

GENETIC, PETROGRAPHIC, MINERALOGIC, ELEMENTAL, AND PARAGENETIC CLASSIFICATIONS AND MODELS OF MAGMATOGENIC-VOLCANOGENIC ORES AND HOST-ROCKS

"Every science begins by accumulating observations, and presently generalizes these empirically; but only when it reaches the stage at which its empirical generalizations are included in a rational generalization does it become developed science."
    Herbert Spencer (in: *The Data of Ethics,* 1879)

"To avoid this grave danger (of a single hypothesis becoming a 'ruling theory' and master of its author) the method of working hypotheses is urged where the mutual conflicts of hypotheses whet the discriminative edge of each."
    T.C. Chamberlin (in: *The method of multiple working hypotheses, J. Geol.,* 1897)

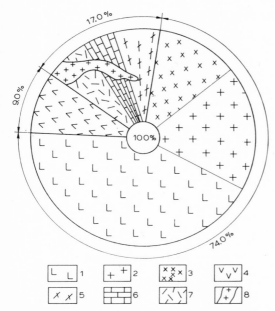

Fig.37. Compositional diagram of the ore-surrounding rocks of the copper-porphyry deposits. Legend: *1* = diorites, syenite-diorites, and monzonites; *2* = granodiorites; *3* = granite-porphyries; *4* = eruptives of intermediate composition; *5* = schists; *6* = limestones; *7* = eruptives of varied composition; *8* = minor intrusions and veins of granodiorite- and granite-porphyries, and diorite-porphyries. (After Smirnov, 1977, vol. 2.)

As an expansion of and to augment the preceding section a number of terminological and classificatory schemes will be presented below that utilize more specific, more narrowly defined, parameters or variables. Consequently, their applications are restricted to particular types of deposits and/or types of investigations. Data will be furnished on:

(1) Subdivision of magmatic—pneumatolytic—hydrothermal processes.

(2) Geochemistry, mineralogy and depth-zonation of magmatogenic deposits.

(3) Distribution of elements in magmatic to hydrothermal (plus supergene) deposits.

(4) Comparison of four classifications.

(5) Development of a granitic magma and its main constituents.

(6) Main chemical elements and minerals of granitic pegmatites.

(7) Geochemical concentrations in a granite batholith.

(8) Groups of accessories in "specialized" granites.

(9) Interrelationships among mica compositions, accessory minerals and granitic rocks.

(10) K/Rb-relationships of "normal" vs. "specialized" granites.

100

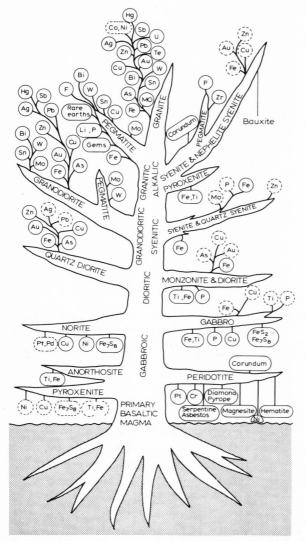

Fig.38. Relation of igneous rocks to mineral deposits. (After Bateman, 1950; courtesy of Wiley, New York, N.Y.)
The elements that branch upward occur in deposits formed from magmatic emanations; those hanging down are magmatic concentrations; fallen ones are weathered products. Full circles indicate major associations.

(11) Averages and dispersion of fluorine and tin in granites (including Clarke value vs. regional values).

(12) Model of tin-bearing granitoid rocks vs. mineralogy, depth-of-intrusion, and lithology of host-rocks.

TABLE XLIV

Classification of magmatic and related ore deposits (freely translated from Schneiderhöhn, 1962)

I. *First magmatic main group: liquid-magmatic deposits*

    1. Chromite
    2. Diamond   } in ultrabasics
    3. Platinum-group (platinoid) metals
    4. Titanomagnetite and ilmenite in gabbros, norites and anorthosites
    5. Fractional crystallization sulfides and their autohydrated alteration products
    6. Pentlandite/pyrrhotite–chalcopyrite deposits in norites and gabbros (type Sudbury)
    7. Platinum- and palladium-containing sulfide–pyroxenite and sulfide–diallage–norite (type Merensky-reef)

I/II. *Intermediate group: liquid-magmatic–pneumatolytic transitional deposits*

    1. Autohydrated chromite and platinum ores
    2. Autohydrated titanomagnetite and rutile ores
    3. Autohydrated pentlandite/pyrrhotite ores
    4. Intrusive magnetite–apatite ores (type Kiruna)
    5. Intrusive apatite–nepheline ores (type Chibine)

II. *Second magmatic main group: pegmatitic–pneumatolytic deposits*

    1. Pegmatite veins and pegmatite zones
    2. Tin, tungsten and molybdenum deposits
    3. Tourmaline-containing gold–copper–lead–silver–bismuth- etc. deposits
    4. Contact-pneumatolytic tin, tungsten and molybdenum deposits
    5. Contact-pneumatolytic-iron and manganese deposits
    6. Contact-pneumatolytic gold, copper, lead, silver, cobalt, zinc deposits
    7. Contact-pneumatolytic platinum deposits
    8. Other contact-pneumatolytic deposits

II/III. *Intermediate group: pneumatolytic–hydrothermal transitional deposits*

    1. Silver–lead–zinc deposits
    2. Tin deposits
    3. Tungsten deposits
    4. Gold–tourmaline–quartz veins
    5. Gold–scheelite deposits
    6. Magnetite–hematite deposits
    7. Copper deposits

III. *Third magmatic main group: hydrothermal deposits*

    (A) Gold and silver deposits
        (a) Plutonic varieties
            1. Plutonic–katathermal gold–quartz veins (type Mother lode)
            2. Plutonic gold-bearing impregnations and replacements in silicate rocks (type Kalgoorlie)
            3. Plutonic gold-bearing replacements in carbonate rocks
            4. Plutonic–mesothermal gold–selenium–lead ores (type Corbach)
        (b) Subvolcanic varieties
            5. Epithermal gold and gold–silver ores
            6. Epithermal silver ores (type Mexico)
    (B) Pyritic (= "Kies") and copper deposits
        (a) Plutonic varieties

(continued)

TABLE XLIV (continued)

---

1. Kata- to mesothermal pyrite ("Kies") stockworks as silicate replacements and dissemitions (type Rio Tinto)
2. Pyritic ("Kies")-chalcopyrite impregnations in silicate rocks (type "disseminated copper ores")
3. Mesothermal chalcopyrite—quartz veins
4. Mesothermal chlorite-bearing chalcopyrite veins and mineralized tectonized zones in metamorphic basic rocks ("chloritic copper formations")
5. Mesothermal copper—arsenic veins (type Butte)
6. Mesothermal copper—arsenic replacement deposits (type Tsumeb-mine)
7. Tetrahedrite—tennantite veins (type Schwaz)

(b) Plutonic—subvolcanic transitional deposits
    8. With copper-ore impregnated eruptive breccias (type Pilares mine)
    9. Zeolitic—chloritic native-copper impregnations (type Lake Superior)
    10. Telemagmatic chalcopyrite deposits (type Kennecott)

(c) Subvolcanic varieties
    11. Meso- to epithermal pyritic "Kies" deposits (type Maiden Pek)
    12. Japanese black ore deposit
    13. Meso- to epithermal copper—arsenic ores (type boron)
    14. Mesothermal silicified (quartzose) subvolcanic rocks with chalcopyrite and pyrite (type Murgul)
    15. Epithermal zeolitic epidotic copper ore in basic rocks (type Nahe)
    16. Mineralization with chalcocite and native copper in red sandstones, slates and tuffites (type Imsbach and Corocoro)

(C) Lead—silver—zinc deposits

(a) Plutonic varieties
1. Kata- to mesothermal silver-rich galena-quartz veins (type Freiberg "precious quartz crystal deposits")
2. Kata- to mesothermal quartz-pyritic (= kiesige)-galena veins [type Freiberg "pyritic (= kiesige) lead deposits"]
3. Mesothermal silver-rich carbonate galena veins (type Freiberger "precious brown Fe—Mn—Ca carbonate deposit")
4. Mesothermal quartz—pyritic—galena—tetrahedrite/tennantite veins (type Granite-Bimetallic Mine)
5. Mesothermal fluorite—barite—galena veins (type Schwarzwald) (= type Black Forest)
6. Sideritic mesothermal galena—tetrahedrite/tennantite veins (type Wood River)
7. Meso- to epithermal quartz—siderite lead—zinc veins (type Rheinisches Schiefergebirge)
8. Meso- to epithermal quartz—calcareous lead—zinc veins (type Upper Harz)
9. Meso- to epithermal quartz—barite lead—zinc veins (type South Harz)
10. Mesothermal mineralized tectonized zones and silicate-rock replacements with galena and sphalerite (type Bawdwin)
11. Kata- to mesothermal replacement deposits with Ag-rich lead—zinc ores (type Leadville)
12. Apomagmatic, meso- to epithermal lead—zinc ores (type Bleiberg)
13. Secondary-hydrothermal layered-metasomatic lead—zinc deposits (type Upper Silesia, Tri-State, Mississippi Valley)

(b) Subvolcanic varieties
    14. Subvolcanic meso- to epithermal lead—zinc ores (type Pulacayo)
    15. Subvolcanic replacements with lead—zinc ores (type Cartagena)

(continued)

TABLE XLIV (continued)

(D) Silver–cobalt–nickel–bismuth–uranium deposits
    1. Calcitic native-silver veins (type Kongsberg)
    2. Calcitic–zeolitic complex silver veins (type St. Andreasberg)
    3. Fluorite–barite–silver- and bismuth-bearing cobalt veins (type Wittichen)
    4. Baritic silver- and bismuth-bearing cobalt–nickel veins (type Schneeberg)
    5. Carbonate-containing bismuth-free silver–cobalt veins (type Cobalt)
    6. Carbonate-containing – quartzitic copper-bearing cobalt–nickel veins (type Dobschau)
    7. Barite cobalt veins (type Richelsdorf)
    8. Katathermal quartzitic uranium veins (type Chingolobwe)
    9. Quartzitic silver-bearing cobalt–bismuth veins with uranium (type St. Joachimsthal, Great Bear Lake)
    10. Calcitic cobalt–nickel veins with uranium (type Schmiedeberg)
(E) Tin–silver–bismuth deposits
    (a) Plutonic varieties
        1. Kata- to mesothermal bismuthinite veins
        2. Colloidal cassiterite deposits (type Vila Apacheta)
        3. Sulphostannate ores with wurtzite (type Monserrat)
        4. Epithermal tungsten veins
    (b) Subvolcanic varieties
        5. Subvolcanic pneumatolitic–hydrothermal tin deposits (type Llallagua)
        6. Subvolcanic meso- to epithermal sulfidic silver–tin deposits (type Potosi)
        7. Subvolcanic meso- to epithermal wurtzite–teallite veins (type Carguaicollo)
(F) Antimony–mercury deposits
    1. Stibnite veins (Chinese type)
    2. Antimony–mercury deposits (type Ferghana)
    3. Mercury deposits: (a) sub-type Almaden
                        (b) sub-type Amiata
    4. Realgar–orpiment deposits
    5. Selenium veins
(G) Oxidic iron–manganese–magnesium deposits
    1. Siderite/manganite/rhodochrosite veins (type Siegerland)
    2. Metasomatic siderite deposits (type Erzberg)
    3. Hematite veins and specularite veins
    4. Manganese veins
    5. Platinum-bearing hematite–quartz veins
    6. Metasomatic specularite deposits
    7. Metasomatic magnesite deposits
(H) "Ore-free" deposits
    1. Fluorite veins
    2. Metasomatic fluorite deposits
    3. Barite veins
    4. Metasomatic barite deposits
    5. Witherite veins
    6. Quartz veins and hydrothermal silicifications

IV. *Fourth magmatic main group: exhalative deposits*

    1. Exhalites
    2. Solfatoric sulfur deposits
    3. Exhalative boron deposits

V. *Transitional formations between submarine exhalative-sedimentary deposits*

    1. Submarine exhalative-sedimentary specularite deposits ("Keratophyr-iron ore") (type Lahn-Dill)
    2. Exhalative-sedimentary manganese deposits
    3. Submarine exhalative-sedimentary sulfide deposits

(13) Characteristic elements of granite-pegmatites and nepheline-syenite-pegmatites.

(14) Approximate element-content of pegmatitic solution.

(15) Distribution of chemical elements in the pegmatite-to-pneumatolytic range of rocks.

(16) Mineralogic types of pegmatites and elemental distribution.

(17) Pegmatite types according to temperature of formation.

(18) Direct association of non-pegmatitic veins with pegmatites.

(19) Phases of origin of and elemental distribution in pegmatites.

(20) Temperature, phase and mineral-paragenetic diagram of pegmatites.

(21) Feldspar distribution in pegmatites.

(22) Rhythms in plagioclase formation.

(23) Geochemical diagram of lithium and lithium minerals in pegmatites.

(24) Paragenesis and milieu of origin of boron minerals in pegmatitic—pneumato-lytic—hydrothermal rocks.

(25) Distribution of chlorites in various types of magmatic ore deposits.

(26) Quantitative paragenetic scheme of mineralization phases of sub-volcanic silver veins (example).

(27) Environmental range of origin of calcite-types according to temperature and concentrations of the mineral-formation solutions.

(28) Comparative diagram of mineralogic paragenesis of veins.

(29) Typical paragenetic relationships.

(30) Sequential-genetic classification of a tin deposit.

(31) Characteristic trace elements to identify ortho- and para-metabasites (examples from the published literature).

The prevalently used terminology and classification of ore deposits and host rocks formed by multitudinous igneous processes are listed in Table XLV, including the temperature range and Fersman's phases A to L (Schneiderhöhn, 1961). These phases are the magmatogenic subdivisions upon which Fersman based much of his descriptions and interpretations; they are useful in so far as they categorize the major stages of magmatic evolution and provide the subsequent schemes with a genetic framework for a paraphernalia of mineralogical, elemental, paragenetic, temperature and related data. These models can be explained only briefly.

Table XLVI summarizes the idealized apportionment of the chemical elements, minerals (ore and gangue) (main, minor and accessory) according to depth, temperature and origin (phases A to L), whereas Table XLVII is a distribution diagram of individual elements — thus supplementing the preceding table. Prior to proceeding with the data of Fersman's work (and summarized by Schneiderhöhn, 1961), it must be clarified that other schemes have been offered — and periodic comparisons of new information should be encouraged for both theoretical and practical reasons. One such comparison of the pegmatite-to-pneumatolytic (B—H) range is given in Table XLVIII.

The multiplicity of classifactory approaches provided below are the result of taking

TABLE XLV

Schematic division of the magmatic–pegmatitic–pneumatolytic–hydrothermal processes (based on several tables by A. Fersman, 1952; compiled by Schneiderhöhn, 1961)

A	B	C	D	E	F	G	H	J	K	L
900°	800°	700°	600°	550°	500°	450°	400°	300°	200°	100–50°

Magmatic main crystallization — Epimagmatic, pre-pegmatitic processes — Pegmatitic s.s. — Pegmatic processes s.l. — Pegmatoid supercritical → — High-, intermediate-, low- hydrothermal — Supergene

← Pegmatitic →   ← Pneumatolytic →   ← Hydrothermal →

Granite proto- — Aplite meso- — Pegmatite — s.s. — Pneumatolytic deposits — Hydrothermal deposits

crystallization

("out-of-geological context", if you wish) specific mineralogic, elemental, and/or geological parameters to project or disclose their influence without introducing a maze of complexities as is done in multi-variable schemes. In investigations of this kind, two possibilities exist in the preparation of models: either (a) one commences with a comprehensive type and proceeds to delineate, study and modelize certain parts of it, or (b) one does the reverse by starting with the examination of particular systems, the data of which are then combined to constitute a complex all-encompassing scheme. Here, the first approach was selected, e.g., from Table XLVIII to Table LX.

Table XLVIII lists the main minerals that are formed from a granitic magma, whereas Table XLIX summarizes the predominant elements and minerals of a granite pegmatite, and finally Fig.39 illustrates the zonal distribution of a granite batholith. To continue illustrating the multiplicity in studying the characteristics of one specific lithology, or related rock types, Table L lists the accessory minerals in "specialized" granites (Tischendorf, 1977); Fig.40 exemplifies the interrelations between micas, accessories and granitic rocks; and Fig.41 displays the K/Rb-relationships of normal granites, of the precursors of "specialized" granites as well as of the "specialized" granites. As to Fig.40, Pälchen and Tischendorf (1974, 1978) stated that according to the geological setting and the petrographical/geochemical properties, the postkinematic Variscan granites are divisible into two different intrusive complexes (each with three intrusive phases): (1) "older intrusive complex" (OG), with a special two-mica granite (OGt) facies; and (2) "younger intrusive complex (YG) with preceding phases of fine- to medium-grained, porphyritic (intermediate) phases (IG); the latest stage of evolution is characterized by late- to post-magmatic transformations (especially albitization) forming metasomatized

TABLE XLVI

Geochemistry and mineralogy of the magmatic depth zones (After A. Fersman, P. Niggli, H. Schneiderhöhn) (cf. Schneiderhöhn, 1961)

Phases (After A. Fersman)	Temperature	Distance from source	Macro milieus	Local milieus		Predominant elements	
L	40°		Oxidation zone	Hypergene		Pb, Zn, Au, Cu, Sn, O, CO_2, SO_3	
	50°	>2 km ↑		Antimony—mercury		Hg, Sb, As, S, Se, CO_2	
K	100°			Upper	polymetallic zone	Ni, Co, Ag, Bi, S, As, CO_2	
	150°	2 to 1.5 km ↑	Hydro-thermal	Central		Pb, Zn, Cd, Zn, Ag, S	
J	200°			Lower		Zn, Sb, As, S, Cu, Ag Au	
H	350°			Zone with gold and arsenopyrite		Au, As, Fe, Se, S, W, CO_2	
			Pneumato-lytic	Zone with gold and scheelite			
G	425°	1 km ↑		Zone with chalcopyrite and tourmaline		Cu, Fe, Mo, Co, Au, As, B(F), S, CO_2 (W, Sn)	
F				Zone with wolframite		Sn, W, U, Mo, Bi, As(Cu) (Li), Cl, F	
				Zone with cassiterite			
E	550°			Zone with molyb-denite		Mo, Li, Cl, F	
D	600°	0.5 km	Pegmatitic	Zone with tourmaline and muscovite		K, Al, Be (Mo, Sc), Si, B(F, Zr)	
C	700°			Pegmatites s.s.		K, Na. Al, Ti, Y, Sc, U, Th, heavy metals, Si, Zr, Nb, Ta, Hf, P	
B	800°			Aplite		K, Na, Al, P, Se, earth-metals (Fe, Ca, Mn), Si, Zr	
A	<1000°	Source	Magmatic	Granite		Na, K, (Ca, Mg), Zr, Th, Fe, Si, P, Cl, S)	

Main minerals	Minor minerals	Accessory minerals	Remarks
opal, chalcedony carbonates kaolin	carbonates, sulfates, silicates of heavy metals		
chalcedony carbonates barite	orpiment cinnobar stibnite	alunite zeolites kaolin	secondary quartz of chalcedony
carbonates barite	native silver Co-, Ni-arsenides marcasite		upper boundary (limit) of hematite
carbonates barite (quartz)	sulfides of Pb + Zn	argentite	much
quartz little barite	sulfantimonides and sulfarsenides of Cu	little	barite
quartz pyrite arsenopyrite	gold, scheelite hematite	black tourmaline partly green tourmaline siderites	
quartz, tourmaline (siderite, ankerite)	chalcopyrite, pyrite molybdenite hematite		lower boundary (limit) of hematite
quartz topaz beryl	wolframite scheelite		wolframite tin molybdenum
	cassiterite molybdenite topaz, Li-micas		
microcline quartz	schorl muscovite (beryl)	apatite molybdenite	
microcline, quartz graphic granite, garnet plagioclase	uraninite tantalite, niobite Ti–Nb uranite, zircon xenotime, orthite monazite, apatite	biotite (muscovite) black tourmaline (beryl) magnetite, Pb-, Zn-sulfides	"black minerals"
microcline quartz plagioclase	orthite zircon monazite	magnetitie garnet biotite muscovite	strong influence of the surrounding rocks
quartz feldspar biotite	magnetite pyrite zircon monazite	titanite orthite	

TABLE XLVII

Geochemical distribution of the major elements in magmatic (and supergene) rocks (After Fersman, 1952, cf. Schneiderhöhn, 1961)

Stages	Magmatic		Epimagmatic		Pneumatolytic (fluidal)		
Geophases	A	B	C	D	E	F	G
Temperature 800°		700°	Point Q 600°			500°	poin
						Pneumatolytic and hy
Distance from the source: in the massif		———		at the contact—not more than 100 m	not more than 500 m	not more than 1000	
Mo				←---——molybdenite——→			mo
Li				←——— zinnwaldite			
B and Be			black tourmaline	←—←——— beryl ———→			gr
Sn				cassiterite			
Sc							
W				Fe > Mn ←——wolframite—Mn > Fe —×—			
Bi				←——— native bismuth ———		bismuthini	
Fe			magnetite	pyrrhotite(pyrite)			py
Au							na
As						←arsenopyr	
S			Fe, Cu, Pb sulfide				sulfides
Cu				—			←
Zn			—?				– sphale
Cd							_ _ ↑_
In and Ge							_ _ ↑_
F				←—— topaz —→			fluorine
Mn							
Pb			—?				
Ag							
Ba				(in the micas and feldspars at contacts)			
Se							(
Co							
Ni							
U, Ra							
Tl							
Te							

Hydrothermal			Supergene
H	I	K	L
	300°	200°	100°

nal elements and minerals
not more than 1500 to 2000 m | − more than 2000 m

H	I	K	L
te		powellite	with Pb bearing wulfenite
maline	ludwigite		
nite		(secondary cassiterite) "wood" tin	detrital cassiterite
elite⟶	(wolframite)		W in eluvium, alluvial scheelite
		native bismuth	change into basobismuthite
hotite	hematite siderite	red hematite	Fe hydroxides
I⟶	gold II	gold III Ag	detrital
	arsenides	realgar and orpiment	
	——— sulfosalts ———	sulfates ———	——— sulfates
copyrite ———	⟶ grey copper		malachite, azurite
———	——— complex compounds of Zn		smithsonite, calamine

↑   ↑
↑

- - - - - - -←——— fluorine ———⟶
- - - rhodochrosite - - - - - | maganese oxides
- - - - - galena ——— galena   Pb sulfosalts | cerrusite, calamine, etc.
                    ↑—native Ag argenite
fibrous barite   - - - -barite - - - -
| selenides
danaite ⟶   - - -cobaltite | erythrine
Ni sulfides and arsenides
pitchblende   autunite, chalcocite
in sulfides
(only in effusive rocks) tellurides of Au, Ag

(continued)

TABLE XLVII (continued)

Stages	Magmatic	Epimagmatic		Pneumatolytic (fluidal)			
Geophases	A	B	C	D	E	F	G
Temperature 800°		700°	Point Q 600°			500°	
Sb							
Hg							
SiO$_2$				little	grey	light milky	quartz (d colored)
	Micas possibly present			biotite zinnwaldite		muscovite	chlorite
CO$_2$							
BaSO$_4$							

granites (YGm). The interrelationships depicted in Fig.40 illustrate that a classification founded on specific major minerals, accessory minerals, major and minor chemical elements can be successfully applied in both theoretical and practical (exploration) studies. (See also Rundkvist, 1978.)

In ore petrology and geochemical prospecting, the understanding of the behaviour and distribution, as well as element-interrelationships, are paramount in establishing some descriptive and genetic classifications. For example, Fig.42 demonstrates the correlation between tin (x-axis) versus fluorine (y-axis), giving the averages and dispersion values in each of the eight cases, plus the planetary and regional averages of F- and Sn-values. The study illustrates that the older granite complex (OG) may have been the precursor of the tin-richer and fluorine-richer younger granites (IG, YG). The importance of fluorine on the genesis of tin and rare-element deposits, on the physicochemical character of the magma, and on late-magmatic metasomatic processes, has been briefly mentioned by Tischendorf (1977). Investigations on the interrelationships between elements are a prerequisite to classifications of "indicator elements" (e.g., Boyle, 1974a), which are employed in geochemical prospecting. Further, such detailed geochemical information — in combination with other data — can then allow the grouping of different types of related deposits, as done by Varlamoff (1974).

Varlamoff (1974, 1978) presented a diagrammatic-tabular conceptual model of ore associated with acid intrusives (i.e., granitic—pegmatitic range of lithologies), exemplified by tin mineralization: three major types of groups, divided into six subtypes, are offered,

Hydrothermal			Supergene
H	I	K	L
00°	300°	200°	100°
		stibnite	Sb oxides
(?)	grey copper	cinnabar	
laria quartz		chalcedony	quartz-opal
ertite	sericite	kaolinite	argilite
Fe——Mn	——Mg———Ca	—Fe———Mn	——Ca——Ca
		barite ——→	

which in turn are further subdivided into thirteen geographic examples. Against these, Varlanoff plotted the mineral distribution and the depth-of-formation of these thirteen tin-bearing volcanic—hypabyssal—plutonic rock assemblages (with two additional tin-free abyssal—plutonic rocks). (See also Štemprok, 1978a,b; Taylor, 1978.) Particularly interesting in this model is the significance of depth-of-origin controlling, for example, the mineralogy and mode of occurrence of mineralization; these factors, in combination with the level of present-day erosion, are of prime importance in exploration work. One should especially keep in mind that more than one type of intrusion is to be expected in any specific region, based solely on depth-of-intrusion. Similar "spatial—temporal—mineralogical distribution" classifications should be worked out for other metalliferous and industrial-mineral deposits.

Continuing the above parade on grouping and classifying, Tables LI, LII, and LIII list the characteristic elements of two types of pegmatites, the elemental composition of a pegmatitic solution during magmatic differentiation, and the distribution of elements in the pegmatitic—pneumatolytic range, respectively. Finally, Table LIV depicts the main mineralogy of eleven types of pegmatites (B—G) and the distribution of numerous elements (including hydrothermal material, H—K). The relatively great variety of pegmatites, and their range of formational conditions during phases B—K, are listed in Fig.43. The earlier table already indicated the presence of non-hydrothermal constituents in pegmatite complexes, which is emphasized in Table LV. The four magmatic phases during which pegmatoid components crystallize from a magma are divided according to the main and

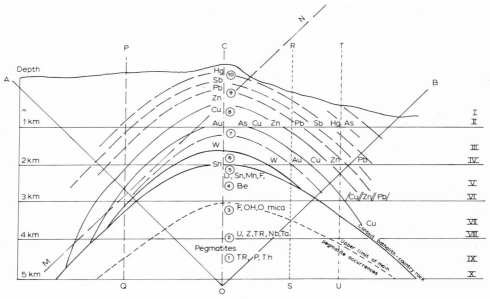

Fig.39. Geochemical concentrations associated with a granitic batholith. (After Fersman, 1952, vol. 2.)

minor elements in Table LVI, which can be compared and correlated with the temperature range and mineralogy in Table LVII.

Further illustrating the procedure of refining the description and classification of rocks, as exemplified by the pegmatites, Table LVIII lists the feldspar varieties, Fig.44 depicts the zoning and cyclicity of plagioclase, Table LIX gives the lithium minerals and Table LX the boron minerals — which concludes the treatment of the pegmatoid and related rocks and associated deposits. However, some supportive data are furnished on mineralogical varieties and their paragenesis as criteria in ore-genesis investigations.

The characteristic features of *altered country rocks* around mineralizations are well-known exploration guides, if the width of the zone is about 10—50 times greater than the orebody. However, these surrounding rocks are of little or no help when their width is excessive (e.g., hundreds or thousands of times larger). Certain mineralogical criteria have been used as demonstrated by the various chlorite types in association with certain ores (Fig.45, Kreiter, 1968, quoting Shilin and Ivanova).

Wilke (1952) provided an exemplar of hydrothermal mineralization of which the paragenetic diagram (Fig.46) lists several generations of each mineral. Calcite has been classified into five groups, with some sub-varieties (i.e., I; IIa, IIb, IIc, IId; IIIa, IIIb; IV; and V), which originated under four hydrothermal and supergene conditions, respectively. The precipitation of the hydrothermal calcite types was controlled by the temperature and the concentration of the solutions (see Fig.47). A paragenetic comparison of

TABLE XLVIII

Evolution of a granitic magma and its main chemical elements (After Fersman, 1931; cf. Schneiderhöhn, 1961)

Temperature	800°	700°	600°	500°	400°	300°	200°	100°
Phases	A	B	C	D–E	F–G	H	J	K
	magmatic	epimagmatic	pegmatitc	pegmatoid	supracritical-pneumatolytic	high-hydrothermal	intermediate hydrothermal	lower hydrothermal
Mineral composition	quartz feldspars hornblende magnetite	graphic granite orthite monazite	graphic granite with Ti-Nb-Ta-rare earth	tourmaline mica / topaz beryl	Li-minerals / manganese minerals; phosphates	cryolite / fluorite	sulfides	zeolites
Types of deposits	acidic granites		pegmatites		pneumatolytes	hydrothermalites		
Type number		1	2	3 / 4	5 / 6	7	high / intermediate	lower
Characteristics of the solutions	magmatic-melt solutions			fluid solutions		superheated aqueous solutions		

TABLE XLIX

Main chemical elements and major minerals of granite pegmatites and their percentages (After Schneiderhöhn, 1961, p. 464; courtesy of Gustav Fischer, Stuttgart)

Frequency	Elements	%	Minerals	%
Always present	H, B, O, F, Na, K, Al, Si, P, Ca, Fe, Mg	12	albite (plagioclase), microcline, quartz, apatite, topaz, magnetite, biotite, muscovite, black tourmaline, garnet, fluorite	12–13
Frequent	Li, Be, S, Cl, Ti, Mn, Zr, REE, Y. Nb, Sn, Cs, Rb	12–13	lepidolite, spondumene, beryl, pyrite, titanite, zircon, rutile (ilmenite), orthite monazite, xenotime, columbite, cassiterite, polluzite	13–14
Rare	C, Cu, Zn, Mo, W, Pb, Bi, Sc, Th, U	10	calcite, chalcopyrite, sphalerite, molybdenite, wolframite, galena, bismuthinite, bismuth, euxenite, samarskite, uraninite	11
		34–35		37–38

four mining districts is offered in Fig.48, based only on the major minerals.

Generalized or idealized paragenetic relationships of typical mineral associations have been established empirically – and many, but not all, explained theoretically – to assist the ore petrologist. One scheme is presented in Table LXI listing the parageneses of two, three and four or more members. Such "averaged-out" schemes serve only as guides; they do not set the rule for specific studies as numerous exceptions are to be expected.

Paragenetic or sequential classifications or models are more complete if the total geological history can be schematized, because each event, process and product is prop-

TABLE L

Schematic grouping of accessories present in "specialized" granites as reported in the literature (Tischendorf, 1977)

Very important accessories, typical for specialized granites	Topaz, fluorite, tourmaline, cassiterite, columbite–tantalite, beryl
Less frequent accessories, typical for specialized granites	Orthite, titanite, thorite, xenotime, monazite, apatite
Accessories appearing in specialized and normal granites often influenced by processes of assimilation	Andalusite, cordierite, sillimanite, dumortierite, garnet
Non-typical accessories appearing in specialized and normal granites	Magnetite, ilmenite, pyrite, rutile, anatase, zircon, epidote, hornblende

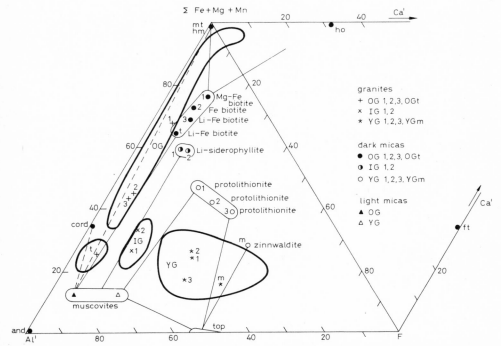

Fig.40. Interrelations between the composition of micas, accessory minerals and the granitic rocks in the Erzgebirge. (After Pälchen and Tischendorf, 1974, 1978.)
The parameters are: $Al' = Al - (Na + K + 2 Ca)$; $\Sigma Fe^{3+} + Fe^{2+} + Mn + Mg$; $Ca' = Ca - [\{Al - (Na + K)\}/2]$; F (in atom-6).

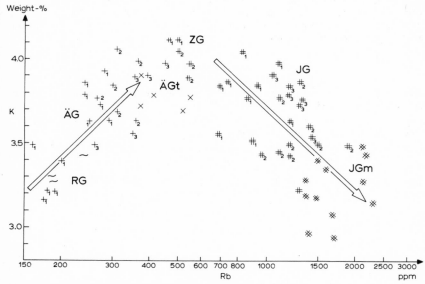

Fig.41. K/Rb-relationship of normal granites (RG) and precursors of specialized granites (AG) as well as specialized granites (ZG, JG) in the Erzgebirge. (From Tischendorf, 1977.)

116

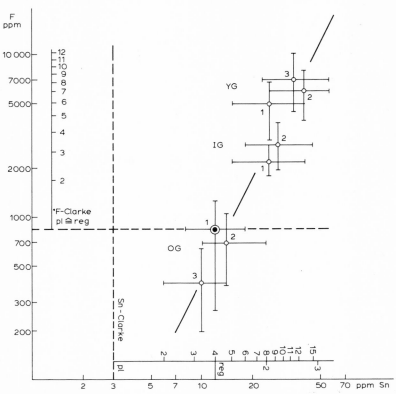

Fig.42. Averages and dispersion of fluorine and tin in postkinematic granites of the Erzgebirge and the planetary and regional average values. (After Tischendorf, 1977; Tischendorf et al., 1978.)
(*OG* ≙ Variscan Older Intrusive Complex ≙ precursors of stannigene granites, *IG YG* ≙ Variscan Younger Intrusive Complex ≙ stannigene granites.)

TABLE LI

The characteristic elements of the granite pegmatites and the nepheline—syenite pegmatites (After Fersman, 1952; cf. Schneiderhöhn, 1961)

Granite pegmatites	Nepheline—syenite pegmatities
—	Sr, Ba, Ca
P, Hf	P, Zr
Nb, Ta	—
yttrium earth	cerium earth
F, B	—
	Co_2, S, Cl
K > Na	K < Na
U < Th	U < Th
Li, Be	—
—	V present
—	Ti
—	Fe

TABLE LII

Approximate content of chemical elements in pegmatitic solutions (After Fersman, 1952; cf. Schneiderhöhn, 1961)

Contents in g/tonne	Elements
>10 000	O, Na, Al, Si, K
±1000	H, Li, Be, Mg, P, Ca, Fe
±100	Be, C, F, Ti, Mn, Rb, Y, Zr, Mo, Sn, Se, Er
±10	Cl, Se, Cu, Zn, Nb, Cs, Hf, Th, U
±1	Ho, S, Ta, W, Pb, Bi
~0.1	Ge, Sb
0.01	Au, Tl
~0.001	Ga
~0.0001	Pd, Ra, Po, Ac, Pa

TABLE LIII

Distribution of chemical elements of the "pegmatitic" late-stage (remaining) magma in the pegmatitic, pegmatitic–pneumatolytic and pneumatolytic phases (After Fersman, 1952; cf. Schneiderhöhn, 1961)

Pegmatitic	Pegmatitic – pneumatolitic	Pneumatolytic
H ————————————	H ————————————	H
Li ———————————	Li	—
Be ——————————	Be	—
	B	
O ————————————	O	
	F	
Na ———————————	Na	
	Mg	
Al		
Si ————————————	Si	CO_2
P		
K	Cl	S
Ca ——————————	Ca	
	Sc	
Ti		
Rb		Mn
Y		Fe
Zr	Mo	Cu
Nb		Zn
Cs	Ga	
rare earth metals		Ge
Hf	Sn	As
Ta		Sb
		Ba
Th		W
		Au
		Pb
		Bi
	U ————————————	U

TABLE LIV

Types of pegmatites (upper table) and distribution of main chemical elements and minerals in these pegmatites (After Fersman, 1952; cf. Schneiderhöhn, 1961)

Types of pegmatites → Minerals of pegmatites ↓	Aplitic sahlband	1 Orthite–monazite	2 Titano-niobotantalates	3 Black tourmaline, muscovite	4 Beryl–topaz	5 Albite–lepidolite	6 Li–Mn–Fe phosphates	7 Cryolite	8 Fluoro-carbonates	9 Sulfides Fe, Cu, Zn	10 Zeolites Ca, Na (K)	
Quartz — Si	greyish	light smoked	transparent	dark smoked	smoked	pink	milky	transparent "Dauphiné"-twinned quartz, elongated and radiating		chalcedonic amethyst	short prisms	—— accumul. SiO_2
feldspars — Si, Al, Na, K, Ca		microcline		microcline orthoclase	orthoclase	albite	albite	albite	adularia I	albite	zeolites and adularia II	kaolinization
Micas — Si, Al, Fe, K, Li, H–	biotite	lepidolite		muscovite (zinnwaldite)		green muscovite	lepidolite	"gilbertite"		sericite nacrite	kaolinite	argilite

Principal chemical elements and minerals of pegmatites

Phases of processes	B	C	D	E	F	G	H	I	K	L
P	monazite	monazite	apatite	amblygonite	apatite	phosphates	phosphates			detrital monazite
TR, Y, Th	monazite orthite	cerium—yttrium—silicates		(monazite)	hematite					detrital monazite
Fe (Mg)	magnetite	biotite—magnetite								"limonite"
Ta, Nb, Ti, Zr	sphene	tantalo-niobates with Ti, Zr, Nb	titano-tantalates Mn, Fe (in biotite)	niobates, oxides of titanium						
U, Ra		uraninite		uraninite						autunite
B			black tourmaline	blue tourmaline	polychrome tourmaline		black tourmaline			
Mo, Re			molybdenite (Re)				molybdenite			
K		microcline	perthitic microcline	orthoclase			adularia			
Sc, Ce		(euxenite)	thortveitite							
Be		gadolinite chrysoberyl (in tantalite)		beryl phenacite	aquamarine	vorobievite		emerald	green sphene	
Sn					cassiterite					detrital cassiterite
Na					albite					
W, Bi				wolframite native Bi bismuthinite	Se in cassit. and wolfr.	scheelite				rarely in eluvium
Li				spondumene	lepidolite	Li-phosphates	Li-phosphates	Li-thermals and sources		
Mn			titano-tantalates Mn, Fe	Mn-silicates	Mn-silicates	(Fe)-Mn-phosphates		Mn-carbonates		
Cs						pollucite (helvite)				
S	asphaltites		MoS_2							
F					topaz	fluorine	cryolite—topaz yellow and pink	fluorine		
CO_2 + C	graphite					carbonates of the	fluorocarbonates	sulfides Zn, Fe, Cu, Pb, Mo	. carbonates Ca, Mg, Fe, Me –	carbonates Ca (Mg)
H_2O						progressively occurring			hydro-alumino-ferri-silicates and phosphates	

Note: Important. The original of Fersman's geochemical "table" as well as Varlamoff's French translation (with reinterpretations) contain ambiguities and possibly disputable information. Full lines and dashed lines indicate with more-or-less certainty the extension of minerals into the different geospheres, and the commas denote a succession. In Fersman's original paper it is sometimes difficult to understand all of

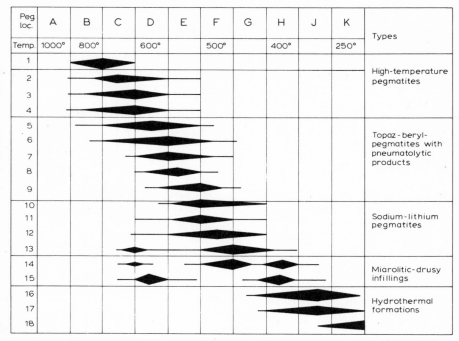

Peg. loc.	A	B	C	D	E	F	G	H	J	K	Types
Temp.	1000°	800°		600°		500°		400°		250°	
1											High-temperature pegmatites
2											
3											
4											
5											Topaz-beryl-pegmatites with pneumatolytic products
6											
7											
8											
9											
10											Sodium-lithium pegmatites
11											
12											
13											
14											Miarolitic-drusy infillings
15											
16											Hydrothermal formations
17											
18											

Fig.43. Specific phases characteristic of the pegmatite types. (After Fersman, in Schneiderhöhn, 1961.)

TABLE LV

Close association of pegmatites with non-pegmatitic mineral veins (After Fersman, 1952; cf. Schneiderhöhn, 1961)

400°	K J H	Zeolitic veins Alpine-type veins	Post-pegmatitic filling of miarolitic vugs	Mineralized hydro-thermal veins ↑
500°	G F E	quartz veins		pneumatolytic mineral veins, especially with gold ↑
600°	D		pegmatites	pneumatolytic quartz veins pneumatolytes
700°	C			silexite ↑
800°	B A			
			granitic magma	

TABLE LVI

Genetic phases and chemical element distribution of the pegmatites (After Schneiderhöhn, 1961, p. 622; courtesy of Gustav Fischer, Stuttgart)

Stages	Main elements	Minor elements	
I. Older, almost purely magmatic	K, Na, Al, Si, Fe	Be, Sc, Y, La, Ti, Zr, Se, Er	In addition much boron, which was partly also derived from the slaty marine host-rocks
II. Pegmatitic—pneumatolytic	K → Na → Li	Nb, Ta, Sn, Mo, P, U, Mn, Bi, F, Cl	
III. Pneumatolytic	K, Li	Be, W, Rb, Cs	
IV. Hydrothermal	Fe, Au, Cu, Zn, Pb, S, As, CO_2 (+Ca, Mg, Fe), H_2O	Replacement of oxygen prevailing in stages I—III by sulfur and in the younger stages by CO_2	

TABLE LVII

Temperature, phases and parageneses of the pegmatites (After Schneiderhöhn, 1961, p. 626; courtesy of Gustav Fischer, Stuttgart)

Temperature	Phases	Parageneses
1000—700°	magmatic-melt phases	biotite → plagioclase → microcline perthite → high-quartz
700—600°	pegmatite phase *per se*	microcline perthite, high-quartz → low-quartz, muscovite
600—570°	inversion temperature high-quartz → low-quartz	beryl, tourmaline, Y-La-minerals, zircon, columbium—tantalum-oxides, cassiterite
600—370°	pneumatolytic phase	cleavelandite, lithium silicates, lithium phosphates, berylium minerals, boron minerals, manganese phosphates, columbium—tantalum—titanium oxides, casserite, wolframite, cesium—rubidium—fluorine minerals
370°	critical temperature of water	
370—50°	hydrothermal phase	hydrolysis of the silicates, origin of fine-grained clay minerals, zeolites, carbonates of Ca, Mg, Fe, Mn, sulfides of Fe, Cu, Zn, Pb

TABLE LVIII

Feldspar distribution in pegmatites (After Fersman, 1952; cf. Schneiderhöhn, 1961)

phase	>800°		700°	600°	500°		400°
	A	B	C	D	E	F	G

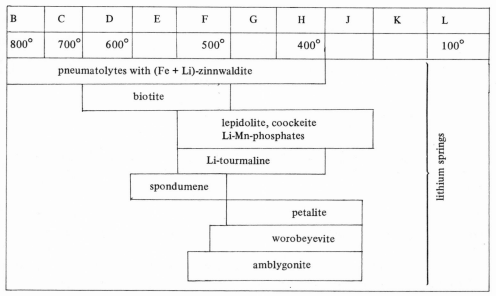

erly put in its natural context. For instance, the detailed structural, mineralogical and chemical work by Čada and Novák (1974) permitted the reconstruction of the phases of alterations (=alteration paragenic sequence) of the greisen-type mineralization of the Cinovec-South tin deposits, summarized in Table LXII. In particular, the need for precise

TABLE LIX

Geochemical diagram of lithium and lithium minerals in the pegmatites (After Fersman, 1952; cf. Schneiderhöhn, 1961)

B	C	D	E	F	G	H	J	K	L
800°	700°	600°		500°		400°			100°

pneumatolytes with (Fe + Li)-zinnwaldite

biotite

lepidolite, coockeite
Li-Mn-phosphates

Li-tourmaline

spondumene

petalite

worobeyevite

amblygonite

lithium springs

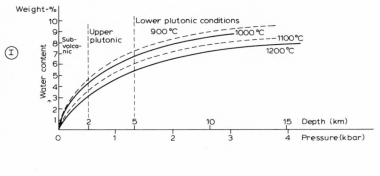

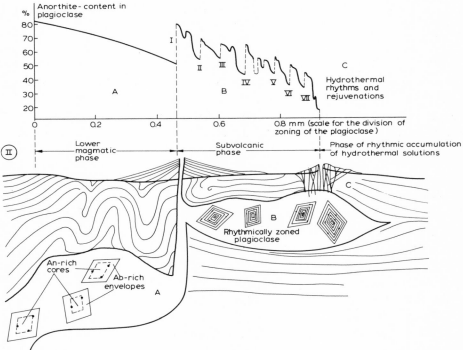

Fig.44. I = Solubility of water in an albite melt. (After R.W. Goranson, 1936.) II = Plagioclase as a manometer under plutonic and subvolcanic conditions. (After Borchert, 1964; courtesy of Glückauf, Essen.)

thin-section work of hundreds of samples (over 1000) is emphasized in such investigations.

The unequivocal recognition of metamorphosed volcanic (flow rocks and pyroclastics) and sedimentary sequences has posed considerable difficulties with which every

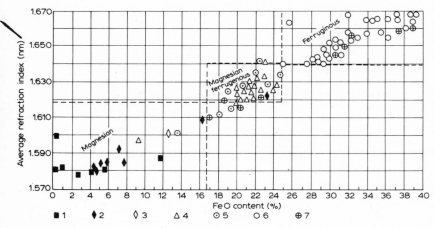

Fig.45. Diagram of the distribution of chlorites of various types in magmatic deposits. (After Shilin and Ivanova, in Kreiter, 1968.) Legend: *1* = iron ore and chromites; *2* = lead–zinc and copper–zinc; *3* = copper–bismuth; *4* = copper and copper–pyrite; *5* = piezoquartz; *6* = sulfide–cassiterite; *7* = primary gold.

geologist working in Precambrian terrain is familiar. In unaltered rocks, the whole spectrum of "mineralogy–texture–fabric–structure–stratigraphic relations" can be utilized for genetic reconstructions, and in many cases only one of these criteria is required to distinguish, for example, sedimentary from volcanic rocks (although one can recall instances where this is not easy). With an increase in the grade of metamorphism, a progressive obliteration or destruction of the primary features occurs; hence in the early stages the "mineralogy–texture–fabric" range is obliterated, and at a higher metamorphic grade the "mineralogy–texture–fabric–structure". In the latter situation only the gross stratigraphic relations may offer a clue to the rock's origin. However, the stratigraphic setting can be similar or identical for both sedimentary and volcanic lithologies, which makes proper interpretation impossible. In such instances, geochemical methods have been used in the attempt to find criteria to differentiate metasediments from metavolcanics, e.g. in Table LXIII. It is assumed, then, that metamorphism was *iso*chemical (i.e., no chemical changes took place) and the composition of the metamorphic rock mirrors the original lithology; this is in contrast to metasomatism where the reaction is *allo*chemical (with "transvaporisation" or an exchange – i.e., addition or removal – of chemicals taking place). The geochemical criteria, such as those listed in Table LXIII, appear to the based on empirical methods; the data is applicable only in one particular region (or indicate an average) after a positive correlation has been found by trial-and-error. In other words, geochemical criteria found to be useful in differentiating metasediments from metavolcanics at one locality may not necessarily be applicable elsewhere. Such comparative work is being continued in many laboratories. In particular bulk composition has proved to be unreliable in distinguishing between the protogenes (=precursors) of metamorphic rocks,

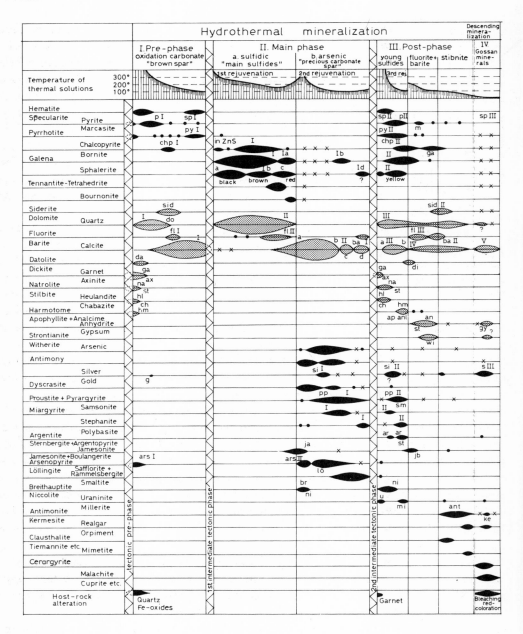

Fig.46. Quantitative paragenetic scheme of the phases of mineralization of the subvolcanic silver veins of St. Andreasberg, Harz. (After Wilke, 1952; courtesy of Glückauf, Essen.)

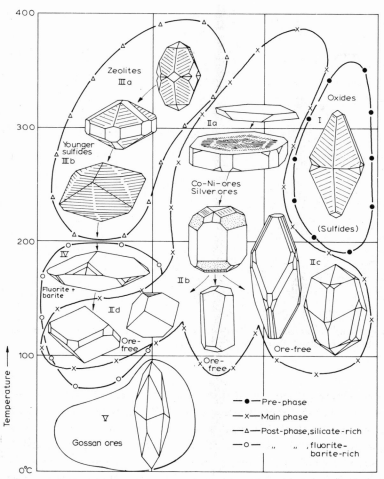

Fig.47. Environment of formation of the calcite-types depending on temperature and concentration of the solutions (the concentration increases to the right). (After Wilke, 1952; courtesy of Glückauf, Essen.)

because the primary chemical composition of many types of igneous rocks can be matched by sedimentary lithologies of equal make-up (see for example fig.9.1, p.227 and fig.9.2, p.229, in Garrels and Mackenzie, 1971).

"SEDIMENTARY" ORE AND ROCK CLASSIFICATIONS

"Science is built up of facts, as a house is built up of stones; but an accumulation of facts is no more science than a heap of stones is a house."
    Lucien Poincaré (in: *Science and Hypothesis*)

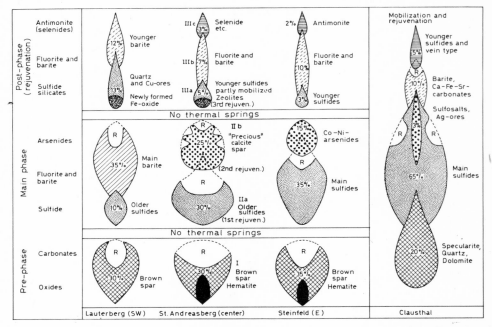

Fig.48. Comparative diagram of the precipitation sequence of the vein minerals of the Central- and Upper-Harz region. (Courtesy of Glückauf, Essen.)

Among the mineralizations and ore deposits collectively called "sedimentary" belong numerous genetic types that range widely in chemical and mineralogic composition as well as in texture and structure. This class of accumulations encompasses the total spectrum of syngenetic–clastic (cf. Hails' Chapter 5, Vol. 3; Pretorius Chapters 1 and 2, Vol. 7), syngenetic–physicochemical (numerous chapters in earlier volumes), syngenetic–biochemical (cf. Trudinger's Chapter 6, Vol. 2 in this Handbook); and various weathering (cf. Wopfner and Schwarzbach's Chapter 2, Vol. 3; Lelong et al.'s Chapter 3, Vol. 3 in this Handbook), and infiltration (supergene) (cf. Rackley's Chapter 3, Vol. 7 in this Handbook) deposits of the early- and late-diagenetic to early-epigenetic types. (For related terminologies see, for example, Wolf and Chilingar, 1976, and Wolf, 1976c). The investigation of these fall into the realm of sedimentary geology — comprising metalliferous, non-metalliferous and industrial minerals plus hydrocarbon accumulations (cf. Table LXIV).

Among the "sedimentary" ores no clearly defined boundaries exist as they grade into each other, i.e., they are all transitional as the following brief oversimplified list indicates:

(1) syngenesis   → diagenesis

     syngenesis   → supergenesis

TABLE LX

Parageneses and milieus of origin of the boron minerals in the pegmatitic–pneumatolytic–hydrothermal environment (After Kunitz, 1929, cf. Schneiderhöhn, 1961; courtesy of Gustav Fischer, Stuttgart)

Prevailing chemical condition	SiO₂		Al₂O₃		MgO		CaO	
Rocks	granite gneiss sandstone		granite clay-slates phyllite		amphibolite dolomite		gabbro, diabase limestone amphibolite	
	with boron minerals	without boron minerals	with boron minerals	without boron minerals	with boron minerals	without boron minerals	with boron minerals	without boron minerals
Pegmatitic ±600°	schorl	biotite muscovite high-quartz orthoclase albite beryl	schorl (dumortierite)	biotite muscovite oligoclase–andesine spinel (corundum) garnet	schorl ludwigite	hornblende biotite labrador cordierite	(schorl) danburite	diopside amphiboles (biotite) anorthite apatite wollastonite grossularite
Pneumatolytic ±400°		low-quartz orthoclase albite	kornerupine	chrysoberyl topaz muscovite gilbertite	serendibite dravite fluoborite	phlogopite pargasite humite	wiluite uvite (axinite) (datolite)	vesuvianite scapolite fluorite
High-hydrothermal ±250°	sassoline		dumortierite	sericite	sussexite	chlorite actinolite talc dolomite	axinite	epidote zoisite fluorite apatite scheelite calcite
Low-hydrothermal ±100°		flaser quartz chalcedony opal		nacrite pyrophyllite kaolinite wavellite	szajbelyite camsellite	steatite serpentine talc magnesite dolomite	datolite	prehnite calcite aragonite chabazite Ca-zeolites

(continued)

TABLE LX (continued)

Prevailing chemical condition	FeO + Fe₂O₃		Li₂O		Na₂O		Sn, Ti, Zr, rare earth	
Rocks	iron ore deposits		granite pegmatite		foyaite		granite syenite pegmatite	
	with boron minerals	without boron minerals	with boron minerals	without boron minerals	with boron minerals	without boron minerals	with boron minerals	without boron minerals
Pegmatitic ±600°	schorl ludwigite	hedenbergite grunerite lepidomelane magnetite andradite pyrrhotite	schorl (rhodizite)	spodumene biotite petalite worobleffite spessartite		aegirite barkevikite arfvedsonite nepheline leucite cancrinite		ilmenite titanite titanaugite mica hornblendes zircon melanite perovskite Y-garnet orthite gadolinite andalyte

	(cut)	(cut)	(cut)	(cut)	(cut)	(cut) sodalite	(cut) mastic	(cut) zircon
(cut) **±400°**	wolframite arsenopyrite	(axinite) ludwigite	rubellite	spondumene lipedolite	homilite	aegirine arfvedsonite cancrinite melinophane	painite warwickite melanocerite tritomite norden- skioldine	rutile monazite samarskite columbite cassiterite eudialyte kataplite astrophyllite mosandrite
High- hydrothermal ±250°	hematite lievrite pyrite	(rhodizite) manadonite hambergite		amblygonite triphylite	axinite avogadrite	cryolite fluorite villiaumite analcime calcite	kariocerite cappellinite	sphene bratkite euxenite fluocerite Ce-apatite benitoite
Low- hydrothermal ±100°	hematite goethite siderite rhodonite manganite Fe-phosphate sulfides			cookeite	datolite	analcime natrolite Na-zeolites		wood tin malacon anatase tschffkinite hatchettolite

TABLE LXI

Typical paragenetic relationships (After Thomas, 1973)

Two members	Three members	Four or more members
galena−sphalerite (lead−zinc)	galena−sphalerite−pyrite	galena−sphalerite−pyrite−quartz
pyrite−chalcopyrite (iron−copper)	pyrite−chalcopyrite−quartz	pyrite−chalcopyrite−galena−sphalerite−quartz
gold−quartz	gold−quartz−pyrite	gold−quartz−pyrite−sphalerite−galena−arsenopyrite
gold−tellurium	gold−tellurium quartz	gold−tellurium−quartz−pyrite
cobalt−nickel	cobalt−nickel−pyrrhotite	cobalt−nickel−pyrrhotite−quartz (iron)
cassiterite−wolframite (tin−tungsten)	cassiterite−wolframite−quartz	cassiterite−wolframite−quartz−tourmaline
molybdenite−bismuth	molybdenite−bismuth−quartz	molybdenite−bismuth−quartz−wolframite−pyrite
cinnabar−pyrite (mercury-iron)	cinnabar−pyrite-quartz	cinnabar−pyrite−quartz−calcite
magnetite−chlorite	magnetite−chlorite−garnet	magnetite−chlorite−garnet−pyroxene−pyrite
chromite−serpentine	chromite−serpentine−olivine	chromite−serpentine−olivine−pyroxene

diagenesis	→ epigenesis
diagenesis	→ supergenesis
diagenesis	→ infiltration
supergenesis	→ infiltration
infiltration	→ epigenesis
epigenesis	→ low-grade metamorphism

There are also gradations and transitions between:

(2) "sedimentary"	→ weathering-produced ores (supergenesis)
"sedimentary"	→ volcanic-exhalative/sedimentary ores
volcanic-exhalative	→ hypabyssal (subsurface)-exhalative ores
(3) clastic	→ physicochemical ores
supergene in situ	→ clastic (reworked) ores
physicochemical	→ biochemical
biochemical	→ physicochemical.

TABLE LXII

Sequential-genetic classification table (After Čada and Novák, 1974)

Stage	Process	Space
Magmatic	Intrusion of granite batholith	
	Solidification of magma	Origin of contraction fissures in the apical part of massif
		Origin of structural pattern in the southern part for the development of greisen mass
	Granitization phenomena	Some granitic facies in relatively deep parts of the massif
	Origin of "Stockscheider"	
Late magmatic	K-feldspathization I albitization I	Local effects

I. tectonic interval — structures mostly trending NW—SE-ward

I. metasomatic	Greisenization I (— silicification I —zinnwalditization I — topaz I)	W(-Mo) mineralization Mostly diffusion origin of greisens I and greisenized granites in endocontact zone
	Regeneration (— intensive albitization II and adularization)	Throughout apical part

II. tectonic interval — structures trending NE—SW-ward
— rejuvenation of structures trending NW—SE-ward

Magmatic	Intrusion of aplitic granites	Sills, locally developed
II. metasomatic	Greisenization II (— silicification II — muscovitization and/or zinnwalditization II — fluoritization I — topaz II)	Sn mineralization Flow of solutions along structural lines mostly striking NE—SW, with diffusion especially in greisen I

III. tectonic interval?

Hydrothermal-metasomatic	Sulphidization I—II sericitization low-temperature silicification III fluoritization II	Impregnation in metasomatites

TABLE LXIII

Characteristic trace elements for the identification of ortho- and para-metabasites given in various publications (bibliography in Lange, 1965)

References	Characteristic trace elements		
	Ortho-metabasite	para-metabasite	unspecific
Engel and Engel (1951)	higher contents of Co, Cr, Cu, Ni, Sc	corresponding lower values	
Engel (1956)	Co, Cr, Ni only locally suited		
De Widt (1957)	Cr (>170 ppm)	Cr (>170 ppm)	
Lapadu-Hargues (1958)	higher Ti contents		
Wilox and Poldervaart (1958)	differentiation according to content of Sr		Ba, Co, Cr, Cu, Ga, Ni, Pb
Hahn-Weinheimer (1959)		in eclogites: $0.1-0.5\%$ C, $^{12}C/^{13}C = 90.38-91.07$	
Walker et al. (1960)	with increasing metamorphism and metasomatosis, the contents of trace elements change slightly higher contents of Cr, Co, Cu, Ni		
	$TiO_2 > 1.0\%$ (only locally suited)	$TiO_2 < 1.0\%$ (only locally suited)	
Leake (1963)	negative correlation of Cr and TiO_2 also of Ni and TiO_2	positive correlation of Cr and TiO_2 also of Ni and TiO_2	
	Cr + Ni >250 ppm further the relations of certain main elements (or Niggli values) and their distribution trends (e.g. C/mg, K/mg, Alk/mg, Ba/mg, Zr/mg, Ti/mg)		Cr + Ni <250 ppm
Lange (1965)	Cr, Ni, Co; negative correlation of Cr/Ti; Ni/Ti; positive correlation of Cr/Ni; the Ni–Co relation; higher Cr and Ni contents; gabbroid macrochemism	positive and negative correlations, respectively	Ci, V; Ti (as individual value), Cr <150 ppm, Cr + Ni < 200 ppm

One of the best styles of summarizing, depicting and classifying such complex interconnexions is by resorting to flow-chart-like and/or diagrammatic conceptual models (e.g., Wolf's Chapter 2, Vol. 1 in this Handbook). The above list provides only two-variables

TABLE LXIV

Modelized sequence of formation of the sedimentogenic deposits (After Baumann and Tischendorf, 1976)

Depth range \ Genetic types (physico-chem.)	Weathering products		→ Exogenic dynamic (different. caused by solubility and transportation) Origin of sediments s.s.			
	physical	chemical	clastic deposits	chemical precipitations oxid. hydrolys.	carbonates	silicates, sulfides, evaporates
Terrestrial-limnic	fanglomerates (breccia/rubble-type of deposits eluv. and aeolian placers (Pt, Au, gemstones among others)	siallites weathered kaolinites weathering-type phosph.	alluv. placers (Au, Sn, Ti, RE, etc.) glass-sands gravel, clays	swamp- and lake-formations (Fe, Mn); limnic coal formations; chalc		
		allites weathered: laterites (Al, Fe, Mn, Mg, Ni)	terrestrial depressions: molasse form. fanglom., sands, clays	red bed (Fe, Mn) →	infiltration-b. → arid concentration-b. (Cu, Pb, Zn, Ag, U, S) ↑	→ terrestrial saltb. (Ca, Ba, Na, K, N, etc.)
Shallow-marine (shelf)		halmyrolyses (glauconite)	marine depression molasse form. (congl., quartz-sands, clays) marine rubble-breccia deposits marine placers (Ti, Sn, Zr, RE, among others)	oil deposits ↑ red bed → "Rote Fäule" (Kupferschiefer) oolite form. (Fe, Mn) glauconite-phosphate form. paral. coal formations	oil-source-bed ↑ bitum. marl form. (Cu, Pb, Zn, U, V, C, etc.) → anhydr., marine salt deposits (Ca, S) (Na, K, Mg) limestone–dolomite form. (partly with Pb, Zn)	
Marine — Labile			flysch form. (congl. sandstone, clay-slate)	oil-source-rock ↑ bitumen. black slate (with FeS₂, Cu, Mo, U, V, C among others)	limestone–dolomite form. (partly with Pb, Zn, P)	siliceous/cherty form. (partly with Fe, Mn)
Marine — Stable			red clay form.		pelagic Globigerina ooze form. (partly with Pb, Zn)	diatom. and radiol. ooze form. (partly with Mn, Fe, Mo among others)

Climatic range: ☐ -humid   ⬚ -arid   〜 -biogenic influences

TABLE LXV

Genetic classification of sedimentary ores (After Schneiderhöhn, 1962)

A. *Weathering products over rocks and ores*

  1. Eluvial heavy mineral placers
  2. Oxydation- and cementation-zones of ores

B. *Terrestrial, clastic (= mechanical) sediments*

  I. Gravity (mass movement) deposits
    1. Breccias and conglomerates of ores   Mineralogic placer types:
    2. Eluvial heavy mineral placers   Gold
  II. Wind deposits   Platinum
    1. Eolian heavy mineral deposits   Tin
  III. Glacial and fluvio-glacial deposits   Ilmenite, magnetite, chromite,
    1. Ore-rich moraines   rutile, zircon
  IV. River deposits   Monazite
    1. Fluvial heavy mineral placers   Diamond
  V. Lake deposits (humid climate)
    1. Mud
    2. Aluminum-rich mud
  VI. Deposits of arid climates
    1. Fanglomerate placers
  VII. Human dump deposits
    1. Many of original uneconomic value became economic at a later stage due to technological
      advances

C. *Eolian–atmospheric deposits*

  No known economic sediments

D. *Terrestrial, chemical deposits (weathering products)*

  I. Deposits formed in close proximity of sources:
    1. Carbonate-bauxite and silicate-bauxite
    2. Clay deposits of various types (montmorillonite, kaolinite, etc.)
    3. Magnesite ("gel-magnesite")
    4. Ni-silicate
    5. Laterite iron ores
    6. Basalt- and serpentine-iron ores (± Ni, Co, Cr)
    7. Weathering phosphates
    8. Zeolites
  II. Deposits formed at greater distance from sources:
    1. Crust or concretion forming iron ores
    2. Iron and manganese weathering ores on carbonates and shales-slates (Type Hunsrück)
    3. Concretionary to oolitic iron ores (1, 2, and 3 may be transitional, as well as with E and
      "descending" ores)
    4. Weathering products of other heavy metals (2nd to nth cycle)
    5. Eluvial-arid Pb–Cu deposits (Cap Garonne-type)
    6. Arid Cu deposits (red-bed type)
    7. Arid Ag deposits
    8. Arid U–V deposits (Coloardo Plateau–Wyoming or Western States type)
    9. Pb–Zn in sandstones
    10. Terrestrial evaporites

E. *Terrestrial, chemical deposits in swamps, lakes and rivers*

  1. Ores in lakes, lake-shore, and lake-bottom
  2. Ores in swamps and in vegetation-rich and peat-rich environments

(continued)

TABLE LXV (continued)

F. *Organic sediments in swamps and lakes*

   1. Silica-rich organic deposits (diatomite)
   2. Freshwater black muds, bituminous sediments, and ores within black shale (ores partly related to the sulfur cycle (see GIII6)
   3. Coal
   4. Ores associated with peat and coal

G. *Marine Ores*

  I. Shore- and coastal deposits
    1. Conglomeratic and breccia iron ores
    2. Marine heavy mineral placers
    3. Bird guano and guano-phosphate
  II. Shallow, marine inland lake deposits
    1. Oolitic iron-oxide and iron-silicate ores, Mn-ores
  III. Hemipelagic (shelf) deposits
    1. Marine black shale and associated ores [Alum-(=alaun)rich, pyritic and cupriferous shales—slates, and SiO_2-ores part of the sulfur deposits, oil shales]
    2. Exhalative-sedimentary iron-oxide (e.g., Lahn—Dill type)
    3. Exhalative solfata deposits
    4. Exhalative bor deposits
    5. Exhalative manganese deposits
    6. Exhalative sulfide deposits
    7. Marine phosphates and manganese
  IV. Deep-sea deposits
    No known economic deposits.
  V. Marine evaporites
    1. Marine chemical carbonates
    2. Gypsym and anhydrite
    3. Halite
    4. Kali- and magnesium minerals
    5. Rarer evaporitic minerals.

in contrast to the multi-parameter natural systems (see for example fig.1, Chapter 2, Vol.1, on soil genesis and supergene ores).

Most classifications that are comprehensive and attempt to incorporate *all* major types, give some fleeting consideration to the so-called sedimentogenic ores. However, for the specialist a more detailed approach becomes necessary, such as those in Tables LXIV and LXV. In the former, the depositional environments (terrestrial—limnal, shelf or shallow-water marine, and marine) are plotted against the type of deposits; the influences of climate and biological processes are superimposed on the table. Table LXIV is merely a list of the sedimentogenic accumulations.

The principles of clastic and chemical sedimentology have found their application in the interpretation of strata-bound ores, e.g., placer and supergene concentrations, respectively (cf. Hails' Chapter 5, Vol. 3; and Lelong et al.'s Chapter 3, Vol. 3). As this

TABLE LXVI

Relative mobility of various elements in weathering processes (After Rösler and Lange, 1972)

Type of rock	Main elements	Trace elements
1. Gabbroid rocks	Ca > Na > Mg > Si > K > M = Fe	Mn > Ni > Ti > Cr > V
2. Porphyrites	Ca > Mg > Na > K > Si > Al = Fe	Co > Mn > Ni=Cr=V > Cu > Zr > P=Ti > Ga

TABLE LXVII

Transport of elements in aqueous solutions (After Gundlach, 1964, in Rösler and Lange, 1972)

Kind of transport	Acid solutions (pH < 6)	Neutral solutions (pH 6–8)	Alkaline solutions (pH > 8)
Simple ions, hydratised ions, O-complexes (e.g. SbO$^+$ or WO$_4^-$)	Ag, Al, As, Ba, Be, Bi, Ca, Cd, Co, Cr, Cu, Fe, Hg, Mg, Mn, Mo, Ni, Pb, Sb, Sn, Sr, Ti, Tl, U, V, Zr	As, Ba, Ca, Mg, Mo, Sb, Se, Sr, V, W	Al, As, Ba, Ca. Cr, Mo, Nb, Sb, Sn, Sr, Ta, Ti, Tl, U, V, W, Zn, Zr
Isopoly and heteropoly acids	As, Mo, Nb, Sb, Sn, Ta, Te, V, W		
Halide and oxyhalide complexes (e.g. FeCl$_4^-$, VOF$_4^-$)	Ag, Au, Be, Bi, Cd, Cu, Fe, Ga, Ge, Hg, In, La , Mo, Nb, Pb, Pt , Re, Sb, Sn, Ta, Th, Ti, Tl, U, V, Zn, Zr	as in acid solution (soluble in brines: CaCO$_3$, CaSO$_4$, CuS, PbS, SrSO$_4$, ZnS)	
Other anion complexes (above all sulfate, phosphate, carbonate)	Cr, La, Th, Ti, Tl, U, V, Zr	Ag, Be, Cr, Cu, La, Mn, Mo, Sb, Th, Ti, U, V, Zr	Ag, Be, Cu, La, Mo, Th, U
as hydrocarbonate		Ba, Ca, Co, Fe, Mg, Mn, Sr	
Sulfur hydride complex (acid) sulfide complexes (alkaline)	Ag, Ca, Cu, Fe, Hg, Ni, Pb, Zn	Ag, Cu, Fe, Hg, Ni, Pb, Zn	Ag, As, Au, Bi, Cu, Ga, Ge, Hg, In, Mo, Ni, (Pb), Pt, Re, Sb, Se, Sn, Te, Tl, W, Zn
Thiosulfate and polythionate complexes		Ag, Au, Cu, Hg, In, Pb, Zn	Ag, Au, Cu, Hg, In, Pb, Zn
Colloidal	Au, Fe (oxide sol) Ni (sulfide sol), Sn (oxide sol)	Au, Fe (oxide sol), Ni (sulfide sol), Sn (oxide sol)	Au (SiO$_2$ as protective colloid), Fe (oxide sol), Ni (sulfide sol)

TABLE LXVIII

Geochemical classification of sediments (After Goldschmidt, 1964, in Rösler and Lange, 1972)

Type of sediment	Conditions of formation	Material	Examples
Resistates	chemically unchanged residues from rocks subject to weathering	from blocks of rock to fine detritus, mainly quartz	sandstones, conglomerates, heavy mineral placers
Hydrolysates	hydrolysed decomposition products, finest particles transported and deposited by water	clay minerals, Al hydroxides	bauxites, soils (partly)
Oxidates	products of oxidation formed in weathering	Fe, Mn oxides	oxidic iron- or manganese-containing sediments or ores
Reductates	deposition under reducing conditions	carbon-rich substances	sapropel with sulfides
Precipitates	precipitation of ionically dissolved elements	Ca, Mg, Fe compounds (carbonates and others)	carbonate, mud and silica sinter and ooze, phosphates (partly), borates (partly)
Evaporates	precipitation products of compounds easily soluble in water due to oversaturation	sulfates, carbonates, chlorides of Na, K, Mg and Ca	rock salt, potassium salts, gypsum
Biolites	deposition of animal and plant remains	liquid and solid hydrocarbons	peat, coal, petroleum, part of natural gases

Handbook testifies, the multidisciplinary realm of sedimentology has become specialized and complex, and to conclude this section merely four self-explanatory schemes are provided (see numerous recent books on sedimentology that include sections on ore genesis, e.g. Conybeare, 1979).

Weathering is the first stage of the origin of components that eventually constitute sediments, some of which are of economic viability. In ore petrology, e.g., in the study of red-bed (sandstone- and shale-hosted) copper, uranium, vanadium and silver concentrations as well as iron, manganese, bauxite and magnesite deposits, to mention only a selected few, the source of the metals has been under debate for some time and the problem is left unresolved although we can theoretically visualize numerous alternatives. One of the possible sources are trace- and minor-element containing rocks that, upon weathering (or subsurface leaching), release them into solutions which, in turn, can precipitate the elements either in dispersed or concentrated form. Petrologists have been interested, therefore in (1) the amounts of chemical elements in various lithologies and the elements'

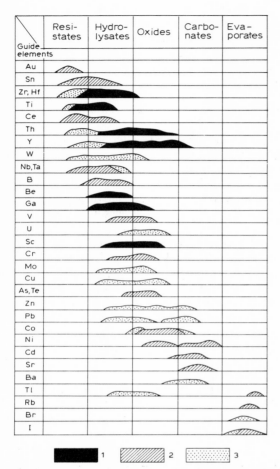

Fig.49. Sequence of precipitation of some chemical elements in sedimentary milieus. (After Ginsburg, 1963.)

relative mobilities under various natural settings, (2) their mode of transportation, and (3) the type of deposits formed. These three aspects are exemplified, respectively in Table LXVI (for more details see Lelong et al.'s Chapter 3, Vol. 3), Table LXVII, and Table LXVIII plus Fig.49.

CLASSIFICATIONS OF SPECIFIC COMMODITIES

"Every great advance in science has issued from a new audacity of imagination."
    John Dewey (in: *The Quest for Certainty* 1929)

The substantial amount of exacting information on specific groups of mineral deposits cannot be condensed meaningfully in all-inclusive classifications; consequently, each type of ore is best comprehensively treated by establishing particular schemes for particular chemical elements and/or mineralizations, the selected parameters varying according to the needs of the research performed. Above, we have already undertaken one step of specialization — at least two groups were classified separately, namely, the "sedimentary" and "magmatic/volcanic" types. In the present section, a further step in refining classifications will be offered.

The specific-commodity schemes can take different forms, either as tables or as pictorial/diagrammatic classifications in which either the ore minerals are listed or the chemical element(s), or both. Most of these models are hybrid in nature because they are both descriptive and genetic. In addition to the data provided in this section, the reader may wish to refer to Routhier (1963, Vol. 2) who has listed the natural occurrences of the major metals according to host-rock type, genesis, mineralogy, and other pertinent factors. Štemprok (1977, pp.128—132) discussed the geochemical classification of elements, and pointed out one important distinction to be made between "cyclic" and "dispersed" elements — the former take part in "reversible" processes within the earth's rock and geochemical cycles, in contradistinction to the latter that become dispersed rather than concentrated in minerals or accumulations. Classifications based on *both physicochemical* (thermodynamic) properties and *empirical* observations are required, for the geochemical cycles are governed in many instances by totally different rules than suggested by simple calculations and arm-chair reasoning.

The following classifications or models will be provided below:

(1) Tabular classification of iron-ore deposits.

(2) Iron resources diagram (example: Europe and adjacent localities, 1977).

(3) Characteristic features of different types of iron-formations (especially Precambrian types).

(4) Diagram of major types of chemically precipitated iron-formations.

(5) Classification of gold ores.

(6) Classification triangle of phosphorites.

(7) Classification table of phosphorites.

(8) Pictorial diagram of kaoline deposits of magmatic/volcanic associations.

(9) Idealized pictorial model of porphyry copper deposits.

(10) Conceptual model of "exotic"-subtype copper concentrations.

(11) Model of lead—zinc ores.

(12) Classificatory tables of zinc ores based on geologic environments and quantitative estimates.

(13) Tabular scheme of bismuth deposits.

(14) Conceptual model of bismuth geochemical distribution ("cycle").

(15) Geological-genetic classification of beryllium ores.

(16) Model of beryllium geochemical distribution ("cycle").

TABLE LXIX

Classification of iron-ore deposits of the USSR (After Smirnov, 1977, vol. 1)

Genetic group	Class (Association)	Deposit [*]
Magmatic	Low-titanium magnetite, in intrusives of the dunite—pyroxenite—dunite association	Kachkanar, Gusevogorsk, Pervoural'skoe (Urals); Lysansk (East Sayan)
	Titanomagnetite—ilmenite, in gabbroic and gabbro—amphibolitic intrusives	Kusinsk, Kopansk (Southern Urals)
	High-titanium titanomagnetite, in gabbroic and gabbro—diabase intrusives	Pudozhgorsk, Koikar (Karelia); Kharlovo (Altai)
	Perovskite—titanomagnetite and apatite—magnetite, in alkaline—ultramafic intrusives with carbonatites	Afrikanda, Kovdor (Kola Peninsula)
Contact-metasomatic	Magnetite calc-skarn	Magnitogorsk, Vysokøgorsk, Lebyazhinsk, Goroblagodat, North Peschansk, etc. (Urals); Adaevo and other deposits of the southern half of the Turgai iron-ore province, Dashkezan (Azerbaidzhan), Atansor (Central Kazakhstan); Belorets and Kholzun (Gornyi Altai); Tashtagol', etc. (Altai-Sayan district); Chokadam-Bulak (Tadzhikistan)
	Magnetite magnesian-skarn and magnesian—calc-skarn	Teya (Kuznets Alatau); Kazsk, Sheregesh (Gornaya Shoriya); Zheleznyi Kryazh (Iron Ridge) (Eastern Transbaikalia); Taezhnoe, Pionersk (Southern Yakutia)
	Scapolite—albite and scapolite—albite—skarn magnetite	Kachar, Sarbai, Sokolovsk (Turgai province); Goroblagodat (Urals); Anzas (West Sayan)
	Magnetite and hematite, hydrosilicate	West Sarbai (Turgai province); Abakan (Khakassia); individual sectors of deposits of preceding classes)
Hydrothermal	Magnomagnetite, associated with traps	Korshunovsk, Rudnogorsk, Tagar, Neryuda, etc. (Eastern Siberia)
	Magnetite specularite, intensely metasomatic	Paladaur (Georgia); Kutimsk (western slopes of the Northern Urals)
	Iron—carbonate vein- metasomatic	Bakal (Southern Urals); Abail (Southern Kazakhstan)

(continued)

TABLE LXIX (continued)

Genetic group	Class (Association)	Deposit *
Marine sedimentary (weakly metamorphosed and unmetamorphosed)	Sideritic (brown-ironstone in the zone of oxidation) layered, in marine terrigenous-carbonate sediments	Komarovo-Zigazinsk, Katav-Ivanov, and other groups (Southern Urals)
	Hematitic, in marine carbonate-terrigenous sediments	Nizhe-Angara (Eastern Siberia)
	Hematitic and magnetite-hematitic, in eruptive-sedimentary sequences	Atasu group (Central Kazakhstan); Kholzun (Gornyi Altai)
	Siderite–leptochlorite–hydro–goethite, pisolite–oolitic, in marine carbonate-terrigenous sediments	Kerchen (Crimea); Ayat (Turgai province); Bakchar (Western Siberia)
	Magnetite, partially titaniferous marine placers	Modern 'black' beach sands of the coasts of the Black, Caspian, and Japan seas; fossil beach sands in Azerbaidzhan, etc.
Continental sedimentary	Hydrogoethite, pisolite–oolite, lacustrine–paludal	Large number of small deposits on the Russian Platform and other parts of the Union
	Siderite–leptochlorite–hydrogoethite, pisolite–oolite, naturally alloyed with chromium and nickel, lacustrine–paludal, associated with weathering crust of ultramafic rocks	Orsk-Khalilova group (Southern Urals); Serovo (Northern Urals); Malka (Northern Caucasus)
	Siderite (brown-ironstone in zone of oxidation) hypergene-meta-somatic, in littoral–lacustrine coarsely-clastic, predominantly carbonate sediments	Berezovo (Eastern Transbaikalia)
	Siderite–leptochlorite–hydro-goethite, in ancient fluvial sediments	Lisakoyo (Turgai province); Taldy-Espe, etc. (Northern Aral region)
	Predominantly martite eluvial–deluvial (cobbly)	Vysokogorsk (Central Urals)
Weathering crusts (residual and infiltration)	Goethite–hydrogoethite (brown-ironstone) and martite–hydro-goethite zones of oxidation of deposits of sideritic and skarn–magnetite ores	Bakal, etc. (Southern Urals), Berezovo (Eastern Transbaikalia); Vysokogorsk (Urals)

(*continued*)

TABLE LXIX (continued)

Genetic group	Class (Association)	Deposit *
	Goethite—hydrogoethite, ocherous, naturally alloyed with chromium and nickel, in weathering crust of ultramafic rocks	Yelizavetinsk (Central Urals)
	Hydrogoethite, in eluvial—deluvial sediments in karst limestones	Alapaevo (eastern slopes of Urals)
	Martite and hydrohematite, in ferruginous quartzites	Yakovlevsk, Mikhailovsk, etc. (KMA); Saksagan' group (Krivoi Rog)
Metamorphic (metamorphosed)	Precambrian ferruginous quartzites	Krivol Rog, Kremenchug, Belozero, Mariupol' (Ukraine); Olenegorsk (Kola Peninsula); Kostamukshsk (Karelia); Karsakpai (Central Kazakhstan); Malyi Khingan, Ussuri (Soviet Far East)
	Magnetite and magnetite—specularite contact-metamorphosed sedimentary (with relicts of sedimentary iron ores)	Kholzun (Gornyi Altai)

* In this table, almost all the significant deposits are shown that possess proved reserves of over 100 M tonnes. Smaller deposits are indicated in those cases when significant deposits are not known for comparison.

(17) Six conceptual models of the geochemical distribution (or "cycles") of the following: barium, fluorine, tungsten, boron, strontium and lithium.

Three iron-ore classifications are considered first: Table LXIX by Smirnov (1977) is a comprehensive overview of seven genetic varieties of the USSR, subdivided into classes of mineral and lithologic associations. Synopses of this kind covering every continent should eventually be synthesized, compared and contrasted to establish a world-wide usable scheme, and to delineate similarities and differences between regions, geological ages and genetic ore types. Fig.50 (Zitzmann, 1977) depicts the iron resources of Europe and adjacent areas according to ten genetic types, geologic epochs and regions of mineralization (=metallogenic provinces). Noteworthy is that in the latter scheme three additional genetic varieties have been introduced, implying that — as consistently maintained in this chapter — the choice of nomenclature and classifications is a paramount first step in such investigations.

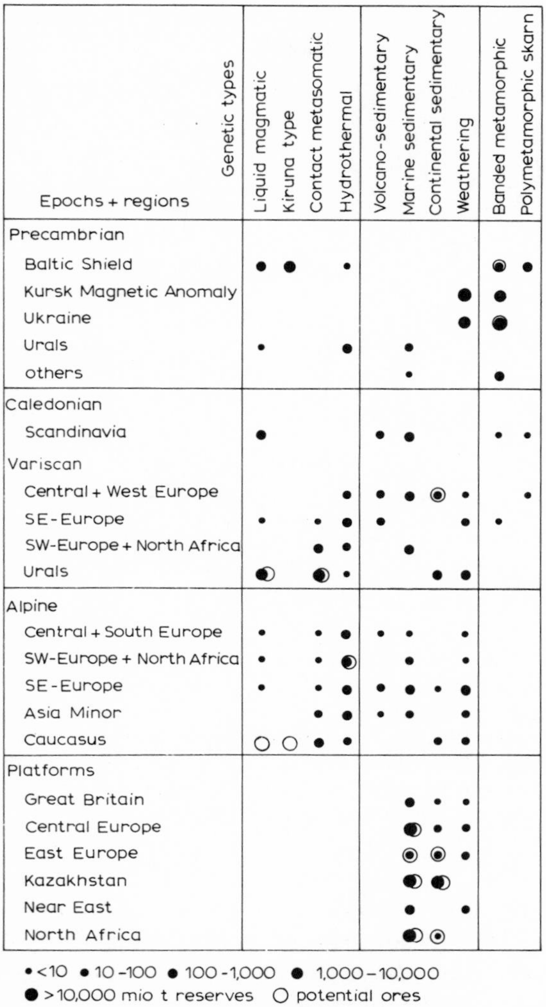

143

Epochs + regions \ Genetic types	Liquid magmatic	Kiruna type	Contact metasomatic	Hydrothermal	Volcano-sedimentary	Marine sedimentary	Continental sedimentary	Weathering	Banded metamorphic	Polymetamorphic skarn
Precambrian										
Baltic Shield	●	●	•						◉	●
Kursk Magnetic Anomaly							●		●	
Ukraine							●		◉	
Urals	•		●		•					
others						•			●	
Caledonian										
Scandinavia	●				●	●			•	•
Variscan										
Central + West Europe			•		•	•	◉	•		•
SE-Europe	•		•	●	•			•	•	
SW-Europe + North Africa			●	●	•					
Urals	◑	◑	•					•	●	
Alpine										
Central + South Europe	•		•	●	•	•		•		
SW-Europe + North Africa	•		•	◉	•			•		
SE-Europe	•		•	●	•	•	•	●		
Asia Minor			•	●	•	•		•		
Caucasus	○	○	●	•		•		•		
Platforms										
Great Britain					●	•		•		
Central Europe					◑	•		•		
East Europe					◉	◉		•		
Kazakhstan					◑	◑				
Near East					•			•		
North Africa					◑	◉				

• <10   ● 10 –100   ● 100 –1,000   ● 1,000 –10,000
● >10,000 mio t reserves   ○ potential ores

Fig.50. Iron resources in Europe and adjacent areas corresponding to the genetic types of iron ore deposits, their epochs and regions of mineralization. (After Zitzmann, 1977; courtesy of Bundesanstalt für Geowissenschaften, Hannover.)

The next classification is a further refinement, as especially the so-called "iron-formations" (so well known from the Precambrian terrains, but not restricted to these) are divided and then compared with similar ones to auspicate distinguishing/characteristic features. Gross (1965) proposed a sixfold division of these iron ores (Fig.51 and Table LXX) founded on several variables: mineralogy of ore and gangue, textures and structures, facies, associated lithologies, depositional environments, and tectonic setting,

144

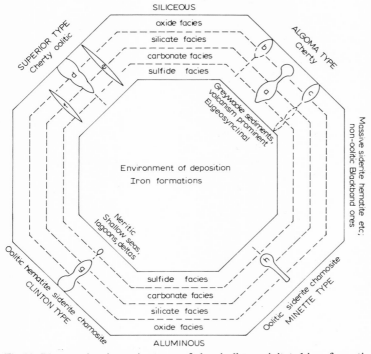

SILICEOUS

SUPERIOR TYPE
Cherty oolitic

ALGOMA TYPE
Cherty

oxide facies

silicate facies

carbonate facies

sulfide facies

Greywacke sediments,
volcanism prominent,
Eugeosynclinal

Massive siderite hematite etc;
non-oolitic Blackband ores

Environment of deposition
Iron formations

Neritic
Shallow seas;
lagoons, deltas

Oolitic hematite siderite chamosite
CLINTON TYPE chamosite

sulfide facies

carbonate facies

silicate facies

oxide facies

Oolitic siderite chamosite
MINETTE TYPE

ALUMINOUS

Fig.51. Diagram showing major types of chemically precipitated iron-formation with their sedimentary facies and depositional environment. (After Gross, 1965; courtesy of Geological Society of Canada, Ottawa.) Examples: a = Michipicoten, Ontario; b = Moose Mountain, Ontario; c = Timagami, Ontario; d = Knob Lake, Labrador and Quebec; e = Iron River, Michigan; f = Gunflint and Biwabik iron-formation, Ontario and Minnesota; g = Wabana, Newfoundland; h = Clear Hills, Alberta.

including presence or absence of volcanism. Such classifications have proved to be useful — but inasmuch as they are highly interpretive (if not conjectural), more descriptive systems should be utilized simultaneously with them: *descriptive* schemes and terminologies *must underlie, or be correlative with, the genetic classifications.*

A detailed examplary classification of igneous-rock-associated gold has been offered by Smirnov (1977, Vol. 3) in Tables LXXI and LXXII, who recognized two major groups: near-surface and medium-depth gold concentrations. The near-surface deposits are divided into (1) gold—silver, (2) gold—quartz, and (3) gold—quartz—sulfide associations, and then further subdivided according to additional mineralogical subassociations. The medium-depth gold ores, on the other hand, are primarily specified by their geotectonic environment (mio- and eugeosynclines, ancient shields, activation zones), then classified as to ore-generating magmatic complexes, structures, minerals and morphology. Although both schemes are recommendable, a reader who may be particularly interested in the differences of magmatic complexes and geotectonic milieus between the near-surface and

TABLE LXX

Characteristic features of the different types of iron-formation (After Gross, 1965; courtesy of Geological Society of Canada, Ottawa)

1. *Algoma type*: Thin-banded chert or quartz and iron oxides, silicates, carbonates and sulfides, associated with volcanic rocks, and greywacke; usually in thin lenses with streaky banding; oolitic textures absent or inconspicuous; relatively rare but in rocks of all ages, especially early Precambrian; eugeosynclinal; e.g., Moose Mountain and Michipicoten area, Algoma district, Ontario.
2. *Superior type*: Interbanded chert or quartz and iron minerals with prominent granular or oolitic textures; associated with dolomite, quartzite, black carbon-bearing shales and slate, chert breccias and volcanic rocks; usually well-defined formations of broad regional extent; form in continental shelf and miogeosynclinal environments. Extensively developed in late Precambrian rock groups, e.g., Lake Superior and Labrador iron ranges.
3. *Clinton type*: Hematite–chamosite–siderite formations; interbedded with carbonaceous shale, carbonate rocks and sandstones; high in phosphorus compared to cherty formations; oolitic to granular texture; miogeosynclinal and shallow-basin environments; lenticular beds restricted to thin formations of broad regional extent as in the Appalachian belt. Most common in rock groups of Cambrian to Devonian age, e.g., Wabana, Newfoundland; Birmingham, Alabama.
4. *Minette type*: Siderite–chamosite–iron silicate, and goethite beds; usually high in phosphorus; oölitic textures; lenticular beds containing clastic material, sideritic mudstones, and concretionary siderite layers; associated with sandstones, carbonaceous shales, siltstone and limestone; cyclic sedimentation of sandy beds, shale and ironstones, and black shale, common; formed in transition zones between marine- and brackish-water environments. Most common in Mesozoic and younger rocks, e.g., Peace River area of Alberta, Canada; Jurassic beds in England; Minette ores of Lorraine, France, and in Luxembourg and Belgium.
5. *Non-oolitic*: A heterogeneous group of non-oolitic, lensy, ferruginous beds; vary in composition and type from the blackband sideritic carbonaceous claystones associated with coal; to the Lahn and Dill hematite and goethite beds locally rich in manganese and/or silica; to siderite–iron silicate–goethite lenses in shale. Associated with fine-grained clastic or volcanic rocks; occurring mainly in late Palaeozoic and younger rocks, e.g., Lahn and Dill River area in Hesse, Germany; Vares, Yugoslavia; Gross Ilsede and Peine in Lower Saxony.
6. *Clastic*: Placer deposits of black sands; mainly magnetite with siderite or hematite in sandstones, usually contain titanium and rare earth elements; thin beds of lensy distribution, varied grade, and quality; deposited along beaches or near shore marine environments, e.g., iron-formation near Burmis, Alberta, in the Belly River sandstone.

medium-depth gold deposits, may have to extract the data piecemeal for comparisons. Noteworthy is the absence of certain varieties of gold accumulations in other parts of the world, especially the sediment-associated deposits (e.g., Witwatersrand and Western States/Colorado Plateau types described, respectively by Pretorius, Chapters 1 and 2 and Rackley, Chapter 3, in Vol. 7 in this Handbook).

The sedimentary phosphorites (cf. Cook's Chapter 11, Vol. 7) can have textures that resemble limestones, so that the classification proposed by Folk (1959) has been employed after appropriate modification, as demonstrated in Table LXXIII and Fig.52. The latter triangular diagram is merely another scheme supplementing the more utilitarian Table LXXIII. Folk's carbonate classification has also been modified for the study of Precambrian iron ores, as illustrated by Dimroth in Chapter 5, Vol. 7. These investiga-

TABLE LXXI

Classification of near-surface gold-ore deposits (After Smirnov, 1977, vol. 3)

Subassociation	Fineness of gold	Mineral type	Typomorphic minerals	Magmatic complexes	Geotectonic environment	Examples of deposits
Gold–silver association						
Silver with gold (Au : Ag = 1 : 100–1 : 10)	<500	Argentite	Argentite, pyrargyrite, stephanite, polybasite, argentiferous pyrite, native silver	Volcanics of andesite–dacite–rhyolite series (necks, plugs, and subvolcanoes)	Young volcanic belts	Pachucha (Mexico), Tonopah, Comstock (USA) Konomai, Titosi, Motikosi, etc. (Japan), Khakandzha (USSR)
Gold–silver (Au : Ag = 1 : 10–1 : 1)	500–750	Pyrite–argentite	Arsenopyrite, pyrite, sulfosalts of silver, electrum, argentite	Volcanics of andesite–dacite–rhyolite series	Young volcanic belts	El Oro (Mexico); gold shows in Kamchatka of Miocene age (USSR)
		Pyrite–arsenopyrite	Arsenopyrite, fahlores, sulfosalts of silver, argentite, electrum, gold tellurides	Volcanics of andesite–dacite–rhyolite series	Alpine orogenic belts	Kremnica (Czechoslovakia), Sasar (Romania)
					Young volcanic belts	Karamken (USSR)
					Hercynian fold belts	Chadak (USSR)
Gold with silver (Au : Ag = 1 : 1–1 : 0.1)	750 and above	Pyrite–bismuthinite	Bismuthinite, cinnabar, gold tellurides, proustite, polybasite, enargite	Volcanics of andesite–dacite–rhyolite series	Young volcanic belts	Goldfield (USA)
					Activation zones of regions of terminated orogeny	Dzhalinda (USSR)
		Pyrite–pyrargyrite	Arsenopyrite, tetrahedrite, pyrargyrite, electrum, gold and silver tellurides	Volcanics of rhyolitic composition	Activation zones of regions of terminated orogeny	Balei (USSR)

Gold–quartz association

Deposit type	Temperature (°C)	Subtype	Mineral association	Intrusive facies	Tectonic setting	Examples
Gold–quartz without tellurides	750 and above	Gold–pyrite	Pyrite, fahlores, arsenopyrite	Minor intrusions of subvolcanic and hypabyssal facies	Activation zones of platforms	Kuranakh (USSR)
					Belts of late Hercynian orogeny and young volcanic belts	Mnogovershinnoe (USSR)
Gold–quartz with tellurides		Tetrahedrite–telluride	Tetrahedrite, tennantite, arsenopyrite, jamesonite, boulangerite, and other sulfosalts, krennerite, sylvanite, nagyagite, and other tellurides, and cinnabar	Volcanics of andesite–dacite–rhyolite series (necks, plugs, and subvolcanoes)	Young volcanic belts	Cripple Creek (USA)
				Minor intrusions of subvolcanic and hypabyssal facies	Alpine orogenic fields	Sekeri and other deposits (Romania), Zod (USSR)

Gold–quartz–sulfide association

Deposit type	Temperature (°C)	Subtype	Mineral association	Intrusive facies	Tectonic setting	Examples
Gold–quartz–sulfide with sulfosalts and tellurides of gold and silver	850–900		Sulfosalts (boulangerite, freibergite, etc.), tennantite and tetrahedrite, tellurides of gold and silver	Volcanics and subvolcanics of andesite–dacite series	Late Hercynian orogenic belts	Kochbulak (USSR)
				Subalkaline series of minor intrusions of subvolcanic and hypabyssal facies	Activated platforms	Lebedinskoe (USSR)
Gold–quartz–sphalerite	from electrum to 850		Pyrite, sphalerite, galena, chalcopyrite; little or no tellurides	Volcanics of andesite–dacite–rhyolite series	Alpine orogenic belts	Stemnica (Czechoslovakia), Ilba, Baya-Sprie, Kapnik, etc. (Romania), Beregovo (USSR)
Gold–pyrite–polymetallic		Gold–pyrite–polymetallic	Pyrite, galena, sphalerite, chalcopyrite, tennantite, argentite, bornite, gold, hessite, sylvanite, arsenopyrite, freibergite, enargite, pyrargyrite	Volcanics of andesite–dacite–rhyolite series	Eugeosynclines of different ages	Zmeinogorsk (USSR)

TABLE LXXII

Classification of medium-depth gold-ore deposits (After Smirnov, 1977, vol. 3)

	Geotectonic environment	Ore-generating magmatic complexes	Ore-controlling structures	Ore association	Mineral type	Structural–morphological features of ore bodies
Miogeo-synclines	With large batholithoid intrusions, and a small number of minor intrusions and dykes	Granitoids of the granite–porphyrite series (orogenic and post-orogenic)	Extensive zones of crushing and shearing, mainly in anticlinal structures	Gold–quartz	Pyrite–galena–chalcopyrite / Pyrite–arseno-pyrite	Saddle-like, complexly branching, concordant and discordant veins
				Gold–sulfide	Pyrite / Pyrrhotite / Pyrite–pyrrhotite	Quartz veins, vein zones, and zones of sulfide segregation
	With extra-geosynclinal multiphase granitoid intrusions and several generations of dykes and minor intrusions	Granitoids predominantly of the grano-diorite series and minor intrusions of variable composition (post-orogenic)	Major faults zones, combined with dome-like and brachyformal structures		Pyrite–arseno-pyrite / Chalcopyrite–galena–sphalerite / Gold–sulfo-antimonite / Antimony	Quartz veins and vein systems / Vein–stockwork / Mineralized zones of crushing and shearing
	With widespread development of minor intrusions and dykes, without manifestations of plutonic granit-oid magmatism	Minor intrusions of variable composition (post-orogenic)	Fields of minor intrusions	Gold–quartz	Pyrite–arseno-pyrite / Pyrite–sphale-rite–chalcopy-rite–galena	Systems of concordant and discordant veins
Eugeo-synclines	With a complete cycle of develop-ment	Minor intrusions of the gabbro–plagio-granite and gabbro–syenite series of the late geosynclinal phase	Bands of develop-ment of minor intrusions		Pyrite–chalcopy-rite–galena	Quartz veins
			Shear zones amongst ultra-	Gold–carbonate–	Pyrite–chalcopy-rite–millerite	Sinuous veins

Cycle of development	Intrusion	Structural setting	Ore formation	Mineral assemblage	Morphology
		ment of skarns at the contact with diorite stocks	quartz–sulfide	…rite–galena	…ntact segregations of types in skarns
With a complete cycle of development	Granitoids of batholithic intrusions of the orogenic stage	Faults zones in exo- and endo-contacts of granitoid massifs	Gold–copper	Pyrite–chalcopyrite–chalcocite	Mineralized shear zones
			Gold–quartz	Pyrite–chalcopyrite–rite	Veins and vein systems
		Faults and dyke belts within and outside granite batholiths		Pyrite–arsenopyrite–sulfosalt	
	Late and post-orogenic granitoid dykes	Dyke belts and faults	Gold–quartz–sulfide	Pyrite–galena–aikinite	Stockworks in dykes and veins
	Complex of minor intrusions of the gabbro–diorite series of the pre-orogenic stage	Faults and discordant segregations of diorites		Pyrite–pyrrhotite with fahlores	Stockworks and quartz veins
With an incomplete cycle of development	Batholithic intrusions of granite–granodiorite series of orogenic stage	Faults and exo-contact zones of intrusions	Gold–sulfide	Pyrite–arsenopyrite–galena–sphalerite	Veins and vein systems
				Pyrite Pyrite–pyrrhotite Pyrite–chalcopyrite–rite with fahlores	Irregular segregations in exocontacts of intrusions with limestones and in skarns

(continued)

TABLE LXXII (continued)

Geotectonic environment	Ore-generating magmatic complexes	Ore-controlling structures	Ore association	Mineral type	Structural-morphological features of ore bodies
Ancient shields	Minor intrusions of granitoids of the post-orogenic stages of development of ancient fold systems	Deep-seated faults, defined by minor intrusions	Gold–quartz	Arsenopyrite–pyrite with pyrrhotite	Veins and vein systems
		Graben–syncline zones, filled with volcanogenic–sedimentary deposits		Pyrite–arsenopyrite–pyrrhotite with tellurides	Veins, metasomatic segregations, and vein systems
			Gold–quartz–sulfide	Pyrite–arsenopyrite with sulfo-salts	
Zones of activation	Minor intrusions of granitoid series	Bands of minor intrusions and faults in arched uplifts		Quartz–arsenopyrite–tetrahedrite–chalcopyrite Quartz–pyrite with tourmaline	Veins, vein stockworks, systems, and stock-works

TABLE LXXIII

Classification of Jordanian phosphorites (After Folk's, 1959, 1974, terminology, originally for limestones; extensively modified; courtesy of *Geol. Jahrb.*)

Classification of Jordanian phosphorite classification pattern according to Folk (1961) substantially modified B. Beerbaum (1975)		>10% allochems allochemical phosphorite (Type I-II)		<10% allochems microcrystalline phosphorite (Type III)		
		Sparry allochemical phosphorite (Type I) — Sparry calcite (quartz) cement > microcrystalline ooze matrix	Microcrystalline allochemical phosphorite (Type II) — Sparry calcite (quartz) cement < microcrystalline ooze matrix	Most abundant allochems	1–10% allochems	<1% allochems
Volumetric allochemical proportion	Volume ratio (1:3 · 1:3/3:1 · 3:1)					
Detritus to skeletons	1:3	Skeletal detrisparite	Skeletal detrimicrite		<u>Detritus</u>	
	1:3/3:1	Skeldetrisparite	Skeldetrimicrite			
	3:1	Detrital skelsparite	Detrital skelmicrite		Detrital micrite	
Detritus to pellets	1:3	Pelletal detrisparite	Pelletal detrimicrite		<u>Skeletons</u>	
	1:3/3:1	Peldetrisparite	Peldetrimicrite			
	3:1	Detrital pelsparite	Detrital pelmicrite		Skeletal micrite	
Pellets to skeletons	1:3	Skeletal pelsparite	Skeletal pelmicrite		<u>Pellets</u>	
	1:3/3:1	Skelpelsparite	Skelpelmicrite			
	3:1	Pelletal skelsparite	Pelletal skelmicrite		Pelletal micrite	Phosphatic / Calcitic / Siliceous → Micrite

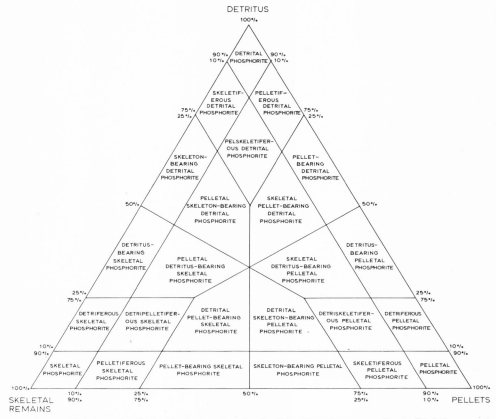

Fig.52. Classification triangle for Jordanian phosphorite (triangle pattern according to Füchtbauer and Müller, 1970). (After Beerbaum, 1977; courtesy of *Geol. Jahrb.*)

tions serve well as examples of cross-fertilization of ideas from one geological subdiscipline to another.

An interesting overview of the genesis of "primary" kaolin has been presented by Seyhan (1971) which opens numerous avenues for future research on igneous and volcanic rock-associated industrial mineral (non-metallic) concentrations. Fig.53 shows that many of these kaolin deposits are also associated with metalliferous ores, and in contrast to these so-called primary clays, the kaolinization of feldspars by descending water is to be considered as "secondary" in origin (=supergene types). A total of five kaolin deposits is described by Seyhan to be briefly outlined in the succeeding paragraphs.

(1) Hypomagmatic (=plutonic) kaolin (type No. 1). When plutonic rocks crystallize at some depth, the contact facies are finer grained (felsitic) and the hydrothermal gases and acidic solutions left as a result of magmatic differentiation may be prevented from escape due to the presence of the fine-grained lithology, absence of tectonic fracturing

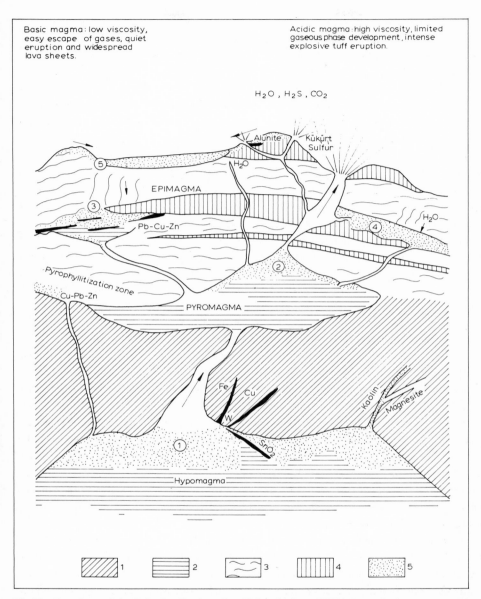

Fig.53. Genetic model of the kaolin deposits. (After Seyhan, 1971.) Legend: *1* = metamorphic and sedimentary rocks; *2* = acidic rocks (plutonic and subvolcanic); *3* = lavas; *4* = tuffs; *5* = kaolin.

and the comparatively low to intermediate pressure, can cause lateral kaolinization of feldspar in porphyries and granite mass (kaolinite type No. 1, Fig.53). This variety of kaolin, which forms a "cover" along the upper parts of subvolcanic (hypabyssal) and plutonic rocks, has been interpreted as weathering crusts, which may be true in some cases but not in others. In Fig.53, types 1 and 2 are associated with numerous high-temperature minerals, especially with cassiterite veins, indicating an autometamorphic origin of these kaolin concentrations.

(2) Pyromagmatic (subvolcanic) kaolin (Type No. 2). The depth at which subvolcanic conditions occur varies from as little as 0.5 km to as much as 3–4 km, depending on the chemistry of the magma and the petrographic and tectonic conditions of the "hanging rocks" (=roof country rocks). During periods when the conduits of the volcanoes are plugged or closed and the eruption is interrupted, both the pressure and temperature increase, which results in the dissolution of the alkali-silicates. When then the volcanic eruption reoccurs, the lavas are at first alkali-rich. The "dryness" of the subvolcanic material and the change in alkali-content during eruption result in anorthite-richer plagioclase and the precipitation of metal-sulfide. The kaolin formed from these subvolcanic rocks are rich in Ca and have Fe-mineral disseminations which are the secondary product after pyrite — thus making the kaolin economically unviable. When the eruptions are less frequent and the dormant periods are longer, the plagioclase are richer in alkali, so that the kaolin derived from them are higher in Na_2O and K_2O. Should the magma become "dry" as a consequence of eruption, or repeated eruptions, the formation of kaolin is prevented — only the reactive circulating fluids under high temperature and strongly acidic pH-values will form subvolcanic kaolin deposits (type No. 2).

(3) Epimagmatic (volcanic) kaolin (types Nos. 3, 4 and 5). The above-described influence of the water-content (or degree of "dryness") of the magma on the origin of kaolin, controlled by the frequency of volcanic eruption, can be extended to the near-surface and surface deposits. As a consequence of these varying volcanic conditions, there are several environments conducive to kaolin formation. Most likely, the feldspars surrounded by solutions will undergo kaolinization already during magmatic transportation. This is indicated by white-speckled andesites and dacites where sometimes the feldspar phenocrysts, but not the feldspar of the groundmass, were kaolinized. This is due to the presence of water (constituting 90–95% of the volcanic gases), which is required for the process of kaolinization, and which was present prior to the consolidation of the groundmass material but had escaped thereafter.

In many andesites and dacites, which solidified under thick lavas and tuffs of previous eruption(s) or were injected into sedimentary sequences, kaolinization has often occurred along specific horizons and they are also rich in lead–zinc and copper minerals. This indicates that the fluids and gases required for kaolinization were trapped in older country rocks to continue then to differentiate and react with the host lithologies. Heated subsurface water probably played an important role too in the kaolinization of the feldspar-rich tuff units (type No. 3).

It has been frequently observed that hot thermal fluids and the acidic solutions of the last volcanic phase penetrated some tuffs and have kaolinized them (type 4). As the solutions are of limited quantity and since the pH changes irregularly, these kaolin-tuffs are in places silicified. These deposits are composed of "hard" kaolin, difficult to wash during mining. Only subsequent weathering by descending water has changed them into a "soft" kaolin.

The telethermal solutions of the magma are, generally, alkaline. However, in sub-volcanic deposits the telethermal sulfide-solutions may become oxidized in the ground-water, and the released acidic sulfur-containing solutions can produce a strongly acidic milieu near the earth surface. Under such conditions, the prophylitization changes into alunitization and leads to the origin of kaolin deposits with a paragenesis of sulfur—alunite—kaolin.

The volcanic activity alters the surface geomorphology; hence in the newly created basins for example lakes are formed. Into these lakes thermal fluids may pass, tuffs may accumulate, hot lava tongues may reach them, etc. — with the consequence that the pH will be conducive to the lateral kaolinization of fine-grained feldspar in the tuffs and tuffites. Also, it is known that volcanic gases may be trapped for longer periods in the lavas, and that the degassing can take several years. Consequently, it is probable that in sedimentary—volcanic basins the water seeping through the lavas becomes heated and reacts with the gases, then causing kaolinization of the flow rocks and pyroclastics (type No. 5). During kaolinization, the pH becomes alkaline; the alkali released from the feldspar may react with the silicic acid of the already formed kaolin to cause "bauxitization" of the kaolin. At the same time, such basins are recorded to have received an influx of clastic material so that the alkali solutions may alter to an organic matter-rich "swampy" milieu, which in turn results in the well-known lithologic paragenesis of kaolinitic clay—brown-coal—clay—bauxite and brown-coal accumulations.

Although the main theme of the present multi-volume treatise is related to strata-bound ores, the *gradational nature* of volcanic-exhalative mineralization (through various near-surface hypabyssal into truly magmatic and deep-subsurface hydrothermal ores), as well as the *tight genetic relationships* and the consequent possible associations in time and space, all necessitate that at least a fleeting reference should be made to some non-strati-form and non-strata-bound ore types. The porphyry-copper types are an example. Sillitoe (1973) proposed a model of a porphyry-copper *system* (Fig.54) which has been assembled or reconstructed by combining information from many areas, i.e., it is a conceptual or idealized model allowing the classificatory grouping of seemingly unrelated prophyry-copper sub-varieties into a more comprehensive scheme.

By using several different (genetically unrelated) ore deposits from geographically widely separated areas, Sillitoe (1973) gave examples of natural occurrences that have features characteristic of a specific level or depth below surface, as shown in Fig.54. The lower-level-type ore is illustrated by the deposits of Chuquicamata and the highest by Volcan Aucanquilcha, with four others in between. The following characteristics apply to the model:

(1) The porphyry copper is genetically related to porphyritic-textured stocks containing a large proportion of the Cu—Mo mineralization; the porphyry (e.g., dacite por-

156

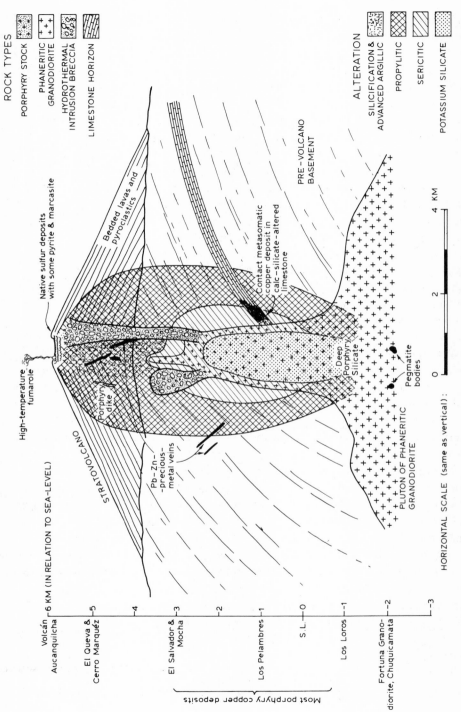

Fig.54. Idealized cross-section of a typical, simple porphyry copper deposit showing its position at the boundary between plutonic and volcanic environments. (After Sillitoe, 1973; courtesy of *Econ. Geol.*) Vertical and horizontal dimensions are meant to be only approximate.

phyry) changing in depth to a phaneritic rock (e.g., quartz diorite) or similar composition. At this boundary, the diameter of the stock is envisaged to increase, and mineralization appears to die out soon. Some pegmatite bodies may be present in the pluton (Fig.54) formed from trapped fluids that, higher up in the system, give rise to the porphyry-type mineralization.

(2) The central economic section of a typical porphyry-copper deposit is characterized by concentric shells of K-silicate, sericite, argillic and propylitic alteration. In deeper parts of the system, the K-silicate alteration may be the major alteration type, and in the basal section may grade into a modified deep K-silicate alteration variety, with biotite less common and consisting of the assemblage quartz—K-feldspar—sericite—chlorite. Upward, the typical porphyry-copper's sericite—argillite alteration increases in importance at the expense of K-silicate alteration. Near the upper margin of the economic hypogene mineralization, any intrusives are probably smaller and less regular, and large volumes of hydrothermal breccias may occur.

(3) The uppermost comagmatic volcanic superstructure tends to exhibit a less regularly distributed alteration consisting of propylitic and argillic alteration with localized intense silicification and advanced argillic alteration — possibly the result of ascending hydrothermal solutions. Pyrite (and marcasite) is ubiquitous, but other sulfides are rare.

(4) "Epithermal" Cu, Pb, Zn, and precious metal veins and replacements are present as fringe accumulations in the propylitic-altered sections as well as accompanying advanced argillic alteration, silicification or propylitic alteration in the supra-adjacent volcanic edifice (Fig.54).

(5) The evidence used by Sillitoe (1973) suggested that the tops of typical porphyry copper was at 1.5—3 km beneath the summits of stratovolcanoes and that the entire porphyry-copper systems have a vertical extent of up to approximately 8 km.

(6) Most of the deposits were emplaced within much older and, therefore, genetically unrelated country rock. Also, the normal location for the economic portions of porphyry ores is in rock formations that underlie the coeval volcanic pile (see below, however).

(7) The porphyry-copper system proposed by Sillitoe spans the boundary between plutonic and volcanic environments.

(8) In cases where the Cu—Mo mineralization is exposed, the overlying volcanics have been eroded.

(9) During the final phases of stratovolcanism, fumarolic and hot-spring activity are the surface expression of the efflux of metal-bearing magmatic solutions freed during retrograde boiling, leading to an interaction of these solutions with various possible types of connate fluids and resulting in chemical alteration and mineralization.

(10) Sillitoe also believed (p.801) that a porphyry-copper deposit is overlain by a column of pyritic alteration which transects a calc-alkaline volcanic pile, commonly surmounted by an andesitic stratovolcano with native sulfur deposits as well as small base-metal concentrations, particularly copper, in high-temperature fumarolic sublimates near the central vent.

158

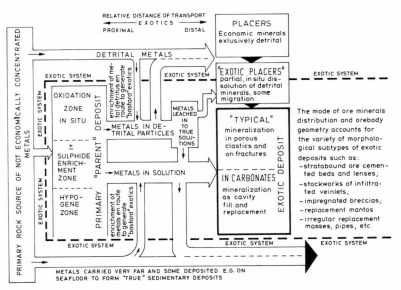

Fig.55. The exotic mineralization system. (After Laznicka, 1978; courtesy of *Northern Miner*.)

Laznicka (1978) supplied a fine example to demonstrate that an exploration and/or research geologist must be familiar with the whole gamut of types and sub-types of genetic mineralization, as this basic "recipe" knowledge serves as a prospecting guide and assists in the formulization of study programs. He described the 200 million-ton 1.65% copper Exotica orebody about 3 km downslope from the Chuquicamata copper porphyry in Chile. The latter intrusion served as a source, supplying copper to surface and ground-waters that, in turn, precipitated the metal to form — what Laznicka termed — an "exotic" mineralization. He defined this type as "a down-the-slope displaced oxidation zone of a primary deposit in which the ore minerals represent the entrapped mobile phase in contrast to residual deposits where the immobile phase constitutes the valuable component." According to Schneiderhöhn's classification, the exotic deposits are a sub-type of the "descendant" group of mineralizations. The exotic accumulations are a *transitional* type, with the residual deposits per se on one end of the spectrum and a truly "sedimentary" mineral concentration on the other. The so-called "transported" bauxites and "pseudo-gossans" may be cited as intermediate-type deposits between the two end-members just referred to. Laznicka discussed the possible misinterpretations of exotic ores in cases where no source is known (i.e., he would call these "bastard exotic" deposits), as well as the influences on the exploration philosophy. In so far as other metals (in addition to copper) can be concentrated in specific climatic and tectonic settings (e.g., in non-marine basins with arkose-associated red-bed sediments), this sub-type mineralization serves as a good example of the importance of formulating and utilizing plausible

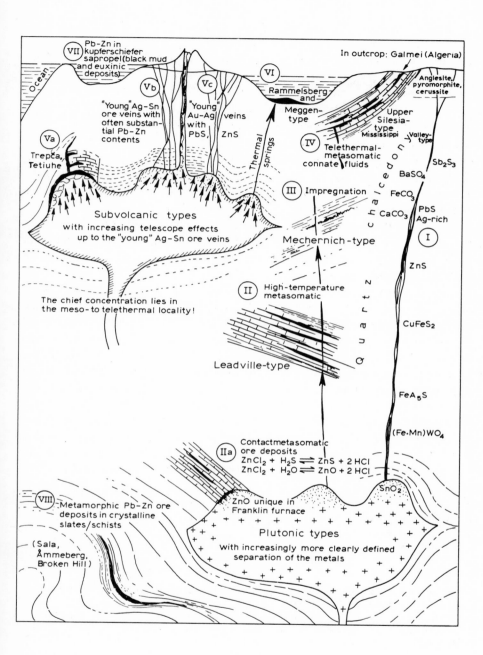

Fig.56. Genetic overview of the lead–zinc ore deposits. (After Borchert, 1950, unpublished; 1978; courtesy of Glückauf, Essen.)

classifications and conceptual models (Fig.55).

Fig.56 presents in a simplified style the genetic types of lead—zinc mineralizations, listing eleven varieties (i.e., I—VIII, with three sub-types, such as Vb) — all of which depict genetic mechanisms, already in voque twenty to thirty years and longer ago. Although not explicitly connotated in the diagram, the ore types range from those syn- genetically formed along the earth's surface through diagenetic varieties originating shortly after host-rock accumulation to those produced very late-epigenetically. As to the source of the cations and anions of the ore minerals, the plutonic, subvolcanic (i.e., hypabyssal), volcanic-exhalative, sedimentary-exhalative, seawater, and continental sources have been illustrated; whereas the physicochemical and biological (e.g., bacteriological) precipitation processes can be envisaged while examining the surface milieu in Fig.56.

As in some other instances, the stratiform/strata-bound lead—zinc deposits are treatable as a separate entity, which results in specialized classifications as exemplified in Table LXXIV. A further refinement in this has been offered through the publications on zinc by the Bundesanstalt für Bodenforschung Hannover, FRG, who have made a much- needed comprehensive summary of the geology, metallurgy and economics of specific commodities (see also their bulletins Nos. I—IV on lead, copper, aluminum and fluorite, respectively). Six tables of the zinc ores (Tables LXXV—LXXX) have been reproduced here to serve as models; the schemes being based on mechanisms of formation, tectonic setting, geologic time distribution and geographic distribution. Table LXXV presents the genetic types of zinc accumulations, each one exemplified by specific ore districts or mines. Table LXXVI depicts a geotectonic division of zinc mineralizations associated with

TABLE LXXIV

Classification scheme of stratified lead—zinc deposits (After Bogdanov and Kutyrev, 1973)

A — Sedimentary (sedimentary-katagenetic deposits)

    I. The Mississippi Valley type (shallow-water marine type): the Tri-States deposit in the USA, Mirgalimsai deposit in the USSR.
    II. The Sumsar type (coastal-marine type): Uch-Kulach, Sumsar, Dzhergalan in Central Asia.

B — Volcanogenic-sedimentary deposits (associated with the submarine acid volcanism)

    I. The Altai type: near volcanic foci (the liparite, andesite-dacite-liparite, liparite-tuffaceous- shaly formations etc.): Rudny Altai.
    II. The Atasu type: laterally (or chronologically) separated from volcanic foci: Zhairem, Atasu and other deposits in Central Kazakhstan, Filizchai deposit in Azerbaijan, Rammelsberg and Meggen in Germany.

C — Metamorphosed deposits

    I. Regionally-metamorphosed: Broken Hill and Mount Isa deposits in Australia; Karagaily, Culshad in Kazakhstan.

TABLE LXXV

Genetic classification of economically important zinc deposits (After Bundesanstalt für Boden-forschung, 1974; courtesy Bundesanstalt für Geowissenschaften, Hannover)

(A) *Siamic volcanogenic–sedimentary deposits,*
which occur partly as massive coarse bodies, partly as impregnations. To this group belong:
  (a) *Deposits in geosynclinal sequences* (partly with volcanites)
  Occurrences of this type are present in old *shields,* in *Paleozoic* as well as in *Mesozoic–Tertiary orogenic* areas.
  (a_1) *Shield deposits*
  Characteristic for deposits in association with Archean volcanites (partly tuffites) is the metal-association Zn–Cu; contents: 4–6% and more Zn; 1–2% Cu; Pb is absent or occurs as minor amounts, mostly <1% (e.g., Kidd Creek, Ontario).
  In deposits of shields in Proterozoic slates/schists without known volcanites, the Cu-content decreases; Pb increases and Zn reaches values of over 10% (e.g., Sullivan Mine, B.C., Canada; Mt. Isa, Qld., Australia, Broken Hill, N.S.W., Australia).
  (a_2) *Deposits in Paleozoic orogenic areas*
  In this category belong:
  Deposits in New Brunswick (No. 6- and No. 12-units), Canada; Rammelsberg, Harz and Meggen, Sauerland, West Germany; Red Roseberry, Tasmania, Australia.
  (a_3) *Deposits in Mesozoic–Tertiary orogenic areas*
  To these belong:
  Anvil mine, Yukon Territory, Canada; Kuroko ores, Japan.
  (b) *Deposits in geosynclinal carbonate sequences without volcanites*
  Exemplified by:
  (b_1) *Occurrences in Paleozoic orogenic areas*
  Tynagh, Navan, Ireland; mines along the Mascot–Jefferson City trend, Eastern Tennessee, USA.
  (b_2) *Occurrences in Mesozoic–Tertiary orogenic areas*
  Bleiberg, Austria and Reocin, Spain.

(B) *Siamic-hydrothermal or marine-sedimentary deposits*
To these belong:
  (a) *Deposits in Precambrian carbonate sequences of shields*
  (often chimney/pipe-shaped)
  Kipushi, Zaire; Kabwe, Zambia; Tsume, SW-Africa
  (b) *Deposits in carbonate platform sediments* (from Paleozoic and younger areas)
  Pine Point, N.W.T., Canada; occurrences along the Carthage trend ("New Zinc Belt"), Central Tennessee, USA; Old Lead Belt and New Lead Belt, Missouri, USA; occurrences in Upper Selisia near Beuthen.

(C) *Sialic vein-, impregnation- and replacement-type deposits*
Here belong the following varieties
  (a) *Shield deposits (contact-metasomatically modified)*
  Sterling mine, Franklin Furnace; New Jersey, USA.
  (b) *Deposits in Paleozoic and Mesozoic orogenic areas*
  Bad Grund/Harz; Montevecchio, Sardinia/Italy and deposits in the Western USA; Idaho (Coeur d'Alene district), Colorado (Leadville district), Utah (Bingham district).
  (c) *Deposits in Mesozoic–Tertiary orogenic areas*
  Deposits in Mexico and Peru as well as Trepca, Yugoslavia.

TABLE LXXVI

Proportion of geotectonic units plus the simatic and sialic deposits supplying zinc-ore (in %) to the "Western World" during 1973 (After Bundesanstalt für Bodenforschung, 1974; courtesy of Bundesanstalt für Geowissenschaften, Hannover)

Geotectonic unit	Mining %	Simatic deposits (1 to 4)				Sialic deposits (1 to 3) hydrothermal veins and replacements		
		volc.-sedim.	volc.-sedim.	hydrotherm. or marine-sedim.			late-orogenic	subsequent
			without recognizable magmatism			± contact-metasomatic		
		± volcanics geosynclinal series shields Paleoz./Mesoz.-Tert. orogenic regions	geosyncl. limestone sequences Paleoz./Mesoz.-Tert. orogenic regions	Precambr. limestone sequences in shields (platform sediments)	limestone sequences and sandstones of Paleozoic platforms	late-orogenic and subsequent shields	Paleoz./Mesoz. orogenic regions	Mesoz./Tert. orogenic regions
		(1)	(2)	(3)	(4)	(1)	(2)	(3)
Shields								
Canadian shield	19.9	19.1	–	–	–	0.8	–	–
Fennoscandian shield	3.7	3.6	–	–	–	0.1	–	–
	38.4	32.9		4.6				
African shield	4.5	0.4	–	4.1	–	0.9	–	–
Australian shield	8.9	8.9	–	–	–	–	–	–
Remaining Western World	1.4	0.9	–	0.5	–	–	–	–
Paleozoic orogenic regions								
Appalachian orogenic regions	7.9	5.0	2.9					
European orogenic	6.?						1.8	

region	1.6	–	1.4	–	–	–	–	0.2	–
Remaining Western World	0.1	–	0.1	–	–	–	–	–	–
Mesozoic–Tertiary orogenic regions									
Cordilleran-Andian orogenic regions	24.8	–	4.5	–	–	–	–	4.1	16.2
Alpine orogenic region	6.6	–	–	4.3	–	–	–	–	2.3
Nord-African orogenic region	1.0	39.5	7.7	0.5	4.8	–	0.5	6.6	20.4
East-Asiatic orogenic region	7.1	–	3.2	–	–	–	–	2.0	1.9
Remaining Western World	–	–	–	–	–	–	–	–	–
Platform regions (from Paleozoic)									
North American platform	5.8	–	–	–	–	5.8	–	–	–
European platform	0.4	6.2	6.2	–	–	0.4	–	–	–
Remaining Western World	–	–	–	–	–	–	–	–	–
Proportion of the mining	100.0	100.0	49.4	9.8	4.6	6.2	0.9	8.7	20.4

Subtotals: columns 3–6 = 70.0; columns 7–9 = 30.0

TABLE LXXVII

Proportion of the ore types in supplying zinc-ore (in %) to the most important countries of the "Western World" during 1973 (After Bundesanstalt für Bodenforschung, 1974; courtesy of Bundesanstalt für Geowissenschaften, Hannover)

Countries	Simatic deposits				Sialic deposits		
	volc.-sedim.	volc.-sedim.	hydrotherm. or marine-sedim.		hydrothermal veins and replacements		
		without recognizable magmatism					
	± volcanics geosynclinal series shields Paleoz./Mesoz.-Tert.orogenic regions	geosyncl. limestone sequences Paleoz./Mesoz.-Tert. orogenic regions	Precambr. limestone sequences in shields (platform sediments)	limestone sequences and sandstones of Paleozoic platforms	± contact-metasomatic late-orogenic and subsequent shields	late-orogenic Paleoz./Mesoz. orogenic regions	subsequent Mesoz./Tert. orogenic regions
Canada	82.6	–	–	13.8	–	3.6	–
Australia	98.2	–	–	–	–	1.8	–
USA	12.7	29.8	–	17.1	8.0	32.4	–
Peru	10.0	–	–	–	–	–	90.0
Mexico	–	–	–	–	–	–	100.0
Japan	54.5	–	–	–	–	33.5	12.0
West Germany	69.8	–	–	–	–	30.2	–
Yugoslavia	–	8.7	–	–	–	–	91.3
Sweden	93.5	–	–	2.2	4.3	–	–
Zaire	–	–	100.00	–	–	–	–
Spain	2.0	88.0	–	–	–	10.0	–
Italy	–	75.9	–	–	–	24.1	–
Rep. Ireland	–	100.0	–	–	–	–	–
Finland	100.0	–	–	–	–	–	–
Zambia	–	–	100.0	–	–	–	–
Iran	54.5	45.5	–	–	–	–	–
Bolivia	82.0	–	–	–	–	–	18.0
South Korea	–	–	–	–	–	–	100.0
Denmark/Greenland	100.0	–	–	–	–	–	–
Total of	49.4	9.8	4.6	6.2	0.9	8.7	20.4
Western World	70.0				30.0		

TABLE LXXVIII

Zinc reserves in Canada divided according to geographic provinces and ore-deposit type-data of 1974 (in 1000 mt and %) (After Bundesanstalt für Bodenforschung, 1974; courtesy of Bundesanstalt für Geowissenschaften, Hannover)

	Simatic deposits					Sialic deposits			
	volc.-sedim.			hydrotherm. or marine-sedim.		hydrothermal veins and replacements			
	All provinces		± volcanics geosynclinal series shields Paleoz./ Mesoz.-Tert. orogenic regions		without recognizable magmatism limestone sequences and sandstones of Paleozoic platforms		contact metasomatic late-orogenic and subsequent shields		late-orogenic Paleoz./ Mesoz. orogenic regions	
	1000 t	%	1000 t	%	1000 t	%	1000 t	%	1000 t	%
Newfoundland	800	2.1	800	2.1	–	–	–	–	–	–
Nova Scotia	10	<0.1	10	<0.1	–	–	–	–	–	–
New Brunswick	10,200	27.2	10,200	27.2	–	–	–	–	–	–
Quebec	2,530	6.8	2,530	6.8	–	–	–	–	–	–
Ontario	7,615	20.3	7,615	20.3	–	–	–	–	–	–
Manitoba	1,590	4.2	1,590	4.2	–	–	–	–	–	–
Saskatchewan	120	0.3	120	0.3	–	–	–	–	–	–
Northwest Territories	7,140	19.0	4,780	12.7	2,360	6.3	–	–	–	–
Yukon	4,395	11.7	3,735	9.9	–	–	–	–	660	1.8
British Columbia	3,160	8.4	2,980	7.9	–	–	130	0.4	50	0.1
Canada as a whole	37,560	100.0	34,360	91.4	2,360	6.3	130	0.4	710	1.9
	$36.72 \ 10^6 t = 97.7\%$						$0.84 \ 10^6 t = 2.3\%$			
Canadian shield	19,535	52.0	19,535	52.0	–	–	–	–	–	–
Appalachian orogenic areas	11,090	29.5	11,090	29.5	–	–	–	–	–	–
Cordill.-Andian orogenic area	4,575	12.2	3,735	9.9	–	–	130	0.4	710	1.9
North American platform	2,360	6.3	–	–	2,360	6.3	–	–	–	–
Total:	37,560	100.0	34,360	91.4	2,360	6.3	130	0.4	710	1.9

[1] Data as of 1.1.1974.

siamic and sialic magmatism/volcanism; an absence of the influence of igneous activity being explicitly mentioned in three columns. Table LXXVII gives the countries which produce zinc and it is noteworthy that some mine this metal from several genetic varieties

TABLE LXXIX

Zinc reserves and zinc ore mined in Canada based on ore-deposit type (in 1000 mt and %) (After Bundesanstalt für Bodenforschung, 1974; courtesy of Bundesanstalt für Geowissenschaften, Hannover)

Deposit type	Reserves [1]		Mining [2]	
	1000 t	%	1000 t	%
Volcanic-sedimentary in geosynclinical sequences in shields and orogenic regions, ± volcanites	34,360	91.4	1,021.2	82.6
Hydrothermal or marine-sedimentary, without recognizable magmatism, in limestone sequences and sandstones of Paleoz. platforms	2,360	6.3	169.9	13.7
Simatic deposits as a whole	36,720	97.7	1,191.1	96.3
Hydrothermal veins and replacements in shields, ± contact-metasomatic	130	0.4	23.9	1.9
Hydrothermal veins and replacements in orogenic regions	710	1.9	21.2	1.8
Sialic deposits as a whole	840	2.3	45.1	3.7
Total	37,360	100.0	1,236.2	100.0
Geotectonic unit				
Canadian shield	19,535	52.0	709.0	57.4
Appalachian orogenic region	11,090	29.5	198.2	16.0
Cordilleran-Andian orogenic region	4,575	12.2	159.1	12.9
North American platform	2,360	6.3	169.9	13.7
Total	37,560	100.0	1,236.2	100.0

[1] Data as of 1.1.1974.
[2] Mining in the year 1973 of recoverable zinc.

(e.g., USA, Canada and Japan), whereas others only from one type (e.g., Ireland and Zaire). In Table LXXVIII, considering only one particular country, namely Canada, the zinc reserves are listed according to geographic provinces and types of ore deposits, whereas Table LXXIX provides data on both zinc reserves and mining arranged in accordance with the types of mineralizations of the same country. Table LXXX gives an overview of the world zinc reserves and supply from shields, orogenic areas and platforms. In all tables one should note the percentage distribution of the zinc obtained from the various genetic varieties — the volcanic-sedimentary ores making the major contribution.

The bismuth geochemical cycle is represented in Fig.57 (Feiser, Borchert and Anger, 1965, 1978) with numerous types of bismuth concentrations depicted by the names of specific mining districts. Table LXXXI lists seven genetic varieties of bismuth deposits with a twenty-fivefold mineralogic sub-classification. Ten of the *endogenic* bis-

TABLE LXXX

Comparison of the proportions of different geotectonic units in regard to reserves (1974) and supply (1973) of zinc to the "Western" world (in %) (After Bundesanstalt für Bodenforschung, 1974; courtesy of Bundesanstalt für Geowissenschaften, Hannover)

Geotectonic unit	Reserves	Mining	Types of deposits [1]
Shields			
Canadian shield	14.8	19.9	SIM-1, SIA-1
Fennoscandian shield	3.8	3.7	SIM-1, SIA-1
African shield	3.0	4.5	SIM-1, SIM-3
Australian shield	9.7	8.9	SIM-1
remaining Western world	3.5	1.4	SIM-1, SIM-3
Paleozoic orogenic regions			
Appalachian orogenic region	12.8	7.9	SIM-1,2
European orogenic region	11.6	6.3	SIM-1,2, SIA-2
Australian orogenic region	1.6	1.6	SIM-1, SIA-2
remaining Western world	0.3	0.1	SIM-1
Mesozoic—Tertiary orogenic regions			
Cordilleran-Andian orogenic regions	14.3	24.8	SIM-1, SIA-2,3
Alpine orogenic region	5.6	6.6	SIM-2, SIA-3
North-African orogenic region	0.8	1.0	SIM-2, SIA-2
East-Asiatic orogenic region	6.1	7.1	SIM-1, SIA-2,3
Platform regions *(from Paleozoic)*			
North American platform	11.6	5.8	SIM-4
European platform	0.5	0.4	SIM-4
Total	100.0	100.0	

[1] Deposits after classification in Table LXXVI. SIM-simatic, SIA-sialic.

muth deposits listed in Table LXXXI have then been outlined in detail in Table LXXXII, classified according to "bismuth mineralization" and "Bi-bearing mineral association" and its component minerals (in order of paragenesis); the examples of deposits in combination with the mineral associations permit easy correlation between the two tables.

The occurrence of beryllium is illustrated in Fig.58 (Borchert, 1950), together with some of the host rocks and geological settings. A rather complete geological-genetic classification has been provided by Smirnov (1977) in Table LXXXIII. By comparing Table LXXXII of the bismuth with Table LXXXIII of the beryllium, some fundamental differences in the mode of dividing ores are obvious. The latter scheme is first classified into three genetic groups, namely into hydrothermal, pneumatolytic—hydrothermal and pegmatitic; then follows a sevenfold division (I—VII) according to (1) mineral associations

TABLE LXXXI

Types of bismuth mineralization, characteristic of various genetic classes of ore deposits (After Smirnov, 1977, vol. 2)

Class	Type of bismuth mineralization	Practical importance
Magmatic	bismuth—platinoid	nil
Pegmatoid	bismuth—pyrrhotite—chalcopyrite	nil
	bismuth—arsenosulfide	nil
	bismuth—polymetallic	nil
Carbonatite	bismuth—polymetallic	nil
Skarn	bismuth—pyrrhotite—chalcopyrite	great
	bismuth—arsenosulfide	individual deposits
	bismuth—polymetallic	great
	bismuth—bornite—chalcopyrite	small
Albitite	bismuth—polymetallic	nil
Greisen	bismuth—cassiterite—wolframite	great
	bismuth—pyrrhotite—chalcopyrite	great
	bismuth—polymetallic	small
Hydrothermal:		
plutonogenic	bismuth—pyrrhotite—chalcopyrite	individual deposits
	bismuth—polymetallic	great
	bismuth—arsenide	small
	bismuth—arsenosulfide	small
	bismuth—chalcopyrite	great
	gold—bismuth	small
volcanogenic	bismuth—chalcopyrite	medium
	bismuth—polymetallic	medium
	bismuth—cassiterite—wolframite	great
telethermal	bismuth—polymetallic	small
Pyritic	bismuth—puritic	medium
	bismuth—polymetallic	medium
Eluvial and deluvial placer		small

plus mode of origin and/or host rock, and (2) depth of formation (e.g., acroabyssal through hypabyssal and medium-depth to deep-seated) and range of temperature. The above parameters are plotted against two parent rocks and six country rocks. (Note the different meanings of host-, parent- and country-rock assemblages.)

To bring this section to conclusion, additional pictorial models of the following six chemical elements are provided: barium, fluorine (question: where, how and when do the cycles of Ba and F overlap with those of Pb and Zn to give rise to the Mississippi Valley-type deposits?), tungsten, boron, strontium, and lithium (Figs.59—64).

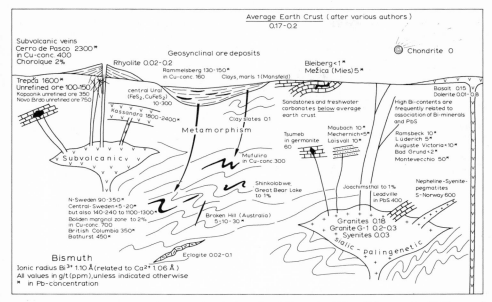

Fig.57a. Conceptual model of the geochemical cycle of bismuth. (After Feiser et al., 1965, unpublished; 1978; courtesy of Glückauf, Essen.)

CLASSIFICATIONS OF CHEMICAL ELEMENTS RELATED TO ORE GENESIS AND/OR ORE INDICATORS

"Prudens quaestio dimidium scientiae — to know what to ask is already to know half."
    (Unknown source)

"Although this may seem a paradox, all exact science is dominated by the idea of approximation."
    Bertrand Russel

"Those who refuse to go beyond fact rarely get as far as fact ... Almost every great step (in the history of science) has been made by the 'anticipation of nature', that is by the invention of hypotheses which, though verifiable, often had very little foundation to start with."
    Thomas H. Huxley

The chemical elements may be considered to be the smallest constituents of ores and host rocks (if one conveniently ignores for the present purposes the sub-atomic particles!); and many classifications are founded on elements as numerous schemes summarized in this contribution testify. Many data have been accumulated on the behaviour and distribution of elements under natural conditions. Nevertheless, many questions remain unanswered and the problems await future attention. The data on chemical elements cross-cut many geological subdisciplines; e.g., the physicochemical characteristics

170

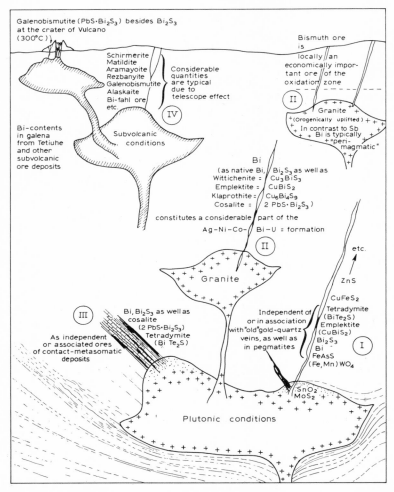

Fig.57b. Genetic overview of the bismuth ore deposits. (After Borchert, 1950, unpublished; 1978; courtesy of Glückauf, Essen.)

of elements apply to low- and high-temperature ore and gangue minerals alike, and transgresses sedimentary, igneous as well as metamorphic rock genesis.

From the information available in the published literature, only a few selected classificatory schemes are abstracted below in support of and to augment several sections:

(1) Crustal abundances of elements.

(2) Ore-grade versus crustal abundances.

(3) Relation between total world reserves and crustal abundances.

(4) Domestic reserves of elements compared to their earth's crustal abundances.

TABLE LXXXII

Types of bismuth mineralization of endogenic deposits (After Smirnov, 1977)

Type of bismuth mineralization	Bismuth-bearing mineral association and its component minerals (in order of precipitation)	Bismuth minerals	Examples of deposit
Bismuth–platinoid	chalcopyrite–bismuthide chalcopyrite–(cubanite)[1]–(pentlandite)–(millerite)–bismuth minerals	bismuthides of palladium and platinum, tellurobismuthite, tetradymite	copper-nickel deposits of USSR, Canada, and South Africa
Bismuth–cassiterite–wolframite	wolframite–cassiterite–bismuthinite–bismuth quartz–muscovite–wolframite–cassiterite–molybdenite–pyrite–bismuth minerals	bismuthinite, native bismuth, ikunolite, joseite A, galenobismuthite[2], cosalite[2]	greisen, molybdenum–tin–tungsten deposits of Central Kazakhstan (Kara-Oba, Akchatau, Taishek); Transbaikalia (Bukuka); Shansi province in China; Queensland, Australia; Barralia deposit in Portugal, etc.; volcanogenic tin and tin–silver deposits of Central Bolivia; volcanogenic copper–lead–zinc deposits of Japan
Bismuth–pyrrhotite–chalcopyrite	pyrrhotite–chalcopyrite–bismuthinite–bismuth pyrrhotite–chalcopyrite–bismuth minerals	bismuthinite, native bismuth, joseite A, joseite B, hedleyite, tetradymite, cosalite[2], kobellite[2], galenobismutite[2]	skarn–scheelite deposits of Middle Asia, Primor'e and Yakutia; greisen deposits of Central Kazakhstan (Upper Kairakty, Ak-maya, and Saran); Transbaikalia (Belukha); Portugal (Panaskeira); cassiterite–sulfide deposits of Soviet Northeast; tungsten–tin deposits of Bolivia (Tasna, and Colavi); bismuth deposit of Middle Asia (Ustarasai)
Bismuth–chalcopyrite	chalcopyrite–bismuthinite–sulfobismuthite pyrite–chalcopyrite + bismuth minerals–(sphalerite)	bismuthinite, emplectite[2], wittichenite, berryite lindströmite, aikinite	copper–molybdenum deposits, gold–quartz–sulfide deposits of Eastern Transbaikalia; some copper–bismuth deposits of Eastern Karamazar (Andrasman, Kaptar-Khona, Dzhuzum) and region of Val d'Annivier
Bismuth–polymetallic	sulfobismuthite sphalerite–chalcopyrite–(fahlore)–bismuth minerals(I)–galena with isomorphous addition of bismuth and/or inclusions of bismuth minerals(II)–bismuth minerals(III)	aikinite, emplectite, wittichenite tetradymite aikinite[3], mamildite[3], tetradymite[3], native bismuth[3], giessenite, emplectite, wittichenite	polymetallic and pyritic–polymetallic, deposits of the world

(continued)

TABLE LXXXII (continued)

Type of bismuth mineralization	Bismuth-bearing mineral association and its component minerals (in order of precipitation)	Bismuth minerals	Examples of deposit
Bismuth—arsenosulfide	bismuthinite—bismuth—arsenosulfide pyrite—(pyrrhotite)—minerals of the arseno-pyrite series—cobaltite with subdispersed gold—(löllingite)—bismuth minerals	bismuthinite, native bismuth, wehrlite, cosalite, galenobismutite [2]	skarn—magnetite—polymetallic deposits of Northwestern Balkhash region (USSR); Moravia (Czechoslovakia); Madan region (Bulgaria); skarn—scheelite deposits of Zeravshan and Altai Ranges; skarn—magnetite—bismuth deposit of Middle Asia (Chokadam-Bulala)
Bismuth—arsenide	bismuth—arsenide native bismuth associated with pyrite, tri-, di-, and monoarsenides of nickel (Ni-skutterudite, rammelsbergite, and niccolite) and löllingite	native bismuth	deposits of 'five-element' (Co—Ni—Ag—Bi—U) association in DDR and Czechoslovakia (Erzgebirge); Canada (in the region of Great Bear Lake and Ontario); USA (Monte Cristo deposit in Arizona)
Bismuth—pyritized	pyrite—bismuth—telluride pyrite—chalcopyrite—bismuth minerals—(sphalerite)—bornite—fahlore	tellurobismuthite, tetradymite	chalcopyrite deposits of the Urals, Caucasus, and Bulgaria, and pyritic—polymetallic deposits of the Altai
Bismuth—bornite—chalcopyritic	chalcopyrite—bornite—sulfobismuthite sphalerite—emplectite—chalcopyrite—bornite—wittichenite	wittichenite [3], emplectite	skarn—polymetallic deposits of Kazakhstan (Akchagyl, Karagaily); copper—skarn deposits of Kazakhstan (Sayak) and Middle Asia (South Yangikan)
Gold—bismuth	gold—bismuth—telluride quartz—carbonates—bismuth minerals—gold minerals	bismuthinite, joseite A tetradymite, tellurobismuthite, joseite B, native bismuth	gold—rare metal deposits of Upper Indigirka region; gold—sulfide, low-temperature deposits of Eastern Transbaikalia (Golgotai), Transcaucasia (Zod), etc.

[1] Minerals not ubiquitously present are placed in parentheses.
[2] Sometimes a breakdown product of more complex minerals.
[3] Product of breakdown or decomposition.

(5) Ionic potential ($=Z/r$) of the elements.
(6) Geochemical subdivision of the elements.
(7) Minero-genetic significance of the geochemical groups.

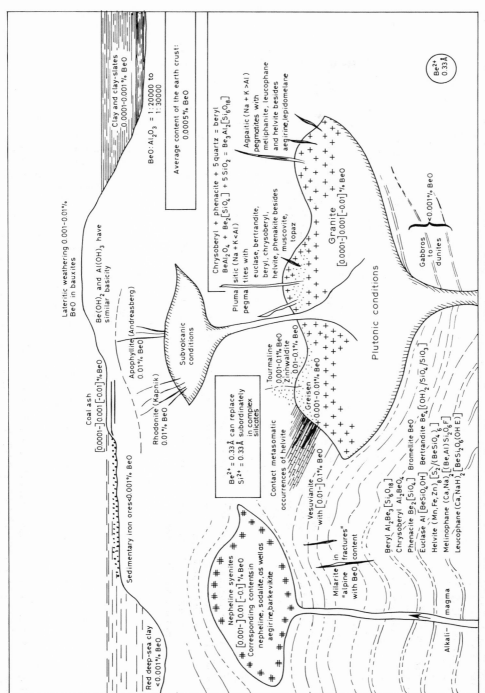

Fig.58. Conceptual model of the geochemical cycle of beryllium. (After Borchert, 1950, unpublished; 1978; courtesy of Glückauf, Essen.)

TABLE LXXXIII

Geological–genetic classification of beryllium deposits (After Smirnov, 1977, vol. III, table 14, pp. 321–323)

Genetic group	Parent rocks					
	Granites, normally leucocratic and with increased alkalinity			Subalkaline granites (granosyenites, quartz syenites, and syenites)		
	Country rocks					
	Acid silicate rocks with increased alkalinity	Carbonate rocks; interstratified carbonate and aluminosilicate rocks; silicate rocks with increased calcium content	Basic and ultramafic rocks	Acid and subalkaline silicate rocks	Carbonate rocks; interstratified carbonate and aluminosilicate rocks	Basic and ultra-mafic rocks

Associations and mineral types of beryllium deposits

Hydro-thermal
I. Association of bertrandite-bearing, hydrothermally altered volcanic and subvolcanic rocks (bertrandite association)
Acro-abyssal, and less frequently hypabyssal (0.5–1.5 km), low-temperature (150–200°C), hydrothermal formations

Quartz–adularia–bertrandite type (in eruptives)

Quartz–bertrandite type (in eruptives)

Hydro-thermal
II. Association of bertrandite–phenakite–fluorite metasomatites (fluorite–bertrandite–phenakite association)
Hypabyssal (1–2 km), mainly medium-temperature (200–350°C), hydrothermal formations

	dolomites	(in granosyenites and alkaline granites)	stones and skarns)	epidote–actinolite schists)
	Fluorite–phenakite type (in limestones) (a) mica–fluorite–phenakite subtype (b) mica–tourmaline–fluorite–chrysoberyl subtype (c) fluorite–tourmaline–cassiterite–phenakite subtype Topaz–fluorite–chrysoberyl type (in dolomites and limestones)		Britolite–thorite–fluorite–phenakite type (in limestones) (a) fluorite–leucophane subtype (b) britolite–chrysoberyl–phenakite subtype (c) thorite–fluorite–phenakite subtype	Feldspar–phenakite type (in andesite-basalts)

Hydrothermal   III. Association of beryl–fluorite–mica metasomatites (fluorite–mica–beryl association)
Medium-depth (2–3 km), medium- and high-temperature (200–400°C) hydrothermal formations

Fluorite–beryl type (a) muscovite–fluorite–beryl subtype (in calc-silicate rocks) (b) biotite–fluorite–beryl subtype (in granodiorites)	Phlogopite–oligoclase–beryl type (in serpentinites and talc schists) Fluorite–phenakite–beryl type

Hydrothermal   IV. Association of beryl–molybdenite–cassiterite–wolframite greisens and quartzose-vein formations
(quartz–molybdenite–wolframite–beryl association)
A. Hypabyssal, shallow-depth, medium-temperature subassociation
Shallow-depth (2–3 km), medium- and high-temperature (200–300°C) hydrothermal formations

Molybdenite–wolframite–cassiterite–beryl type (in granites) .

(continued)

TABLE LXXXIII (continued)

Genetic group	Parent rocks					
	Granites, normally leucocratic and with increased alkinity			Subalkaline granites (granosyenites, quartz syenites, and syenites		
	Country rocks					
	Acid silicate rocks with increased alkalinity	Carbonate rocks; interstratified carbonate and aluminosilicate rocks; silicate rocks with increased calcium content	Basic and ultramafic rocks	Acid and subalkaline silicate rocks	Carbonate rocks; interstratified carbonate and aluminozilicate rocks	Basic and ultra-mafic rocks
Tourmaline–cassiterite–beryl type (in acid alumino-silicate rocks)						

B. High-temperature, medium-depth subassociation
Medium-depth (3–4 km), pneumatolytic–hydrothermal (250–450°C) formations

Wolframite–molybdenite–beryl type (in granites and hornstones)
Bismuthinite–cassiterite–beryl–wolframite type (in granites)
Beryl–molybdenite–wolframite type

matol-
ytic–
hydro-
thermal

Medium-depth (3–4 km), less frequently deep (4–6 km), and shallow-depth (2–3 km), pneumatolytic–hydrothermal formations

Mica–feldspar–beryl type (in gneisses)

Mica–feldspar–genthelvine type
Feldspar–magnetite–phenakite type
Feldspar–gadolinite type

Amphibole–albite–leucophane type

Pneu-
matol-
ytic–
hydro-
thermal

VI. Association of beryllium-bearing skarns
Medium-depth (3–4 km), pneumatolytic–hydrothermal (350–500°C) formations

Magnetite–fluorite–chrysoberyl–helvine skarns (in limestones)
Taffeite–chrysoberyl–hsianghualinite skarns (in dolomites)

Pegma-
titic

VII. Association of beryllium-bearing pegmatites
Deep-seated (4–6 km), pegmatite, high-temperature (350–500°C) formations

Tantalite–beryl pegmatitic type
Beryl–spodumene pegmatitic type
Beryl–spodumene–lepidolite pegmatitic type
Quartz–feldspar–beryl pegmatitic type

Gadolinite pegmatitic type

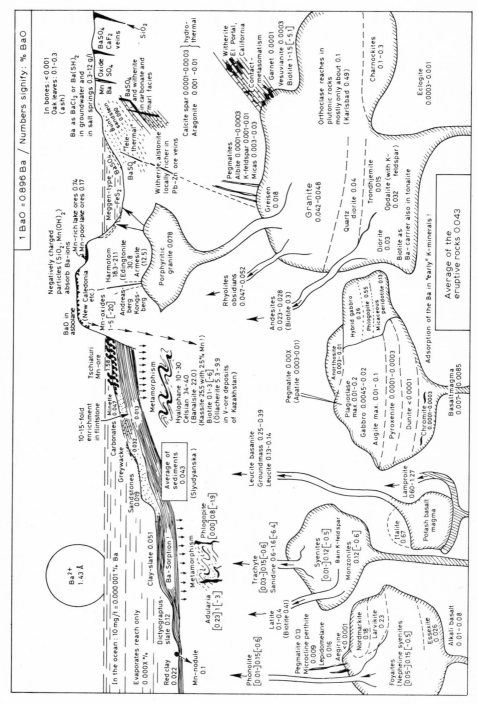

Fig.59. Conceptual model of the geochemical cycle of barium. (After Borchert, 1950, unpublished; 1978; courtesy of Glückauf, Essen.)

179

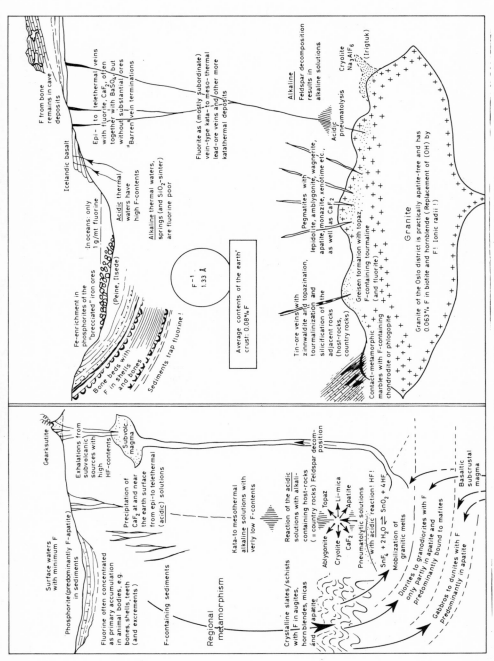

Fig.60. Conceptual model of the geochemical cycle of fluorine. (After Borchert, 1950, unpublished; 1978; courtesy of Glückauf, Essen.)

180

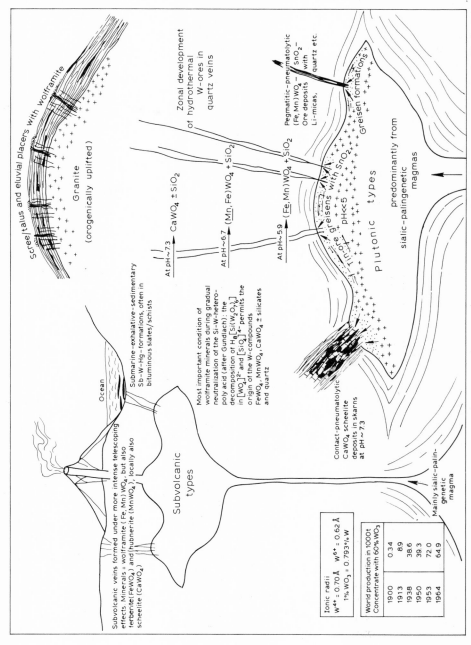

Fig.61a. Genetic overview of the tungsten ore deposits. (After Wiendl and Borchert, 1968, unpublished; 1978; courtesy of Glückauf, Essen.)

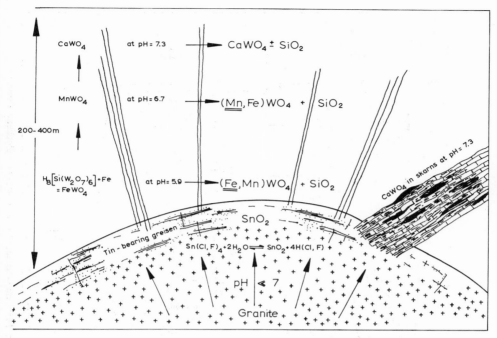

Fig.61b. Zonal sequence of tungsten minerals in the roof section of a granite pluton. (After Borchert and Wiendl, 1967, unpublished; Borchert, 1978; courtesy of Glückauf, Essen.)

(8) Summary of the geochemical distribution of the ore-forming elements and minerals.

(9) Association of metals with magmas.

(10) Indicator elements.

(11) Hydrochemical elements.

(12) Four types of mineralizations in the Cu—Zn—Pb triangle.

(13) Correlation between different Cu—Zn—Pb classifications.

(14) Dispersed elements in minerals of different ore types.

(15) Concentratability of metallic elements by thermal geological mechanisms.

(16) Minerals of pneumatolytic and hydrothermal alteration processes.

(17) Gain and loss of constituents during carbonatization and granitization.

(18) Energy requirements for the recovery of metals at different grades from various sources.

Erickson (1973) (cf. also Booth, 1974) offered data on the connections between the crustal abundance of elements and mineral reserves plus resources — exemplified by several metals. As depicted in Table LXXXIV, the total amounts in the earth's crust, in various crustal segments, in the United States' crust, down to a 1-km depth of the latter, and the world's reserves and resources, are provided. (For the mode of calculation, see the

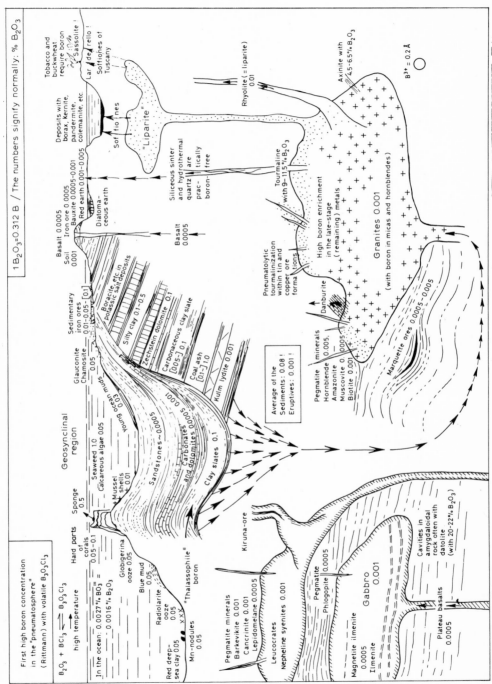

Fig.62. Conceptual model of the geochemical cycle of boron. (After Borchert, 1950, unpublished; 1978; courtesy of Glückauf, Essen.)

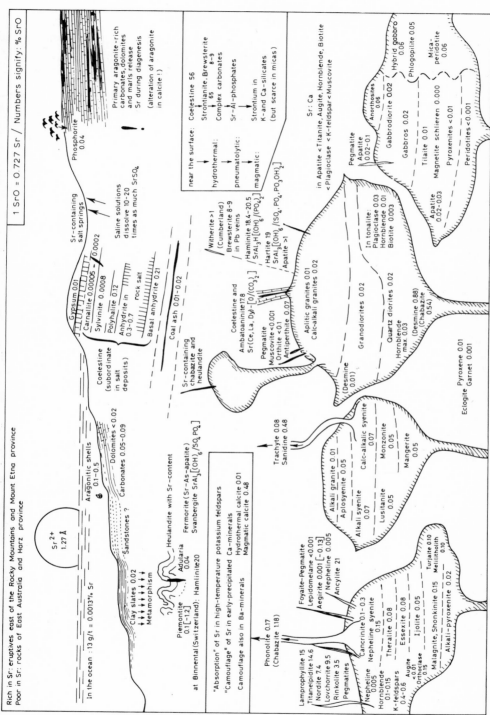

Fig.63. Conceptual model of the geochemical cycle of strontium. (After Borchert, 1950, unpublished; 1978; courtesy of Glückauf, Essen.)

TABLE LXXXIV

Abundance, mass, reserves, and resources of some metals in the earth's crust and in the United States crust (After Erickson, 1973; courtesy of U.S. Geological Survey, Washington)

[Abundance in grams/metric ton (g/mt); mass and reserves in metric tons (mt.). Calculations = mass (metric tons) × abundance (decimalized) = total content of element]

Element	Total earth's crust			
	Goldschmidt [6] (g/mt)	Vinogradov [7] (g/mt)	Lee and Yao [8] (g/mt)	Mt × 10^{12}
Antimony	1	0.5	0.62	14.9
Beryllium	6	3.8	1.3	31.2
Bismuth	0.2	0.009	0.0043	0.1
Cobalt	40	18	25.0	600
Copper	70	47	63	1,510
Gold	0.001	0.0043	0.0035	0.084
Lead	16	16	12	290
Lithium	65	32	21	500
Mercury	0.5	0.083	0.089	2.1
Molybdenum	2.3	1.1	1.3	31.2
Nickel	100	58	89	2,130
Niobium	20	21	19	460
Platinum	0.005	—	0.046	1.1
Selenium	0.09	0.05	0.075	1.8
Silver	0.02	0.07	0.075	1.8
Tantalum	2.1	2.5	1.6	38.4
Tellurium	0.0018	0.001	0.00055	0.013
Thorium	11.5	13	5.8	140
Tin	40	2.5	1.7	40.8
Tungsten	1	1.3	1.1	26.4
Uranium	4	2.5	1.7	40.8
Zinc	80	83	94	2,250
	G/mt	G/mt	G/mt	Mt × 10^{15}
Aluminum	81,300	80,500	83,000	1,990
Barium	430	650	390	9.4
Chromium	200	83	110	2.6
Fluorine	800	660	450	10.8
Iron	50,000	46,500	58,000	1,392
Manganese	1,000	1,100	1,300	31,2
Phosphorus	1,200	930	1,200	28,8
Titanium	4,400	4,500	6,400	153.6
Vanadium	150	91	140	3.36

ceanic crust		Continental crust		Continental crust segments	
				Shield areas	
G/mt	Mt × 10^{12}	8G/mt	Mt × 10^{12}	8G/mt	Mt × 10^{12}
0.91	8.1	0.45	6.8	0.46	4.9
0.83	7.4	1.5	23.8	1.5	16.7
0.0066	0.059	0.0029	0.041	0.003	0.029
·7	330	18	270	19	190
·5	760	50	760	52	550
0.0035	0.032	0.0035	0.052	0.0034	0.035
·0	90	13	200	13	0.140
·0	180	22	320	21	220
0.11	0.9	0.08	1.2	0.078	0.81
1.5	14.6	1.1	16.6	1.1	11.6
·0	1,200	61	920	64	0.680
8	160	20	300	20	210
0.075	0.67	0.028	0.43	0.031	0.30
0.1	0.89	0.059	0.91	0.054	0.64
0.091	0.82	0.065	0.98	0.067	0.70
0.43	3.8	2.3	34.7	2.3	24.3
0.00088	0.0078	0.00036	0.005	0.00038	0.0036
4.2	37	6.8	0.100	6.6	68
1.9	16.8	1.6	24	1.5	16.3
0.94	8.3	1.2	18.1	1.2	12.7
1	7.8	2.2	33	2.1	22.6
·0	1,030	81	1,220	83	870
'mt	Mt × 10^{15}	G/mt	Mt × 10^{15}	G/mt	Mt × 10^{15}
·,000	747	83,000	1,242	84,000	869
370	3.3	400	6.1	400	4.3
160	1.4	77	1.2	81	0.84
420	3.74	470	7.1	470	5
·,000	667	48,000	725	49,000	508
·,800	16	1,000	15.2	1,000	10.6
·,400	12.5	1,200	16.3	1,200	11.4
·,100	72.1	5,300	81.5	5,500	57.1
170	1.51	120	1.85	120	1.3

(continued)

TABLE LXXXIV (continued)

Element	Continental crust segments – Con.				
	Folded belts		United States crust		United States crust to 1-km depth
	^{8}G/mt	Mt $\times$ 10^{12}	^{8}G/mt	Mt $\times$ 10^{12}	Mt $\times$ 10^9
Antimony	0.43	1.9	0.45	0.41	11.2
Beryllium	1.6	7.1	1.5	1.4	38
Bismuth	0.0025	0.012	0.0029	0.0025	0.07
Cobalt	16	80	18	16	440
Copper	46	210	50	45	1,230
Gold	0.0038	0.017	0.0035	0.003	0.085
Lead	13	60	13	12	330
Lithium	23	100	22	20	550
Mercury	0.086	0.39	0.08	0.072	2.0
Molybdenum	1	5	1.1	1	27
Nickel	53	0.240	61	55	1,500
Niobium	19	90	20	20	550
Platinum	0.022	0.13	0.028	0.026	0.71
Selenium	0.071	0.27	0.059	0.055	1.5
Silver	0.062	0.28	0.065	0.059	1.6
Tantalum	2.4	10.4	23	2.1	57.5
Tellurium	0.00031	0.0016	0.00036	0.00031	0.0085
Thorium	7.1	32	6.8	6	0.160
Tin	1.7	7.7	1.6	1.4	38
Tungsten	1.2	5.4	1.2	1.1	30
Uranium	2.3	10.4	2.2	2	55
Zinc	77	350	81	73	2,000
	^{8}G/mt	Mt $\times$ 10^{15}	^{8}G/mt	Mt $\times$ 10^{15}	Mt $\times$ 10^{12}
Aluminium	82,000	373	83,000	74.5	2,000
Barium	390	1.8	400	0.37	10
Chromium	68	0.36	77	0.070	1.92
Fluorine	480	2.1	470	4.30	11.8
Iron	4,000	217	48,000	43.5	1,200
Manganese	930	4.6	1,000	0.9	24.9
Phosphorus	1,100	4.9	1,200	0.98	26.8
Titanium	5,000	24.4	5,300	4.9	1.30
Vanadium	110	0.55	120	0.11	3

ited States				World		
serve [1]	Recover-able resource potential [2]	Ratio of potential to reserve	Reserve [3]	Recover-able resource potential [4]	Ratio of potential to reserve	Grade [5]
$\times 10^6$	$Mt \times 10^6$		$Mt \times 10^6$	$Mt \times 10^6$		
.10	1.1	11	3.6	19	5	Unknown
.073	3.7	50	0.016	64	4,000	
.013	0.007	0.5	0.081	0.12	1.5	
.025	44	1,760	2.14	763	360	
8	122	1.6	200	2,120	10	0.86 percent
.002	0.0086	4.1	0.011	0.15	14	
.8	31.8	1	0.54	550	1,000	3 percent
.7	54	12	0.78	933	1,200	
.013−0.028	0.20	15−6.8	0.11	3.4	30	
.83	2.7	1	2	46.6	23	Unknown
18	149	830	68	2,590	38	1.5 percent
known	49	Unknown	Unknown	848	Unknown	
.00012	0.07	560	0.009	1.2	133	
.025	0.14	6	0.695	2.5	36	
.05	0.16	3.2	0.16	2.75	18	
.0015	5.6	4,000	0.274	97	354	
.0077	0.0009	0.11	0.054	0.015	0.3	
.54	16.7	31	1	288	288	Unknown
	3.9	^{9}?	5.8	68	12	0.6 percent
.079	2.9	37	1.2	51	42	
.27	5.4	20	0.83	93	112	
.6	198	6.3	81	3,400	42	4 percent
$\times 10^6$	$Mt \times 10^6$	Ratio of potential to reserve	$Mt \times 10^6$	$Mt \times 10^9$	Ratio of potential to reserve	
8.1	203,000	24,000	1,160	3,519	3,000	
30.6	980	32	76.4	17	223	
1.8	189	387	696	3.26	47	
4.9	1,151	235	35	20	600	
00	118,000	65	87,000	2,035	23	
1	2,450	2,450	630	42	67	
31	2,940	3	15,000	51	34	
25	13,000	516	117	225	2,000	
0.115	294	2,560	10	5.1	500	

(*continued*)

TABLE LXXXIV (continued)

[1] U.S. Bureau Mines (1970); 1 short ton = 0.91 mt.
[2] Recoverable resource potential = 2.45 A × 10^6 (abundance A expressed in g/mt).
[3] U.S. Bureau Mines (1970); 1 short ton = 0.91 mt; does not include United States reserve.
[4] Recoverable resource potential = 2.45 A × 17.3 × 10^6 (abundance A expressed in g/mt; land area of world is 17.3 times United States land area).
[5] U.S. Bureau Mines (1970); data on world basis.
[6] Goldschmidt (1954, p. 74–75).
[7] Vinogradov (1962, p. 649–650).
[8] Lee and Yao (1970, p. 778–786). All calculations are based on this work.
[9] Very high.

Calculation of mass of crustal segments

Total earth's crust	24 × 10^{18} mt.
Oceanic crust	8.9 × 10^{18} mt (37 percent of total crust).
Continental crust	15.1 × 10^{18} mt (63 percent of total crust).
Shield areas	10.6 × 10^{18} mt (30 percent of continental crust or 43.8 percent of total crust).
Folded belts	4.54 × 10^{18} mt (30 percent of continental crust or 19.1 percent of total crust).
United States crust	0.90 × 10^{18} mt (based upon United States as 1/17 of land area of world's continental crust).
United States crust to 1-km depth	24.6 × 10^{15} mt (based upon average thickness continental crust = 36.5 km; therefore 1 km is 2.74 percent of United States crust).

original publication.) The metals, with known reserves most closely near the estimated potential recoverable resource, are the metals that have been most intensely explored for the longest time (e.g., Pb, Cu, Zn, Ag, Au, Mo). On the other hand, nickel and tin are two metals of which the reserves are known to be very much less than the calculated potential recoverable resources, and are located in geological settings uncommon in the United States. Also, the metals with "reserve—potential resource" relationships are intermediate between the crustal abundance and reserve abundance, which have up to recently either not been vital to the economy, or their prices fluctuated greatly, or they were available chiefly as industrial by-products. Erickson concluded that the scheme of predicting resource potentials in relation to crustal abundance is valid, i.e., if one searches for an element hard enough, one will discover it eventually in approximately the quantities as expected. He stresses, however, that resources whose feasibility of economic recovery is not known, are not included in this calculation (p.21): " . . . recent discoveries of several *new* types of mineral deposits indicate that we need to critically examine our criteria for where and how to look for mineral deposits. Certainly, significant mineral deposits remain undiscovered because exploration efforts are commonly confined to the classic environment of ore deposition."

Although the work summarized by Erickson (1973) is partly confined to the United States, and the data was obtained principally from English-language publications,

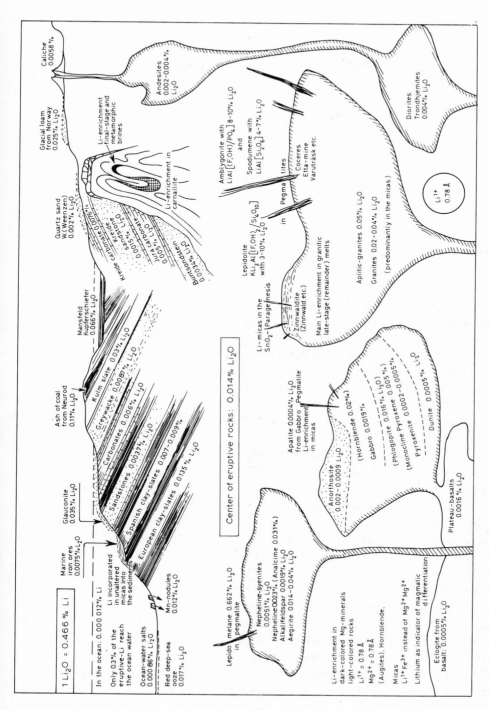

Fig.64. Conceptual model of the geochemical cycle of lithium. (After Borchert, 1950, unpublished; 1978; courtesy of Glückauf, Essen.)

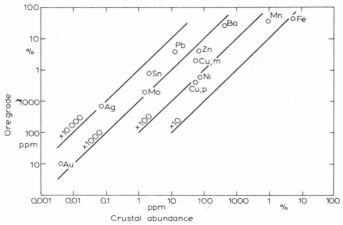

Fig.65. Ore grades versus crustal abundance of selected elements. (After Garrett, 1977; courtesy of *J. Int. Assoc. Math. Geol.*)

modifications and expansions are to be expected. An additional exemplar has been provided by Garrett (1977) (Fig.65), who stated that an ore deposit can be considered as a point or locality at which one or more elements have been concentrated many times its average crustal abundance, thus permitting economic extraction and benefication. The "concentration factor" varies from element to element (excluding here, the *minimum* concentration required to make the deposit economically viable at a *particular time* and for a *specific* set of conditions, which has been discussed by Garrett). Fig.65 serves as an example of the range of concentrations required to transform the average crustal distribution to an average ore grade. Non-ferrous and rare elements require a concentration factor of about 100–10,000.

All industrially used chemical elements as well as minerals could be plotted similarly, although in considering "ore grades" one has to account for "all the sociological, political, economic, and technological factors . . . " (Garrett, 1977, p.248).

The status of the controversy regarding the remaining supplies of mineral resources has been examined by Govett and Govett (1972), and they stated that all exploitable mineralizations within the presently mineable thickness of the earth's crust have not been discovered as yet, and that it is difficult to establish within reasonable limits the probable additional reserves to be expected. They calculated the world reserves for the common elements plotted in Fig.66, and their mathematical treatment of the data indicated that nine of the thirteen elements have a linear relation. This correlation is useful only if a reasonable upper limit can be established. The lower the acceptable grade, the higher the apparent reserve will be (based on both economic and technological variables). This uncertainty related to reserves is exemplified by chromite for the years 1948, 1965 and 1970. Although the enormous reserve increase of iron during twenty years has been cited as a possibility for other elements, this is not justified geologically because lead, copper

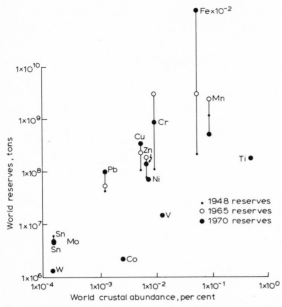

Fig.66. Relationship between total world reserves and crustal abundances of some elements. (After Govett and Govett, 1972.)

and zinc reserves will hardly be increased similarly. The geochemical behaviour and thus the concentration processes in the crust (and mantle?) are different for various elements. It is interesting to follow Govett and his co-worker's reasoning in pointing out that the 1948-reserves of iron were considerably *below* the calculated trend – and that based on these data major additional discoveries should have been predicted. Tungsten, cobalt and vanadium also lie considerably below the established trend.

Fig.66 (in combination with the mathematical calculations) suggests that at the present level of mining and extractive technology, the reserves will probably not be increased by more than a factor of five. Although this makes little difference to the depletion dates, as calculated by Govett and co-worker, it predicts an increase in exploration success: for every copper deposit now known, five new ones of the same size and grade will be found. (However, see their qualifying comments on the importance of "economic grade" which has changed in some commodities from higher to "low-grade" deposits. The ultimate limit is the mining of "whole rock" for all elements!)

One of the key points to be remembered is the relatively shallow depth to which mining has penetrated the earth's crust. There is no absolute shortage of any one element, and any crises that may exist in the future is related to the limitations of our exploration (geological, geophysical, geochemical) and exploitation (economic, engineering, metallurgical) methodologies.

The chemical behaviour or properties of elements (especially when in an ionic state)

192

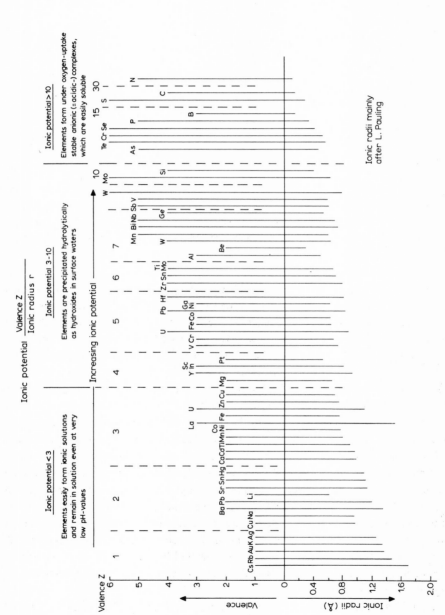

Fig.67. Linear diagram of the ionic potential of the most important elements present in rocks and ore deposits, divided into three groups according to their characteristic chemical behaviour. (After Borchert and Kleinevoss, 1969, unpublished; Borchert, 1978; courtesy of Glückauf, Essen.)

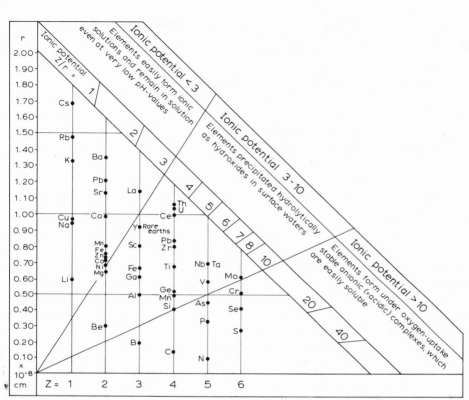

Fig.68. Triangular diagram of the ionic potential of the most important elements present in rocks and ore deposits, divided into three groups according to their characteristic chemical behaviour. (After V.M. Goldschmidt, in Schneiderhöhn, 1962; courtesy of Gustav Fischer, Stuttgart.)

have been depicted by the ionic potential, which is valence (Z) divided by the ionic radius (r). Fig.67 gives the Z and r values of the more common elements in rocks and ore deposits; Z above the zero-line and r below it. Another mode of presentation, familiar from Schneiderhöhn's book (1962, fig.10, p.187) as well as from various English-language publications (e.g., Mason, 1958), is the triangular diagram with Z and r plotted as the horizontal and vertical axes, respectively (Fig.68). From the intersection of these axes, i.e., the zero point, lines radiate out to constitute the ionic potential values from 1 to 40. In Fig.67, however, these values increase to the right to about 30. In both diagrams, three fields are recognized; ionic potentials <3, 3 to 10 and >10, respectively, representing (1) elements easily forming ionic solutions and remaining in this ionic form even under very low pH-values, (2) elements which are precipitated hydrolytically as hydroxides under near-surface natural waters; and (3) elements forming under conditions of oxygen-uptake stable anionic (=acidic) complexes, which are easily soluble.

Here again, as in earlier sections, in considering the distribution of elements — either in pure form or as compound/mineral constituents — one has to take into account that some elements are part of a geochemical cycle or cycles, and can become "concentrated", in contrast to those elements that become "dispersed" either within the earth's crust or near or along the earth's surface. The physicochemical characteristics of the elements, as theoretically calculated or experimentally determined in the laboratory, are sometimes insufficient in estimating the elements' behaviours in nature with its numerous environmental variables or parameters acting simultaneously. Pragmatically collected data are indispensible to supplement the theoretically and experimentally deduced information — if there are discrepencies a plausible explanation has to be offered. Some pragmatic data are to be found in the following — in most cases easily explainable by the physicochemical properties of the elements.

Tables LXXXV and LXXXVI present the chemical elements according to their tendency to be associated with specific, naturally occurring compounds and/or minerals. (Although these two tables were quoted from Baumann and Tischendorf, 1976, similar data have been presented in many other publications.)

A particularly useful mode of presenting the distribution of elements, as mineral constitutents versus geological environment and host rock, has been provided by Cissarz (1951) (Figs.69 and 70). In this regard, the interested reader may wish to refer also to the tabular classification by Routhier (1963, vol. 2), where each element has been treated according to natural chemical or geologic environmental/lithological association, thus providing a comprehensive classification of ore deposits with "type"-exemplars.

McKelvey (1973) has similarly discussed at length the problem related to the potential resources from several view-points prior to turning to the geological side of the situation by examining the extent of *undiscovered* reserves and resources. He stressed estimates of the amounts of numerous types of ores associated with various geological environments, because such data offer present and future exploration targets. (See McKelvey's papers for other methods less likely to succeed.)

The "area" and "volume" method is described by McKelvey as one exploration philosophy that has paid many dividends to both the metal-mining and oil industries, of which he cites several published examples. This approach of estimating undiscovered resources involves the extrapolation of mineral-abundance data from explored to unexplored localities, reasoning that the area or volume of similarly favourable lithologies should contain similar amounts of economic deposits. A related technique is founded on calculations that establish the amount of drilling required to explore sufficiently a favourable area as indicated by the reserves already discovered by drilling (several case histories and theoretical discussions are available from the literature).

McKelvey offered as a variant of the area-method for estimating reserves of non-fuel minerals an approach based on the observation that the tonnage of mineable reserves of well-explored elements in the United States is approximately equivalent to their crustal abundance in percent times a billion or 10 billion, as shown in Fig.71. (Note: the American "billion" equals a thousand million $= 10^9$.) This relationship depends on the extent of past exploration, for only reserves of the long-sought and well-explored commodities

TABLE LXXXV

Grouping of elements according to geochemical characteristics (After Baumann and Tischendorf, 1976)

Geochemical character	Decreasing tendency $\longrightarrow$													
Atmophile	noble gases	H_2	O_2	N_2	O_2	H_2	CO_2	Hg	CH_4	NO_2	SO_2	Cl	Br	J
Hydrophile	H_2	O_2	Cl 114 000	Br 32 000	S 2560	N 815	Na 430	B 370	J 160	C 122	Mg 69	Sr 24	K 15	Ca 14
Biophile	C 780	H_2	N_2 240	O_2	P 7.5	Cl 6.4	S 5.4	J 5	B 3	V	Br			
Litho- phile / sedimento- phile	O_2	C 43	N 36	Se 12	B 8.3	S 8.1	Hg 4.8	As 3.9	Br 2.9	Li 1.9	Ga 1.6	V 1.4		
granito- phile	U 1.6	Th 1.39	K 1.33	Cs 1.30	Li 1.25	Rb 1.20	Sn 1.20	Be 1.19	W 1.16	F 1.14	Pb 1.13	Si 1.11		
inter- mediate	Na 1.09	Ge 1.07	Zr, Hf 1.06	O_2 1.04	Y, La 1.02	B 1.0	Nb, Ta 1.0	Al 0.96	Ba 0.92	Mo 0.91	Sr 0.88	P 0.78	Zn 0.72	
basalto- phile	Ni 2.8	Co 2.5	Cr 2.4	Mg 2.4	Ca 2.3	V 2.2	Cu 2.1	Ti 2.0	Fe 1.85	Mn 1.75	Sc 1.67	Zn 1.57		
Sulfophile	Ni 540	Cu 232	Bi 220	Cd 154	Ag 143	Co 116	Zn 102	Re 29	Mo 18	Hg 13	Fe 11	Pb 6	In	Tl
Chalcogenic	S	Se	Te	As	Sb	Bi								
Siderophile	Fe	Co	Ni	Mo, Re	platinoids	Au								

(Arrows in the original connect Si 1.11 → P 0.78 / Zn 0.72 and Zn 1.57 → Zn 0.72.)

TABLE LXXXVI

The minerogenetic significance of the geochemical groups of elements (After Baumann and Tischendorf, 1976)

Element	Properties and chemical characteristics	Minerogenetic significance
Atmophile	non-metals, volatile easily soluble	mineralizers, i.e. carriers of less-soluble components in gases and fluids
Hydrophile	non-metals, alkalies and alkali earth metals, easily soluble	main constituents of aqueous solutions, a prerequisite for the transportation of elements necessary for the subsequent concentration of these elements, concentration in exogenic late-stage solutions (saline and subsaline)
Biophile	non-metals	main components of organic substances, which actively influence the enrichment of numerous elements (especially non-ferrous metals).
Sedimento-phile	predominant non-metals, easily ab-/adsorbable, biophile	enrichment in "black slates" among other types of sediments, thereby causing sedimentary geochemical specialization
Granitophile	non-ferrous and rare metals, alkalies	enrichment in epigenetic deposits from late-stage (remaining) solutions of acidic specialized magmatic melts
Intermediate	accessory/trace elements alkali earth metals	enrichment of epigenetic often pegmatoid deposits from late-stage (remaining) solutions of syenitic specialized magmatic melts.
Basaltophile	elements of the ferrous group, non-ferrous group, alkali earth metals	enrichment in syngenetic deposits by gravitational differentiation or liquid segregation of basic and ultra-basic, specialized magmatic melts
Sulfophile	non-ferrous and rare metals	enrichment under reducing condition and high sulfide availability (intramagmatic, hydrothermal, sedimentogenic)
Siderophile	elements of the ferrous group, precious metals	enrichment presumably in the earth's core and during unnatural (smelting) processes

display the correlation to crustal abundance. This feature, however, lends the methods its practicality in making potential-resource estimates of elements that have been explored for only a relatively short period.

This reserve—abundance interrelationship is only an approximation, because crustal abundance is merely one variable that determines each element's concentration. Nevertheless, as demonstrated by examples of metal distributions empirically established, McKelvey provided a connection between crustal abundance and the latter's influence on both the magnitude of reserves and the geochemical mode of occurrence. "For example, of the 18 or so elements with crustal abundances greater than about 200 parts per mil-

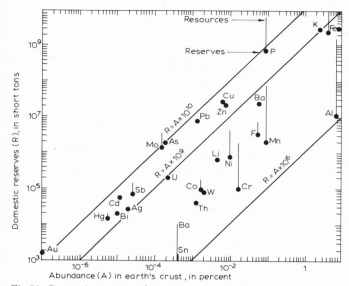

Fig.71. Domestic reserves of elements compared to their abundance in the earth's crust. Tonnage of ore minable now is shown by a dot; tonnage of lower-grade ores whose exploitation depends upon future technological advances or higher prices is shown by a bar. (After McKelvey, 1973; courtesy of U.S. Geological Survey, Washington.)

lion, all but fluorine and strontium are rock forming in the sense that some extensive rocks are composed chiefly of minerals of which each of these elements is a major constituent. Of the less abundant elements, only chromium, nitrogen, and boron have this distinction. Only a few other elements, such as copper, lead, and zinc, even form ore bodies composed mainly of minerals of which the valuable element is a major constituent, and in a general way the grade of mineable ores decreases with decreasing crustal abundances. A similar gross correlation exists between abundance of the elements and the number of minerals in which they are a significant constituent."

McKelvey continued his elegant arguments and requested that we continue to refine our geological methods, e.g., study the effects of the natural fractionation of chemical elements (like the one described and illustrated by the reserves-versus-abundance data), as the information helps to comprehend the geochemical cycles, and through this eventually the precise mechanism of concentration. In turn, such understanding assists in estimating resources in various size- and grade-categories. At present our knowledge of the genesis, mode of occurrence, and totally integrated geological setting of many metalliferous minerals is still insufficient for meaningful predictions. On the other hand, enough data have been accumulated of certain types of deposits to ascertain some success in applying predictive techniques, so that practical "model studies" are now feasible.

Returning to immediate considerations of mineralizations and their component

TABLE LXXXVII

Association of metals with magmas (After Thomas, 1973)

Magma type	Rock type	Associated metals
Ultrabasic	peridotite	platinum, chromium, nickel, diamond, asbestos, corundum, iron
	pyroxenite	nickel, copper, iron, titanium
	anorthosite	titanium, iron
Basic	gabbro	iron, titanium, copper
	norite	platinum, palladium, nickel, copper
Intermediate	quartz-diorite	gold, silver, lead, zinc, iron, copper, arsenic
	monzonite-diorite	titanium, iron, copper, gold, arsenic
	granodiorite	mercury, antimony, gold, iron, silver, lead, zinc, copper, arsenic, molybdenum, tungsten, bismuth, tin
	syenite	iron, molybdenum, zinc
	nepheline-syenite	iron, titanium, zinc, gold, copper, zirconium
Acidic	granite	silver, cobalt, nickel, antimony, uranium, tellurium, tungsten, lead, zinc, gold, tin, molybdenum, iron, arsenic, bismuth, copper
	pegmatite	molybdenum, copper, tin, tungsten, bismuth, beryllium, rare earths, uranium, lithium, iron, tantalum, columbium, titanium

TABLE LXXXVIII

Indicator elements of mineral deposits with different compositions (After Beus and Grigorian, 1977)

Type of deposit	Major indicator elements identifiable by the semi-quantitative emission spectrographic method	Additional indicator elements identifiable by special methods
Copper−nickel in basic and ultrabasic rocks	Cu, Ni, Co, Zn	S
Rare-metal aplogranites and pegmatites	Be, Sn, Nb, Ta, W	Li, Cs, Rb, F
Tin−tungsten of quartz−greisen origin	Be, Sn, W, Mo, Bi	Li, Rb, F
Rare-metal carbonatites	Nb, Zr	U, P
Tungsten−molybdenum in skarns	W, Mo, Sn, Cu, Be, Bi	−
Bismuth in skarns	Bi, Cu, Pb, Zn	−
Tin-sulfide	Sn, Sb, Pb, Cu, Ag, Zn	−
Polymetallic	Pb, Zn, Cu, Ag, Ba	−
Gold ore	Sb, As, Ag	Au
Porphyry copper	Cu, Zn, Mo, W, As, Sb	−
Copper	Cu, Pb, Zn, Ag, Mo	−
Hydrothermal uranium	Pb, Mo, Zn, Ag, Cu	U
Antimony−mercury and mercury	As, Sb, Ba	Hg
Celestite	Sr	−
Phosphorites	−	P
General list	Be, Sn, Mo, W, Nb, Zr, Bi, Ta, Ni, Co, Cu, Pb, Zn, Ag Sb, As, Ba, Sr	Li, Cs, Rb, Au, Hg, U, F, S, P

constituents, Table LXXXVII (Thomas, 1973) lists the four magmatic types which gave rise to twelve host lithologies and their respective associated metals (similar data have been provided in different format elsewhere in this chapter). In geochemical exploration for metallic deposits, socalled "indicator-elements" (Boyle, 1974a), which are commonly associated with the sought-for mineralization(s), are utilized to locate ores (e.g., Sb—As anomalies may indicate presence of gold ore; and Hg-vapour-techniques have been employed in general exploration). Table LXXXVIII outlines the major and some additional indicator-elements, whereas Table LXXXIX summarizes the two groups of hydrogeochemical indicator-elements (strongly and weakly oxidizing).

In the above tables of elements, no attention has been given to their precise mode of occurrence, e.g., the minerals with which the elements are associated. The questions can be raised: are the trace elements within major ore-forming minerals, or have they formed trace amounts of rarer minerals; are they absorbed on the surfaces of material, etc? It appears that the classifications for routine geochemical prospecting are often insufficient for detailed studies and need to be supplemented, if not superseded, by meticulously researched data, such as exemplified in Table XC. To accomplish the latter, *specific*

TABLE LXXXIX

Hydrogeochemical indicator elements for different types of deposits (After Boyle, 1974)

Type of deposit	Indicator elements of ore bodies	
	strongly oxidizing	weakly oxidizing
Copper—pyrite	Cu, Zn, Pb, As, Ni, Co, F, Cd, Se, Ge, Au, Ag, Sb	Zn, Pb, Mo, As, Ge, Se, F
Polymetallic	Pb, Zn, Cu, As, Mo, Ni, Ag, Cd, Sb, Se, Ge	Pb, Zn, As, Mo, Ni
Molybdenum	Mo, W, Pb, Cu, Zn, Be, F, Co, Ni, Mn	Mo, Pb, Zn, F, As, Li
Tungsten—beryllium	W, Mo, Zn, Cu, As, F, Li, Be, Rb	W, Mo, F, Li
Mercury—antimony	Hg, Sb, As, Zn, F, B, Se, Cu	Ag, As, Zn, B, F
Gold ore	Au, Ag, Sb, As, Mo, Se, Pb, Cu, Zn, Ni, Co	Ag, Sb, As, Mo, Zn
Tin ore	Sn, Nb, Pb, Cu, Zn, Li, F, Be	Sn, Li, F, Be, Zn
Titaniferous magnetite	Ti, Fe, Ni, Co, Cr	Ni, Fe
Spodumene	Li, Rb, Cs, Mn, Pb, Nb, Sr, F, Ga	Li, Rb, Cs, F
Copper—nickel	Ni, Cu, Zn, Co, Ag, Ba, Sn, Pb, U	Ni, Zn, Ag, Sn, Ba
Beryllium—fluorite	Be, F, Li, Rb, W	Be, F, Li
Baritic—polymetallic	Ba, Sr, Cu, Zn, Pb, As, Mo	Be, Sr, As, Mo

TABLE XC

Average amounts of dispersed elements in minerals of different types of ore deposits ($\bar{x}_i$ in ppm) (Smirnov, 1977, vol. III)

Type of deposit	Mineral	Re	Se	Te	In	Cd	Tl	Ga	Sc
Magmatic, pegmatitic, and greisen deposits									
Magmatic copper−nickel	chalcopyrite	<0.1	60	11	5	−	−	−	−
Magmatic titano-magnetite	titano-magnetite	−	−	−	−	−	−	−	20
Pegmatitic rare-metal	cassiterite	−	−	−	3	−	−	50	160
Greisen rare-metal	cassiterite	−	−	−	15	−	−	75	390
	wolframite	−	−	−	6	−	−	−	121
	molybdenite	51	−	−	−	−	−	−	−
Skarn, hydrothermal, pyritic, and other deposits									
Skarn, galena−sphaleritic	galena	−	124	39	−	−	2.7	−	−
	sphalerite	−	−	−	51	4720	−	26	−
Cassiterite−sulfide	cassiterite	−	−	−	18	−	−	−	10
	sphalerite	−	−	−	1470	3450	−	7	−
	chalcopyrite	−	−	−	300	−	−	−	−
Molybdenite−wolframite−sulfide	molybdenite	40	−	−	−	−	−	−	−
Molybdenite−chalcopy-rite−porphyric	molybdenite	660	−	−	−	−	−	−	−
	chalcopyrite	1.6	153	44	−	−	−	−	−
	pyrite	1.3	60	24	−	−	−	−	−
Galena−sphalerite in altered silicate rocks	galena	−	35	16	−	−	1.4	−	−
	sphalerite	−	−	−	116	2900	−	10	−
Galena−sphalerite in altered carbonate rocks	galena	−	3	1.8	−	−	20	−	−
	sphalerite	−	−	−	50	2000	−	34	−
Pyritic galena−sphalerite−chalcopyrite	galena	−	550	85	−	−	14	−	−
	sphalerite	−	−	−	6	2040	8	18	−
	chalcopyrite	−	150	15	14	20	5	−	−
	pyrite	−	90	12	−	−	5	−	−
Stratiform galena−sphalerite	galena	−	28	6	−	−	8.5	−	−
	sphalerite	−	−	−	9	2500	45	90	−
	pyrite	−	−	−	−	−	23	−	−
Antimonite−cinnabar	cinnabar	−	1312	−	−	−	2	−	−
	antimonite	−	125	−	−	−	3	−	−
Cinnabar	cinnabar	−	478	−	−	−	6	−	−
Antimonite	antimonite	−	34	−	−	−	3	−	−
Copper−pyritic	chalcopyrite	−	91	64	11	20	5	−	−
	sphalerite	−	−	−	23	1600	8	−	−
	pyrite	0.1	70	29	−	−	5	−	−

single minerals were obtained from *particular genetic* deposits and analyzed to determine their minor- and trace-element contents for a number of practical (e.g., possible by-products and influence on metallurgical properties of the ore) and theoretical (e.g., variation of trace elements in pyrite provides clues to mode of origin) purposes.

The three-element ratios (plotted in triangles) of the major elements in polymetallic deposits have also been used to establish certain descriptive and genetic ore varieties (cf. Voke's Chapter 4, Vol. 6, fig.6; Jung and Knitzschke's Chapter 7, Vol. 6, fig.12; Sangster and Scott's Chapter 5, Vol. 6, fig.26; and Stanton, 1972, figs.15-12, 15-15, 16-17 and 17-12). The classificatory scheme provided by Saager (1967) consists of four types of sulfide ores (i.e., his "Kieslagerstätten") in a Cu—Pb—Zn triangle (Fig.72), which are then compared in Table XCI with the divisions of other researchers.

Saager suggested the following types: (1) the Mofjell-type deposits which are banded pyritic Zn—Pb ores probably formed by submarine exhalative-sedimentary processes in a geosynclinal milieu; (2) the Tretthammer-type deposits which are less regular and poorly banded pyrrhotite Cu—Zn ores presumed to have a similar origin as the ore type (1); and (3) the Hauknestind-type accumulations that are epigenetic pyrrhotite Zn—Pb replacement ores in marble beds and tectonic breccia zones formed from originally primary syngenetic sulfide ores, mobilized and emplaced in their position during the later stage of orogenesis. According to his non-genetic, descriptive classification, Saager recognized the pyrite-type which corresponds to (1) above; the pyrite—pyrrhotite- and Cu—pyrrhotite-types, as represented by (2); and the Zn—pyrrhotite-type of which No. 3 above is an example. These three types of ore mineralizations (a fourth one is a sub-type) have been plotted in a Cu—Zn—Pb triangle (Fig.72) and collectively form an area near the Zn-pole (compare with fig.15-12, p.520 in Stanton, 1972). The unsymmetrical distribution along the Zn—Cu side of the triangle is characteristic of the natural Cu—Zn—Pb occurrences. Zn occurs in all deposits and, in general, predominates among the three metals —

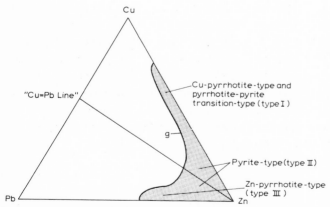

Fig.72. The position of the four types of sulfide ore deposits in the Cu—Zn—Pb—triangle and in the sulfide ore field. (After Saager, 1967; courtesy of *Nor. Geol. Tidsskr.*)

TABLE XCI

Correlation of the various classifications based on Zn–Pb–Cu composition (After Saager, 1967)

	Zn ≫ Pb > Cu	Cu ≧ Zn > Pb		Zn > Pb ≫ Cu
Main ore	pyritic ore	pyritic ore and chalcopyritic-pyrrhotitic ores	chalcopyritic-pyrrhotitic ores	sphaleritic-pyrrhotitic ores
Main ore minerals	pyrite	pyrite and pyrrhotite	pyrrhotite	pyrrhotite and sphalerite
Texture	banded	poorly banded-compact		often brecciated
Saager	pyritic type	pyrite-pyrrhotite transition type	Cu-pyrrhotite type	Zn-pyrrhotite type
Mofjell-area	Mofjell type	Tretthammern type		Hauknestind type
Vokes	pyritic type		pyrrhotitic type	
Carstens	Rödhammer type			
	Leksdal type			
		Röros type		
	(Flöttum type?)			Flöttum type
Foslie	Lökken-Grong subtype	Röros-Sulitjelma subtype		
Stanton	banded pyritic Zn–Pb–Cu ores		chalcopyrite-pyrrhotite ores	

except for the extreme Cu-rich ores of the Cu–pyrrhotite and pyrrhotite–pyrite transitional type (type I) in which Cu is somewhat more abundant than Zn. Also, high Pb-contents are also associated with low Cu-contents, and vice versa. Ore occurrences with more than 20% Cu and 20% Pb have not been recorded, hence the absence of plots near the Cu-pole. The antipathetic relationship between Cu and Pb is clearly expressed by the boundary-line "g", which forms a curved line between the Cu-rich (type I) and the Pb-rich (type III) ores. At the intersection between the Cu–Pb line, the curve shows a distinct embayment towards the Zn-corner and cuts the Cu–Pb line at about 60%.

Based on the above, as well as on the Cu-, Zn- and Pb-data discussed by Saager, the "Kieslagerstätten" have the following features:

if $Cu > 20\% \rightarrow Pb \leqslant 20\%$,

if $Pb > 20\% \rightarrow Cu \leqslant 20\%$,

and fall into the area delineated by $Zn > 30\%$, $Cu < 70\%$, $Pb < 40\%$, as shown by the

shaded area in the triangle in Fig.72. In Table XCI, Saager summarized the different subdivisions of the Norwegian sulphide deposits and demonstrated that he has offered a more
comprehensive approach.

It has been stated loosely that "ore" can be formed by any mineral (or element)
that is sufficiently concentrated to be viably mined (this *not* to be taken as the definition
of ore!) — which leads one to the problems related to the numerous possible mechanisms
resulting in mineral concentrations. This is partly implicit in many tables and figures
where no reference to "ore-grade" has been made (see below for further details). Let us
examine a number of publications that have discussed processes of elemental and mineral
congregations or assemblages.

Sullivan (1970) presented in Fig. 73 the relationships between the composition of
igneous parent-rock and the metal content (log-scale) of associated ore bodies: (1) Ni
occurs from peridotite to gabbro as concentrations of only 10–100 times; (2) Cu from
gabbro to monzonite in the range of 100–1000 times concentration; and (3) As, Sb and

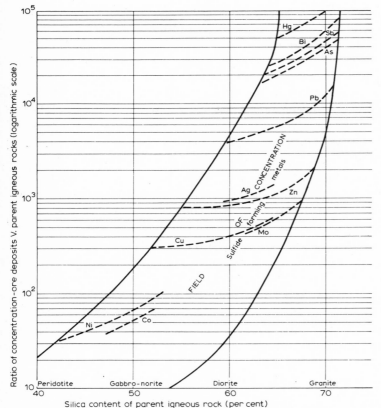

Fig.73. Concentratability of metallic elements by thermal geological processes. (After Sullivan, 1970;
courtesy of *Can. Min. Metall.*)

Hg, of which the ratio of concentration may range from 10^3 to 10^5. The order of concentratability of metals by igneous thermal processes corresponds to the well-known zoning sequences — the higher concentrations occur among the metals that are mobile at the lowest temperatures. According to Sullivan, contrary to popular assumptions, there is no relationship between the concentration and thermal zoning and the metals' relative solubilities.

Fig.73 is useful in determining the *total* amount of metal within a specific district with a certain amount of particular rock types; the data in the figure provide the cut-off point for exploration when areas have been well-mined and have supplied a known amount of metals. The concepts of "ultimate reserves" and "relative discovery potential" were discussed which are intimately related to metallogeny and exploration philosophy, as exemplified by Sullivan in several instances (see his sections on Cu, Zn, Ag, Sn, Mo, U and Au).

Sullivan maintains that "any metal which occurs in veins can also occur in disseminated form — in some cases in rocks which do not look like hosts for mineral deposits" (pp.773–774, 783). This statement needs some explanation, for (1) there are at least two genetic mechanisms of vein formation as represented by two concepts described in the literature, and (2) these two hypotheses must be further tested in the future to establish not only their general validity, but the precise geological conditions under which they are applicable. The results will influence exploration thinking.

The first vein type is formed by the precipitation of igneous or other ore (±gangue) material from a variety of hydrothermal fluids that moved over a considerable distance into a fractured rock. The second type of veins is produced by "lateral secretion" processes, the components having been derived from the immediate host or country rock(s) in response to compaction, tectonic squeezing, metamorphism and others that consequented remobilization and movements of metal-bearing solutions into open spaces. The *most important* difference in genesis of the two types of veins lies, therefore, in the *source* of the metals: in one case it was relatively remote, derived possibly from differentiated igneous hot gases and fluids (=igneous-hydrothermal solutions); i.e., "syngenetic–igneous" processes produced the metalliferous solutions from a hot igneous mass. This is the classical theory. The second vein variety is subdivisible into two, namely (1) "diagenetic–sedimentary" and (2) "epigenetic–lateral secretion" in origin. Both of these vein types derive their constituents from their host rock that is either (a) unconsolidated or poorly lithified so that mere compaction can mobilize metal-bearing fluids, or (b) from a solid rock that undergoes leaching and consequent concentration and removal of their trace/minor components by circulating ground/connate waters; the latter precipitating the ore and gangue into earlier-formed spaces. No further details can be provided here of this more complex phenomenon (cf. Wolf, 1976c for discussions); some arguments were given here merely in support of Sullivans statement that "any of the metal which occurs in veins can also occur in disseminated form ... " Is this necessarily true for the veins formed from igneous hydrothermal solutions? The reply is neither a simple "yes" nor an easy "no". It is "yes" only in so far as magmatic/volcanic processes give rise to both solid rocks that may contain metal disseminations upon crystallization of a magma and through differentiation to the metal-bearing hydrothermal fluids that can subsequently precipitate minerals into fractures. The reply is "no", however, because the presence of a solid igneous mass with trace metals does not necessarily ensure that hydrothermal solutions were originally produced.

Several practical questions arise as to the connections between host/country rocks, on one hand, and the origin of veins of type one, on the other.

(1) What are the conditions under which metals are released from host rocks through metamorphism, for example, to form veins, in contrast to the conditions that result in the disseminations being left unaffected?

(2) What are the investigative procedures that can be utilized to determine the type(s) of host lithologies which supplied material (anions and cations) that was subsequently concentrated in veins?

(3) If an answer can be supplied to point (2), can certain quantitative procedures be developed (beyond those described by Sullivan, 1970, and others) to calculate the amount of material released from the country rock and, therefore, the quantity of originally disseminated constituents? If a high proportion has been precipitated in veins, there would be no economic incentive to prospect for disseminated metals in this particular case and locality.

In particular the granitoid group of rocks and ores are associated by pneumatolytic and/or hydrothermal mineral concentrations, which have been treated elsewhere in this chapter. However, the alterations of the country/host rocks have to be well understood in the exploration for these metasomatic-type mineralizations. Consequently, the diagrammatic model of Schneiderhöhn's (1962, fig.8, p.109) can be considered as another example of an element/mineral classification; note his pH versus temperature plotting of the range of mineral formations.

The enigma of the ultimate and intermediate sources of the compounds of ores has been a longstanding challenge, which will be met only by a continual investigation of, for example, the gains and losses of the elements. These studies on specific lithologies in particular areas, should eventually permit the formulation of generalized conceptual models (qualitative as well as quantitative, the latter being the ultimate aim) of (1) the geochemical and rock cycles of the elements, (2) specific segments of the cycles, and (3) the particular part(s) of the cycles which most likely will result in ore concentrations of the elements. Boyle (1976) has worked out the gain and loss of some compounds during granitization and carbonitization, which can serve as exemplars (see his tables 6 and 7).

To conclude this section on the conceptualization or classification schemes of ore elements and minerals, let us examine an approach of direct industrial application. Page and Creasey (1975, fig.12, p.12) offered curves showing the energy requirements for the recovery of Fe, Ti and Al at different grades from various sources, i.e., how much coal or oil (in kilowatt-hours) are needed to produce one ton (Fig.74).

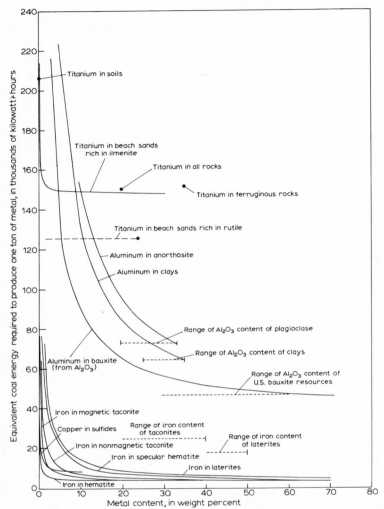

Fig.74. Energy requirements for recovery of iron, titanium and aluminium at different grades from various sources. (After Page and Creasey, 1975.)

COMPARATIVE CLASSIFICATIONS–GENETIC MODELS OF SPECIFIC COMMODITIES

"The recording of facts is one of the tasks of science, one of the steps toward truth; but it is not the whole science. There are one-story intellects, two-story intellects, and three-story intellects with sky lights. All fact collectors, who have no aims beyond their facts, are one-story men. Two-story men compare, reason, and generalize, using the labors of the fact collectors as well as their own. Three-story men idealize, imagine, and predict."

Justice O.W. Holmes

As already pointed out on several occasions (Wolf, 1973, 1976a, 1976b), the ultimate goal of each geological study should be a *comparison* of the data obtained from one district with those of other localities. The present writer also has gone as far as saying — and retains this "philosophical" position — that investigators should be given the opportunity, whenever possible, to culminate their reports by a comparative investigation of the topics considered (may it be a regional field, geochemical, petrologic, structural, or an orebody study). Both *observational/descriptive* and *genetic/interpretive* information can be compared—contrasted on any scale (thin-section to regional—continental; e.g., see Nasseef and Gass, 1977), to establish the *similarities* and *differences* [1]. This comparative approach has been a powerful tool in the sciences and arts — and could find much more application in geology. As maintained elsewhere in this chapter when discussing the "five levels of capacities" (or types of methodologies and studies when extrapolated to the scientific method), nomenclatures and classifications are utilized during *all* stages of investigations, from the lowest, simplest to the highest, most complex. The comparative style of reasoning is also applied during all these levels, much of the comparison being done *unconsciously* and forming an integral part of the fundamental tools in learning and communication. What is urgently required, however, is the *deliberate* or *calculated* (i.e., *conscious*) use of the comparative method applied to specific commodities and specific problems. That is, we have to expand and refine this technique — one first step would be establishing conceptual models (Wolf, 1973, 1976b) as well as comparative tables (e.g., Tables XCII and XCIII on "Cu—Ag shale" and "red-bed sandstone" deposits; Boyle, 1968). The ultimate style of handling any geological information is provided by the "Systems Science" or "System Analysis" techniques (see introductory comments by Lehman, 1977). Summary and comparative tables are needed in hanspecimen, thin-section, polished-section, structural unit, etc., investigations to add one more dimension to the data.

One more example must suffice, namely Lambert/McKerrow's (1978) investigation on comparing the San Andreas fault with the Highland Boundary fault system of Scotland; thus delineating both differences and similarities between the two systems.

MULTI-CLASSIFICATION APPROACHES TO SPECIFIC COMMODITIES: EXAMPLES

"Science concerns itself with the co-existence and sequences among phenomena; groups these at first into generalizations of a simple or low order, and rising gradually to higher and more extended generalizations."

Herbert Spencer (in: *First Principles,* 1862)

---

[1] Arthur Koestler in *The Act of Creation,* for example, explains creativeness as a consequence of "biosociation", i.e., combining two or more unconnected facts or ideas (=making a comparison) to form a single new idea or concept.

TABLE XCII

Characteristics of the copper–silver shale deposits (After Boyle, 1968; courtesy of Geological Society of Canada, Ottawa)

Name	Age	Rock types	Degree of folding	Nature of faulting in ore beds	Grade of metamorphism
Kupferschiefer, Germany	Permian	Shales, siltstones, sandstones	Gentle	Minor	Diagenetic effects only
White Pine, Michigan, U.S.A.	Precambrian	Shales, siltstones, sandstones	Gentle	Minor	Low grade (greenschist)
Zambian (northern Rhodesian) Copperbelt	Late Precambrian	Shales, siltstones, argillite, impure dolomite, micaceous quartzite, greywacke	Complex series of anticlines and synclines; some orebodies exhibit complex folding	Minor	Low grade (greenschist) to medium grade (epidote amphibolite)
Boléo, Mexico	Pliocene	Clay tuffs, tuffaceous sandstones	Relatively flat-lying	Minor	Low grade, mainly diagenetic changes
Creta area, Oklahoma, U.S.A.	Permian	Shales and siltstones	Flat-lying	Minor	Low grade, mainly diagenetic effects
Redstone River area, N.W.T., Canada	Late Precambrian or Early Cambrian	Argillaceous limestone, silty limestone, siltstone	Folded	Faulted	Low grade, mainly diagenetic effects

The multiple requirements of researchers and explorationists (whether field geologists, ore petrologists, geochemists, geophysicists, economists, or computer specialists) have led to various types of classifications of the same commodity. This is illustrated here by a number of schemes applied to uranium and copper. In most instances, *all major* varieties of ores are represented in the classifications, including high-temperature and

ype of ore shoots (Minerals)	Late veins	Principal elements	Grade Cu (%)	Grade Ag (ppm)	Remarks
ds, layers, elongate nses, dissemina- ons (Bornite, chal- cite, chalcopyrite, rite, galena, halerite)	"Rucken" (Barite, calcite, niccolite, pyrite, chalcopy- rite, etc.) (Secretion veins?)	Cu, Zn, Pb, Ag, Mo, Co, Cd, As, Sb, S	2.5	150	Carbonaceous mat- ter is abundant in some beds. Silver is recovered
ds, layers, and sseminations Chalcocite, native pper, pyrite)	Veinlets and veins with associated halos (Calcite, quartz, chalcocite, native copper. etc.) (Secretion veins)	Cu, Ag, S	1–3	Up to 150	Carbonaceous mat- ter is abundant in some beds. Silver is recovered
ds, layers, elong- e lenses, dissemina- ons (Chalcopyrite, rnite, chalcocite, rite, linnaeite, rrollite)	Metamorphic veins. (Quartz carbonate, feldspar, pyrite, chalcopyrite, bornite, chalcocite) (Secretion veins)	Cu, Co, S	3.0	~1–10	Carbonaceous mat- ter is abundant in some beds. Silver is not recovered
ds, ribs, elongate nses (Chalcocite, alcopyrite, rnite, covellite, tive copper, py- te, galena)	None	Cu, Mn, Zn, Pb, Cl, Co, S	5.0	9	Carbonaceous mat- ter is abundant in some beds. High Mn and Cl content in ore. Silver is recovered
ds	None	Cu, S	Up to 4	1	
ven mineralized ds	Veins of quartz, calcite, bornite, etc. in general area	Cu, Ag, Co, S	2.5	~10	At exploration stage

high-pressure varieties because, as in earlier discussions, it is much more informative to do so instead of selectively choosing the stratiform ores for the present purpose of this Hand-book. Preferential treatments prevent the reader from obtaining a "bird's-eye" or syn-optic view and making comparisons.

TABLE XCIII

Characteristics of the "Red Bed" Sandstone Deposits (After Boyle, 1968; courtesy of Geological Society of Canada, Ottawa)

Country and location	Rock types and age	Form of deposits	Primary ore minerals and gangue
Canada: New Brunswick Dorchester, New Horton Goshen	Relatively flat lying Carboniferous arkose, grit, conglomerate, and sandstone	Disseminated lenses and layers, nodules, replaced plant fragments. Irregular veins	Chalcocite, bornite, pyrite. No gangue
Canada: Nova Scotia	Relatively flat lying Caboniferous grit, sandstone, and conglomerate	Disseminated lenses and layers, nodules, replaced plant fragments. Irregular veins	Chalcocite, bornite, pyrite. No gangue
Canada: Walton, Nova Scotia	Carboniferous sandstones, shales, limestone, and evaporites	Raking pipe-like deposit in brecciated zone. Small disseminated zones with replaced plant fragments	Barite, pyrite, chalcopyrite, galena, sphalerite, tennantite, proustite, siderite, hematite
Canada: Nova Scotia, Salmon River, Cape Breton Island	Gently dipping Carboniferous sandstones	Lenses and layers of disseminated galena	Galena, pyrite, minor barite
Southwestern U.S.A.: Texas, Oklahoma, New Mexico, Arizona, Colorado, Wyoming Utah, and Idaho	Arkosic sandstone, conglomerate and shale of Permo-Carboniferous, Triassic, Jurassic or Tertiary age	Lenses and layers of disseminated ore minerals, nodules, replaced plant fragments. Irregular veins	Chalcocite, bornite, chalcopyrite, covellite, pyrite, uraninite; silver sulfides at Silver Reef, Utah
U.S.A.: New Jersey (Newark Series)	Triassic sandstones and shales. Extrusive basalt and intrusive diabase sills	Lenses and irregular bodies of disseminated ore minerals near faults and fractures	Native copper, chalcocite, bornite, chalcopyrite, pyrite
U.S.A.: Michigan	Precambrian Copper Harbour conglomerate and sandstone and Nonesuch shale and sandstone	Irregular tabular lenses mainly in sandstone. Disseminations in sandstone	Native silver, native copper, solid hydrocarbons, chalcocite, etc.
U.S.A.: Colorado. Cashin mine, Montrose county	Jurassic flat-lying sandstones and shales	Pods and lenses in fissure cutting La Plata sandstone	Argentiferous chalcocite, covellite, bornite, native copper, argentite, native silver, kaolin, calcite, and barite

Secondary ore minerals	Grade and tonnage	Reference	Remarks
Malachite, azurite	Cu–0.5–5% Ag–0.5–2.0 ppm	Papenfus (1931); Brummer (1958); Smith (1960, 1963)	Not economic. Dorchester deposit was mined in the past
Malachite, azurite	Cu–0.5–1.0% Ag–0.5–2.0 ppm	Papenfus (1931); Brummer (1958)	Not economic
Manganese oxides, malachite, azurite, limonite, cobalt sulfates	Pb–5–25% Zn–1–20% Cu–0.5–5% Ag–20 oz/ton	Boyle (1963a)	Economic for barite, lead, zinc, copper, and silver
None	Pb–1–4.5% Ag–1.5 ppm Large tonnage	Carter (1965)	Exploration stage
Malachite, azurite, carnotite, autunite; cerargyrite at Silver Reef, Utah	Variable Ag content generally low in most deposits. Ag content at Silver Reef, Utah averaged 20 oz/ton	Proctor (1953); Wyman (1960); Fischer and Stewart (1961)	Economic uranium and vanadium deposits. Generally uneconomic for copper. The Silver Reef deposits were economic for silver and uranium
Malachite, azurite, brochantite, chrysocolla, cuprite, tenorite, and secondary native copper and chalcocite	Low grade. Ag content very low	Woodward (1944)	Not economic. Some of the deposits contain small amounts of uraninite and other uranium minerals
	Ag content highly variable, assays from 1 oz Ag/ton to 70 oz Ag/ton or greater recorded	Rominger (1876); Nishio (1919); White and Wright (1954)	Mined for silver prior to 1900 (see also description of White Pine mine)
Azurite, malachite, cuprite, iron sulfate, argentite, native silver	Very high grade copper and silver, but small ore shoots. Some assays show up to 512 oz Ag/ton	Emmons, W.H. (1905)	Exhausted. Worked for copper and silver

(continued)

TABLE XCIII (continued)

Country and location	Rock types and age	Form of deposits	Primary ore minerals and gangue
U.S.A.: Wyoming Silver Cliff mine, Lusk	Gently dipping Cambrian (?) calcareous sandstone	Pods and irregular replacement deposits along a fault zone	Native silver, chalcocite, and pitchblende. Calcite is the principal gangue. There is also some clinozoisite
U.S.A.: Emery country, Utah	Near base of Shinarump conglomerate Upper Triassic conglomerate and sandstones	Lenses and tabular bodies of mineralized sandstone, conglomerate, and shale. Numerous fossil trees and plant fragments	Uraniferous asphalt and woody fragments
England: Alderly Edge	Relatively flat lying Triassic sandstones and conglomerates	Lenses and irregular bodies of disseminated ore minerals. Irregular veins	Chalcocite, chalcopyrite, bornite, pyrite, galena, barite, manganese oxides, cobalt minerals
Germany: Maubach-Kaller Stollen area	Mesozoic flat-lying Bunter sandstone and conglomerate	Layers, lenses, and irregular bodies of disseminated ore minerals. Irregular veinlets	Galena, sphalerite, nickeliferous pyrite, pyrite, chalcopyrite, boulangerite, tetrahedrite, linnaeite. "Knotten" of galena are characteristic. Veinlets of barite and chalcopyrite
Sweden: Laisvall, Maiva, Dorotea, and Vassbo	Horizontal Cambrian quartzites and quartzitic sandstones	Layers, lenses, and elongate bodies of disseminated ore minerals	Galena, sphalerite, pyrite, calcite, barite, fluorite
Norway: Dalane area	Folded and metamorphosed Precambrian quartzites and conglomerates with associated greenstones and quartz porphyries	Impregnation zones in quartzite near contact of younger greenstone	Native copper, native silver, cuprite, calcite
Europe: Various countries, particularly U.S.S.R.	Permian and Triassic sandstones, shales, and conglomerates	Layers, lenses, irregular bodies of disseminated minerals	Chalcocite, bornite, chalcopyrite, barite

econdary ore minerals	Grade and tonnage	Reference	Remarks
₁alachite, azurite, uprite, native copper, hrysocolla, limonite, ₁ummite, uranophane, ₁etatorbernite	High grade silver and low grade uranium deposit	Wilmarth and Johnson (1954)	Silver ore shoots exhausted
₁utunite (?), gum-₁ite (?), carnotite, ₁alachite, azurite, ₁hrysocolla, cobalt ₁loom, native sulfur	Ag uniformly distributed throughout uraniferous zones. Ag 0.44–1.64 oz/ton Ag (average) 0.70 oz/ton	Reyner (1950)	
₁alachite, azurite, yromorphite, anglesite, ₁anadinite, mottramite, ₁erussite, asbolite	Low grade. Cu–1.5–2.0% Ag very low Ni and Co present	Warrington (1965)	Exhausted. Lead ore contained from 0.09 to 7.5 oz Ag/ton
₁alachite, cerussite, yromorphite, anglesite, ₁tc.	Pb–1–3% Zn–1.5% Ag–3.5 ppm Relatively large tonnages	Behrend (1950); Fritzsche (1951); Schachner (1958)	Economic
	Pb–4% Ag–10 ppm 50,000,000 tons	Grip (1960)	Economic
	2.6 to 6.7 parts Ag per 100 parts copper	Neumann (1944a)	Mined for copper and silver prior to 1917
Malachite, vanadinite, ₁zurite, volborthite, etc.	Cu–0.5–1.0%		Generally not economic

(continued)

TABLE XCIII (continued)

Country and location	Rock types and age	Form of deposits	Primary ore minerals and gangue
U.S.S.R.: Kazakhstan, Dzhezkazgan area	Permo-Carboniferous sandstones, siltstones, argillites, minor thin limestone and flintstone. Abundant plant remains. Eight ore-bearing horizons are present	Beds, layers, sheets, seams, blankets, lenses, ribbons, and irregular bodies of disseminated ore minerals; also veins. Veins are not economic	Chalcocite, bornite, chalcopyrite, pyrite, domeykite, tetrahedrite, galena, sphalerite, molybdenite, betechtinite, arsenopyrite, native copper, native silver; minor barite, calcite, etc.
Bolivia: Corocoro	Steeply dipping, faulted Tertiary sandstones and conglomerates	Lenses, elongated lenses, and beds of disseminated ore minerals	Chalcocite, native copper, galena, native silver, domeykite, pyrite, chalcopyrite
China: Yunnan	Folded Mesozoic sandstones and shales	Lenses, layers, beds, and irregular bodies of disseminated ore minerals; also veins	Chalcocite, bornite, chalcopyrite, pyrite; minor calcite, quartz, and barite. Chalcopyrite, Co-bearing tennantite, barite, and siderite in veins

Uranium deposits

In the first part of this section the following uranium classifications are presented:

(1) Diagrammatic model of the geochemical cycle of uranium (two figures).

(2) Tabular classifications of uranium deposits based on mode of origin, host rocks, with examples (two tables).

(3) Diagram depicting genetic types of uranium occurrences.

(4) Supplementary tabular and flow-chart-like uranium classifications with additional variables.

(5) Metallotectonic classification of uranium ores.

(6) Triangular classification of the principle uranium deposits, augmented by a detailed tabular scheme founded on genesis, host rock, environment, mineralogy and other features.

econdary ore minerals	Grade and tonnage	Reference	Remarks
alachite, azurite, native ver, native copper	Cu−1.0−3.5% (?) Large tonnages. Ag follows copper and is present mainly in chalcocite	Narkelyun (1962); Davidson (1962, 1965)	One of main sources of copper in the U.S.S.R.; deposits also produce lead, silver, and some zinc. Narkelyun considers the bedded deposits to be of sedimentary origin later modified by diagenesis, others think an epigenetic origin more likely. Narkelyun concludes that the veins are of secretion origin
alachite, azurite	Cu−3−5% or higher Ag−7−16 ppm Large tonnages	Ljunggren and Meyer (1964); Berton (1937)	Economic. Continued development work. Local concentrations of native silver (Singewald, 1933)
lalachite	No data. Chalcocite and other copper minerals contain traces to small amounts of silver and gold	Li et al. (1964)	At exploration stage. Li et al. consider that the deposits are essentially sedimentary, later modified by post-depositional processes. The veins are thought to owe their origin to secretion processes

(7) Special-purpose uranium classification.

(8) Flow-chart diagram of the world's present uranium reserves.

(9) Table of uranium grade and potentiality.

(10) Table and diagram depicting geochronologic and geologic setting of uranium host rocks and ores.

The two cartoon-like conceptual models (Figs.75 and 76) illustrate the geochemical cycle of uranium in conjunction with or as part of the rock cycle (Borchert, 1950, unpublished; Schneiderhöhn, 1962, fig.9, pp.156−157). Fig.75 provides an overview of the various igneous and sedimentary environments in which uranium is found, whereas Fig.76 depicts especially the sedimentary environments, plus some igneous and metamorphic milieux, of the uranium cycle.

The tabular Figs.77 and 78 by Dahlkamp (1978, figs.1 and 2) offer, respectively, a simple "mode of origin-host rock" division of the uranium occurrences and a more

220

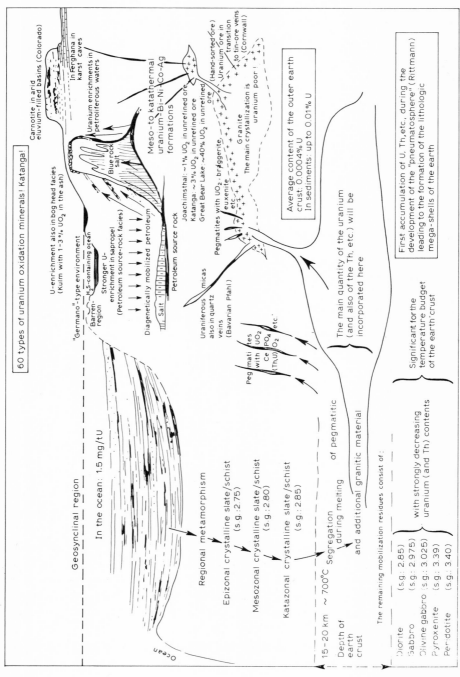

Fig.75. Conceptual model of the earth-crust's geochemical cycle of uranium. (After Borchert, 1950, unpublished; 1978; courtesy of Prof. Borchert.)

221

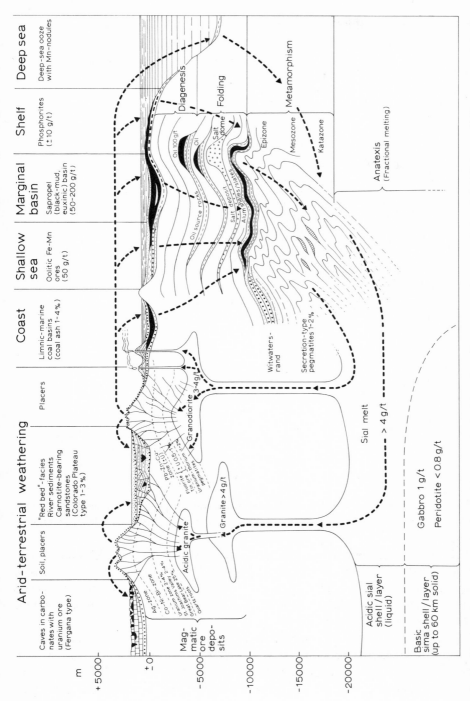

Fig.76. Distribution of uranium in the earth's crust, emphasizing the sedimentary environment. (After Schneiderhöhn, 1962; courtesy of Gustav Fischer, Stuttgart.)

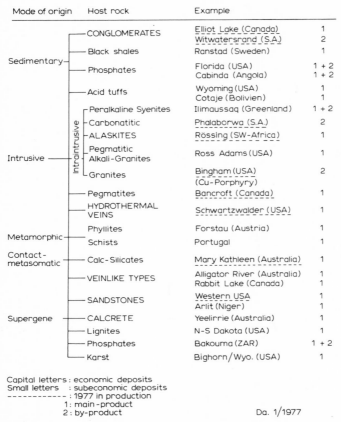

Mode of origin	Host rock	Example	
Sedimentary	CONGLOMERATES	Elliot Lake (Canada)	1
		Witwatersrand (S.A.)	2
	Black shales	Ranstad (Sweden)	1
	Phosphates	Florida (USA)	1 + 2
		Cabinda (Angola)	1 + 2
	Acid tuffs	Wyoming (USA)	1
		Cotaje (Bolivien)	1
Intrusive	Peralkaline Syenites	Ilimaussaq (Greenland)	1 + 2
	Carbonatitic	Phalaborwa (S.A.)	2
	ALASKITES	Rössing (SW-Africa)	1
	Pegmatitic Alkali-Granites	Ross Adams (USA)	1
	Granites	Bingham (USA) (Cu-Porphyry)	2
	Pegmatites	Bancroft (Canada)	1
	HYDROTHERMAL VEINS	Schwartzwalder (USA)	1
Metamorphic	Phyllites	Forstau (Austria)	1
	Schists	Portugal	1
Contact-metasomatic	Calc-Silicates	Mary Kathleen (Australia)	1
	VEINLIKE TYPES	Alligator River (Australia)	1
		Rabbit Lake (Canada)	1
Supergene	SANDSTONES	Western USA	1
		Arlit (Niger)	1
	CALCRETE	Yeelirrie (Australia)	1
	Lignites	N-S Dakota (USA)	1
	Phosphates	Bakouma (ZAR)	1 + 2
	Karst	Bighorn/Wyo. (USA)	1

Capital letters : economic deposits
Small letters   : subeconomic deposits
– – – – – – – – – : 1977 in production
1 : main-product
2 : by-product

Da. 1/1977

Fig.77. Classification of uranium occurrences, I. (After Dahlkamp, 1978; courtesy of *Miner. Deposita.*)

detailed "multi-genetic—host-rock—structural" classification. These are supported by a diagrammatic model of numerous uranium-containing host rocks and environments (Fig.79).

The next three tabular classifications (respectively by Little, 1970; Barnes and Ruznicka, 1972; and Ziegler, 1974) are again easy to utilize — they all incorporate both petrographic and genetic parameters. In particular Table XCIV summarizes the genetic aspects. An additional parameter in Table XCV and XCVI is the tectonic setting, thus making the scheme a metallotectonic one.

Ruznicka (1971) in his comprehensive review of uranium mineralizations first devised triangular classification schemes consisting of two diagrams (Figs.80 and 81) with the apices occupied by uranium formed through syngenetic—igneous (1.1), sedimentary (1.3), and metamorphic (1.5) processes with transitional genetic mechanisms represented

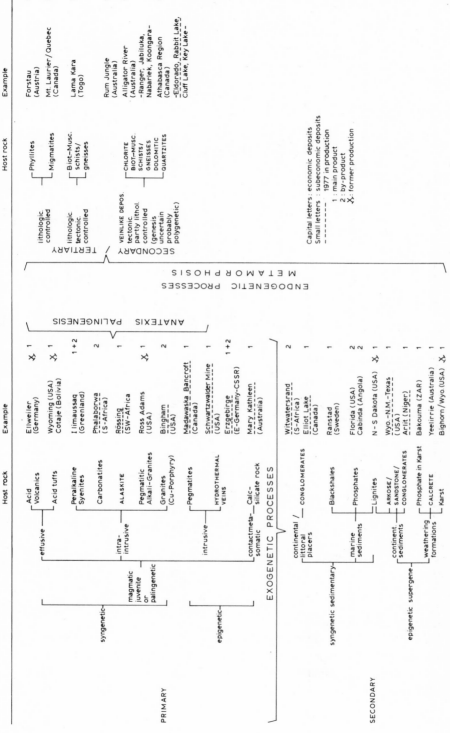

Fig. 78. Classification of uranium occurrences, II. (After Dahlkamp, 1978; courtesy of *Miner. Deposita*.)

TABLE XCIV

Classification of uranium deposits (After Little, 1970, table 1, p. 36; courtesy of International Atomic Energy Agency, Vienna)

Types		Characteristic elements	Characteristic uraniferous minerals	Other characteristic minerals	Example(s)
Igneous and related types	Granitic (large plutonic bodies)	Th, U, Zr, Si	Allanite, zircon, thorite, uraninite	Magnetite, sphene, apatite	St. Pierre Uranium property, Johan Beetz area, Quebec
	Pegmatite red pegmatites	Th, U, Zr (Si, F, S, P, Ce)	Uranothorite, uraninite, thorite, allanite, cyrtolite	Uralite, hematite, pyrite, pyrrhotite, magnetite, sphene	Faraday and Bicroft mines, Bancroft area, Ontario
	unzoned white pegmatites	Mo, U, Th, Si (Nb, RE^3)	Uraninite, uranothorite, betafite, euxenite, allanite	Biotite, magnetite, molybdenite, hornblende	Grandroy prospect, Johan Beetz area, Quebec
	zoned	Li, Be, Nb, Mo, Si (Nb. Sn, RE^3)	Pyrochlore, betafite, allanite, columbite tantalite	Molybdenite, cassiterite, calcite	Richardson mine, Wilberforce, Ontario
	Anhydrite	Th, U, Zr (Si, S, P)	Uranothorite, uraninite, cyrtolite	Uralite, calcite, pyrite, pyrrhotite, magnetite	Lower levels of Faraday mine, Bancroft, Ontario
	Carbonatite (with associated fenite)	Nb, U, Th	Pyrochlore, betafite, perovskite, niocalite	Calcite, soda pyroxene and amphibole, apatite, biotite, magnetite	Newman property, Lake Nipissing, Ontario
	Metasomatic	U, Th, P, Si (F, Mo, Fe, S)	Uraninite, thorite, thorianite, monazite	Calcite, pyroxene, quartz, feldspar, hema-	Charlebois Lake, Saskatchewan

		As, S, Se (Au, Pt)	blende, thucolite	ite, chlorite, quartz	Bear Lake, District of Mackenzie
Hydrothermal	Type IIa (simple mineralogy)	U, C, Fe, Se	Massive pitchblende, thucolite	Hematite, pyrite, calcite	Fay-Ace-Verna mines, Beaverlodge area, Saskatchewan
	Type IIb (simple mineralogy)	U, C	Sooty pitchblende, thucolite(?)	Kaolin, ankerite, pyrite (hematite)	Rabbit Lake prospect, Wollaston Lake area, Saskatchewan
Sedimentary	Placer (modern)	Th, U, P, Zr, Nb, RE[3])	Monazite, uraninite, pyrochlore, zircon	Magnetite, ilmenite, pyrite, garnet	Bugaboo deposits, British Columbia
Conglomerate	Syngenetic (Blind River type)	U, Th, Ti, Fe, S, Zr, RE[3]	Brannerite, uraninite, monazite	Pyrite, monazite, zircon	Denison mine, Elliot Lake, Ontario
	Epigenetic	U, Fe, C	Autunite, gummite, etc.	Limonite	Dear Creek prospect, British Columbia
Sandstone	Lignitic	U, C	Urano-organic compounds (thucolite)	Jarosite	Cypress Hills area, Saskatchewan
Dolomite		Th, U, Fe	Monazite, uraninite	Hematite, zircon	MacLean Bay occurrence, Stark Lake, Dist. of Mackenzie
Phosphorite		U, Ca, P, C	Unknown	Collophanite, bitumen	Fernie area, B.C.
Shale		U, C, H	Unknown	Bitumen, lignite	Fernie Formation, Alta and B.C.
Supergene	Cappings and secondary deposits formed by percolating water	U, Si, S	Gummite, uranophase, etc.	Barite, gypsum	Parts of Bolger and Gunnar deposits, Sask.

Notes: 1. Types with major production have solid underline; types with minor production have broken underline. 2. Elements in brackets are present in minor amounts or in certain deposits only. 3. RE means rare earth elements.

TABLE XCV

A genetic classification of uranium deposits (After Barnes and Ruznicka, 1972, p. 160)

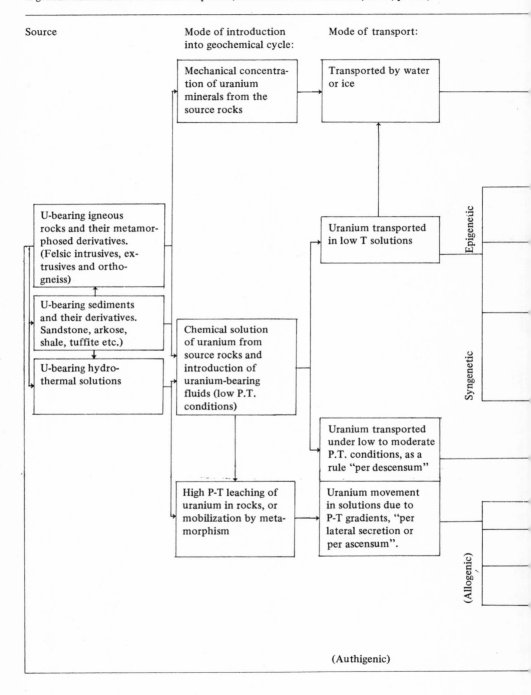

Mode of concentration and deposition:		Type	Example
Alluvial and marine deltaic and littoral syngenetic	Oxidation on site	Modern uranium–thorium bearing placers	Indus River, Pakistan Bugaboo Cr. Can.
	Reduction on site	Pyritic quartz–pebble conglomerates	Huronian, Can. Rand, S. Afr.
		Conglomerate	Ningyo-toge, Japan Crooks Cap. U.S.A.
Eh and pH changes (chemical cells)		Sandstone	Colorado Plateau, U.S. Hamr, C.S.S.R.
Adsorption or complexing	by carbon	Coal, lignite, bitumen	Dakotas, U.S.A. Stachanov, C.S.S.R.
	by other agents	Supergene precipitates and evaporites	Tyuya Myuyun, U.S.S.R.
Adsorption or complexing	by organic material	Organic shales	Chatanooga shales, U.S. Kolm shales, Sweden
	by other agents	Phosphorites, limestones, etc.	Phosphoria Formation U.S.
Metamorphic reconcentration		Sedimentary metamorphics	Ronneberg basin, E. Germany
		Mineralized water (potential source)	Closed basins, central Asia, oceans, mine, mine waters
Deposition by Eh, pH changes	Authigenic	Zones in granitic rocks (supergene)	Massif Central, Fr. Rossing, S.W. Afr.
	Allogenic	Shear zones in metamorphic rocks	Rabbit Lake, Can. Nabarlek, Austr.
Precipitates in tuffites		Pyroclastics	Huta-Muran, C.S.S.R.
Eh, pH, P-T change	simple solutions	Monomineralic veins	Beaverlodge, Can. Pribram, C.S.S.R.
Eh, pH, P-T changes	complex solutions	Polymetallic veins (volcanic environments)	Great Bear, Can. Jachymov, C.S.S.R.
Metasomatism in porous rocks		Pyrometasomatic deposits	Rexspar, Can. Bur Region, Somalia
Differentiation and crystallization high P-T	Diatremes	Carbonatites	Newman and Oka deposits, Canada
	Dykes	Pegmatites	Faraday, Canada
	Plutons	Granites and syenites (commonly alkaline)	Ross-Adams, Alaska Rossing, S.W. Afr.

TABLE XCVI

Metallotectonic classification of uranium (After Ziegler, 1974, pp. 666–669)

Structural units	Components of structural units	Deposits of pyritic conglomerates (auriferous or not) (1)	Deposits in sandstone argillite continental assemblages (2)	Vein deposits almost monometallic in composition (3)	Finely disseminated deposits being stratiform controlled, sometimes associated with veins (4)	Vein deposits associated with polymetallic mineralizations (5)	Vein deposits associated with magnetite (6)	Deposits associated with molybdenum and with acidic to intermediate volcanism (7)	Deposits in epimetamorphic peribatholithic schists (8)
Archean cratonic nuclei (older than 2500 MA)	Archean complex (basement) (A)								
	cover-rock formations (B)	▲ Elliot Lake (CDN) ▲ Witwatersrand (ZA)	★ Wyoming (USA)						
Folded belts (mobile belts of the Proterozoic)	geosynclinal stages (C)			▲ Beaverlodge (CDN) ▲ Nabarlek (AUS) ▲ Koongarra (AUS)	▲ Singbum belt (IND) ▲ Ranger I (AUS)	▲ Port Radium (CDN) ▲ ? Shinkolobwe (CGO)	Krivoy Rog (SU) Vasterik (S)		
	cover-rock formations (D)	▲ Jacobina (BR)	★ Niger (with MO)						

(continued)

TABLE XCVI (continued)

Structural units	Components of structural units	Deposits in pegmatoids and skarns (9)	Deposits in pegmatite zones (10)	Deposits of U and Th in alkaline complexes of carbonatites associated with intracontinental rifts (11)	Deposits of uraniferous calcretes (12)	Deposits in black schists (13)	Deposits in marine phosphates (14)	Deposits in karsts (15)
Archean cratonic nuclei (older than 2500 MA)	Archean complex (basement) (A)		Ontario (CDN) Tete (Mozambique)	▲ Palabora (ZA)				
	cover-rock formations (B)				• Yeelirrie (AUS)		★ Cabinda (Angola)	
Folded belts (mobile belts) of the Proterozoic	geosynclinal stages (C)	▲ Rossing (SWA) ▲ Bancroft (CDN) ▲ Mary Kathleen (AUS) ▲ Fort Dauphin (RM)	▲ Central Plateau (Madagascar)	▲ Illimoussaq (Greenland) ★ Pocos de Caldas (BR)				
	cover-rock formations (D)				• Namib Desert (SWA)	• Aluminous schists of Sweden Schists of Chattanooga (USA)		★ Bakouma (RCA) Le Vigan (F)

(continued)

230

TABLE XCVI (continued)

Structural units	Components of structural units	(1) Deposits of pyritic conglomerates (auriferous or not)	(2) Deposits in sandstone argillite continental assemblages	(3) Vein deposits almost monometallic in composition	(4) Finely disseminated deposits being stratiform controlled, sometimes associated with veins	(5) Vein deposits associated with polymetallic mineralizations	(6) Vein deposits associated with magnetite	(7) Deposits associated with molybdenum and with acidic to intermediate volcanism	(8) Deposits in epimetamorphic peribatholithic schists
Lower Paleozoic orogenies (Caledonian type)	geosynclinal stages (E)							♦ ? URSS	
	cover-rock formations (F)							♦ ? URSS	
Hercynian orogenies or others of the Upper Paleozoic	geosynclinal stages zone (G)					■ Jachymov Erzgebirge (CS and GDR) (Ni-Co-Bi-Ag-Cu-Pb-Zn, As-Sb)			Portalegre (F) Ciudad-Rodrigo (E)
	Cordilleran zone (H)			Central French Massif Iberian Meseta		■ ? Pribram (CS) (Pb-Zn-Ag)			
	cover-rock formations (I)		■ Hérault (F) ★ Hamri (CS) ★ Gulf Coast (USA)						
Alpine orogenies	geosynclinal stages (J)		★ Suleiman-Range (PAK)			★ Isérables (CH) (Pb-Cu-Sb) ★ Talmessi (IR) (Ni-Co-Cu-Bi)			

(continued)

TABLE XCVI (continued)

Structural units	Components of structural units	Deposits in pegmatoids and skarns (9)	Deposits in pegmatite zones (10)	Deposits of U and Th in alkaline complexes of carbonatites associated with intracontinental rifts (11)	Deposits of uraniferous calcretes (12)	Deposits in black schists (13)	Deposits in marine phosphates (14)	Deposits in karsts (15)
Lower Paleozoic orogenies (Caledonian type)	geosynclinal stages (E)							
	cover-rock formations (F)							
Hercynian orogenies or others of the Upper Paleozoic	geosynclinal stages zone (G)							
	Cordilleran zone (H)							
	cover-rock formations (I)						* Morocco * Florida (USA)	
Alpine orogenies	geosynclinal stages (J)							

(continued)

TABLE XCVI (continued)

Structural units	Components of structural units	Deposits of pyritic conglomerates (auriferous or not) (1)	Deposits in sandstone argillite continental assemblages (2)	Vein deposits almost monometallic in composition (3)	Finely disseminated deposits being stratiform controlled, sometimes associated with veins (4)	Vein deposits associated with polymetallic mineralizations (5)	Vein deposits associated with magnetite (6)	Deposits associated with molybdenum and with acidic to intermediate volcanism (7)	Deposits in epimetamorphic peribatholithic schists (8)
	fragments of (K) Precambrian basement incorporated in the orogen		■ Trentin (I) ■ Mts Messek (H) ■ Spisbasses Tatra (CS) ■ Novazza (I) (with Zn)	Kalna (YU)		★ Macedonia (YU) (Pb-Zn)			
American cordilleras (Nevado-Laramian assemblage)	geosynclinal stages (L)					★ Numerous occurrences with Pb-Zn-Ag; Cu-Ag; Ni-Co-Cu-Ag; auriferous pyrite			★ Spokane (USA) ★ Austin (USA)
	fragments (M) of Precambrian basement incorporated in the orogen		★ Colorado (USA) ★ Salta (RA) ★ Malargue Mendoza (RA)			★ Front Range (USA) (Pb-Zn-Cu-Mo)		★ Maryvale (USA)	

(continued)

TABLE XCVI (continued)

Structural units	Components of structural units	Deposits in pegmatoids and skarns (9)	Deposits in pegmatite zones (10)	Deposits of U and Th in alkaline complexes of carbonatites associated with intracontinental rifts (11)	Deposits of uraniferous calcretes (12)	Deposits in black schists (13)	Deposits in marine phosphates (14)	Deposits in karsts (15)
	fragments of (K) Precambrian basement incorporated in the orogen							
American cordilleras (Nevado-Laramian assemblage)	geosynclinal stages (L)						▪ Phosphoria	
	fragments of (M) Precambrian basement incorporated in the orogen							

☐ Types consisting of districts with more than 50,000 t U (in economic minerals).
⌐⌐⌐ Types consisting of districts with more than 200,000 t U (in subeconomic minerals).
Age of the first deposition of the mineralization: ▼ Archean, ▲ Proterozoic ◆ Paleozoic, ▪ Upper Paleozoic, ★ Cenozoic, ● Recent.
Note: The countries in which the deposits occur, are designated by the symbols of the FIA.

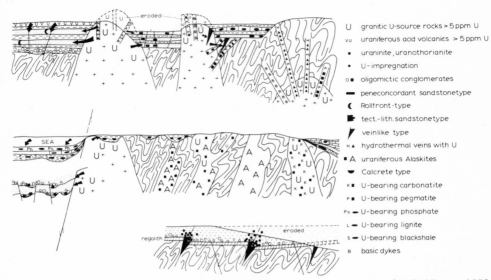

Fig.79. Diagram depicting the occurrence and formation of uranium deposits. (After Dahlkamp, 1978; courtesy of *Miner. Deposita.*)

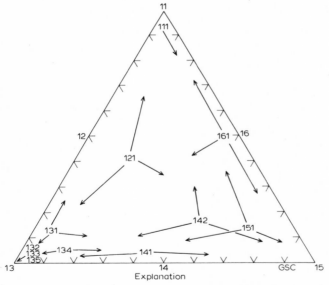

Fig.80. Schematic illustration of the principle of classification of uranium deposits formed by syngenetic magmatic and sedimentary, and metamorphic processes. (After Ruznicka, 1971; courtesy of Geological Society of Canada, Ottawa.)

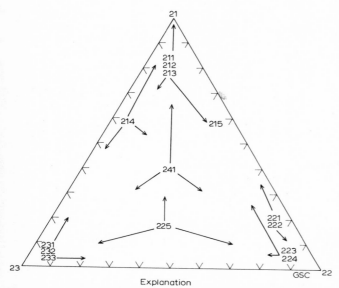

Fig.81. Schematic illustration of the principle of classification of uranium deposits formed by epigenetic processes. (After Ruznicka, 1971; courtesy of Geological Society of Canada, Ottawa.)

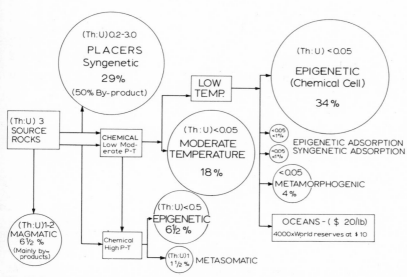

Fig.82. World's present low-cost uranium reserve by genetic types (Barnes and Ruznicka, 1972).

TABLE XCVII

Classification of uranium deposits (After Ruznicka, 1971; courtesy of Geological Society of Canada, Ottawa)

Part 1

Classification			Characteristic		Example
Class	Group	Type	Geological environment	Mineralization	Other distinct features	(reference)
a	b	c	d	e	f	g
...entary, and metamorphogenetic.	(1.1.) Deposits formed mainly by igneous processes.	(1.1.1.) Deposits in pegmatitic granites and syenites (Grenville type).	Mainly pegmatite bodies which are complex mixtures of syenitic, granitic and quartz rich phases.	Uraninite, uranothorite, allantite, betafite, zircon, fergusonite, microcline–microperthite, antiperthite, sodic plagioclase, quartz, pyroxene, hornblende, minor biotite and muscovite; magnetite; fluorite, calcite.	Ore mineralization is mainly concentrated along dyke margins, adjacent to inclusions and in zone near internal fractures. Metamorphic (sedimentary) or epigenetic features may be present in varying degrees.	Deposits in the Grenville Province of Canadian Shield as mentioned in this paper, radioactive mineralization in granites and syenites mainly of Precambrian age. Lang et al. (1962), competent references in this paper.
	Deposits formed by ...us and sedimentary processes.	(1.2.1.) Effusive-sedimentary deposits (Huta type)	Uranium mineralization is related to quartz–porphyry volcanic events. Uranium mineralization is confined mainly to tuffites and related rocks originating through volcanic and sedimentary processes.	Sooty pitchblende, pitchblende, secondary uranium minerals, molybdenite, chalcopyrite, tenantite-tetrahedrite, galena, sphalerite, arsenopyrite, ilmenite, magnetite, hematite, covellite.	The uranium mineralization is syngenetic due to differentiation through volcanic processes, accumulation during sedimentation with participation of hydrogen sulfide exhalations. Metamorphism may	Novoveska Huta and Muran deposits in the Czechoslovakian West Carpathians. Rojkovic (1968). Zhukov (1963). Drnzik (1965). Adamek (1967).

(1) Syngenetic magmatogenetic	(1.3) Deposits formed mainly by sedimentary processes.			
	congomerates (Elliot Lake type).	tion in many confined to quartz–pebble conglomerates comprised in basal sedimentary complexes unconformably overlying Precambrian complexes, in which uranium enriched rocks were weathered. The igneous processes and metamorphism could influence the formation of deposits.	rite, brannerite, allanite, xenotime, pyrite, hematite, magnetite, titanium minerals, gold and other minerals characteristic of placers.	mous deposit in the U.S.S.R., Witwatersrand, Jacobina. Roscoe (1969), Shcherbin (in Vol'fson, editor, 1968).
	(1.3.2.) Uranium-bearing placers.	Unconsolidated placers of Recent age.	The main radioactive minerals are pyrochlore, allanite, monazite, uraninite, zircon; nonradioactive: magnetite, garnet, ilmenite, pyrite.	Bugaboo Creek, British Columbia; Haggart Creek, Yukon Territory. Lang et al. (1962). Robinson (1958).

(continued)

238

TABLE XCVII (continued)

Part 2

Classification			Characteristic			Example (reference)
Class	Group	Type	Geological environment	Mineralization	Other distinct features	
a	b	c	d	e	f	g
		(1.3.3.) Uranium-bearing phosphatic sediments.	Marine sediments with uranium content proportionally related to phosphate content (roughly) and inversely proportional to CO_2 content.	Uranium is as a rule a part of phosphate minerals, other compounds: collophanite, bitumen or organic matter, eventually iron, cobalt, molybdenum, rare earths and gold.	A specific example of deposit with uranium and rare earths sorbed by phosphate compounds of fish-bone detritus is known from U.S.S.R.	Fernie group, Rocky Mountains (Canada) very low grade mineralization; anonymous deposit from U.S.S.R.; some deposits in U.S.A. Robinson (1958), Lang et al. (1962), Gotman and Zubrev (1963).
		(1.3.4.) Caustobioliths, organic shales and related rocks with syngenetic uranium mineralization (Chatanooga type).	Pre-orogenic deposits in marginal part of platform adjacent to geosyncline mostly of Paleozoic age.	Uranium is finely distributed and mainly proportionally related to carbonaceous matter; clay minerals, pyrite, quartz, feldspar, etc.		Marine shales in Saskatchewan, mineralization in graptolitic shales in Barrandien (Czechoslovakia): from other countries: Chattanooga shale and Kvarntorp and Ranstad (Sweden). Klepper-Wyant (1957).

...ned mainly by sedimentary processes.

...nentary, and metamorphogenetic.

(1.3) Depos...	deposits in (lime-stones), (clays) and sandstones.	ments.	not appear in mineral forms.	Surazhskiy (1960) the syngenetic deposits in marine sandstones are similar to the deposits in black shales, but they do not contain phosphatic material. They are rich on quartzose siltic material.	by Surazhskiy (1960) as representants of the tenth type in his classification.
(1.4). Deposits formed mainly by sedimentary and metamorphic processes.	(1.4.1.) Uranium-bearing graphitic shales and graphitic schists (Iron Ranges type).	Syngenetically deposited uranium in marine sediments was further locally concentrated by moderate metamorphis processes.	Uraninite, secondary uranium minerals; uraniferous turquoise, kolovratite, volborthite, vanadium minerals.		Mineralization from Middle Asia (U.S.S.R.). Iron Ranges (U.S.A.). Klepper and Wyant (1956).
	(1.4.2.) Sedimentary-metamorphic deposits in argillaceous, micaceous and black shales.	Argillaceous, micaceous and black shales and dolomitized limestones, graptolithic and other material of organic origin, pyrite containing rocks.	Pitchblende, sooty pitchblende; quartz, sericite, clay minerals, calcite, dolomite, pyrite, chalcopyrite, sphalerite, galena, bravoite, graphite, pyrophyllite, marcasite, chloanthite, niccolite, chlorite, pyrobitumen, anthraxolite.	Uranium mineralization occurred syngenetically with sedimentation and was concentrated during the metamorphic processes.	Anonymous deposit mentioned by Getseva (1958) and by Surazhskiy (1960).

(1) Syngenetic magmatogenetic a...

(continued)

TABLE XCVII (continued)

Part 3

Classification			Characteristic			Example (reference)
Class	Group	Type	Geological environment	Mineralization	Other distinct features	
a	b	c	d	e	f	g
(1) Syngenetic magmatogenetic and sedimentary, and metamorphogenetic.	(1.5) Deposits formed mainly by metamorphic processes.	(1.5.1) Metamorphogenic disseminations (Aillik type).	Uranium contained in original sedimentary or igneous rocks was further concentrated by metamorphic processes.	Pitchblende, quartz, feldspar, biotite, hornblende, pyrite, hematite.	The arrangement of uranium minerals follows usually the metamorphic structural and textural features of the host rocks.	Several localities within the Makkovik-Seal Lake uranium-bearing area, Beavan (1958), Gandhi et al. (1969), Little and Ruznicka (1970).
	(1.6) Deposits formed mainly by igneous and metamorphic processes.	(1.6.1) Granulites and/or lithologically analogous uraniferous metamorphic rocks (Makkovik type).	Metamorphic rocks,	Pitchblende, hematite, plagioclase, clinopyroxene, apatite, quartz, calcite, (soddyite).	Ore-mineralization may be affected by epigenetic processes.	Occurrence in the Makkovik-Seal Lake uranium-bearing area. Steacy (1969), written communication).

(continued on p.242)

by 1.2, 1.4 and 1.6, respectively. The second triangle (Fig.81) represents the epigeneti-cally produced uranium concentrations, i.e., impregnation and metasomatic (2.1), vein and related (2.2), and infiltration (2.3) types. Both diagrams have numerous hydrid ura-nium deposits located within the figure. Based on this approach, Ruznicka (1971) sub-divided the deposits into classes, groups and types, utilizing the one-, two- and three-digit numbers of the two triangles for his detailed tabular classification (Table XCVII).

Some profound investigations may require specific schemes that are unique, and may or may not be generally applicable to the particular study for which the scheme has been developed. The unusual classification of Bostick served a specific purpose in his evaluation of the uranium favourability of a basin on the basis of a few selected geological criteria to allow rapid data input during computer calculations. His scheme (Table XCVIII) applies to the Western States-type (also known as "Colorado Plateau–Wyoming type") of uranium mineralizations, i.e., to the epigenetic-infiltration, roll-shaped variety of ores (cf. Rackley's Chapter 3, Vol. 7). Future research will show as to whether a similar approach can be used for genetically related deposits, e.g., ores formed under "red-bed" conditions. Among several interesting controls, the most outstanding is the "average dip" variable, which (as far as the writer is aware) has not been systematically or statistically investi-gated as to its influence on the rate and mode of epigenetic fluid movements and, conse-quently, the degree and localization of mineralization. The data in Table XCVIII indicates that the dip of the host rock is of prime significance.

The approximate percentage of the world's uranium reserves according to genetic types are presented in Fig.82 (Barnes and Ruznicka, 1972), whereas the uranium-grade (in percent) and the potential (in tons) of seven ore-types form the basis of Table XCIX.

To terminate this section on uranium classifications, two further schemes are offered — an approach that can and has in several cases (see separate treatment below) been applied to other metals and non-metal concentrations. As can be deduced from Table C and Fig.83, specific genetic uranium mineralizations are more prevalent in the host rocks of certain geologic eras or periods.

Copper deposits

The following classifications of copper serve as additional examples in the present section of "Multiple-Classifications":

(1) Diagrammatic model of the copper geochemical cycle.

(2) Classification of six stratified copper deposits.

(3) Terminology and classification of copper ores according to temperature (T) and pressure (P).

(4) $T–P$ classification of copper deposits with examples.

(5) Division of thirteen copper-ore types based on major and minor elemental and mineralogical composition.

TABLE XCVII (continued)

Part 4

Classification			Characteristic			Example (reference)
Class	Group	Type	Geological environment	Mineralization	Other distinct features	
a	b	c	d	e	f	g
		(2.1.1.) Uranium-bearing migmatites and general metasomatic deposit.	Migmatites in gneisses.	Uraninite, thorianite, thorite, monazite, rare earth silicates, biotite, apatite, pyrite, fluorite, molybdenite, magnetite.		Charlebois Lake, Beaverlodge migmatites, Cardiff Mine, Normingo. Robinson (1958).
		(2.1.2.) Deposits formed by alkalic metasomatism in ferriferous rocks (Krivoy Rog type).	Quartzites and related rocks of Precambrian age with iron-ore mineralization, migmatitized and altered (by aegirinization, rhodusitization, albitization and carbonatization).	Uraninite, magnetite, hematite, aegirine, rhodusite, malacon, pitchblende, uranium silicate (nenadkevite), aragonite, graphite, pyrite, galena, marcasite, quartz, carbonates.	Uranium mineralization is related to the alkaline–silicate and carbonate metasomatism.	Uranium deposit at the edge of the Krivoy Rog iron ore-deposit. Reference of such type: Kotlyar (1961). Petrov et al. (1969).
		(2.1.3.) Pyrochlore-bearing fenites (Oka type).	Carbonate and alkaline rocks intruded into gneisses or other rocks.	Calcite, pyrochlore, betafite, niocalite, niobian perovskite. Uranium occurs as a minor constituent of	Replacement of original minerals in rocks by calcite, soda pyroxene, soda amphibole or chlo-	Quebec Columbium property near Oka, Quebec. Lang et al. (1962).

d metasomatic types of deposits

Epigenetic uranium deposits

			sulphides, garnet, diopside, scapolite, microcline, albite.		
(2.1) Impregn[ation]		carbonate rocks at the contact with intrusives (skarn type).			
	(2.1.5.) Pyrometasomatic deposits in intrusive or extrusive rocks.	Granite-porphyries, lava flows and tuffs with epigenetic impregnations of uranium mineralization mainly of metasomatic character.	Uraninite, uranothorite, pitchblende, rutile, phlogopite, feldspar, sulfides, carbonates, fluorite, etc. or some of them.	"Porphyry-uranium", "liparite formation", uranium mineralization in trachytic rocks etc.	Anonymously described deposits in U.S.S.R., Rexspar in British Columbia (Canada). Vol'fson (1968), Lang et al. (1962), Wambeke (1967).
(2.2) Veins and related types of uranium deposits	(2.2.1.) Davidite type.	Amphibolites, granites, gabbro. Lower structural level.	Davidite, diopside, scapolite, chert, carbonates, plagioclase, tourmaline, ilmenite, rutile, pyrite, chalcopyrite, pyrrhotite, molybdenite, apatite, etc.		Radium Hill, Australia. Nininger (1954). Tishkin (1966).
	(2.2.2.) Brannerite type.	Gneisses, amphibolites, granites, quartzites, albitites. Lower structural level.	Brannerite, zircon, apatite, ilmenite, pyrite, chlorite, rhodizite, magnetite, etc.		Deposits mentioned by Tishkin (in Vol'fson, 1966). Zavarzin (1961).

(continued)

TABLE XCVII (continued)

Part 5

Classification			Characteristic		Example
Class	Group	Type	Geological environment	Mineralization	Other distinct features	
a	b	c	d	e	f	g
		(2.2.3.) Pitchblende (with simple mineral associations) type (Pribram-Beaverlodge type).	Veins and related bodies in igneous, metamorphic and rarely unmetamorphosed sedimentary rocks.	Pitchblende, quartz, carbonates, iron oxides, sulfides, uranoan anthraxolite or some of these minerals.	Mainly pitchblende mineralization in carbonate, quartz-carbonate or quartz gangue. Alterations of wall-rocks. The boundaries between this and the next type are not sharp.	Most deposits in the Beaverlodge uranium-bearing area, Pribram uranium deposit etc. Robinson (1955), Klepper and Wyant (1957). Hruby-Sorf (1968). Harlass and Schuetzel (1965). Vol'fson, editor (1966, 1968).

nium deposits.

		Lake type).	rarely unmetamorphosed sedimentary complexes.	Ni, sulpharsenides, selenides, native silver, bismuth, arsenic, copper.	alterations.	in the Krusne hory (Erzgebirge) Mountains, and many others. Robinson (1958), Dymkov (1960), Mrna (1961, 1963, 1967), Sattran (1966) Ol'fson (1968), Campbell (1955), Jory (1964).
(2) Epigenetic uranium depos	(2.2) Veins and related types	(2.2.5.) Pitchblende (often in the sooty form) with hydrous aluminum silicates in shear zones (Rozna type).	Shear zones (often of regional character) in metamorphic rocks.	Sooty pitchblende, pitchblende, uranoan anthraxolite, hydrous aluminum silicates, graphite, fluorite, sulphides or some of these components.	Irregular orebodies, often structurally controlled. Wall-rock alterations developed as argillization, fluoritization, graphitization. Hematitization as a rule absent. (Various grades of alterations.)	Rozna, Zadni bhodov, Hermanicky and some other deposits in Czechoslovakia, deposits mentioned by Tananaeva (in Vol'fson, 1968), probably Rabbit Lake deposit at Wollaston Lake in Canada. Little, Ruzicka (1970). Vol'fson, editor (1968).

(continued)

TABLE XCVII (continued)

Part 6

Classification			Characteristic			Example (reference)
Class	Group	Type	Geological environment	Mineralization	Other distinct features	
a	b	c	d	e	f	g
igenetic uranium deposits.	f infiltration types of deposits.	(2.3.1.) Blanket type.	Deposits formed in porous sedimentary rocks from prevalently downward migrating uranium containing solutions.	Uranium oxides, silicates, uranyl-hydroxide, uranium vanadates, carbonates, sulphates, phosphates, sulphides, or various assemblages of those and other components with carbon, iron, vanadium, selenium, etc.		Mecsek deposit in Hungary. Barabas and Virag (1966). Saum and Link (1969).
		(2.3.2.) Roll type.	Deposits formed in porous sedimentary rocks from the solutions migrating pene-	Sooty pitchblende, coffinite, fluorapatite; vanadium, copper, silver, lead, zinc min-	Uranium is precipitated at the oxidation/reduction front.	Hamr deposit in Czechoslovakia, many epigenetic uranium deposits in sandstone

(2.3.) The group	(2.3.3.) Infiltration deposits in lignites or coals (Dakota type).	Deposits formed in lignites or coal similarly as the two (2.3.1.) and (2.3.2.) types or one of both.	phates; carbon-rich matter. Autunite, carnotite, torbernite, zeunerite, lignite or coal; material with small amounts of molybdenum, vanadium, copper, phosphorus, arsenic, germanium, beryllium, etc.	Uranium was introduced in the coal after coalification by waters. The participation of syngenetic processes may not be excluded.	editor (1963). Cypress Hills in Saskatchewan (Canada), uranium mineralization in Tertiary lignites and Permo-carboniferous coals within Bohemian massif. Denson et al. (1959), Cameron and Birmingham (1969), Skocek (1967), Lepka (1967).
(2.4) Transitional group.	(2.4.1.) Pipe type (Orphan pipe type) of collapse origin.	Ore solutions thought to be hydrothermal which used the permeability and porosity of rocks and possibly reacted with them.	Minerals containing uranium, copper, lead, zinc, molybdenum, cobalt, silver and nickel.	This type of deposit can be classified within a transitional group of epigenetic uranium deposits.	Orphan pipe at Grand Canyon (U.S.A.), perhaps an anonymous deposit mentioned by Velichkin (in Vol'fson, editor, 1968). Saum and Link (1969).

TABLE XCVIII

Highly specialized classification of Colorado Plateau ("Western States"-type) uranium deposits in sedimentary rocks (After Bostick, 1970; courtesy of American Institute of Mining and Engineering)

←———— Less favourable			
Average dip	70° (or any intensely deformed beds)	70–50°	50–30°
Depositional environment	tidal flat, marine shale, turbidite, dune, geosynclinal	overbank river, playa, "marine", "eolian"	alluvial plain, flood plain, coal swamp, lacustrine, nearshore marine
Rock class	limestone, dolostone, salt, anhydrite	shale, mudstone, siltstone, marl, sandy limestone, bituminous	sandy siltstone, "coal"
Sandstone/shale ratio of beds, *not* sorting of grains	<1/4 clastic/nonclastic <1/1	1/4–1/8 shale/nonclastic <1/1	>8/1
Colour (as reported on fresh surface if possibly)		red, purple	orange, reddish brown
Grain composition	limestone, shale, salt, evaporite	quartz sandstone, quartz arenite, orthoquartzite	quartz wacke coal
Grain size	nonclastic limestone evaporite	clay	silt
Sorting		very poor	poor
Organic matter	petroleum		silicified woody material
Uranium		known through exploration yet no uranium found	
Adjacent or included pyroclastics (tuff; ash; tuffaceous-bentonitic beds)	absent in unit and adjacent rocks	some present under unit	some present in overlying, but not contiguous units
Overlying beds		absent, gypsum, permeable sandstone, volcanic flows, limestone	
Underlying beds		absent, gypsym, permeable sandstone, volcanic flows, limestone	

		More favourable ⟶	
-20°	20–10°	10–5°	5°
ch, bar, lagoon, vaporite = "non-rine"	swamp, alluvial fan, shallow marine, delta, "continental"	lagoon (sandy), bog, river channel, non-marine or lacustrine delta	braided stream
areous sandstone, ite, low-grade coal	muddy sandstone	conglomerate peat	sandstone
–1/1 dstone/nonclastic 1	1/1–2/1 4/1–8/1	8/1–4/1	
ow, plish pink	brown, pink greenish yellow	olive, yellow, green tan, gray-brown	green, gray (except lime-stones)
osic wacke, osic arenite, aceous sandstone	lithic wacke, lithic arenite gray-wacke	feldspathic wacke, sub-feldspathic wacke, sub-graywacke	feldspathic sand-stones, feldspathic arenite, lithic sand-stone, subarkose
fine sandstone	fine sandstone	very coarse sandstone, conglomerate	medium sandstone, coarse sandstone
well	well	moderate	
seams		carbonized plant fossils	carbonaceous trash
e tests for $O_8 > 0.01\%$	some uranium mineralization or prospects with ore grade rock	some low-grade deposits or deposits: <1000 tons	deposits mined on large scale
e in unit	some over unit, some under	some over unit, some in unit	abundant over unit (unless in some younger beds)
ermeable, shale, areous sandstone, marl		semi-permeable, sandy shale	
ermeable shale, areous sandstone, marl		semipermeable sandy shale	

250

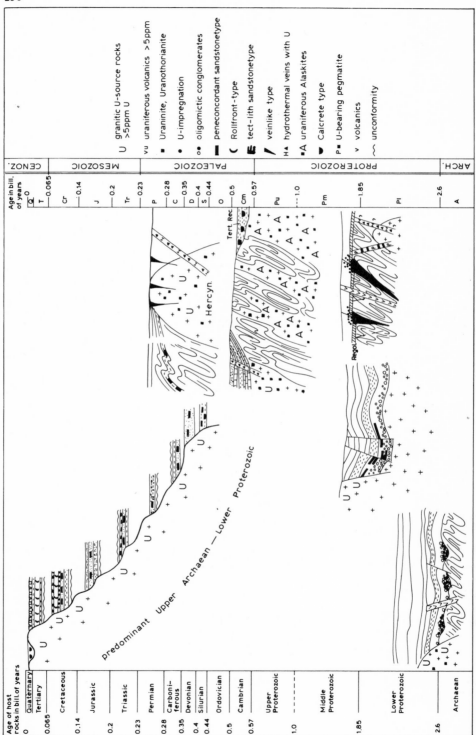

Fig.83. Geochronologic–stratigraphic distribution of uranium-host rocks and types of uranium deposits. (After Dahlkamp, 1978; courtesy of *Miner. Deposita*.)

TABLE XCIX

Uranium grade and potential of types of uranium deposits (After Dahlkamp, 1978; courtesy of *Miner. Deposita*)

Type of deposit	Symbol	Average U-grade in %	Total U-potential (mined reserve) ≤ S 15 p 1b		Mined as	
			Individual Deposit in tU	Uranium District up to max tU	Main product	Secondary product
Conglomerate-type	●	0.025–0.15	15 000–100 000	200 000	1 U 2 Au	– U
Sandstone-type	⊂~	0.2	5 000–25 000	150 000	U	Cu, V Mo, Se
Veinlike type	▼	0.2–2	10 000–250 000	450 000	U	Ni
Hydrothermal veins	▲	0.1–1	100–25 000	50 000	1 U 2 Co, Ni, Ag, Bi	U
Intraintrusive type	◆	0.04	10 000 ≥ 100 000	100 000	1 U 2 Cu	U
Calcrete-type	⊃	0.1–0.2	40 000	40 000	U	V
Black Shale Phosphorite-type	=	0.02–0.08	10 000–70 000	300 000	1 U 2 P	U

(6) Comprehensive classification of eighteen copper deposits founded on host rock and mineralogy, with examples.

(7) Metallogenic (or summary) classification of copper deposits based on geotectonic setting, host rock, structure and form of distribution, zonation and alteration products.

As in earlier volumes and in this chapter, the diagrammatic classifications find their application in offering an easily remembered and conveniently presentable "cartoon" or "pictorial model", of which Fig.84 depicts four plutonic, sub-volcanic and igneous-exhalative copper ores (A–C), six "magmatic" types (I–VI; with several sub-types, i.e., IVa, IVb, Va, Vb, Vc), and two "sedimentary" deposits (1 = weathering or infiltration; 2 = marine syngenetic–diagenetic varieties) (Borchert, 1950, 1978). A number of transitional and/or hybrid types have not been considered in such simplified cartoons, although it is usually much easier to present them pictorially than in pigeon-hole-type tables.

It has become clear from several sections in this contribution that numerous types of terminologies and classifications are the consequence of selectively or preferentially

TABLE C

Types of Uranium deposits (After Dahlkamp, 1970; courtesy of *Miner. Deposita*)

Era		Orogeny	Absolute age in mio years	Types of uranium deposits in order of their economic significance			
				1	2	3	further
Cenozoic	Phanerozoic	Alpine	0–70	C	⊃	—	=/p
Mesozoic		Laramide Cimmerian	70–225	—		C/	=/p
Paleozoic		Hercynian	225–600	▼		▲	
		Caledonian			—		=/s
Upper Proterozoic	Precambrian	Assyntian	600–900	◆	▲		
Middle Proterozoic		Grenville	900–1750				
Lower Proterozoic		Hudsonian	1750–2400	▼			
Archeozoic		Kenorian Laurentian	≥2400	●			

Legend
Type of uranium deposit

- ● Conglomerate - Type   ▼ Veinlike - Type   =/s Black shales
- — Sandstone - Type peneconcordant   ▲ Hydrothermal Veins
- C Sandstone - Type rolls   ◆ Intraintrusive Type   =/p Phosphorite
- C/ Sandstone - Type stack   ⊃ Calcrete - Type

chosen parameters. For example, if one takes the "concordancy" as the *fundamental* feature characterizing a large class of deposits, then a viable pragmatically established new type has been recognized in the past, namely the "stratified" or "strata-bound" copper ores. Table CI presents a sixfold division of these, with two sub-types.

The copper ores have been classified by Pélissonnier (1962) and Pélissonnier and Michel (1972) according to their depth (D) [or pressure (P)] (see Figs.16 and 17) and temperature of formation as depicted in Fig.85, which is divided into areas delineating theoretical "normal" conditions in the earth's crust (and along the surface) and the limit of crystallization of magmas (cf. two thick lines), as well as into zones ranging from igneous-telethermal to hypothermal and magmatic-segregation. In Fig.86 several ore districts have been plotted to reflect their inferred $T–D(P)$ of origin, whereas Fig.87 outlines the types of parageneses. The numbers in the latter diagram were then utilized in Table CII, which presents thirteen ore varieties according to composition founded on major elements (S,

253

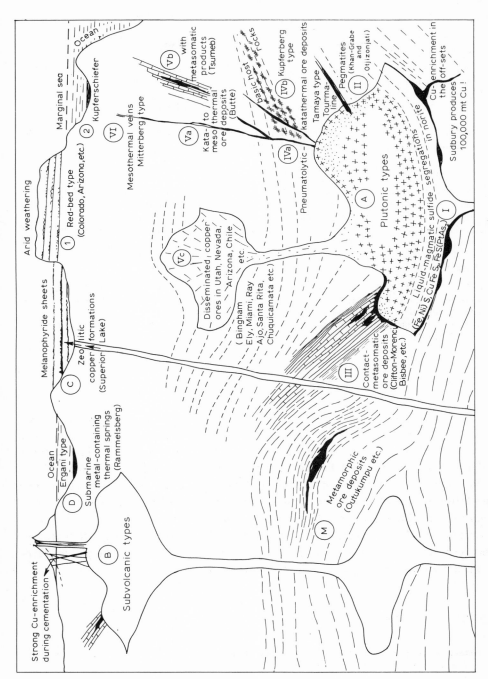

Fig.84. Genetic overview of the copper ore deposits. (After Borchert, 1950, unpublished; 1978; courtesy of Glückauf, Essen.)

TABLE CI

Classification scheme of stratified copper ore deposits (After Bogdanov and Kutyrev, 1973)

A — Sedimentary (sedimentary-diagenetic) deposits

   I. The copper slate type (shallow-water marine type): the Mansfeld deposits, Sudets piedmont monocline in Poland.

  II. The copper-sandstone type:

     1. The Dzhezkazgan sub-type (lagoon-deltaic sub-type): Dzhezkazgan, Sary-Oba, Itauz, etc., in Kazakhstan.

     2. The Pre-Urals subtype (lacustrine-alluvial sub-type): Naukat, Varzyk, etc., in Central Asia, the Colorado Plateau in the USA, the Pre-Urals region.

B — Volcanogenic—sedimentary deposits

   I. Related to underwater acid and basic volcanism (spilite—keratophyre, andesite—dacite formations, etc.): the uppermost parts of the section of some pyrite deposits in S. Urals, N. Caucasus, the "Kuroko"-type deposits in Japan.

  II. Related to terrestrial basic and acid volcanism (paragenesis of basaltic, liparite and variegated formations): Lake Superior and White Pine in the USA, the Minusinsk basin in the USSR.

C — Metamorphosed deposits

   I. Regionally-metamorphosed: Copper sandstone of the Olekmo-Vitim highland (Udokan, etc.), copper slates and quartzites of the Katanga-Zambia copper belt, pyrite deposits of Karelia, fahlbands of Norway and Canada.

  II. Contact-metamorphosed: the Krasnoe, Burpalinsk, and other deposits in the Olekmo-Vitim highland.

Fe, O), primary "marqueurs" (Sn, Mo, S, As, Ni, etc.), and minor elements (Zn, Pb, Ag, Au, As, Se, Te, Ba, Co, Bi, W, etc.). The seventeen types of copper ores are summarized in Table CIII — noteworthy is the combination of variables used, including tectonic setting and varieties of host lithology, in addition to the composition (Routhier, 1963). On the other hand Pélissonier and Michel (1972) utilized mainly the host-rock characteristics in the scheme offered here in Table CIV, presenting thirteen types (excluding metamorphically produced varieties). Interesting to note that Pélissonnier and Michel (1972) proposed a scheme that related the thirteen copper ore types to the orogenic (=geosynclinal) evolution in both space and time, while simultaneously dividing the ores into syngenetic and epigenetic varieties (see their fig.72).

MISCELLANEOUS CLASSIFICATIONS

"Sciences arise from the discovery of identity amidst diversity."
    W.S. Jevons (in: *The Principles of Science,* 1874)

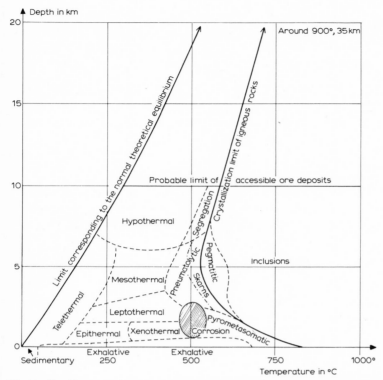

Fig.85. Classification of ore deposits according to their temperature and pressure of formation. (After Pélissonnier, 1962; courtesy of Bureau de Recherches Géologiques et Minières, Paris.)

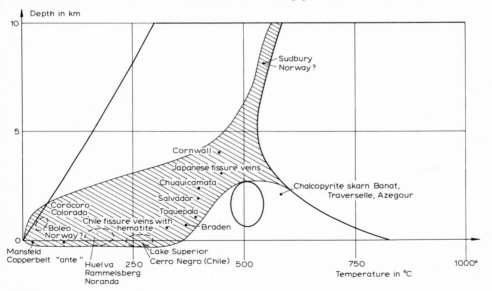

Fig.86. Classification of the copper ore deposits according to their temperature and pressure of formation. (After Pélissonnier, 1962; courtesy of Bureau de Recherches Géologiques et Minières, Paris.)

TABLE CII

Division of the types of copper deposits based on chemical elements (major, minor, "marker") and minerals, with examples (After Pélissonnier and Michel, 1972; courtesy of Bureau de Recherche Géologiques et Minières, Paris)

Type No.	Type of copper deposit	Major elements S, Fe, O	Primary marker elements	Frequent minor elements	Primary characteristic minerals	Characteristic deposit
1	Massive pyrites associated with volcanic rocks	I		Zn, Pb, Ag, Au, Se, Te, Ba	Pyrite, chalcopyrite, sphalerite	Huelva, Flin-Flon, Timmins, Cyprus Mount Lyell, Murgul, Röros, Lökken ...
2	Stratiform impregnations associated with volcanic rocks poor in pyrite	II–III		Ag, Zn	Bornite, chalcosine ...	Kosaka, Hanaoka, Cavallo, Bagacay, Mantos Blancos, Diablo ...
3	Native copper deposits associated with basic lavas	III		Ag	Native copper, chalcosine	Lake Superior
4	Stratiform disseminated deposits in sediments and poor in sulfur and iron	III		Ag, Co	Chalcosine, bornite, chalcopyrite. rite, native copper	Mansfeld, Dzhezkazgan, Zambia, Katanga, Boléo, White Pine, Timna, Corocoro ...
5	Cupriferous tin deposits	II	Sn	Bi, W, Ag	Cassiterite, stannite, chalcopyrite	Cornwall, Akenobé, Ashio, Mount Pleasant ...
6	Porphyrite deposits with molybdenum	II	Mo	Au, Re	Pyrite, chalcopyrite	Braden, Chuquicamata, Cananea, Bingham, Bisbee, Kounrad. Agarak, Majdanpek, Medet ...
7	Deposits with enargite	II	S, As	Ag, Pb, Zn	Enargite, pyrite	Butte, Cerro de Pasco, Morococha, Bor ...
8	Deposits with iron oxide and associated with basic intrusives	II	O (metallic oxides)	Co, Ni, Au	Chalcopyrite, oligiste, magnetite	Craigmont, La Africana, Tamaya, Messina, O'Okiep, Cornwall ...

TABLE CII (continued)

Type No.	Type of copper deposit	Major elements S, Fe, O	Primary marker elements	Frequent minor elements	Primary characteristic minerals	Characteristic deposit
9	Deposits associated with nickel-rich basic and ultra-basic intrusions	I	Ni	Pt, Co	Pyrrhotite, chalcopyrite, pentlandite	Sudbury, Norilsk, Petchenga, Kambalda, Empress ...
10	Deposits of quartz with chalco-pyrite	II	?	Co, Au	Chalcopyrite, pyrite, quartz	Allihies, Bou Skour ...
11	Deposits with siderite	II	Co_3Fe, Sb	Ni, Ba, Ag, Bi, Hg	Siderite, chalco-pyrite, tetrahedrite, barite	Mitterberg, Banca, Alzen ...
12	Deposits with arseno-pyrite or pyrrhotite and gold	I–II	Fe, As	Au, Ag, Bi	Mispickel, pyrrho-tite, chalcopyrite	Cobar, Boliden, Salsigne, Rossland ...
13	Deposits with tennantite containing germanium	III	Ge, As	Pb, Zn, Co, U, V, Mo, Ag, Cd	Tennantite, chal-cosine, renierite	Tsumeb, Kipushi ...

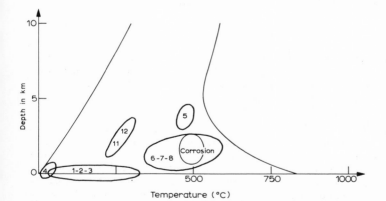

Fig.87. Position of the "paragenetic" types of copper deposits in the pressure–temperature diagram (types 9, 10 and 13 are not shown as their precise $P-T$ conditions of formation are uncertain). (See Table CIV.) (After Pélissonnier, 1962; courtesy of Bureau de Recherche Géologiques et Minières, Paris.)

TABLE CIII

Classification of copper ore deposit (After Routhier, 1963; cf. Pélissonnier and Michel, 1972, table V; courtesy of Bureau de Recherches Géologiques et Minières, Paris)

		Country rocks	Examples	Mineralogic association
In sedimentary rocks without visible relationship to any plutons. Dominantly stratiform.		1. Type in coarse-grained detrital formations, sandstones or schistose sandstones, especially red sandstone (red beds).	Corocoro (Bolivia). Oudokan (USSR) Zambia, in part: Mufulira (quartzites).	native copper, chalcocite, domeykite (CuAs), galena, barite, gypsum, celestite. Oxidative and cementation: native copper, cuprite, native Ag, covellite.
		2. Type in shales, more or less bituminous, more or less carbonated, with possibly Pb, Zn, Co, U.	Mansfeld (Germany): Kupfer-chiefer. Poland. Zambia, in part: Roan Antelope, Nkana, Chambishi. Very large deposits.	Chalcopyrite, chalcocite, bornite, galena, sphalerite, traces of Co, Mo, Ni, V, etc. Linneite, chalcopyrite, chalcocite, bornite, pyrite, sphalerite, minerals of U, quartz, sericite, chlorite, carbonates, tourmaline (rare).
		3. Types in carbonated rocks or schistosed dolomitic sandstones. *N.B.* Combinations and intermediates between the three types are very frequent. The vein variations are very frequent, examples: (Kipushi) Katanga. Oxidation and cementation are playing an important part, notably in Zambia and Katanga.	Katanga. Mindouli-M'Passa.	Cementation: chalcocite, covellite. Oxidation: hydroxides of Co.
		4. Type associated with volcanic tuffs, and other sediments, in volcanic regions (with andesites, liparites, etc.). (cf. 11)	Japan (Honshu). Le Boléo (Mexico).	"kuroko" ore: chalco-pyrite, galena, sphalerite, barite, Au, Ag.
Associated with granitic plutons which are frequently monzonitic.	Intraplutonic	5. Type as veins, with Cu arsenate, Zn (mesothermal); often lattice-like vein systems, crowded and intricate.	Butte (Montana). Very large deposits; with significant cementation. Chuquicamata (Chile).	Chalcopyrite, chalcocite, bornite, enargite, tennan-tite, tetrahedrite, pyrite, galena, sphalerite, Ag and Au in minerals of Cu and of Pb. Quartz, carbonate and silicate of Mn (dialogite, rhodonite). Cementation: chalcocite, covellite.

(continued)

TABLE CIII (continued)

		Country rocks	Examples	Mineralogic association
Associated with granitic plutons which are frequently monzonitic.	Intraplutonic	6. Porphyry copper type (or disseminated): reticulated veinlets mineralized in altered monzonite, also mineralized in apicale position. No arsenate.	Globe, Miami (Arizona). Ely (Nevada), etc. Bingham (Utah), in part. Kourad, etc. (USSR). Very large deposits especially if cementation is significant.	Chalcopyrite, bornite, chalcocite, pyrite.
	Marginal and periplutonic	7. Vein types, "sub-types": (a) Veins with chalcopyrite, grading with depth into stanniferous veins (hypothermal); (b) Veins with chalcopyrite, scheelite, tourmaline (transition between hypothermal and acidic extensions); (c) Veins of chalcopyrite, pyrrhotite, magnetite, molybdenite, tourmaline (transition between hypothermal and acidic extensions); (d) Veins of tetrahedrite-tennantite minerals dominate, very often with Ni–Co, with carbonate gangue.	Cornwall. Japan, in part. Japan (Yakuoji). Cobar (N.S. Wales). Very frequent but small in size.	Chalcopyrite, galena, sphalerite, cassiterite.
		8. Massive type is dominant, in carbonate rocks, in pyrometasomatic zones or external to it.	Bingham, in part. Clifton Morenci (Arizona). Tourinsk (Ural). Conception del Oro (Mexico). very large deposits.	Realgar, orpiment, chalcopyrite, bornite, chalcocite, enargite, tennantite, tetrahedrite with Au, pyrite (with Au), pyrrhotite, oligiste, cinnabar, molybdenite, quartz, calcite.
Associated with volcanic or subvolcanic rocks.		9. Disseminated type (cf. porphyry copper) at the apix of andesitic subvolcanic intrusions = apical subvolcanic.	Majdanpek (Yugoslavia).	Pyrite, cupriferous pyrite, pyrrhotite, magnetite, sericite, calcite, quartz.

(continued)

TABLE CIII (continued)

	Country rocks	Examples	Mineralogic association
Associated with volcanic or sub-volcanic rocks.	10. Volcano-sedimentary type in lenses and beds in tuffs overlying rhyolite flows. Probably related to 13.	Rio Tinto, province of Huelva (Spain). Mount Lyell (Tasmania). Large deposits. West Shasta County (California), etc.	Pyrite, chalcopyrite, tetrahedrite-tennantite, enargite, galena, sphalerite, mispickel. Ag, Au. Small amount of gangue: quartz, chlorite. Cementation of little importance.
	11. Veins or framework of veins in lavas, especially andesitic, post-orogenic varieties. The lavas are very frequently propylitized. Some stratiform deposits are tuffs associated with lavas (cf. 4).	Teniente = Braden (Chile). Bor (Yugoslavia). Cerro de Pasco (Peru). Japan, in part. Algeria, example Tell oranais (Cu and Pb). Large deposits. Le Boleo (Mexico).	Chalcopyrite, bornite, enargite, tennantite, panabase, quartz, tourmaline (volcanic hypothermal!). Cementation: chalcocite, covellite, bornite. The other deposits are of a lower thermal degree.
	12. Type associated with basic lavas (often spilitic), or in basic coarse-grained lavas of the geosynclinal phase (ophiolites).	Montecatini (Toscane). Norway, example: Lökken. Corsica. Saint-Véran (Hautes-Alps). Greece, etc. Very frequent, but small.	Chalcopyrite, bornite, chalcocite, tetrahedrite-tennantite, native Cu with zeolites. Cementation is important after post-dating mechanical processes.
	13. Type associated with acidic to intermediate lavas (dacites, keratophyres, rhyolites) and with tuffs of the geosynclinal phase. Metasomatism of siliceous wall rocks in rocks with sericite, chlorite, andalusite, cordierite, staurolite, almandine; of calcareous lenses in skarn. Without any doubt related to 10.	Boliden (Sweden), Falun (Sweden), Rouyn-Noranda (Canada).	Mispickel, pyrrhotite, galena, pyrite, chalcopyrite, gold (15 g/t at Boliden), silver (49 g/t); a little of Bi, Te (tetradymite, tellurobismutite) and Se.

(continued)

TABLE CIII (continued)

	Country rocks	Examples	Mineralogic association
Associated with volcanic or sub-volcanic rocks.	14. Type of native copper, with zeolites, disseminated in basalts (often spilitic) = "Cupriferous zeolitic formation", and concentrations in veins extending into sedimentary terrain. (Concentration during the cooling of the lavas, or during folding and metamorphism).	Lake Superior (U.S.A.). Type is frequent, but rarely important, the Lake Superior is exceptionally huge in quantity.	Ag, native copper, chalcocite, panabase, tennantite, quartz, zeolites, zoisite, chlorite, orthose.
	15. Volcano-sedimentary type ("exhalative"), associated with basic lavas (often spilitic).	Ergani (Turkey). Much of the pyritic deposits are more or less cupriferous. Norway: Leksdal, Stordö, etc. Corsica (Noceta).	
Associated with coarse-grained basic rocks: norites, gabbros-diabases, sometimes in "traps" (Siberia).	16. Type Ni–Cu–Pt, at the base or along the border of complexes dominated by norites, locally gabbros (with olivine), frequently transformed into amphibolites. Especially in Precambrian shields.	Sudbury (Ontario, Canada). Bird River (Manitoba, Canada). Petsamo (USSR). Montchegorsk (USSR). Norway, many small deposits. Insiawa (South Africa). Varallo (Piemont, Italy). In general important deposits.	Chalcopyrite, cubanite, pyrite, pyrrhotite, pentlandite, polydimite, magnetite, sperrylite, tellurides of Au and Ag. 14 recovered elements: Cu, Ni, Pt, Pd, Au, Ag, Rho, Ru, Ir, Co, Se, Te, iron, SO_2 (sulfuric acid).
Associated with ultra-basic rocks.	17. Type with a little Ni (and Co) at the proximity of serpentines or gabbros? Possibly grading to Ni–Co as at Bou-Azzer (Marocco). *N.B.* It is possible that the types 12, 13, 15 and 16 could be less obviously	Outokumpu (Finland)? (see 17').	

(continued)

TABLE CIII (continued)

	Country rocks	Examples	Mineralogic association
	separated as it appears in the table. The cupriferous deposits in or at the contact of ultrabasic rocks (perido-tites, serpentinites) seem to be very rarely present in economic quantities.		
In metamorphic terrains without visible relation to plutons.	Many of the above-cited types can be included in metamorphic sequences. For example, some Zambian deposits (types 1 and 2) are indicating weak meta-morphism. The lavas of types 12, 14, 15, with or without coarse-grained basic rocks are frequently metamorphosed as amphi-bolites and chloritic rocks ("chloritized cupriferous formations" of Schneider-höhn, 1962), those of type 13 in leptites. Type 16 is frequently metamor-phosed. It is obvious that a great number of cuprifer-ous mineralizations in metamorphic terrains pose problems of their associa-tion: either with granitic plutons, or with lavas, or with metamorphism.	12' ? Norway (Lökken, etc.), Japan,, New Caledonia, 13' Boliden, Falun (Sweden), Rouyn-Noranda (Canada) 14' Lake Superior, etc. 15' 16' Norway 17' Outokumpu (Finland)	Chalcopyrite (Cu : 3.7%); sphalerite (Zn : 1.07%) cubanite, pyrrhotite (with "linneite"); pyrite. Rare: stannite, magnetite, chromite. Very rare: pentlandite, galena, gold, 0.11% Ni, 0.20% Co (in pyrite and cobalt-pentlandite). Miner-alization in quartzites grading into metamor-phosed graphitic schists. In the quartzites, inter-calations of "skarns" with chromiferous minerals: diopside, tremolite, uvarovite, chromite. Very disputable origin; the most recent accepted hypotheses: — primarily sedimentary concentration — then selective mobiliza-tion, especially of Cu and Zn, through dynamic metamorphism (M. Saksela, 1957). It is never-theless difficult not to accept at least the source of chrome from the nearby serpentine rocks. *N.B.* Akjoujit (Mauritia) has not been sufficiently investigated (no publica-tions) as to its setting in the country rock to be included in this tabulation.

When perusing the literature it becomes obvious that a gamut of adjunct and auxiliary classifications and terminologies could be reviewed here, but space has set certain restrictions confining the treatment below to a few augmentary and supportive schemes as a sequel to the other sections. In the succeeding sector the listed schemes are pursued:

(1) Classification of endogenous deposits according to structures and shapes.

(2) Classification of ore deposits founded on shapes, host rocks, mineralogy, variability and exploration methods.

(3) Morphogenic (geomorphologic) classification.

(4) Metal versus lithofacies (host-rock) classification.

(5) Stratigraphic (unconformity) control on ore concentration (three schemes).

(6) Climatic subdivisions of ore-forming environments.

(7) Zonation as a basis for classification of ores and ore-forming milieux.

(8) Geochronological divisions of ore-rich periods.

Structural geology has maintained to be an essential, often pivotal, discipline in economic and ore geology, and thus has constituted the basis for some trenchant classifications; it developed into a complex and irreplaceable field of investigation consisting of numerous sub-disciplines that utilize nearly all fundamental physical and chemical sciences plus mathematics and computer technology. Adler (1970a, b) classified these disciplines and disclosed their interrelationships in tabular—diagrammatic form, simultaneously offering a summary of the nomenclature commonly employed. (The range of scale in structural studies has already been noticed above, cf. Kirchmayer, 1961.) The motivated reader is referred to the above-cited original publications — the classifications are equally applicable to structural and petrofabric analyses of ores. An auxiliary reference to be consulted is that by Lovering and McCarthy (1978) who outlined a classification of ore targets in order of increasing complication of concealment to be considered in geochemical exploration, i.e., overburden—structure relationships (cf. their fig.2, p.126).

Kreiter (1968) proffered a summary table of the major structures and shapes (=forms) of endogenous deposits, classified into six principal classes with a total of twenty-one types (position of metal concentrations in relationship to structure; Table CV). The augmentary Table CVI then lists the accumulations by shape, size, host rock, mode of occurrence, mineralogy, variability and chief exploration techniques.

The *morphogenic (geomorphologic) classification of ore* has proved to be expedient in specific cases, in both metalliferous and non-metalliferous (industrial and fuel) exploration and research; e.g., the schemes specify the properties of the reservoir rocks of hydrocarbons (oil and gas), i.e., beach-, dune-, shoestring-, fluvial-, blanket- or sheet-sandstones, and various carbonate reef types. The same classifications should prove valuable to ore explorationists in the search for mineralized clastic sediments and pyroclastics, and for Mississippi Valley-type concentrations in limestones/dolomites — not to ignore the placer deposits in recent to Precambrian epiclastic accumulations (see pertaining chapters in Volumes 1 to 7 and subsequent issues of this Handbook).

Surface-weathering results in mineral concentrations with a morphology that lends

TABLE CIV

Summary of the principal characteristics of the copper ore types (After Pélissonnier and Michel, 1972, table XXIII; courtesy of Bureau de Recherches Géologiques et Minières, Paris)

	Geotectonic framework	Depth of emplacement with respect to the surface of the lithosphere's surface	Characteristic rock association and/or country/host rocks	Shape of assemblage	Detailed shape	Hypogene alterations	Zonation
1	Pre-orogenic ("initial" volcanism during the geosynclinal evolution)	Zero	Volcanics (acidic and basic); tuffs, agglomerates and marine sediment associations	Concordant to semi-concordant lenses in horizons of lavas and sediments. Strong influence of superimposed folded tectonism	Massif, often layered, sometimes veins or disseminated	Variable (chlorite, quartz, sericite, ...); the strong alteration is never very extensive	Often strongly present and on various scales
2	Syn-orogenic	Zero	Andesitic to rhyolitic lavas, tuffs and sedimentary associations	Concordant to semi-concordant lenses in horizons of lavas and sediments	Very finely crystalline disseminations (Kuroko) or in irregular clots; sometimes massif (oko)	Strongly developed (quartz, chlorite, sericite, albite) and voluminous	Very strong in the kurokos [Cu, Pb, Zn, Ba—Cu Fe(Au)—SiO$_2$ (Cu Fe)]
3	Late-orogenic	Zero	Vesicular basalts, sometimes conglomerates	Concordant to semi-concordant lenses in horizons of lavas and sediments. Sometimes veins	Essentially disseminated (with exceptions in certain parts of the deposits of the Lake Superior native copper massif)	Deuteric alterations of the lavas; formation of adularia, epidote, pumpellyite, datolite, zeolite ... Decolorations	Lake Superior: vertical regional zonation cutting the stratigraphy (Cu—As: associated minerals)

No.	Orogenic relation	Alteration	Host rocks	Form	Habit	Wall-rock alteration	Zonation / typical ores
			dolomites, bentonitic tuffs, marls (in the substratum of volcanics)	form (Kupferschiefer), peneconcordant (red beds) in horizons of epi-continental sediments	disseminated	(decolorations)	(= typical), horizontal: (a) Cu–Pb–Zn (Mansfeld); (b) Chc–Bnite–Chpyrite (Zambia, Lubin) and vertical (chalcopyrite at margin, bornite and chalcosite at the center)
5	Late-orogenic	Weak	Batholiths, stocks, granitic subvolcanic massifs	Veins (stockwork, massive), intra- and extra intrusif	Various mixtures in gangue	Strongly developed, pneumatolytic to epithermal products (tourmaline, greisen, kaolin ...)	Spectacularly present in Cornwall: Sn–Sn, W–Sn(Cu)–Cu(Sn)–Zn(Pb)–Pb(Zn, Ag)–Fe and Mn carbonates. Telescoping in subvolcanic ores
6	Late orogenic	Weak	"Granodioritic" subvolcanic intrusions	Irregular massive in multifissured zones; breccia pipes (massive deposits, skarns and veins)	Disseminated in veinlets and associated fine clots	Very intensely developed and very large deposits ("secondary" quartzites, sericite, chlorite, kaolin ...)	Spectacularly present regionally and locally: (Cu–Mo)–Pb, Zn Ag, barite .... – hypogene alteration zonation
7	Late orogenic	Weak	"Granodioritic" subvolcanic intrusions	Area of veins; replacement, breccia pipes	Massive to disseminated mixed with gangue	Often intensely developed (silica, sericite, chlorite, barite, carbonates, ...)	Like type No. 6, sometimes complicated by telescoping
8	Late to post-orogenic	Weak	Dioritic to gabbroic intrusions; diorite veins; diabase sills	Massive in skarns; veins	Various mixture in gangue	Rather strongly developed, but only locally (chlorite, talc)	No clear observations

(continued)

TABLE CIV (continued)

	Geotectonic framework	Depth of emplacement with respect to the surface of the lithosphere's surface	Characteristic rock association and/or country/host rocks	Shape of assemblage	Detailed shape	Hypogene alterations	Zonation
9	Non-orogenic (at the boundary between the cratonic and disturbed zone)	Poorly defined, probably weak	Basic and ultrabasic intrusions	Lenses and beds, tabular concordant, in marginal zones and at the base of intrusions; veins, massive, intra- and extraintrusive discordant breccia	Massive, veinlets, spherulites, disseminated	Average (or intermediate?): deuteric alterations of the basic or ultrabasic intrusions (uralite, serpentine)	Lateral and vertical irregular changes. No recognizable systematic zonation
10	Post-orogenic	Weak	Schists or carbonate rocks. Relationships with basic intrusive	Veins with mineralized columns	Various mixture in gangue	Very weak	Grading into type no. 11

	probably weak	country rocks	eralized columns	massive to disseminated) in gangue	chlorite, silica	into type no. 12, grading vertically from tetrahedrite–barite association at the surface to the siderite–chalcopyrite association at depth into type no. 10.
12	Post-orogenic	Poorly defined, probably weak	Schists and sandstones, sometimes lavas	Veins, replacement near fractures	From massive to interstitial in gangue	Rather intense, but poorly developed: biotite, chlorite, kaolin, silica	Grading into type no. 11.
13	Post-orogenic	Poorly defined, probably weak	Carbonate series	Veins, replacement near fractures or pipes	Mainly massive, sometimes layered without gangue	Average: silica, calcite, graphite	Increase of copper at depth with respect to Pb–Zn

TABLE CV

Principal structures and shapes of endogenous deposits (After Kreiter, 1968)

Position of deposits relative to structure	Principal shape of ore bodies	Examples
I. Folded Structures		
1. Deposits in "favourable" horizons	Ore "beds", conformable blanket deposits	Zambia
2. Interstratal deposits in limbs of folds	Sheet veins, mineralized blanket breccias	Aurakhmat (USSR), N.Arkansas (USA)
3. Deposits at crests of anticlines and domes	Saddle veins	New Scotland (USA)
4. Deposits in zones of fracturing of diapir folds	Blanket deposits and stocks	Zyryanovka (Altai, USSR)
5. Deposits in "foliation" zones of block folds	Complex sheet and blanket bodies	Khaidarkan (USSR)
II. Disjunctive Structures		
1. Deposits in major thrust zones	Deep high-dipping veins of the ore-shoot type	Mother Lode (California, USA)
2. Deposits in major fault zones	Vein-like deposits, compound veins, occasionally accompanied by stockwork	El Oro (Mexico)
3. Deposits in lateral shifts and upthrusts of small throw	Veins with complex distribution of ore shoots	Kuludzhun (Altai, USSR)
4. Deposits in faults of small throw	Compound veins, fingering out veins, mineralized zones and crush zones	The Pennines (Great Britain)
III. Jointed Structures		
1. Deposits in shear joints of a single system	Simple and branching veins confined to joints of a single orientation	Coeur-d'Alene (USA)
2. Deposits in shear joints of two systems	Simple and branching veins confined to two systems of joints	Grass Valley (USA)
3. Deposits in shear joints of three and four systems	Simple and branching veins confined to three and four systems of joints	Freiberg (GDR)
4. Deposits in shear zones	Compound veins, often accompanied by elongated stockworks	Tatishan, Hsihuashan (China)
5. Deposits in small tension joints	Gash veins, simple veins	Gottlieb, Morgengang, Freiberg (GDR); Mazarron (Spain)

(continued)

TABLE CV (continued)

Position of deposits relative to structure	Principal shape of ore bodies	Examples
6. Deposits in joints associated with linear orientation in intrusives	Branching, occasionally conchiform veins	Zinnwald (Czechslovakia)
IV. Cleavage and microjointed structures		
1. Deposits in zones of schistosity and flow cleavage	Compound vein-like zones	Pyshma-Klyuchi (Urals, USSR)
2. Deposits in areas and zones of microjointing and fracture cleavage	Reticulated veins, stockworks	Tarbaldzhei (Transbaikal, USSR)
V. Contact structures		
1. Deposits at contacts of limestones with intrusions	Blanket deposits and lenses; shape is often complicated by metasomatism	Ingichke (Central Asia, USSR)
2. Deposits near contacts	Veins, short beds, pockets; ore bodies are controlled by fissures and strongly complicated by metasomatism	Koitash (Central Asia, USSR)
VI. Pipe-like and other complex structures		
1. Deposits in simple tubular structures	Pipe-like bodies	Pipes of Rostenberg (Bushveld, South Africa)
2. Deposits in complex pipe-like structures	Compound stocks and stockworks	Braden (Chile), Climax (Colorado, USA)

itself well to grouping, as long as it is realized that numerous transitional/gradational cases exist between the idealized end-member morphological types. Only one scheme is represented here, namely Smirnov's (1977) classification of nickel deposits formed by near-surface supergene processes (Table CVII). The three morphologic varieties (I to III) are listed horizontally at the top; with the definitions, conditions of origin, distributions, zonation, profile, host rocks or associated lithologies, shape, thickness, mineralogy, grades and some examples, along the vertical axis. Supplementing the information on the karst-type nickel concentrations, Smirnov (vol. 2) has provided the data in Table CVIII. (See also Zuffardi's Chapter 4, Vol. 3.)

The controls of stratigraphy, basin structures, paleogeography, depositional structures and unconformities (including diastems and solution features such as stylolites), have been evaluated for a considerable time by geologists, in particular by the petroleum industry. Some classifications of these multi-varied physical parameters influencing ore

TABLE CVI

Distribution of deposits by shape and size, variability and principal exploration methods (After Kreiter, 1968)

The table accepts Kallistov's suggestion of two types of variability: regular and random
The coefficient of variability V relates mainly to the quality of the mineral

Main forms of mineral bodies	Host rocks (and mode of occurrence of deposits)	Types of minerals (elements, minerals, or raw materials)	Variability of deposits	Principal system of exploration	Remarks
1. Sheets and sheet-like bodies, usually large	(a) Usually horizontal, loose, soft, mostly young sedimentary rocks, or mantle of waste	Manganese, iron, nickel, cobalt, sulphur, bauxites, ceramic clay, gypsum, salts, brown coals, magnesite, kaolin, many building stones	Regular	Vertical boring	
	(b) Predominantly alluvium and deluvium	Various placers with gold, platinoids, diamond, cassiterite, wolframite, rutile, ilmenite, monazite, zircon, columbite	Regular and random $50\% < V < 200\%$	Boring as above, with pits as checks	Only certain beach placers, and less often alluvial placers, are large
	(c) Ancient strong sedimentary and sedimentary–metamorphic variously dislocated rocks, and carbonate series (mostly low-dipping) less frequently sandstones	Butiminous coals and anthracites, coaly shales, iron and titanium ores, bauxites, phosphorites, uranium ores, numerous building stones, magnesite, barites (sometimes with sulphides), fluorspar, lead–zinc and copper ores	Regular and random	Chiefly vertical and inclined core drilling	Vein titanium and iron ores also fall under this heading

271

				drilled boreholes	
	basic and alkaline masses	beds with niobium, titanium, zirconium, thorium, rare earths, chrome spinellids, platiniferous rocks			
2. Large isometric and elongated stockworks, and large, irregularly shaped deposits	(a) Small intrusions of various composition and their host rocks	Titanium, copper, molybdenum, tin, tungsten, gold, carbonatites, nepheline syenite	Usually random, occasionally regular 50% < V < 90%	Vertical core and churn (cable) drilling (with check test pits)	Stockworks are explored mostly by cable (churn) drilling
	(b) Skarn and greisen fields; less often serpentinized, graphitized and talcized zones and sections	Iron, copper, lead, zinc, tungsten, molybdenum, tin, lithium, beryllium, asbestos, graphite, talc	Random 60% < V < 120%	Ditto	
3. Small to medium vein-like and lenticular bodies, and medium to small vein zones	(a) Basic and ultrabasic intrusive and effusive rocks; much more rarely strong sedimentary rocks	Copper–nickel ores (usually with platinoids), chrome-spinellids, in part lead, zinc, copper, sulphur, fluorspar, barytes, graphite, corundum, asbestos	Random V ≈ 80%	Drilling and underground workings combined with drilling from the surface	
	(b) Various effusive and strong sedimentary rocks	Tin, tungsten, molybdenum, gold, lead–zinc ores (with Ag, Au, Cd, Bi, Sb, As, In, Ge, Ga, etc.), five-element formation (U, Ni, Co, Ag, Bi)	Random V ≈ 130%	Underground workings with drilling from surface and underground points	In every case exploration begins with mining
	(c) Endo- and exo-contacts of granite intrusions and effusive rocks	Pegmatites with beryllium, tantalum, niobium, cesium muscovite, gold tellurides, phlogopite, etc.	Random V ≈ 200%	Underground workings with drilling from underground points, and underground openings alone	Ditto

(continued)

TABLE CVI (continued)

Main forms of mineral bodies	Host rocks (and mode of occurrence of deposits)	Types of minerals (elements, minerals, or raw materials)	Variability of deposits	Principal system of exploration	Remarks
4. Small, rarely medium-sized, sustained pipelike and branching bodies	(a) Ultra-basic and metamorphosed sedimentary rocks (sustained isolated pipes)	Copper, copper with gold, tungsten, copper–nickel ores (usually with platinoids), platiniferous hortonolites and dunites	Random, V very high	Underground openings, or underground workings with underground drilling	At least the first one and a half to two horizons are explored by underground working alone
	(b) Carbonate rocks skarns (branching, pipe-like bodies)	Tin, lead–zinc ores (with Ag and many other impurities)	Ditto	Ditto. Usually exploration and exploitation are conducted simultaneously	It is even more imperative that not only the initial exploration, but sometimes all of it, should be done by underground workings
5. Small pockets, lenses, veinlets, pipes, cellars, located in definite zones or at random	(a) Endo- and exo-contacts of granite intrusions (pegmatites), skarns, effusive rocks, secondary quartzites, limestones (the bodies are confined to oriented zones)	Beryllium, tantalum, niobium, caesium, tungsten, piezo-quartz, Iceland spar, optical fluorspar, emery, corundum	Random, V cannot be calculated	Underground workings	
	(b) Ultra-basic rocks pegmatites	Platinum, precious, and optical minerals	Ditto	Ditto	

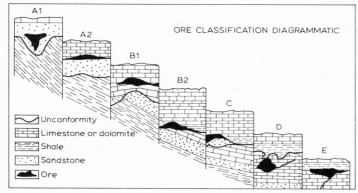

Fig.88. Classification of ores according to mode of association with unconformities. (After Mills and Eyrich, 1966; courtesy of *Econ. Geol.*)

localization have been made available, although more comprehensive systematic work remains to be performed. Wolf (1976c), for example, has reviewed some data indicating similar controls on both the formation of hydrocarbons and metalliferous and non-metalliferous deposits. It has long been appreciated that unconformities control the localization and rate of fluid movements — may they be ore-forming solutions or hydrocarbon-carrying compaction waters. Mills and Eyrich (1966) (cf. also Nicolini, 1970, figs.142 and 143) expounded the modes of controls by unconformities in the origin of ores and they stated that the principal role has been to provide, directly or indirectly, porous and

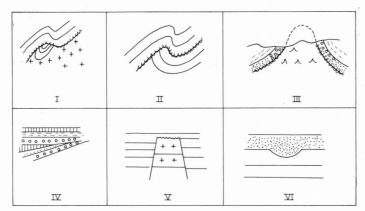

Fig.89. Types of effects along basement — cover-rock unconformities on mineralizations. (After Nicolini, 1970; courtesy of Gauthier-Villar, Paris.) *I* = Cover-rock on crystalline or metamorphic basement; *II* = cover-rock on folded sediments with major discordance; *III* = cover-rock on diapir-type of basement; *IV* = cover-rock on wedge-shaped stratigraphic unit with slight discordance or discontinuity (hiatus); *V* = cover-rock on horst; *VI* = paleo-channel eroded into subjacent formation.

TABLE CVII

Morphogenic classification of nickel deposits of weathering (After Smirnov, 1977)

Morphological type of deposits	I. Areal type		II. Linear type		III. Linear–areal type	
	Ia. Deposits, associated with weathering crust of cerolite–nontronite–ochre profile	Ib. Deposits, associated with weathering crust of cerolite–ochre profile	IIa. Linear-jointed deposits	IIb. Contact-karst deposits	IIIa. Jointed-areal deposits	IIIb. Karst-areal deposits
Brief definition						
Conditions of formation and features of distribution of deposits	On uplifted plateaux. Weathering and ore-formation have occurred mainly in zone of aeration in ultramafic rocks, often at contacts with dykes of basic and acid rocks		In regions with mountainous relief. Weathering and ore-formation have occurred below groundwater level along zones of tectonic faults in ultramafic rocks and at contacts between latter and marbles		Under conditions of hill–valley and low-mountain relief. Weathering and ore-formation have occurred in the zone of aeration and below ground-water level	
Zonation of weathering crust	Vertical		Horizontal		Horizontal and vertical	
Profile of weathering crust	Cerolite-non-tronite-ochre	Cerolite-ochre	Cerolite-ochre	Cerolite-ochre with underlying karst formations	Combination of linear and areal weathering crusts with complete or reduced profile	Combination of contact-karst and areal weathering crusts with com-plete or reduced

rocks in lithological sequence of deposit	...cherous-siliceous formations, non-tronites, and leached serpentinites	...cherous-siliceous formations, and leached serpentinites	...formations, and leached serpentinites	...serpen-tinites and other rocks	...tronites, and leached serpentinites	...and leached serpentinites
Shape of ore bodies	Layer-like and lens-like dimensions	Ore bodies of large size	Wedge- and layer-like, and lensoid	Layer- and nest-like	Layer- and wedge-like and lensoid	Layer-like, pocket-like and lensoid
Thickness of ore bodies	3–20 m	3–20 m	3–10 m	3–30 m	3–10 m	3–30 m
Most common and typical ore minerals	Nontronite, cerolite, hydrochlorite, nickel cerolite, halloysite, goethite, manganese minerals, and prasopal	Goethite, hydrogoethite, cerolite, hydrochlorite, and serpentine	Cerolite, serpentinine, hydrogoethite, goethite, nickel cerolite, and asbolan	Nickel silicates, hydrochlorites, nickel cerolite, ferrihalloysite, goethite, hydrogoethite, asbolan, and chrysoprase	Hydrogoethite, nontronite, goethite, cerolite, nickel silicates, and asbolan	Nickel silicates, hydrochlorites, nickel cerolite, and goethite
Predominant technological grade of ore	Ferromagnesian and magnesian	Ferrous and ferromagnesian	Ferrous and ferromagnesian	Ferromagnesian and silicomagnesian	Ferrous and ferromagnesian	Siliceous and alumino-siliceous
Typical deposits — in the USSR	Aidarbak, Kempirsai, Batamshinsk, Taiketken, Kapitanovo, Devladovo, etc.	Serovo and small sectors of Buruktal	Rogozhinsk	Cheremshan, Sinar, and other deposits of the Ufalei and Polevsk regions of the Urals	Anatol'sk and Yelizavetinsk	Lipovo
Typical deposits — abroad	Cuba		New Caledonia	Lokrida (Greece)		

TABLE CVIII

Principal nickel ore types among karst formations on the basis of composition (After Smirnov, 1977)

1. Karst formations consisting of the eluvium of serpentinites:
   (a) ocherous–siliceous formations
   (b) ocherized leached serpentinites
   (c) leached serpentinites
2. Karst formations consisting of decomposed chlorite, chlorite–tremolite, and talc–chlorite rocks
3. Karst formations of granitic compositions
4. Karst formations of gneissic compositions
5. Karst formations of mixed compositions

permeable plumbing systems for the circulating fluids or for the precipitation of valuable epigenetic mineral concentrations. Callahan (1964) has discussed the influence of unconformities on strata-bound mineralization in carbonate rocks, the premise being that unconformities and diastem systems are of great lateral and vertical extent, very numerous in some stratigraphic sections, offer porosity and permeability along and adjacent to them

TABLE CIX

Six epigenetic deposits associated with unconformities (cf. Fig. 38) (After Mills and Eyrich, 1966; courtesy of *Econ. Geol.*)

A. Epigenetic ore deposits related to indigenous (primary) rock openings of coarse clastic or volcanic rocks above an unconformity.
   1. Ores in clastic or volcanic rocks, especially in ancient stream channels on the old erosion surface, e.g., uranium deposits in clastics with possible leaching of volcanic detritus.
   2. Ores in reactive rocks overlying coarse clastic rocks above the unconformity, e.g., Mississippi Valley-type ores, SE Missouri.
B. Epigenetic ore deposits related ro rock openings (induced or secondary) created by folding or fracturing that has been localized by an unconformity.
   1. Ores near irregularities on the old erosion surface (unconformity), e.g., folding initiated by differential compaction of sediments deposited over old buried hills on the erosion surface and accentuated by later compressive stresses, may create low pressure zones of increased porosity and permeability in the younger strata, e.g., SE Missouri.
   2. Ores in brecciated, fractured, and sheared zones along and adjacent to an unconformity, e.g., Arkansa barite region.
C. Epigenetic ore deposits related to igneous rock bodies intruded along an unconformity.
D. Epigenetic ore deposits related to rock openings (induced), especially subterranean caves, sinkholes, drainage channels, and their associated fractures and breccias, produced by meteoric water dissolution of carbonate rocks below an old erosion surface (unconformity); e.g., East Tennessee zinc district, Upper Mississippi Valley zinc–lead deposits, and Northern Arkansas zinc–lead district.
E. Epigenetic ore deposits related to impermeable barriers above an unconformity; ores in carbonate rocks below the unconformity; e.g., Cave-In-Rock, Illinois, fluorite deposit, Upper Mississippi Valley zinc–lead, Tri-State.
F. Epigenetic ore deposits localized at/near unconformities for as yet unknown reasons.

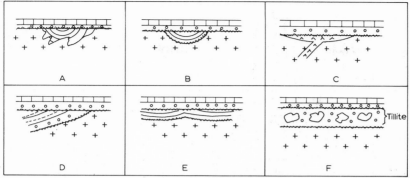

Fig.90. Types of effects along "transitional (=intermediate) basements" – cover-rock unconformities on mineralizations (see figures and A to F explanations for clarification. (After Nicolini, 1970; courtesy of Gauthier-Villar, Paris.) *A* = Intermediate basement constituted of segments of un-granitized units; *B* = intermediate basement constituted of folded sequences which are discordant in the crystalline basement per se; *C* = intermediate basement constituted of volcanic deposits; *D* = intermediate basement constituted of a stratigraphic wedge; *E* = intermediate basement above, which is turn in overlain by a local discontinuity, but regionally concordant with the cover-rock; *F* = intermediate basement as a result of a major climatic change or due to other causes, but without a discontinuity.
* Note: Nicolini uses "intermediate" for those lithologic units that are "intermediate" in some of their properties between the genuine crystalline "basement" and the "cover-rock".

and, consequently, serve as potential conduits for fluids. Every unconformity, either directly or through another unconformity, has connections with the present land surface, so that local unconformities are merely lateral extensions of much more persistent unconformities. Two of the above-cited publications are reviewed here.

Mills and Eyrich (1966) described seven "epigenetic" (according to others some could be "syngenetic–diagenetic") deposits associated with unconformities (Fig.88), with a summary presented in Table CIX.

A broader view-point on the effects of unconformities controlling the localization of ores has been taken by Nicolini (1970), of which the two self-explanatory diagrams (Figs.89 and 90) list twelve "end-member" possibilities (i.e., ignoring transition and combined cases).

Climatic classifications and models related to ore genesis and indirectly, therefore, to geographic (or paleogeographic) variables have been discussed in the present Handbook wherever relevant. (See, for example, Wolf, Chapter 2, Vol. 1; Veizer, Chapter 1, Vol. 3; Wopfner/Schwarzbach, Chapter 2, Vol. 3; Lelong et al., Chapter 3, Vol. 3; Zuffardi, Chapter 4, Vol. 3; Samama, Chapter 1, Vol. 6; Smith, Chapter 8, Vol. 6; Pretorius, Chapters 1 and 2, Vol. 7; Rackley, Chapter 3, Vol. 7.) One supreme three-volume treatise, summarizing the climatic factors in sedimentology and ore-genesis (Mn, Fe, P, and Pb–Zn), has been compiled by Strakhov (1967 to 1970). Let us consider three additional models.

Table CX (Borchert, 1978, p.87) conceptualizes the governances of five climates in

TABLE CX

Types of rock weathering dependent on the climatic belts of the earth (After Borchert and Jackisch, 1966; Borchert, 1978; courtesy of Glückauf, Essen)

		Tropical area			
		North →			Equator
Cold climate (cold humid)	Temperature climate (temperature-humid)	Desert climate (hot-arid)	Subtropical oscillating climate (with seasonal change from rainy to dry times, i.e. monsoonal)		Tropical climate (hot-humid)
Tundra-facies	Forest-facies	Red-bed facies	Steppe facies	Scrub-forest facies	Rainforest-facies (evergreen)
Acidic $Fe(HCO_3)_2$-solutions↓↓↓ ↓SiO_2-residue↓ (often glass sand)	In only weakly acidic soil-solutions, the Al_2O_3 and SiO_2 will not be separated	Predominance of insolation and mainly mechanical weathering at highest oxidation potential	Rhythmically changing oxidizing and reducing conditions as well as rhythmically removed acidic and alkaline soil-solutions (potassium adsorption)		Frequently high groundwater table
Hard-pan formation at high groundwater level, with precipitation of $Fe(OH)_3$ and AlOOH, when groundwater is O_2-containing	Ordinary clay minerals will form groundwater level				Al_2O_3 and SiO_2 will not be separated Formation of ordinary clay minerals from feldspar-rich rocks
Removal of dissolved alkalies, etc.	fluctuation only insignificantly, seasonally and secularly				

Residue of different bauxite types

$Fe(HCO_3)_2$ · $NaHCO_3$, SiO_2

$(Fe, Mn)(HCO_3)_2$ · $NaHCO_3$, SiO_2

oscillating groundwater level

$Fe(HCO_3)_2$ · $NaHCO_3$, SiO_2

Strongly oscillating groundwater level

Under the influence of strong, but often from SiO_2-poor, primate rocks bauxites clay- spar- and clay $NaHCO_3$; SiO_2

Deepest Groundwater reservoir

Iron will be retained, particularly in lateritic residue, under strongly oxidizing conditions. Iron ores of the Conakry- and Cuba-types form on peridotites, often with Ni, but also with Cr-contents

Iron and manganese will be intensely removed in humus-acid-rich solutions near the end of the rainy season, the dissolution of the SiO_2 is impeded

Thus originate Fe-poor, but often still SiO_2-rich bauxites from feldspar- and clay-rich rocks

the origin of six "facies" with varying physicochemical conditions, as reflected by the behaviour of Fe, Mn, Al, SiO_2 as well as by clay minerals' genesis, and laterite/bauxite with possible Ni–Cr concentrations. Particularly relevant to the explanation of the origin of surface-formed ores is Table CXI, of which ten are listed (Borchert, 1978, p.86). The ten ore varieties (i.e., Fe, Mn, Al and SiO_2 types) fall into three climatic zones, namely cold to temperature-wet, temperature-wet to tropical-wet, and subtropical (alternating) "oscillating" (dry-to-wet, monsoonal, savannah). Finally, Table CXII is a contracted diagram of the preceding two and depicts natural restraints under which bauxite originates.

The classification of metallic and non-metallic ore accumulations founded on the *host or country rock plus associated lithologic facies* (the latter either in the immediate vicinity of the ores or somewhat remote, but either genetically related to the mineralizations or at least stratigraphically equivalent), has become a well-accepted practice (e.g., Stanton, 1972; Gabelman, Chapter 3, Vol. 1; Gilmour, Chapter 4, Vol. 1). One auxiliary exemplar will suffice here: Table CXIII (Nicolini, 1970) presents twelve stratiform ore varieties and their respective host rock(s). Such diagrams give a utilitarian overview, in particular if they are detailed and as all-inclusive as convenient. With the availability of new data, occasional up-dating and/or expansion may be necessary; for example, the association of the Homestake-type Au-deposits with the Precambrian Fe-rich sedimentary rocks (Sawkins and Rye, 1974) could be included in Table CXIII.

Classificatory models of zoning (environmental, geochemical, mineralogical, litho-logic-facies, stratigraphic, and textural/fabric) can be useful predictive tools in research and exploration investigations. Again we are restricted to one convenient example (Figs.91 and 92) that depicts zoning as a result of or as a response to (1) environmental factors (shallow to deeper water), (2) geochemical controls (O_2-, CO_2- and H_2S-zones, with corresponding pH, Eh and other changes), (3) mineralogy (e.g., limonite–glauconite–chamosite–chert–pyrite), (4) lithology–facies–stratigraphy (sandstone–ferruginous sandstone/limestone–claystone–siliceous shale–etc.), and (5) textures (oolites-to-bituminous mud, changes in matrix amount, etc.). (See Borchert, 1960a, 1965; also Smirnov, 1976, fig.26; and Stanton, 1972, fig.13-4.) Zoning has been implicitly presented in numerous diagrams in this chapter as well as expounded in several chapters of the earlier volumes (cf. indexes in Vols. 4 and 7).

Classifications reflecting the stratigraphic range and/or distribution of mineralizations are partially augmenting the contribution by Veizer (Chapter 1, Vol. 3) on *evolutionary trends of specific geological deposits* from the Precambrian to the Recent. Such data, which are connected with metallogenesis *per se* considered in a separate discussion below, deal either with particular metallic and/or non-metallic materials of purely physicochemical origin or with organically produced accumulations, or both. As a consequence of the large all-inclusive scope of these studies, investigators may have to confine themselves to narrower fields, i.e. to either one or to a few geological products (e.g., iron, manganese, or iron + manganese, laterite soils, bauxite, or laterite + bauxite, silver, gold, or

TABLE CXI

Behaviour of Fe, Al and SiO_2 in weathering milieu (After Borchert, 1966, 1978; courtesy of Glückauf, Essen)

Climate: cold to temperature-humid

Hard pan	Lake-, swamp- and bog-iron-ores	Pisolitic iron ore	Hunsrück-type	White siderite in peat

Parent-material: Fe-, Mn-, Al- and SiO_2-containing rocks

Humus-acidic solutions:

Fe is dissolved as $Fe(HCO_3)_2$, iron-humate etc.
Mn is dissolved as $Mn(HCO_3)_2$
Al is dissolved as Al^{3+} cation or AlO_2 anion

Bleaching of the layers above the groundwater table (residue: quartz)	O_2-poorer groundwater with $Fe(HCO_3)_2$ and/or protective colloids prevent premature precipitation			carbonic-acidic humus waters with $Fe(HCO_3)_2$
	oxidizing milieu			reducing milieu
O_2-rich ground-water results in precipitation of $Fe(OH)_3$ and AlOOH as hard pan, high ground-water table — removal of alkali-salts	precipitation from stronger O_2-containing waters in exchange with the atmosphere	precipitation during exchange with O_2-surface waters circulating in underlying lime-stones	precipitation as a result of reaction with limestones: $Fe(OH)_3$ and $Mn(OH)_4$ are precipitated — $CaCO_3$ goes into solution, partly also formation of siderite	precipitation as siderite because of an increase in the Eh- and/or pH-values

silver + gold, evaporites, coal, oil, etc.). But eventually with the new information literally "exponentially" expanding, a world-wide data-synthesis will permit the reconstruction of a comprehensive conceptualization of the chronologic distribution. Certain trends (e.g., see Veizer's chapter) are already well-founded, although modifications and refinements are to be expected. With computer techniques now well developed, in the future much of the seemingly fragmental, uncorrelated data will disclose interconnections. For this progress, and for the application of data-processing, geological nomenclatures and classifications (as well as methodologies) more-or-less agreed-upon are to be promoted and implemented. If this is *not* done as a prerequisite, only the *major* geochronological trends based on the *major* geological variables will be considered, because the geological subtleties were ignored in the classifications — thus to discover the effects of minor geological parameters

Climate: temperature-humid to tropical-humid			Subtropical oscillating (monsoonal climate)		
Clay minerals	Kaolinite pH < 7 Montmorillonite pH > 7	Phosphorite	Basalt-iron-ore and Conakry-type	Laterite	Bauxite
Feldspar-containing rocks		apatite-containing rocks	basalt, peridotite	Fe-, Mn- and feldspar-containing rocks	
Decomposition of feldspars in weakly acidic to weakly alkaline soil solutions. *No separation of Si and Al occurs*		mobilization of PO_4^{3-} in carbonic acidic waters weakly acidic or alkaline solutions with low Eh values	*dry season* initially reducing conditions, dissolution of Fe as $Fe(HCO_3)_2$. dessication increases the oxidation potential *precipitation of relatively insoluble Fe(OH)₃* increasing concentration of the soil-solutions increase of the alkali-content; soil solutions react alkaline		
Clay minerals occur as residue and/or originate as weathering-neoformations:				*precipitation of Al(OH)₃ in residue*	
through weakly acidic solutions	through weakly alkaline solutions	precipitation by interaction with carbonates	SiO_2 is being prepared for dissolution *rainy season* dilution of the electrolyte concentration in the soil-solutions, rising of groundwater table, SiO_2 will be partly removed by seepage water → silicification of the surrounding ("desert varnish"). Plant growth creates again reducing conditions through the formation of humus and CO_2.		
Low Al_2O_3-content, higher SiO_2-content, kaolinite pH < 7	higher Al_2O_3-content, lower SiO_2-content, montmorillonite pH > 7				
Intermediate and moderately fluctuating groundwater table					

recorded in the rocks from the Precambrian to Recent we have to refine our present approaches and inaugurate newer techniques and concepts (including classifications).

As mentioned above, both metallic and non-metallic anomalous concentrations (i.e., above the Clarke values) are to be audited in the future preparation of geochronological models — which may reveal hitherto unsuspected genetic inter-dependencies. To mention only two questions to be answered: (1) what are the relationships between evaporites and the occurrence of certain types of mineralizations, and (2) is there a genetic connection between petroleum and ore formations (cf. review by Wolf, 1976c)?

The five ore concentrations and one non-metallic deposit listed below, varying through geologic time, have been choosen selectively from a gamut of available exemplars:

TABLE CXII

Genesis of bauxite deposits (After Borchert, 1959, 1978; courtesy of Glückauf, Essen)

climate: cold (tundra- and swamp-bog facies)	temperate-humid and tropical-humid (forest-facies)	subtropical oscillating (monsoonal) climate of dry- and rainy seasons (savanna-facies)
humus-acidic solutions with $(HCO_3)^{,}$ and low pH-values	soil-solution is weakly acidic to neutral to weakly alkaline	exchange between acidic and alkaline soil-solutions as well as between reducing and oxidizing facies
residue of only quartz sand ↓ ↓ ↓ ↓ ↓ ↓ ↓ solution of Al and $Fe(HCO_3)_2$	residue and/or weathering neoformations: clay minerals (kaolinite, montmorillonite, halloysite)	↑ insignificant mobilization ↕ of Al_2O_3 in acidic solu- ↓ tions
	preferential adsorption of calcium	
precipitation of $Fe(OH)_3$ and AlOOH as hard pan during O_2-containing high groundwater	absence of separation of SiO_2 and Al_2O_3	↑ varying mobilization of ↕ $Fe(HCO_3)_2$ and fixation ↓ of $Fe(OH)_3$
removal of alkali-salts	Si is relatively more mobile than Al	
	moderately oscillating water table	mobilization of SiO_2 in weakly alkaline soil-solutions

bauxite world production in 1000 mt

1903	1913	1923	1939
33	250	560	4350
1947	1949	1951	1955
6306	8548	11 100	16 800
1957			
19 200			

low strongly oscillating groundwater table

removal of SiO_2, silicification of the surrounding rocks required

Intense decomposition in a subtropical oscillating (monsoonal) climate

bauxite-residue on clayey limestones and dolomites (karst- and doline-facies)

reworked bauxites

· reworked bauxites

+   +   +   bauxite ⊤
          as weathering residue from
  +   +   feldspar-containing rocks

+   +   +   +   +   +   +

(1) Bauxites' stratigraphic/chronologic distribution.
(2) Iron-ore: genetic types versus chronologic occurrence.
(3) Manganese deposits; stratigraphic versus host-rock types.
(4) Beryllium mineralization's metallogenic epochs.

TABLE CXIII

Classification or distribution of stratiform metalliferous deposits based on the lithologies of the host rocks (After Nicolini, 1970, table III, p. 60; courtesy of Gauthier-Villar, Paris)

Metals \ Lithofacies	CLASSICAL LITHOFACIES							PARTICULAR TYPE OF LITHOFACIES				
	Sedim. breccias	Conglom.	Sand-stone	Psam-mites, pelites	Clay, argillite	Lime-stones	Dolo-mites	Evapo-rites	Silica	Biochem. rocks	Carbonate rocks	Volcanism
Cu		◇	◇	◇	◇		◇				◇	◇
Pb–Zn	◇ (zn)	◇	◇	◇	◇	◇	◇		◇	◇	◇	◇
Fe (oxide)	◇	◇				◇			◇	◇		
Mn			◇	◇	◇	◇						◇
U		◇	◇	◇	◇					◇	◇	◇
V		◇	◇		◇					◇	◇	◇
Au		◇			◇					◇	◇	◇
Ag											◇	◇
Mo											◇	
Hg			◇		◇	◇						◇
Ni–Co											◇	
Pyrite											◇	◇

N.B. Thickness of the symbols is roughly proportional to the economic importance; cross-hatching = important potential.

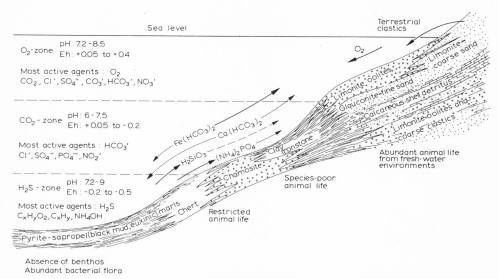

Fig.91. Schematic model of the petrographic and physicochemical-environmental facies of the most important marine iron minerals (cf. Fig.92). (After Borchert, 1960, 1965 and earlier, 1978; courtesy of *Trans. Min. Metall.*)

(5) Chronologic/metallotectonic distribution of ten stratiform deposits.

(6) Chronologic relationship between the earth's mega-tectonic cycles and formation of evaporites.

Reference has been made on several occasions to bauxite (and related concentrations such as laterite, Ni-rich weathering products), so that the geochronological distribution data combined with the "genetic-geological environmental" subdivision of Smirnov's (1977) is a welcome supplement (Fig.93). Simultaneously, the multiformity and multiplexity in classifying the same commodity is demonstrated once more. In numerous instances, the work by Smirnov is restricted to the USSR and adjacent localities, and then compared with the world average; thus illustrating the "norm" concept. We should aim in future data compilations to undertake first intra-continental and inter-continental comparative studies, then establish a world "norm" or average, and finally compare the data of specific regions and continents with this norm.

In addition to merely listing in Fig.93 the bauxite in "relative units" (=quantity), the deposits are described in terms of type in Table CXIV (see Fig.93 depicting a morphogenic classification, which is partly related to mode of genesis) and tectonic milieus. One should also refer to Smirnov's discussion on the typification of bauxites (Smirnov, 1977, vol. 1, pp.275—277) and to Table CXV.

Superimposed on such simplified (mono-, bi- or tri-variant) conceptual models, one should attempt to insert numerous supplementary parameters, e.g., climatic factors, progenitor rock(s) from which bauxite/laterite was derived, mineralogical and geochemi-

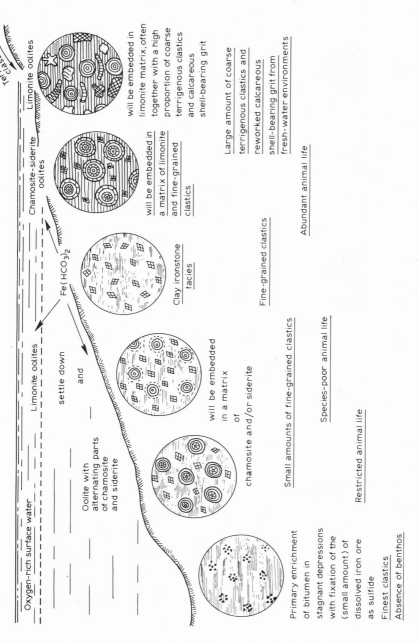

Fig.92. Schematic model of the lithologic, mineralogic and textural variations of the marine iron deposits (cf. Fig.91). (Borchert, 1960, 1965 and earlier, 1979; courtesy of *Trans. Min. Metall.*)

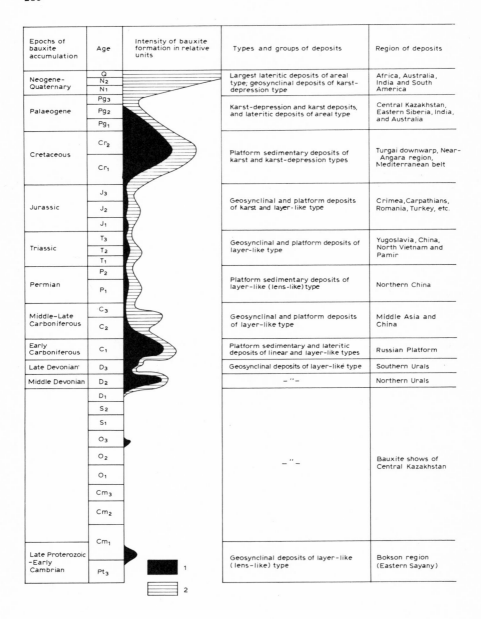

Epochs of bauxite accumulation	Age	Intensity of bauxite formation in relative units	Types and groups of deposits	Region of deposits
Neogene–Quaternary	Q / N₂ / N₁		Largest lateritic deposits of areal type; geosynclinal deposits of karst-depression type	Africa, Australia, India and South America
Palaeogene	Pg₃ / Pg₂ / Pg₁		Karst-depression and karst deposits, and lateritic deposits of areal type	Central Kazakhstan, Eastern Siberia, India, and Australia
Cretaceous	Cr₂ / Cr₁		Platform sedimentary deposits of karst and karst-depression types	Turgai downwarp, Near-Angara region, Mediterranean belt
Jurassic	J₃ / J₂ / J₁		Geosynclinal and platform deposits of karst and layer-like type	Crimea, Carpathians, Romania, Turkey, etc.
Triassic	T₃ / T₂ / T₁		Geosynclinal and platform deposits of layer-like type	Yugoslavia, China, North Vietnam and Pamir
Permian	P₂ / P₁		Platform sedimentary deposits of layer-like (lens-like) type	Northern China
Middle–Late Carboniferous	C₃ / C₂		Geosynclinal and platform deposits of layer-like type	Middle Asia and China
Early Carboniferous	C₁		Platform sedimentary and lateritic deposits of linear and layer-like types	Russian Platform
Late Devonian	D₃		Geosynclinal deposits of layer-like type	Southern Urals
Middle Devonian	D₂		– " –	Northern Urals
	D₁ / S₂ / S₁ / O₃ / O₂ / O₁ / Cm₃ / Cm₂ / Cm₁		– " –	Bauxite shows of Central Kazakhstan
Late Proterozoic –Early Cambrian	Pt₃		Geosynclinal deposits of layer-like (lens-like) type	Bokson region (Eastern Sayany)

Legend: 1 ▨ / 2 ▥

Fig.93. Stratigraphic distribution of intensity of bauxite-formation based on epochs of bauxite accumulation. (After Smirnov, 1977.) Legend: *1* = intensity of bauxite-formation in the USSR; *2* = intensity of bauxite-formation in the world.

TABLE CXIV

Principal types of bauxite deposits based on tectonic setting (After Smirnov, 1977)

(A) On stable sectors of platforms
    (1) In districts of anteclises → structural/erosional, karst/basinal, and structural erosional on lava flows
    (2) In marginal parts of syneclises → structural/erosional, structural/erosional on lava flows, and valley/erosional
(B) On mobile sectors of platforms
    (3) In marginal parts of syneclises, filled with coal-bearing sequences → structural/erosional on lava flows, karst/basinal, and valley/erosional
    (4) In marginal parts of intermontane and pre-montane downwarps → structural/erosional and karst/basinal
(C) In geosynclinal districts
    (5) In regions of mutual transition between geosynclines and platforms → reefogenic/karst and reefogenic/basinal
    (6) In marginal geosynclinal downwarps → reefogenic/karst, and karst/basinal

---

The above list ignores many lateritic bauxites of the weathering crust.
After reviewing several other schemes, Smirnov decided to accept the classification of Table CXV to be best suited for his purposes.

cal variations, textures/fabrics (massive vs. oolitic bauxites, for example, correlated with climatic/hydrologic patterns), presence or absence of Ni, Cr. etc.

The next two exemplars (Figs.94 and 95) depict, respectively, the distribution of iron-ore sub-divided into seven genetic groups and three host-rock types plotted versus geologic epochs, whereas for the manganese a somewhat different graphical approach was utilized with the same results, namely lithologies and ore-mineral composition (indirectly reflecting the origin or genetic types) plotted against stratigraphic age. Instead of varying

TABLE CXV

Principal types of bauxite deposits based on several geological variables (After Smirnov, 1977)

I. Wheathering crusts (lateritic)
    (a) Anteclises
    (b) Shields
    (c) Ridges
II. Sedimentary (redeposited)
    (1) Geosynclinal districts
    (2) Ancient platforms
    (3) Young platforms
Further division:
(a) Nepheline ores
(B) Alunite ores
(C) Kyanite ores
(D) Kaolins and high-Al clays

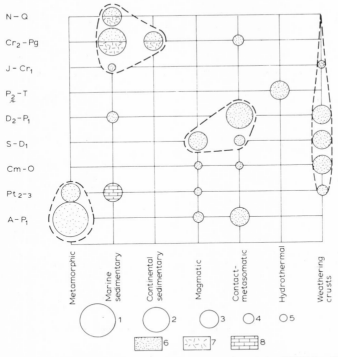

Fig. 94. Distribution diagram of iron-ore deposits of the USSR by different genetic groups based on iron-ore epochs in the Earth's history. (After Smirnov, 1977.) Legend: *1–5*: total reserves of iron ores: *1* = hundreds of billions (10^9) of tonnes; *2* = up to 100 billion tonnes; *3* = up to 10 billion tonnes; *4* = up to 1 billion tonnes; *5* = a few hundred million tonnes; *6–8*: mineral types of ores = *6* = iron-oxide; *7* = carbonate-silicate-oxide; *8* = carbonate.

sizes of circles, the length of the horizontal bars provide the amount in billion and million tonnes (=metric tons), respectively.

Smirnov (1977), in analyzing the variations in intensity of beryllium mineralization (Fig.96) of different mineralogic/genetic associations, plotted the information versus known metallogenic epochs; the latter of which may or may not coincide with the geologic epochs (see also Smirnov, 1976, several chapters; Stanton, 1972, Chapter 19; and several chapters by Baumann/Tischendorf, 1976).

Comprehensive summaries, and at the same time also comparative models, such as exemplified in Fig.97 by Nicolini (1970, fig.XIX, p.520), can serve as research and exploration guides, e.g., in developing metallogenic/metallotectonic concepts (cf. auxiliary sections in this chapter).

The last illustration, evidential of complex but unmistakable correlations between mega-geological factors, is provided by Fig.98 in which Valjaško (1958) (cf. also Baumann and Tischendorf, 1976, fig.6.6) gauged rock-salt (halite), potash salt and Na-sulfate

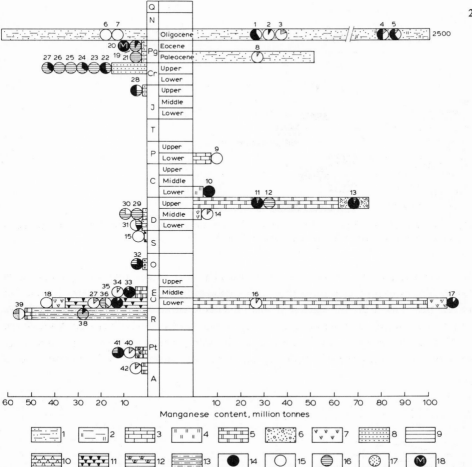

Fig.95. Diagram of the stratigraphical distribution of the manganese deposits and ore-shows of various genetic types in the USSR. (After Smirnov, 1977.) Legend: *1* = sand—silt—clay rocks; *2* = the same rocks with seams of spongolites and gaize; *3* = manganiferous limestones; *4* = cherty shales and gaize; *5* = chert—carbonate rocks (limestones, cherty and clay—chert shales); *6* = conglomerates and coarse-grained sandstones; *7* = jaspers, tuffs, tuffites, cherty shales, and jasperoid quartzites; *8* = porphyrites and their tuffs, tuffites, tuff—sandstones, and limestones; *9* = porphyrites (quartz porphyries), quartzites, cherty shales, and limestones; *10* = jaspers, cherty tuffites, tuff—siltstones, and argillites; *11* = lavas and tuffs of basic composition, jaspers, cherty and chert—clay shales, and limestones; *12* = manganese—iron quartzites; *13* = sand—silt—clay rocks and carbonate sediments; *14—18* = ores: *14* = oxide, *15* = carbonate, *16* = oxidized, *17* = silicate, *18* = mangano-magnetite.

Deposits and ore-shows (figures on diagram): *1* = Nikopol; *2* = Bol'she-Tokmak; *3* = Shkmer; *4* = Chiatur; *5* = Chkhari-Adzhamet; *6* = Laba; *7* = Mangyshlak; *8* = Northern Urals group (Yurka, etc.); *9* = Ulu-Telyak; *10* = Akkermanovsk; *11* = Atasu group (Karazhal, etc.); *12* = Murdzha; *13* = Dzhezda-Ulutauk group (Dzhezda, etc.); *14* = Near-Magnitogorsk group (Niazgulovsk I, etc.); *15* = Dautash; *16* = Usa; *17* = South Khingan; *18* = Bidzhansk; *19* = Aiotsdzor group (Matiros, Karmrashen, etc.); *20* = Near-Sevan group (Chaikenda, etc.); *21* = Tetritskaroi group (Sameb, etc.); *22* = Svarants; *23* = Idzhevan-Noemberyan group (Sevkar, etc.); *24* = Tetritskaroi group (Tetritskaroi, etc.); *25* = Gegechkhor group (Naukhunao, etc.); *26* = Molla-Dzhalda; *27* = Kodmana group (Kodmana, etc.); *28* = Tsedis group (Tsedis, etc.); *29* = West Altai (Zyryanovsk, etc.); *30* = Klevaka; *31* = Sapal; *32* = Mugodzhar group (Kos-Istek, etc.); *33* = Durnovsk; *34* = Kiya-Shaltyr'; *35* = Udsk-Selemdzha group (Ir-Nimiisk, etc.); *36* = Arga group (Mazul, etc.); *37* = Gornoshorsk (Kamzas, etc.); *38* = Nizhne-Uda (Nikolaevo, Kettsk, Arshan, etc.); *39* = Seiba; *40* = Ikat-Garga; *41* = Sosnovyi Baits; *42* = Sagan-Zaba.

Associations	Epochs					
	Proterozoic	Caledonian	Hercynian	Mesozoic	Alpine	
I Bertrandite						
II Fluorite-bertrandite- -phenakite						
III Fluorite-mica-beryl						
IV Molybdenite-wolframite- -beryl						
V Beryllium-bearing alkaline metasomatites						
VI Beryllium-bearing skarns						
VII Beryllium-bearing pegmatites						
Overall distribution						

Fig.96. Intensity of development of beryllium mineralization of different associations in connection with metallogenic epochs of different ages. (After Smirnov, 1977.)

against epeirogenic movements (and consequently also versus regressive—transgressive ocean water movements) during the earth's geologic evolution. A positive correlation between the evaporites and regression was thus demonstrated. An obvious extrapolative query imposes itself on those researchers engaged in, for example, Mississippi Valley-type ores: if it can be accepted that there is a relationship (at least an indirect, if not a direct one) between the presence of evaporites and metal concentrations, is it unreasonable to

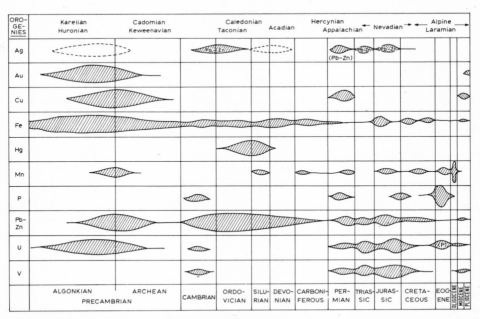

Fig.97. Relative importance of stratiform metal concentrations as a function of age of the host-rocks. (After Nicolini, 1970.)

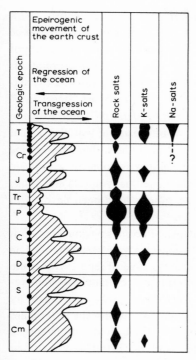

Fig.98. Overview of the relationship between evaporite deposits and tectonic mega-cycles of the earth. (After Valjaško, 1958, in Baumann and Tischendorf, 1976.)

expect and search for a correlation between evaporite "cycles" and cyclic/episodic mineralization? To make the genetic assumption more intricate (and, therefore, relate it to the multivariate natural systems), the next reasonable question is: if an interplay in the origin of hydrocarbon and certain ores exists (*both* may ultimately be founded on the evolution of sedimentary basins), can there be a genetic interplay between the triad "evaporites–hydrocarbons–mineralizations"? Future research may find the answers – and then possibly only when much more data is accessible that can be manipulated by systems-analyses methods to unravel otherwise incoherent and undemonstratable relations (i.e., *not* exposed by simple inspective trial-and-error, pragmatic techniques). This is one practical application of the study of chronological "classifications", "structuring" or "ordering" of all geological causations and products.

CLASSIFICATIONS OF ORE-BEARING PROVINCES AND REGIONS (METALLOGENY/MINEROGENESIS)

"Geology has to choose between the rashness of using imperfect evidence or the sterility of uncorrelated unexplained facts."

    Gregory

Metallogenesis ("minerogenesis" according to Baumann and Tischendorf, 1976, appears to be a term covering both metallic and non-metallic deposits) is considered to be one of the most comprehensive fields of investigation, drawing from the total realm of earth sciences. Although not exactly a new subject, the data collected, the interpretations and extrapolations made, and the hypotheses developed, have merely scratched the surface of what metallogenesis/minerogenesis has to offer in the future. Many "specialists" will make a great input by collecting data, but in particular "generalists", able to metaphorically consume and digest mentally a vast amount of information, will play a vital role in synthesizing the material provided from the various disciplines, thus propagating coherent descriptive and interpretive models for both local and regional to worldwide purposes. For instance, since the paper by Sullivan (1957) numerous symposia and books in English, French, German, Russian, Japanese, among others, have been devoted to metallogenesis (cf. also chapters in the present Handbook).

For those examining metallogenetic reports and maps, it becomes obvious that both descriptive and genetic terminologies, classifications, and concepts have been and must be utilized. Indeed, *objectively* (descriptive) prepared maps and geological reports constitute the foundation upon which the *subjective* (genetic, interpretive, extrapolative)minerogenetic, metallogenic, or metallotectonic maps are prepared. The varying symbols used by different researchers, institutions, companies and government surveys for the latter maps eventually need international standardization (as proposed by the United Nations special committees). The spectrum of metallogenic/minerogenic symbols in itself represents, or rather follows classification schemes, and depicts economic commodities by mode of origin, grade, mineralogy (simple and/or associations), metal-content(s), trace/minor elements, host rock, relative ages, form or shape and/or structure.

The following nine conceptual models on minerogenesis/metallogenesis are furnished here:

(1) Developmental stages of plate tectonics and its ore-forming significance.

(2) Relationships between geosynclinal tectonic stages, magmatism and origin of ores.

(3) Diagram depicting geosynclinal magmatism and environments of ore-formation.

(4) Classification of the associations of major sedimentary and magmatic rocks with the geotectonic-minerogenetic stockworks.

(5) Minerogenetic significance of major tectonic structures.

(6) Minerogenetic evaluation of some major types of regional structures, their geotectonic setting, spatial and depth distribution in crust and mantle.

(7) Archean distribution of volcanic, sedimentary, tectonic, granitization, magmatic and ore-forming periods.

(8) Elemental concentrations during phases of Archean mineralizations.

(9) Synopsis of Archean events related to ore genesis.

In Table CXVI, Baumann and Tischendorf (1976) summarized the six evolutionary stages as envisaged by the hypothesis of plate tectonics, enumerating the regional struc-

TABLE CXVI

The development stages according to the plate-tectonic theory and their minerogenetic significance (After Baumann and Tischendorf, 1976)

Stage	Marginal type	Tectonic structural type	Sedimentation	Magmatism		Types of deposits
				extrusive	intrusive	
1. Pre-stage (rift)	productive	continental rift zone (graben rupture)	terrigenous	trap-rock formation (partly)	ultrabasic-alkaline form.	carbonatites, rare metals, RE; F–Ba–Sr, Cu
2. Young stage (Red Sea)	productive	continental–oceanic rift zone (extended graben zone)	terrigenous pelagic	trap-rock formation (partly)	ultrabasic-alkaline form.	hot saline solutions (evaporites) with Fe, Mn, Cu, Pb-Zn, Ba (oxid., carbon.-silicat., sulfidic)
3. Mature stage (Atlantic)	productive	mid-oceanic ridge	pelagic, (terrigenous)	tholeiitic effusives	ultrabasites, basites	Fe–Mn, Cu, Pb, Zn; liquid-magmat. Cr, Ni–Cu–Pt, Fe–Ti–V
	productive	passive continental slope (shelf), marginal depression	terrigenous, neritic, paralic, evaporites	absent	absent	marine placers, Ti, Zr, Fe$_3$O$_4$ a.o.; P
4. Resorption stage (Pacific)	productive	mid-ocean ridge	pelagic, (terrigenous)	tholeiitic effusives	ultrabasites, basites	Fe–Mn, Cu, Pb, Zn; Cr, Ni, Pt
	destructive	subduction zone	pelagic, neritic, terrigenous	diabase–spilite form., andesite–dacite form.	basic–intermed. and acidic magmatic rocks	Cr, Ni–Cu
5. Terminal stage (Mediterranean Sea)	destructive	subduction zone	(pelagic), neritic, terrigenous	diabase–spilite form., andesite–dacite form.	basic–intermed. and acidic magmatic rocks	FeS$_2$–Cu–Zn–Pb, Fe–Mn, Au, Cu–Mo (porphyries), Ag–Pb–Zn, W, Sn, Sb, Hg
6. Closure stage (Himalayas)	destructive	orogenic zone	terrigenous	absent	absent	red-bed and infiltration-type formations (Cu–Co, U–V), evaporites

TABLE CXVII

Relationship between tectonic development stages, magmatism and ore genesis in the geosynclinal (orogenic) environment (summarized from studies by Biliben, Stille, Schneiderhöhn, Smirnov among others) (After Baumann and Tischendorf, 1976)

Development stage	Magmatism	Magmatic formations		Endogenic types of deposit	
		effusive	intrusive		
(I) Early stage	initial (simatic-juvenile)	diabase–spilite form. keratophyre form. andesite–porphyrite form.	periotite form. gabbro–norite form. plagiogranite form.	Fe, Mn, Cu Fe(Mn), Cu, Pb, Zn	} submarine-volcanic
				Fe, Cu, Pb, Zn Cr, Pt–Os–Ir Ti–Fe, Ni–Cu, Pt–Pd Fe, Cu, Au	} intramagmatic and intracrustal
(II) Orogenic stage (intermediate stage)	syn- and late-orogenic (sialic-palingenic)	dacitic–rhyolitic ignimbrite form.	granodiorite form. granite form.	W, Mo(Au), U, Pb, Zn Sn, W, Mo, Li, Be, rare elements	} intracrustal
(III) Subsequent stage (late stage)	post-orogenic and/or subsequent (sialic-contaminated)	andesite–dacite–liparite form.	"small intrusions" (diorite-porphyritic, granite- and syenite-porphyritic)	Cu, Pb–Zn, Au–Ag; Sn, W–Sb–As–Hg	} intracrustal
(IV) Consolidation stage	final (simatic-contaminated to -juvenile)	alkali basalt form. trachyte form. (basaltic–andesitic–liparitic)	gabbro-syenite-alkali-granite form.	Ti, Fe; rare earth, Zr	} intracrustal

tures, sediments, extrusive and intrusive magmatism and the corresponding types of mineralizations. In Table CXVII, the same authors outline a synopsis of the connection between the four geosynclinal tectonic (orogenic) phases, magmatism and "endogene" ores (including volcanic-exhalative = submarine-volcanic); Fig.99 is a pictorial model of the first three phases. It should be noted that in the latter two models the sedimentary facies have been excluded, which then find their deserved consideration again in Tables CXVIII and CXIX. In the former, Baumann and Tischendorf used the threefold geotectonic-minerogenetic stockwork division, namely, platform-, transitional- and geosynclinaltype stockworks, each characterized by a range of sedimentary, magmatic and metallic, non-metallic plus industrial-mineral concentrations. Table CXIX, then, is similar to the preceding model in its approach, but the earlier threefold structural division has been expanded to deal with numerous important sub-regional orogenic, tectonic and epeirogenic structures, totalling thirteen varieties. The geosynclines have six and the platforms seven kinds of structures. The so-called "transitional"-type stockworks have been incorporated into the platform-type features. The lithologic—petrologic characteristics of the sedimentary rocks as well as the structural and compositional ranges of economic deposits are considered in more detail than in the earlier tabular summaries.

In support of the above-quoted data, Baumann and Tischendorf (1976) advanced in Fig.100 a cross-section of the earth's crust and upper mantle extending idealistically from a marine platform through a eugeosyncline and a miogeosyncline to a continental platform. In each of these four mega-structural settings, specific deformational structures may control the movement of magma and ore-forming fluids as well as the localization of mineralization. The intensity of both the endogenic and exogenic factors influencing oreconcentration, and the level in the crust/mantle at which the processes take place, are in direct response to these structures. The varieties of mineralizations to be expected and the "depth-of-origin" vary within these four mega-environmental settings, as depicted in the diagram.

One problem related to the exploration of metals, industrial minerals, hydrocarbons and water, is the integrated investigation of sedimentary—volcanic *basins*. In every metallogenic/metallotectonic scheme, for example, the *most fundamental* types of basins are mentioned — but often only too superficially. In the future, it is hoped that more comprehensive basin classifications can be offered that appeal explicitly to ore petrologists and explorationists by emphasizing the gamut of economically viable deposits. Accompanying these classificatory models and analogues should be a list of factors and processes that are the prerequisites for the concentration above the Clarke value of each mineral commodity. Most of the progress in this regard has undoubtedly been accomplished by the petroleum exploration industry — see Weeks (1961) and Halbouty et al. (1970), among many others.

Boyle (1976) confined himself in his metallogenic synopses to the Archean and his overviews are presented in Figs.101, 102, and Table CXX. No further comments are advisable here — except to list the deposits conspicuously lacking in Archean terranes:

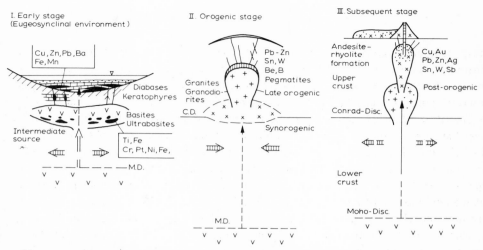

Fig.99; Magmatism and the origin of ore deposits in the geosynclinal (orogenic) environment. (After Baumann and Tischendorf, 1976.)

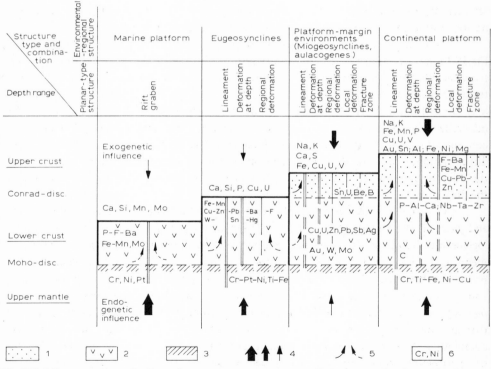

Fig.100. Minerogenetic evaluation of some mega-structures regarding type, mode of combination as well as depth (=level of occurrence) and depth distribution (=range of depth-distribution) of the planar structures. (Baumann and Tischendorf, 1976.) *1* = upper crust (sial s.s.); *2* = lower crust (mainly basic rocks); *3* = upper mantle (sima s.s.); *4* = endogenic and/or exogenic factors (influences are strong, intermediate or weak) controlling precipitation processes in the "spatial" structures; *5* = strong to weak influences of the host-rocks on the concentration of minerals within the planar-type structures.

TABLE CXVIII

Association of the main rock formations and their most important ores and industrial minerals with the geotectonic-minerogenetic units ("Stockwerken") (After Baumann and Tischendorf, 1976)

Geotectonic stockwork = Minerogenetic stockwork	Formations		Main types of industrial deposits (raw materials)
	sedimentary	magmatic	
Platform stockwork	(Glacial form.)	Trap-rock form. (intrusive and effusive)	F, Ba Cr–Pt, Ti–Fe, Ni–Co–Cu, Pb–Zn
	Evaporite formations Carbonate form.		Anhydrite, halite and potash-salts Limestone, bauxite, phosphorite
		Ultrabasic-alkaline rock form.	Apatite, nepheline, carbonatites, T, Nb, Ta, U, rare earth, diamonds, non-ferrous metals
	Sandy formations (partly red-bed form.), marine terrigenous formations		Sands, gravels; non-ferrous metals, Fe, Mn, limn. coals; hydrocarbon deposits Bauxite, kaolin, clay; placer deposits
Transition stockwork (Orogenic stockwork)		Subsequent andesite-rhyolite formation	Non-ferrous metals, rare elements
	Molasse form. (red-bed form.) (flysch form.)		Coal form., hydrocarbon deposits; sands, gravels
		Granite formation, syn- and post-orogenic Granodiorite formation	Non-ferrous and precious metals; W, Li, Be; rare elements; industrial minerals
Geosynclinal stockwork	Flysch form., lagun. form., carbonate formations, siliceous form.		Building materials (consolidated and unconsolidated deposits) Limestones, SiO$_2$-raw material, bauxite, phosphorite
	Clay-slate form. (black slates)	Initial magmatism; diabase-spilite-keratophyre formations	Mn, non-ferrous metals, S, U, V, Mo; hydrocarbons Fe, Mn, non-ferrous metals
	Slate-greywacke formations	Peridotite–gabbro formation	Cr–Pt, Ti–Fe, Ni–Fe

TABLE CXIX

Minerogenetic significance of the most important structure types (After Baumann and Tischendorf, 1976)

Type of structure	Lithologic-petrologic character	Ore depositis	
		structure	composition
(A) *Geosynclinal (orogenic) environment*			
1. Eugeosynclines	pelagic formations	stratiform, impregnating	Mn, non-ferrous metals, S, U, V, Mo; petroleum
	flysch formation		building materials
	diabase—keratophyre—andesite formations	stratiform, impregnating, intracrustal,	Fe, Mn, non-ferrous metals
	ultrabasite—basite formations	intramagmatic impregnating	Cr—Pt, Ni—Cu, Ti—Fe
2. Miogeosynclines	neritic formations		Pb, Zn, S, U; petroleum
	molasse formation		
		stratiform, impregnating-infiltrating	Cu, poly- and precious metals, U, V, evaporites, coal
	granodiorite—granite formations	intracrustal	non-ferrous, precious and rare metals industrial minerals
3. Anticlines	granodiorite—granite formation		non-ferrous, precious and rare metals
	andesite—dacite formations (small intrusions)	intracrustal	(especially Sn, Cu, Mo, rare elements)
4. Fore-deep	molasse formation	stratiform, impregnating-infiltrating (red-bed)	placers, non-ferrous and precious metals, U, V; evaporites; petroleum, coal
5. Intermediary massif	parly analogue to the anticlines being influenced by deep-reaching rupture tectonics and magmatism		
6. Dome- and ring-structures	granodiorite formations	|←——— extensively sterile ———→|
	andesite—dacite formations	intracrustal	non-ferrous, precious and rare metals
(B) *Platform environment*			
1. Syneclise	terrigenous formations		placers; non-ferrous metals, U, V, S; Fe, Mn, P; petroleum, coal
	clayey-calcareous formations	stratiform, impregnating-infiltrating	
	saline formation		anhydrite, halite and potash salts
	trap-rock formation (intrusive and effusive)	intramagmatic intracrustal	Cr—Pt, Ni-Cu, Ti—Fe; F—Ba, non-ferrous metals

(continued)

TABLE CXIX (continued)

Type of structure	Lithologic-petrologic character	Ore deposits	
		structure	composition
2. Anteclise	weathering-type formations ultrabasite—alkaline rock formations	stratiform impregnating intracrustal	placers; Al, Fe, Si, Mg, P apatite, nepheline, carbonates, Ti, Nb, Ta, rare elements, non-ferrous metals; F—Ba; diamonds
3. Platform margin depression, periclines, aulacogen	molasse formation carbonate formation saline formations trap-rock formation	stratiform, impregnating-infiltrating intramagmatic, intracrustal	placers, non-ferrous metals, U, V; Fe, Mn, P; petroleum, coal, evaporites, S Ni—Cu, Ti, F—Ba, non-ferrous metals
4, Horst formations, transposed blocks	partly analogue to the anteclises being influenced by deep-reaching rupture tectonics and magmatism		
5. Graben zone	partly analogue to platform margin depression and aulacogens with additional		
	ultrabasite—alkaline rock formation	intracrustal	carbonatites, rare and non-ferrous metals
6. Dome- and ring-structures	partly analogue to the anteclises with strengthened magmatic influence		
7. Lineaments, deformation at greater depth	ultrabasite—basite—alkaline rock formations	formations in association with the initial- and platform-magmatism	

evaporites and salines (Na, K, Mg, Ca, B, SO_4, N, F, Cl, Br, and I), sedimentary phosphate deposits, sedimentary limestone and dolomite deposits (except for stromatolite carbonates), barium and strontium deposits, mercury deposits, fossil bauxite and laterite deposits, uranium and thorium deposits, carbonatites (Nb, Ta, Zr, Ba, Sr, Cu, etc.), hydrocarbon deposits; tin, lead, zirconium and vanadium deposits.

CLASSIFICATIONS OF MINERAL-RESOURCES REQUIREMENTS

"When you can measure what you are speaking about and express it in numbers, you know something about it, and when you cannot measure it, when you cannot express it in numbers, your knowledge is of a meagre and unsatisfactory kind. It may be the beginning of knowledge, but you have scarcely in your thought advanced to the stage of a science."
    Lord Kelvin

TABLE CXX

Archean volcanic, sedimentary, tectonic, granitization, magmatic, and mineralization events (After Boyle, 1976; courtesy of Geological Society of Canada, Ottawa)

Geological events in chronological order, youngest first		Tectonic events in chronological order, youngest first	Mineralization events	Geochemical features	Remarks
Volcanic events	Sedimentary events	Magmatic and granitization events				
		Diabase dykes and sills.	Late faulting.			Probably most dykes and sills are of Proterozoic age.
		Post-gold–quartz vein basic (lamprophyre) and granitic dykes. Post-vein pegmatites occur in some belts.	Late faulting.	Minor carbonate veins.		Post-gold–quartz dykes are rare in most belts, but are common in some. Post-vein pegmatites generally occur in sedimentary belts.
		Waning stage of granitization and metamorphism. Some late quartz –feldspar porphyry and lamphophyre dykes.	Waning stage of transcurrent shearing and with formation of dilatant zones. Extensive shearing and faulting of ultrabasic bodies in some belts.	Main gold–quartz stage. Main stage of auriferous arsenic and antimony deposits. Formation of some asbestos deposits. Formation of some chromite deposits. Formation of some Fe–Ni–Co sulfide deposits. Formation of dis	Carbonatization of greenstone belts. Concentration of SiO$_2$, Au, Ag, Te, As, Sb, etc. in gold–quartz and arsenic–antimony deposits. Concentration of Fe and Cr in some ultrabasic bodies. Concentration of Fe, Ni, Co	Main gold–quartz stage; main stage of deposition of arsenic and antimony. Main stage of formation of asbestos deposits in most belts; probably main stage of formation of massive and disseminated Fe–Ni–Co sulfide deposits in most

basic and ultra-basic bodies. Concentration of Cu, Mo, etc. in disseminated bodies.	Main pegmatite stage.				of formation of disseminated Cu–Mo deposits.
	Continued granitization. Development of pegmatites mainly in sedimentary terranes; development of quartz–feldspar porphyry dykes and stocks mainly in greenstone belts.	Waning stage of folding and attendant shearing along flow contacts, etc. Extensive transcurrent faulting and shearing across the strata with formation of dilatant zones. Shearing and faulting of ultrabasic bodies in some belts.	Main pegmatite stage; formation of some gold–quartz deposits; formation of some massive Fe–Cu–Zn sulphide deposits; formation of some massive Fe–Ni–Co sulphide deposits in ultrabasic rocks; probable formation of asbestos deposits in some belts.	Concentration of SiO_2, Na, K, B, Be, Li, Nb, Ta, etc. in pegmatites. Carbonatization of greenstone belts. Concentration of Fe, Ni, Co, Pt, Cu, Zn, Pb, Ag, Au, S, etc. in dilatant zones.	
	Extensive granitization; injection of basic and ultrabasic sills, stocks, and irregular bodies into supra-rocks; injection of granitic dykes and stocks.	Extensive folding with intensive shearing, dragging and contortion along flow contacts, sedimentary interflow bands, etc. in greenstone belts. Extensive folding of greywacke–slate sequences. Development of stratiform dilatant	Main period of deposition of massive Cu–Zn sulphide deposits in both greenstone and sedimentary belts.	Extensive granitization and metamorphism of volcanic and sedimentary rocks. Extensive carbonatization of greenstone belts. Concentration of Fe, Ni, Co, Pt, Cu, Zn, Pb, Ag, Au, S, etc. in dilatant zones.	Most massive Cu–Zn sulphide bodies formed during this period in greenstone and sedimentary belts. Some gold–quartz deposits (saddle reefs) formed in stratiform dilatant zones in sedimentary belts. Some

(continued)

TABLE CXX (continued)

Geological events in chronological order, youngest first			Tectonic events in chronological order, youngest first	Mineralization events	Geochemical features	Remarks
Volcanic events	Sedimentary events	Magmatic and granitization events				
			zones in both greenstone and sedimentary belts.			massive Ni–Co sulphide bodies probably formed during this period.
Minor submarine basic volcanism.	Extensive deposition of greywacke, conglomerate, quartzite, and shale; minor impure limestone in some belts. Extensive deposition of iron formation in a few belts; deposition of manganese formation in some belts	Upward and lateral invasive granitization; injection of gabbro and diorite dykes; injection of some granite dykes into supra-rocks.	Continuation of folding and other tectonic activity as detailed below Development of stratiform dilatant zones and breccia pipes in greenstone belts.	Inititation of deposition of massive Cu–Zn sulphide bodies in greenstone belts.	Weathering of volcanic rocks and primordial crust(?) attended by mainly clastic sedimentation with hydrosylates and iron and manganese formations. Granitization and initiation of metamorphic facies in greenstone belts. Concentration of Fe, Cu, Zn, Pb, Ag, Au, etc. in dilatant zones.	Some massive Cu-Zn sulphides bodies formed, especially in breccia pipes. Deposition of some sedimentary iron and manganese formations.
Minor submarine basic volcanism. Acid volcanism in some belts.	Initiation of conglomerate, greywacke, and shale sedimentation.	Granitization at depth; injection of gabbro and diorite dykes into supra-rocks.	Initiation of folding with shearing, dragging and contortion along flow contacts, sedimentary inter-		Granitization at depth; expulsion of basic elements which crystallize as gabbro and diorite dykes in	

		Injection of gabbro and diorite sills. Injection(?) of ultrabasic sills and irregular bodies in some belts. Formation of irregular gabbro and diorite bodies by "dioritization".	Probable minor magmatic segregation of chromite in certain ultrabasic sills; also minor segregation of Fe, Ni, Co, and S in some basic sills.	Magmatic segregation of Cr, Fe, Ni, Co, and S.	Minor chromite and Fe–Ni–Co sulphide deposits formed.
Extensive submarine basic–intermediate–acidic volcanism.	Intermittent deposition of tuffs, agglomerates, carbonates, carbonaceous sulphide shales and minor conglomerate, greywacke, and chert. Extensive deposition of Algoma-type iron formation in many belts; deposition of manganese formations in some belts.	Basic–intermediate–acidic magmatic volcanism. Some ultra-basic flows deposited in some belts.	Major sedimentary deposition of Fe, SiO_2, and CO_3 (carbonate). Sedimentary deposition of Mn in some belts. Minor sedimentary deposition of Mn, S, As, Sb, Cu, Pb, Zn, Ni, Co, Au, and Ag mainly in sulphide sediments. Possible minor magmatic segregation of Fe, Ni, Co, sulphides in thick ultrabasic flows.	Major enrichment of Fe, Mn, SiO_2 and CO_3 in iron and manganese formations. Minor enrichments of Mn, S, As, Sb, Cu, Pb, Zn, Ni, Co, Au, Ag, and other chalcophile elements in sulphidic sediments. Enrichment of Fe, Ni, and Co in sulphide segregations in thick ultrabasic flows.	A major period of formation of hematite and magnetite iron deposits. Deposition of sedimentary manganese formations in some belts.
Minor submarine volcanism with out-pouring of thin basic flows.	Deposition of greywacke and shaly rocks.	Initiation of volcanic period.		Weathering of primordial crust attended by mainly clastic sedimentation	Not observable in most belts. Couchiching of Lawson probably represents this

(continued)

TABLE CXX (continued)

Geological events in chronological order, youngest first			Tectonic events in chronological order, youngest first	Mineralization events	Geochemical features	Remarks
Volcanic events	Sedimentary events	Magmatic and granitization events				
					with minor hydrolysates. Minor basic volcanism.	period. Origin of life manifest by low order forms (coacervates).
		Formation of primordial ocean. Formation of primordial crust of granodioritic (?) composition.			Magmatic(?) differentiation to produce rocks relatively rich in Si, Al, Na, K, and minor Ca, Fe, Mg, etc. Composition of primordial ocean unknown.	Primordial crust not observable; geochemical features mainly speculative.

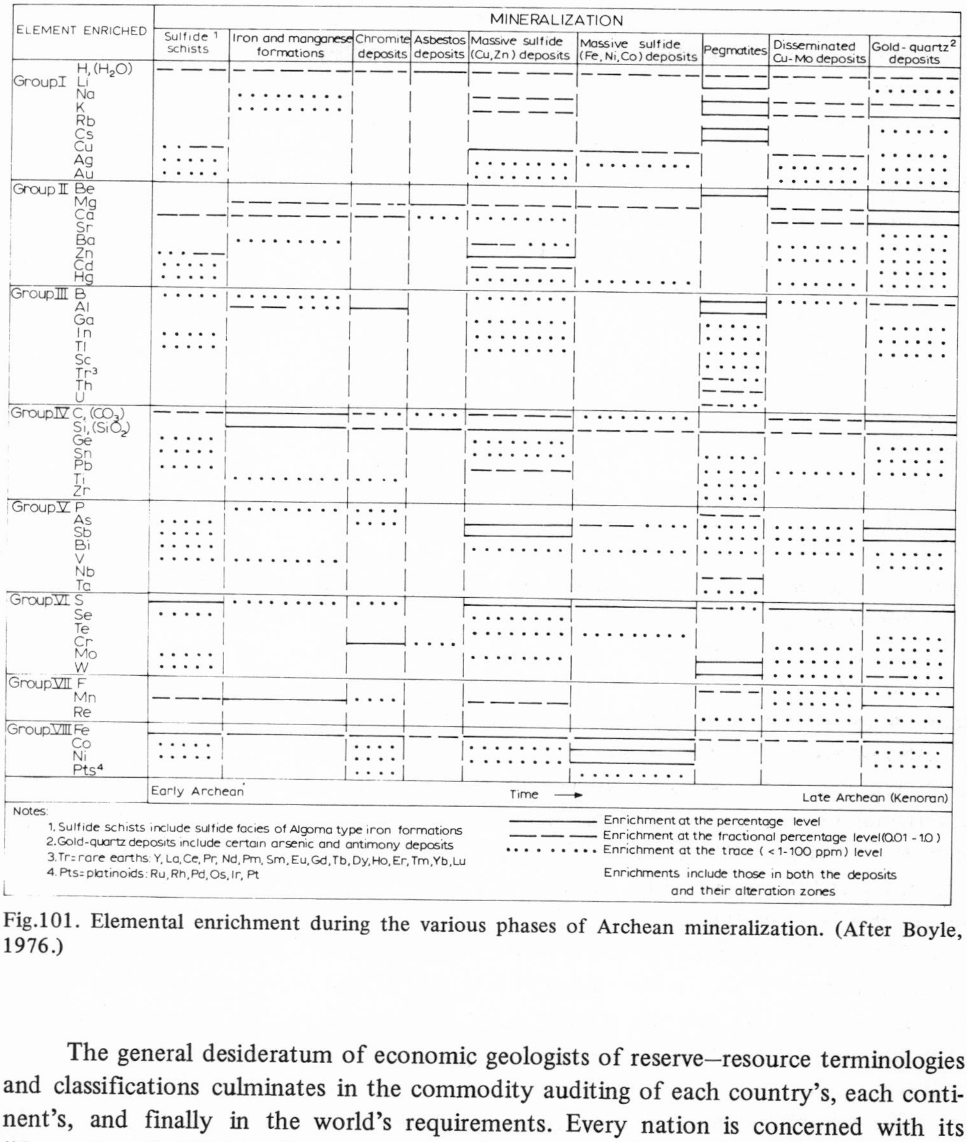

Fig. 101. Elemental enrichment during the various phases of Archean mineralization.

ELEMENT ENRICHED	MINERALIZATION								
	Sulfide [1] schists	Iron and manganese formations	Chromite deposits	Asbestos deposits	Massive sulfide (Cu,Zn) deposits	Massive sulfide (Fe,Ni,Co) deposits	Pegmatites	Disseminated Cu-Mo deposits	Gold-quartz [2] deposits
Group I — H,(H₂O), Li, Na, K, Rb, Cs, Cu, Ag, Au									
Group II — Be, Mg, Ca, Sr, Ba, Zn, Cd, Hg									
Group III — B, Al, Ga, In, Tl, Sc, Tr [3], Th, U									
Group IV — C,(CO₂), Si,(SiO₂), Ge, Sn, Pb, Ti, Zr									
Group V — P, As, Sb, Bi, V, Nb, Ta									
Group VI — S, Se, Te, Cr, Mo, W									
Group VII — F, Mn, Re									
Group VIII — Fe, Co, Ni, Pts [4]									

Early Archean — Time → — Late Archean (Kenoran)

Notes:
1. Sulfide schists include sulfide facies of Algoma type iron formations
2. Gold-quartz deposits include certain arsenic and antimony deposits
3. Tr = rare earths: Y, La, Ce, Pr, Nd, Pm, Sm, Eu, Gd, Tb, Dy, Ho, Er, Tm, Yb, Lu
4. Pts = platinoids: Ru, Rh, Pd, Os, Ir, Pt

——————— Enrichment at the percentage level
— — — — Enrichment at the fractional percentage level (0.01–1.0)
· · · · · · · Enrichment at the trace (<1–100 ppm) level

Enrichments include those in both the deposits and their alteration zones

Fig.101. Elemental enrichment during the various phases of Archean mineralization. (After Boyle, 1976.)

The general desideratum of economic geologists of reserve—resource terminologies and classifications culminates in the commodity auditing of each country's, each continent's, and finally in the world's requirements. Every nation is concerned with its "domestic self-sufficiency" versus the lack thereof, thus directly determining the export—import structure. Crain (1950) preferred "self-adequacy" to "self-sufficiency" because the former is a broader term connotating "overall ability to exploit mineral wealth"; in contradistinction, the latter is the "ratio of production over consumption." The expression "self-adequacy" comprises, in addition to actual production, many other variables, such as reserves, plant capacities (mine, mill, smelter), raw-material distribution, relations between producers versus fabricators, trade agreements, capital availability, success versus

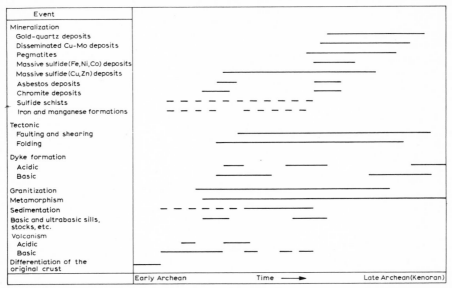

Fig.102. Archean volcanic, sedimentary, tectonic, granitization, magmatic, and mineralization events. (After Boyle, 1976.)

failure in exploration (among many other geological parameters), etc.

Some countries have always been extraordinarily endowed with mineral wealth (e.g., USA, Australia, Canada, USSR), but everyone of them depends on foreign sources either totally or partially for specific metals and/or industrial minerals. The classificatory inventory in Table CXXI (Crain, 1950, fig.3, p.20) is an exemplar of the "relative self-sufficiency" of the USA (1950), divided into "actual and impending" and "potential" — each further subdivided.

Table CXXII (USGS, 1975) is an auxiliary list of commodities obtained from foreign sources divided on a "percentage of liability". Table CXXIII provides data on the United States domestic reserves and resources through to 2000 AD. The latter table illustrates the need for an agreed-upon terminology and a classification in reserves/resources studies.

The mineral supply system of any industrialized country provides an example of the flow of most ore-mineral materials from one resource category to another, progressively from "speculative-undiscovered resources" to refined material production (Sheldon, 1978). To comprehend this (economic) system, one has to examine its components in context. As shown in Table CXXIV, three major phases exist (Research, Exploration, Exploitation), each being divided into two "detailed phases". The table also provides a breakdown of the actual activity carried out, the mineral resource category developed, and the institutions with the prime responsibility.

TABLE CXXI

United States mineral position; relative self-sufficiency of principal metals and minerals (After Lasky and Pehrson, *The Mineral Position of the United States* from Crain, 1950; courtesy of Colorado School of Mines)

RELATIVE SELF-SUFFICIENCY ACTUAL AND IMPENDING
(Based on present technologic and economic conditions and on known "commercial" reserves)

POTENTIAL
(If technological and economic changes permit use of known submarginal resources)

A. VIRTUAL SELF-SUFFICIENCY ASSURED FOR A LONG TIME:

Bituminous coal and lignite	Magnesium	Fluorspar (metallurgical)
	Molybdenum	Helium
Anthracite		Magnesite
Natural gas		Nitrates
		Phosphate rock
		Potash
		Salt
		Sulfur

A. VIRTUAL SELF-SUFFICIENCY:

Bituminous coal and lignite	Aluminum ores	Fluorspar (all grades)
	Copper	Graphite (flake)
Anthracite	Iron ore	Helium
Natural gas	Magnesium	Magnesite
Petroleum	Manganese	Nitrates
	Molybdenum	Phosphate rock
	Titanium	Potash
	Vanadium	Salt
		Sulfur

B. COMPLETE OR VIRTUAL DEPENDENCE ON FOREIGN SOURCES:
 1. Small or remote expectation of improving position through discovery:

Chromite	Industrial diamonds
Ferro-grade manganese	Quartz crystal
Nickel *	Asbestos (spinning quality)
Platinum metals	
Tin	

 2. Good expectation of improving position through discovery:

Cobalt *	Graphite (flake)

B. COMPLETE OR VIRTUAL DEPENDENCE ON FOREIGN SOURCES:

Platinum metals	Industrial diamonds
Tin	Quartz crystal
	Asbestos (spinning quality)

C. PARTIAL DEPENDENCE ON FOREIGN SOURCES, ACTUAL OR IMPENDING:
 1. Good expectation of improving position through discovery:

Petroleum	Arsenic *	Fluorspar (acid grade)
	Bismuth *	
	Cadmium *	
	Copper	
	Iron ore	
	Lead	
	Mercury	
	Tantalum *	
	Titanium	
	Tungsten	
	Zinc	

 2. Little hope of improving position through discovery:

Antimony *	High-grade bauxite
Vanadium	Strategic mica

C. PARTIAL DEPENDENCE ON FOREIGN SOURCES:

Antimony	Strategic mica
Arsenic	
Bismuth	
Cadmium	
Cobalt	
Chromite	
Lead	
Mercury	
Nickel	
Tantalum	
Tungsten	
Zinc	

* *Domestic production chiefly byproduct.*

TABLE CXXII

United States' dependence on foreign sources for some of its minerals (Mining and Minerals Policy, 1973; from USGS, 1975 with permission)

a. Less than half imported from foreign sources:

Copper	Tellurium
Iron	Stone
Titanium (ilmenite)	Cement
Lead	Salt
Silicon	Gypsum
Magnesium	Barite
Molybdenum	Rare earths
Vanadium	Pumice
Antimony	

b. One-half to three-fourths imported from foreign sources:

Zinc	Nickel
Gold	Cadmium
Silver	Selenium
Tungsten	Potassium

c. More than three-fourths imported from foreign sources:

Aluminum	Tantalum
* Manganese	Bismuth
Platinum	Fluorine
Tin	* Strontium
* Cobalt	Asbestos
*Chronium	* Sheet mica
*Titanium (rutile)	Mercury
* Niobium	

* Commodites more than 90 percent imported.

This mineral-supply system constitutes a series of sequential steps, each one necessary for the initiation of the succeeding phase, and each one improving the effectiveness and economic efficiency of the total system. The technological level reached determines the overall economic efficiency and is improved by research at all phases. For example, potential resources are increased by the development of new concepts of mineral deposits, mapping and geological assessment of potentially mineralized districts. Reserves are increased by exploration which itself is made more proficient by improving the techniques of prospecting, extraction and processing.

MINERAL RESOURCES' NOMENCLATURE AND CLASSIFICATIONS

"Economics, the science of managing one's own household."
    Seneca (*Ad Lucilium, epis, sect.* 10, ca. 64)

TABLE CXXIII

General outlook for domestic reserves and resources through 2000 A.D. (From USGS, 1975, with permission)

Group 1: RESERVES in quantities adequate to fulfill projected needs well beyond 25 years.

Coal	Phosphorus
Construction stone	Silicon
Sand and gravel	Molybdenum
Nitrogen	Gypsum
Chlorine	Bromine
Hydrogen	Boron
Titanium (except rutile)	Argon
Calcium	Diatomité
Clays	* Barite
Potash	Lightweight aggregates
Magnesium	Helium
Oxygen	Peat
	* Rare earths
	* Lithium

Group 2: IDENTIFIED SUBECONOMIC RESOURCES in quantities adequate to fulfill projected needs beyond 25 years and in quantities significantly or slightly greater than estimated UNDISCOVERED RESOURCES.

Aluminum	Vanadium
* Nickel	* Zircon
Uranium	Thorium
Manganese	

GROUP 3: Estimated UNDISCOVERED (hypothetical and speculative) RESOURCES in quantities adequate to fulfill projected needs beyond 25 years and in quantities significantly greater than IDENTIFIED SUBECONOMIC RESOURCES; research efforts for these commodities should concentrate on geologic theory and exploration methods aimed at discovering new resources.

Iron	Platinum
* Copper	Tungsten
* Zinc	* Beryllium
Gold	* Cobalt
* Lead	* Cadmium
Sulfur	* Bismuth
* Silver	Selenium
* Fluorine	* Niobium

Group 4: IDENTIFIED-SUBECONOMIC and UNDISCOVERED RESOURCES together in quantities probably not adequate to fulfill projected needs beyond the end of the century; research on possible new exploration targets, new types of deposits, and substitutes is necessary to relieve ultimate dependence on imports.

Tin	* Antimony
Asbestos	* Mercury
Chromium	* Tantalum

[Within each group, commodities are listed in order of relative importance as determined by dollar value of U.S. primary demand in 1971. An asterisk marks those commodities which may be in much greater demand than is now projected because of known or potential new applications in the production of energy].

TABLE CXXIV

Phases of mineral supply system (After Sheldon, 1978)

Major phases	Detailed phases	Activity	Mineral resource category developed		Prime responsibility
RESEARCH	CONCEPTION	Research in geologic processes, i.e. plate tectonics formation of mineral deposits, etc.	UNDISCOVERED RESOURCES	SPECULATIVE	Universities, Government, research organizations, private institutes
	ASSESSMENT	Geologic, geophysical, and geochemical mapping, geostatistical analysis		HYPOTHETICAL	Government Industry
EXPLORATION	DISCOVERY	Prospecting	RESERVES	INFERRED	Industry
		Research on prospecting techniques			Government and Industry
	DELINEATION	Exploration		INDICATED AND MEASURED	Industry
		Research on exploration techniques			Industry and Government
EXPLOITATION	EXTRACTION	Mining and land reclamation	Produced raw material		Industry
		Research and development on extraction	Reserves		Industry and Government
	PROCESSING	Beneficiation reduction and refining	Produced refined material		Industry
		Research and development on processing	Reserves		Industry and Government

"We have a paradox in the method of science. The research-man may often think and work like an artist, but he has to talk like a book-keeper, in terms of facts, figures and logical sequence of thought."
    H.D. Smyth

The countries engaged in mineral exploration and mining have had to establish a terminology and classification scheme of ore reserves in order to assure an unambiguous communication related to economic geology, i.e., to prevent misconceptions and mis-interpretations (cf. McKelvey, 1973, and *Geotimes* Jan. and Sept. 1974, for concepts developed by the United States Geological Survey; Association of Professional Engineers of the Province of Ontario, Canada, 1972, and Zwartendyk, 1972, for proposals in Canada). Two approaches from the English-language geological fraternity are furnished below, consisting of five models, namely:

(1) Triangular diagram depicting the dynamic relation between reserves and resources.

(2) Relation between reserve—resource terminology and basic aspects of the natural stocks.

(3) Reserves—resources classification diagram (USGS).

(4) Two dimensions of resources.

(5) Classification of resource endowment (Canadian Geological Survey).

The interdependency between reserves and resources is illustrated by Govett (1977) in the triangular diagram of Fig.103 with A, B, and C as its apices and D, E and F consti-tuting the sides. The sub-triangle A, F-C, D depicts the measured, indicated and inferred reserves (for definitions, see below) plus the known but non-exploitable resources. The present economic recoverable level is indicated by point F. With an improved technology and/or a cost-decrease, some of the non-exploitable deposits would become reserves. The resources are devided into known and "unknown" groups, the former have been recog-nized as possible future economically viable ores, i.e., they are of a known genetic variety with their grade and localities established also. Point D will be shifted towards apex B, thus expanding the known at the expense of the unknown resources, when an increase in exploration success (e.g., through the recognition of entirely new genetic types of ore deposits hitherto not considered) can be recorded. It must be realized that the "lowest grade considered possible for some defined time period" (represented by apex B) is in this triangle only a qualitative (non-numerical), but useful, concept; this lowest grade will change in particular cases with improvement in technology (new metallurgical methods), accessibility, and price-structure (very low-grade deposits becoming viable) in the national and/or world economy.

Table CXXV summarizes the fundamentally different aspects of the natural mineral stock expressed by the terms "reserves", "resources" and "resource base", whereas Fig.105 shows in an easier to visualize diagrammatic manner the relationship of these vari-ables. The basic framework of this and similar schemes is founded on Fig.104 outlined by

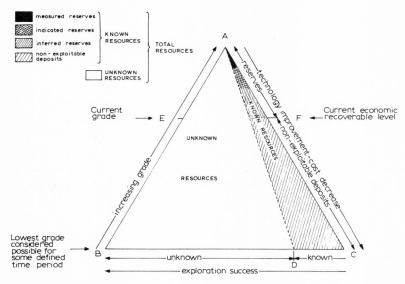

Fig.103. Schematic illustration showing the dynamic relation between reserves and resources. (After Govett, 1977.)

McKelvey (1973). Although the reader must be referred for detailed discussions to the original publications, one should point out that the crucial dimensions of "time" have not been included in conceiving the diagrams.

Zwartendyk (1972) has gone beyond the basic terminology by proposing that the scheme in Figs.105 and 106 must be further refined, otherwise "we will continue to talk on various wavelengths". For example, as to the degrees of certainty, one has to consider as many subdivisions as one can handle or would find useful: (1) "reasonably assured resources" and (2) "estimated additional resources". Zwartendyk stated that one has an

TABLE CXXV

Relation between reserve-resource terminology and basic aspects of the natural stocks (After Zwartendyk, 1972, p. 5; courtesy of the Department of Energy, Mines and Resources, Ottowa)

Terms	Aspects		
	Occurrence	Economic	Technological
Reserves	known	present cost level	currently feasible
Resources	known + unknown	any cost level specified	currently feasible and feasibility indicated in future
Resource base	known + unknown	irrelevant	feasible + infeasible

TOTAL RESOURCES

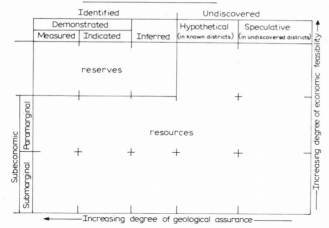

Fig.104. Classification scheme of total resources. (After McKelvey, 1973, for example; to be found in many publications of the USGS.)

intelligible way of depicting a mineral endowment in a relatively uncomplicated fashion by employing two certainty ranges at various prices. The subdivision between these two "certainties" becomes meaningless when moving into the higher price ranges well above the current price at any particular point of time. Not much is known about most deposits

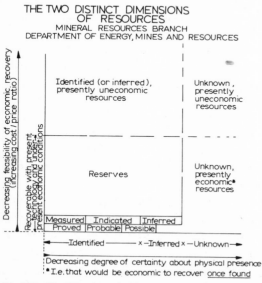

Fig.105. Classification of the two distinct dimensions of resources. (After Zwartendyk, 1972.)

314

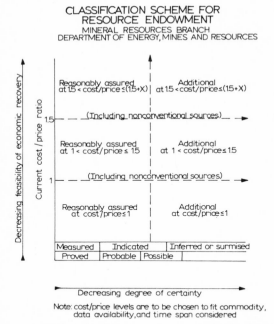

Fig.106. Classification scheme for resource endowment. (After Zwartendyk, 1972.)

in those ranges as it does not seem expedient (economically, at least) to obtain more data.

A number of auxiliary aspects related to possibly sub-categorizing Figs.105 and 106 are discussed by Zwartendyk, and he pointed out that the details are less significant and may be handled differently for different purposes. Important to keep in mind are:

(1) resource estimates must go beyond proved reserves inasmuch as proved reserves have no time dimension and are of little interest beyond the immediate present time — in contrast to the geologist's concern to make extrapolations into the future.

(2) Recourses should be categorized into a sufficient number of groups of "certainty of occurrence".

(3) In order to include the time variable in economic considerations, resource estimates should be further categorized according to how close specific resources are being economically exploitable. Figure 106 was then used by Zwartendyk as an example of possible subdivisions, explaining that "neither the exact number of categories chosen nor the exact names given them are important. What is important is that we see to it that we are covering the spectrum in groups of estimates that make clear that we are doing so".

METHODOLOGIES IN EXPLORATION AND RESOURCE ASSESSMENT

"The aim of science is to seek the simplest explanation of complex facts. We are apt to fall into the error of thinking that the facts are simple because simplicity is the goal of our quest. The guiding motto in the life of every natural philosopher should be, 'Seek simplicity and distrust it'."
Alfred North Whitehead (in: *Concept of Nature*, p.163)

This chapter cannot concern itself with the fundamentals of "Exploration Methodology" (="Exploration Philosophy") to which a separate lengthy and detailed treatment could be devoted to describe and explain the phase-by-phase application of all geological, geochemical and geophysical techniques. Let it suffice to emphasize that in each stage, and during the utilization of each method, several terminologies and classifications are employed. Moreover, the treatment of "Exploration Philosophy" *per se* could itself be based on the outline and discussion of (1) the *classification* of methodologies, (2) their *sequence* of application during research and/or prospecting and (3) the *inter-relationships* and/or *complementary* nature of all geological disciplines. It is possible to draw up numerous tabular or diagrammatic classificatory models — all beyond the scope of this contribution, except for one simple exemplar.

Cargill et al. (1977) listed six major methodologies in resource-assessment investigations, namely: (1) *areal value estimation*: extrapolation of a representative estimated mean unit value to the region of interest; (2) *volumetric estimation*: extrapolation from representative estimated mean concentration of a unit volume to the volume of interest; (3) *abundance estimation*: the estimation of the tonnage of recoverable resource from a representative mean abundance through an empirical function; (4) *deposit modeling*: the estimation of resources in a specific geological environment based on the analysis of characteristics of known deposits in similar environments; (5) *delphi estimation*: and (6) *integrated synthesis*: a resource estimate based upon a combination or integration, or both, of some or all of the above methods. Each of these six methodologies "has different basic requirements in terms of data types and intellectual involvement", shown in Table CXXVI. This table consists then of six methodologies and six requirements; the first five requirements are for data and the sixth is for intellectual involvement. Cargill and co-workers pointed out that as to methodologies "unconstrained and constrained resource assessments require similar, but not identical, methodologies or strategies for their completion. In general, the following initial steps should be taken: (1) inventory of the known occurrences of the resource within the constraints, (2) develop occurrence model(s) reflecting the constraints, and (3) use the model(s) within their scope of applicability to determine an estimate of the amount of the resource available."

Particularly pertinent is the conclusion by Cargill et al. that most resource-assessment failures in the past were the consequence of misunderstanding the basic nature of the methodologies.

As to the *relative or proportionate contributions* of several methods of investiga-

TABLE CXXVI

Relationship between geological methodology and the respective requirements in mineral exploration (After Cargill et al., 1977; courtesy of *J. Int. Assoc. Math. Geol.*)

Requirements	Methodology [a]					
	1	2	3	4	5	6
Mineral deposit data	D	–	D	E	D	E
Geological maps [b]	D	D	D	E	D	E
Geochemical data	D	E	E	D	D	E
Geophysical data	D	D	D	D	D	E
Regional mineral statistics	E	D	E	D	D	E
Scientific background	D	D	D	E	E	E

[a] E denotes an essential item; D denotes a desirable item. The E entries outline areas where specific data requirements must be defined on the basis of the methodology.
[b] Geological maps include metallogenetic maps.

tions made towards the *discovery* of mineralizations, too little information is available — even less on the correlation between the *types of ore deposits* discovered versus the techniques utilized. A quick, non-systematic and non-quantitative review of the literature will indicate that the three major disciplines of geology—geophysics—geochemistry have each their share of ore discoveries; in some cases only one approach led to the find, whereas in other instances two- or three-discipline methodologies were responsible. (One should realize, however, that *geology* with its numerous fundamental concepts and techniques is really never totally excluded from any study as it constitutes the *basis* for all investigations of ore exploration and development. Only the *relative* contribution and the *stage* at which it is utilized may vary! For example, the more geology a geophysicist knows of an area, the greater the likelihood for offering a plausible interpretation.)

Required are more profound, possibly semi- to fully-quantitative studies of the contributions of disciplines or methodologies made in (1) ore discoveries in general and (2) in cases of specific types of mineralizations. These studies will no doubt, commence with simple approaches and progress to more complex ones. As a start I can visualize, for example, the plotting of the *"contributory-values"* of methods in ore discoveries on a triangle with geology, geophysics and geochemistry at its apices; the sides of the triangle (geology—geophysics, geophysics—geochemistry, and geology—geochemistry) ranging from 0 to 100%. Thus, one can plot these "contributory-values" in percent (the technique of estimating this percentage to be worked out may be no easy task, it is realized). *Subsidiary triangles* of a similar nature may well be plausible for the subdisciplines. For example, the major triangle for geology may have at its apices "sedimentary-environmental reconstructions", "stratigraphic interbasinal correlations", and "deformational structural studies". A *second subsidiary triangle* may then be drawn for "structural geology-contributions", namely with macroscale-, megascale- and microscale-structural methods at its apices. Similar refinements may be tried for geophysics and geochemistry.

As an extrapolation of the above, it is interesting to note the needs for graduate training in the earth sciences, that (although it may be partly a subjective judgement) vary from one year to the next. Friedman (1978) in his "The Golden Age of the Geoscientist" notes that at present the fields most required by exploration include sedimentology, stratigraphy, tectonics and geophysics.

EPILOGUE

" ... the quality of a civilization depends upon its ability to discern and reveal truth, and this depends upon the scope and purity of its language."
    Iris Murdoch (in: *Salvation by Words,* New York Review of Books)

"Science is an allegory that asserts that the relations between the parts of reality are similar to the relations between terms of discourse."
    Scott Buchanan (in: *Poetry and Mathematics,* IV, 1929)

"To acquire the first *elementary* principles of any science is unpleasant, but without them one cannot overview *the totality,* which makes the *elementary* members interesting."
    Scharnhorst (freely translated)

"Respect for the word − to employ it with scrupulous care and incorruptible heartful love of truth − is essential if there is to be any growth in society or in the human race."
    Dag Hammarskjöld, Markings

"In every tongue the speaker labours under great inconveniences, especially on abstract questions, both from the paucity, obscurity, and ambiguity of the words, on the one hand, and from his own misapprehensions, and imperfect acquaintance with them, on the other."
    George Campbell (1776)

"A word is not a crystal, transparent and unchanged; it is the skin of a living thought and may vary greatly in color and content according to the circumstances and the time in which it is used."
    Justice O.W. Holmes

"The purpose of classification is not to set forth final and indisputable truths but rather to afford stepping stones towards better understanding."
    L.C. Graton

"It is as important for the purpose of thought to keep language as efficient as it is in surgery to keep tetanus bacilli out of one's bandages."
    Ezra Pound (in: *Literary Essays*)

"It is the mark of an educated man to look for precision in each class of things just so far as the nature of the subject permits."
    Aristotle

"Methodology is half the art of science."
  Karl H. Wolf (March 1979, Passau to Waldmünde)

"Observations never become obsolete."
  Karl Krejci-Graf (in: *Data on the Geochemistry of Oil Field Waters,* 1978)

" . . . what is metaphysics today, will be science tomorrow."
  William Pepperell Montaque (in: *Philosophy as Vision,* 1933)

"Rem tene, verba sequentur." ("Master the subject, and the words will follow.")
  Cato

"Master the terminology and classifications, and the subject matter will unfold itself."
  Karl H. Wolf (Inversion of Cato's above-cited statement)

" . . . all science, even the most rigorous mathematics, is relative in its truth."
  Immanuel Kant (in: Will Durant: *The Story of Philosophy,* Ch. VI)

"The most incomprehensible thing about the world is that it is comprehensible."
  Albert Einstein (in: *Recalled in reports of his death,* April 19, 1955)

"The experiment serves two purposes, often independent one from the other: it allows the observations of new facts, hitherto either unsuspected, or not yet well defined; and it determines whether a working hypothesis fits the world of observable facts."
  René J. Dubos

"To be ignorant of many things is expected.
To know you are ignorant of many things is
  the beginning of wisdom.
To know a category of things of which you are
  ignorant is the beginning of learning.
To know the details of that category of things
  of which you are ignorant is to no longer
  be ignorant."
  Phenella (in: *The Unwritten Comedy*)

"Truth generally lies in the coordination of antagonistic opinions."
  Herbert Spencer (in: Will Durant: *The Story of Philosophy,* ch. VIII)

"Cliché means a stereotype; in its literary sense it is a word or phrase whose felicity in a particular context when it was first employed has won it such popularity that it is apt to be used unsuitably and indiscriminantly."
  Fowler's Modern English Usage (2nd revised edition by Sir Ernest Gowers, 1975)

  The terminologies and classifications audited in this chapter consist of generalized, all-inclusive ore and host-rock schemes for primary and secondary igneous, sedimentary and metamorphic types; encompassing petrographic (mineralogic), geochemical (elemental), paragenetic, metallogenic, exploration-methodological, and resource classifications, plus classifications of specific commodities; and numerous miscellaneous schemes. Multi-classificatory schemes of the same entity have also been exemplified. A total of 25 to 30 major variables with many auxiliary parameters were recognized, of which only a very limited few can be utilized in any one two-dimensional pigeon-hole-type scheme. At least 11 operational factors have been identified that can cause difficulties and confusion in the application of nomenclatures and classifications.

Certain conclusions reached in this chapter could have been obtained without such an extensive survey, but one of the chief aims is to provide concrete information from which a researcher can extract approaches to suit his own work, or can combine and/or modify older data to formulate new systems. Many of the tables and figures will be updated in the future, but in the meantime they will serve specific purposes.

By supplying the large dosage of examples from the published literature with concomitant brief discussions, supported by the writer's own experiences in several geological disciplines, it is concluded that no two persons look at the same object in precisely the same manner. "What makes one writer's description more vivid than another's is the kind of details selected and the way they are arranged" (Corder, 1978). This is true for both scientific and non-scientific observations and is too serious to be comfortably and smugly brushed aside, for the same enigma has been recognized long ago (yes, centuries ago) when *logic* was developed, and more recently when, the *scientific method* has been outlined in innumerable dissertations to offer us an operational method for checking our results. Similar attempts in philosophy—psychology—sociology have been very productive (see, for example, Thouless, 1971, regarding logical communication). (For similar approaches in management and administration, see Gomez, 1978, and Park, 1963.)

Nomenclatures and classifications, as well as scientific methodologies, are a direct outgrowth and integral constituent of logic; logic is indeed the foundation of communication. Our technical deliberations are best when systematized according to an agreed-upon sequence of steps, when our analytical geological—geochemical—geophysical methods are standardized, and when our descriptions—interpretations—extrapolations (including terminologies and classifications) follow a logical pattern.

At present we still appear to have a haphazard, motley collection of nomenclatures and classificatory schemes, each constituting a separate entity and each being reliable to various degrees. In all sections of this compilation, it has been reiterated that both descriptive and genetic [1] nomenclatures and classifications are needed in all types and in all phases of geological studies; that the two should, whenever possible, be kept separate initially (although this may be difficult in the case of "genetically flavoured" terms).

When examined collectively, the schemes do not comprise a coherent self-supporting system — i.e., an underlying unifying principle is absent. This "fragmentation" in terminologies and classifications does not allow one to move smoothly from one geologic study to another; e.g. from the micro- through the macro- to the megascopic scale (or vice

---

[1] It is interesting to note the division of Corder (1978) of the styles of writing into (1) *expository writing* which informs, simply presents information — correlative in geology with descriptive, non-genetic reporting, and (2) *argumentative writing* that convinces, uses data to support or test a belief, and presents opinions, and is comparable to geological interpretive—genetic—extrapolative reporting.

versa), from mineralogical to lithological to textural and structural observations, and from purely descriptive to purely genetic studies.

During later stages of an investigation, when more data are available, more thought should be given to "transitionary/gradational" classification schemes that allow a convenient bridge to move unhindered from one phase to another (e.g., from purely descriptive to the genetic, or from one scale to another). Inherent in this is the requirement to delve deeper into *naturally* occurring transitional/gradational *processes and products* (note: in contrast to *man-inspired* terminologies and classifications) in agreement with Francis Bacon's dictum "Nature, to be commanded, must be obeyed", so that our scientific language has to reflect natural occurrences. *Integrated* nomenclatural/classificatory systems should be developed; several approaches are probably required in each sub-discipline to satisfy the multiple requirements. Descriptions and interpretations are as important as ever and, indeed, are becoming more exacting in accordance with newly evolving methodologies.

The writer's contention that we have as yet not clarified many of the most fundamental terminologies, has been reiterated during the compilation of this contribution. We have accumulated vast amounts of data, are aware of the problems, and have developed plausible theories and hypotheses — but nevertheless we are sometimes stuck when we wish to express our ideas unequivocally and precisely. This confusion (or chaos) is exemplified by "diagenesis" and "epigenesis": when a *clear definition* is not offered; it is impossible to know whether these terms are applied to geologic time, space, source, process, or environment, or a combination of these. This inadequacy is not restricted to the English-language nomenclature: the above-chosen two terms, and the underlying concepts, are similar in many other languages. One more example: what are the precise differences (and/or common features of them) between the French expressions (1) "roche magasin", (2) "roche encaissante", and (3) "roche envelope"? Are they equivalent to the English "host-rock" and/or "country rock" and/or "wallrock" and/or "mother rock"? The same confusion exists in the intra-English-language communication: what are the unequivocal differences between these four?

To return briefly to the absence of the "unifying principle" and the present "state of fragmentation" in the geological nomenclatures, it is concluded that some coherence must be wrangled from the haphazard situation by a systematic analysis of the underlying principles. To mention some specifics, we have to clarify (1) the meanings of individual terms, (2) their applications, limitations, range of utilization, synonyms, overlaps and transitional situations, modifications of terms by prefixes; and (3) the consequences of combining various terms as members of one classification. The meanings of *isolated individual* terms may differ from those meanings of the *same* terms when examined *in context* as members of a classification.

In deliberations such as the above ones, Thoreau's shrewd remark about "Improved means to an unimproved end" reminds one of the futility in improving our descriptive and analytical techniques (including nomenclatures and classifications) without a concomitant attempt to sharpen the genetic/interpretative/extrapolative tools, concepts and models. This problem has not been articulated well enough over a large sector of the geological fraternity — needed are both proscriptions and prescriptions (diagnoses and cures) for this dilemma. What is the utility or relevance of a large volume of descriptive data pouring out of laboratories without proper intellectual digestion, correlation, comparison/contrasting, periodic syntheses, and especially plausible interpretations? The data should always be shown to either fit into an existing genetic concept or, if this cannot be done (where the information may even contradict existing theories), one ought to revise the prevailing ideas or formulate a new one.

For the day-to-day routine work it has been found that the flow-chart conceptual models may be superior to the pigeon-hole-type classifications, for the former depict interrelationships between processes, environments and products, outline the gradational/ transitional cases, and directly point to the numerous combinational possibilities (e.g., multiple reconstructions). In contrast, the pigeon-hole schemes have set boundaries, are inflexible and oversimplified, do not depict transitional cases, and thus do not enhance comparative studies. Nevertheless, the pigeon-hole classifications remain useful — if care is exercised. A combination of flow-chart and pigeon-hole schemes may be advisable, and *the following should be provided* with each system: the parameters upon which the schemes are founded must be clearly defined and the aims, assumptions, limitations, and applications (scopes) must be fully explained. Any internal contradictions, overlaps and other irregularities should be pointed out.

In addition to the well-chosen classification, investigators have used various forms of *"check-lists"* consisting of key words or key statements to compel the researcher (1) to record his observations in an unvarying sequence, (2) to collect all the required data, and (3) to consider all possible interpretations. Each term or phrase in these check-lists represent an idea or concept, and one cannot "accidentally" forget them once they are part of one's repertoire of scientific language — one can agree with Shakespeare who stated in Hamlet (in a different context, however): "Suit the action to the word, the word to the action."

One sometimes wonders about the reasons for the present semi-chaotic state in the geological modes of communication, and before proceeding to find any solutions, one should heed Reiners' (1976, p.365) discussion on the causes that result in vaque statements or muddled nomenclatures because each problem demands a different response: either (1) the writer does not express himself clearly, or (2) the writer has come to an incomplete or incorrect conclusion, or (3) the subject matter is not clear enough or insufficient data has been collected, or (4) a combination of (1) to (3).

When I read in Corder (1978, pp.19–20) about the *"standard of appropriateness"* in writing, it struck me that a similar "standard or rule of appropriateness in geologic report writing" (which includes the appropriate usage of terminologies and classifications) should be stressed. That is to say, the format and content of any report in general, as well as the nomenclature and conceptual models, for example, must be tailored carefully to both the types of problems studied and the predetermined aims set for the investigation. For instance, a "standard of appropriateness" in this latter context would establish the "degree of completion" to which a study should be carried out (e.g., either reconnaissance of very detailed, scales used, ± supplementary microscopic and/or geochemical data). To put this into context with the present chapter: if the "appropriate standard" has been followed in selecting the most suitable terminologies and classifications, then there is little doubt as to the utilization of the proper schemes.

Corder pointed out (p.307) that certain incongruities violating the "principle of appropriateness" may be useful by mixing different kinds of languages to obtain a comic effect in poetry or non-scientific prose! A similar situation must be avoided in scientific (*factual* "expository" and "argumentative") writing through a careful separation of terminologies and classifications that cannot be "mixed". Should one ignore this, the consequences may also be comical — sometimes the comical results are hidden and discoverable only through a more penetrating scrutiny. Just as inappropriate is it to employ abstract words for intellectual discipline — it is equivalent to hiding behind scientific jargon.

Although it is beyond the scope of the present compendium to enter into details, permit me to offer at least a few comments on the types of classifications to illustrate a possible first step in the analysis and division of these schemes, i.e., to consider a *"classification of classifications"*.

(1) *"Rudimentary"* classifications are the simplest or most basic schemes, e.g., in igneous petrology the "plutonic—hypabyssal—volcanic—exhalative" mega-division, and in sedimentology the grain-size-based terminologies of "clay—silt—sand—pebble—cobble—boulder" (or "claystone/shale—siltstone—sandstone—conglomerate") classifications. Such "rudimentary" schemes are applied commonly when only brief superficial examinations are possible (e.g., in the field) and are the result of *"first-appearance"* or *"first-impression"* judgements at the beginning of a study. They are an example of the *"minimal"* terminological and classificatory schemes (see below).

(2) *"Subsumptive"* classifications are those that are the direct logical derivative of a rudimentary scheme and the 2nd- to nth-degree classificatory refinement of them; i.e., one, two, or more subdivisions of the rudimentary classification, each further subdivision being a refinement of the previous one. ("Subsume" means to bring one idea, principle, etc., under another, a rule, or a class.) Examples are: in volcanic nomenclature, the "volcaniclastic—pyroclastic—exhalative" sequence as a 1st-degree subdivision and "volcanic ash—tuff—lapilli—agglomerate" as a 2nd-degree refinement; and in sandstone petrology the 1st-degree subdivision would be "quartzite—arkose—greywacke—calclithite."

(3) *"Auxiliary"* (=*"subsidiary"*) classifications are schemes that, in addition to a major, generally accepted scheme, use a different approach in an attempt to solve the same problems. Consider the various sandstone classifications that have been available in the past and are still in circulation: although a general concensus seems to be emerging on which the best criteria and variables are in adopting the most plausible classification,

enough schemes are still used to have one or more auxiliary systems to augment or support the major classification. (If employed in this fashion, could one call these either "augmentative-auxiliary" or "supportive-auxiliary" classifications?)

It is rather obvious, that at the present state of our knowledge, what one researcher calls his "main" classification, another may consider his "auxiliary" scheme. If a petrographer decides to use only one scheme to the exclusion of all others, then the above distinction is not required. However, from my point-of-view, in more detailed studies it is illuminating to test various terminologies and classifications, and in such instances the differentiation remains valid.

"Auxiliary" systems are to be found on all levels of subdivisions as well as in any type of geologic work: both "rudimentary" and "subsumptive" types of schemes (and of the latter on the 1st, 2nd and nth degree of subdivision) can have auxiliary classifications.

(4) *"Confluence"* classification schemes are those schemes that have been formulated by combining two or more systems. Here, too, several possibilities exist: (a) Two simpler schemes are combined to constitute a more complex one — more complex because the number of variables were increased, or (b) the number of variables remained constant but a different combination or arrangement has been attained, or (c) the relative emphasis on the parameters has been changed.

(5) Elsewhere in this chapter it has been pointed out that geological investigations must follow specific logical steps, the best example being the sequence: field petrography → laboratory petrography → petrology/petrogenesis → environmental reconstructions → extrapolations in space and time. One may call this the *"tandem"* approach practised in geological sciences — an approach that imposes concepts and methodologies to be used "one behind the other" in a certain sequence. If this is accepted, then the consequence is obvious that specific terminologies and classifications are also most logically employed in a sequence of operations, as defined by the petrographic (descriptive) *prior* to the petrologic (genetic—interpretive) nomenclatures, both of which are succeeded by the utilization of an environmental classification.

An investigatory system that prescribes a predetermined sequence consists of several "tandem" nomenclature and classification schemes; each succeeding one is or should be dependent on a preceding scheme. In every subdiscipline in geology, it is hoped that eventually we have such logical tandem systems, each link within them being easily bridged by so-called "nexus" classifications (see next section).

(6) *"Nexus"* classifications are those that afford an easy transition from one scheme to another ("nexus" meaning connection, link, bond). Therefore, one can also speak of "nexus nomenclatures", which offer a transition from one type of terms to a different set. The need for such "nexus" schemes has already been discussed in the earlier parts of this chapter, and the example discussed showed this need by illustrating the *consequences when there is an absence of an easy transition* from the "petrographic—petrologic—environmental" type of investigations. To use another exemplar: during the stages of field work through the laboratory studies (both collecting "concrete" data) to the interpretive

extrapolations (manipulating "abstract" ideas), at least three respective classification systems should be "connected" by two nexus schemes.

A model of a "nexus" scheme might be one composed of both descriptive and genetic terminologies, thus being situated along a spectrum between the purely descriptive and purely interpretive end-members. Only extensive testing in the future will reveal whether, and under what circumstances, such nexus nomenclatures and classifications are useful, or whether instead it is possible to find a more *direct* transition from the descriptive to the interpretive/extrapolative schemes.

(7) Although the above-cited classification of classifications offers only a preliminary superficial attempt, it may show the way for future refinements. I do not think that such a division is purely visionary or idealistic in the sense that it is impractical/academic, because the mere structuring of classificatory schemes may open new avenues of visualizing hitherto unknown approaches. Well-meaning advocates of the "lets-simplify-science-to-make-it-accessible-to-all" school, who have raised objections to the intricacies of the existing terminologies (possibly not realizing that we have not as yet solved the old problems in scientific communication), will expectedly protest against the introduction of further presumed "complexities" (under the misconception that the established nomenclatures and classifications are adequate?). Those who abhor the complexity of science must consider that the needs for further detailed schemes are to assure communication which not any layman can understand, but which a specialist cannot misunderstand. It certainly is not true, as an overzealous "scientific non-believer" will have it, that "today's scientific jargon is tomorrow's hieroglyphics".

(8) The above deliberations result in one further type of system, namely the "*minimal*" and "*optimal*" classification schemes to be used in each specific investigation. The "minimal" terminologies are those that are absolutely necessary to fulfill the set objectives; the "minimal" referring to quality not quantity. Examples: if a study is purely based on field observations, the "size-grade" classification of the sedimentary rocks and the "volcaniclastic—pyroclastic—flow rock" division of the volcanic lithologies may be sufficient. The data would be the simplest, qualitatively speaking, and may be equivalent to the above-described "rudimentary" information.

The so-called "optimal" schemes (i.e., the most detailed, most precise schemes attainable under set conditions and aims) utilizable in any particular study would have to be determined, of course, by the scope of the data required which depends on the time, finances, equipment and expertise made available, as well as on the aims set for the investigation. Already in the planning stages of a study, i.e., prior to its implementation, one should determine when, where and how detailed methodologies are to be employed, and this in turn would consequent the types of "optimal" terminologies and classifications to be used concomitantly.

Certain "philosophical" barriers still exist in the sciences that need to be identified and brought to the attention of researchers in order to deal with them properly — mainly

TABLE CXXVII

Static and dynamic scientific knowledge (K.H. Wolf, 1979, unpublished)

---

(1) Static, stationary, out-of-circulation information in specific areas in the world because of preferential use.

Reason (a): – nation/country-preferential information
  – language-preferential
  – culture-preferential
  – time (in human history)-preferential
  – institutional-preferential (especially if confidential)
  – "school-of-thought"-preferential *
  – topic/problem-preferential * (i.e., no cross-fertilization even within same broad discipline)
  – discipline-preferential * (i.e., no cross-fertilization across boundaries of disciplines)

Reason (b): information proven wrong, hence shelved

[* Interim note: Differences are made between "disciplines" (physics, chemistry, mathematics vs. geology), "school-of-thought" (e.g., drifters vs. non-drifters which is not discipline-confined), and "topic/problem" (i.e., within one discipline, e.g., in geology: sedimentology and biology vs. ore petrology).]

(2) Dynamic, moving, in-circulation information.

Pattern (a): moving *within* above-set (see 1a) boundaries.

Pattern (b): transgressing one, several or all above-set boundaries (most ideal system in a world-wide, across-all-boundaries circulation, except for confidential data).

---

by *consciously* eliminating these barriers! The effects of this on communication in general (including the application of terminologies and classifications) become apparent by considering Table CXXVII of *Static and Dynamic Scientific Knowledge*. Again to keep the discussion in the context of the present chapter it should be pointed out that although this table lists "scientific knowledge" in general, it is also directly applicable to nomenclatures and classifications, as it provides a glimpse of several possible reasons for their *preferential developments* as well as their *patterns of circulation*.

A related situation of the acceptance of scientific concepts through history may be colloquially called the *"roller-coaster"* effect (a phrase adopted from politics where one speaks of "roller-coaster diplomacy"). In geology, this can be illustrated by two examples:

(1) Certain concepts have had their ups and downs to the extent of which they were *developed, accepted and utilized, then rejected, resurrected, possibly updated and once again accepted*. The downtrends in acceptability of an idea may have been the consequence of a new concept, so that the early theory fell victim to either (a) the "Law of Diminishing Appeal" [1] or (b) because the new idea was truly an improvement over the earlier one. With regards to (a) one should hasten to say that indeed there are examples in the history of the sciences when a concept was replaced by another merely because it

"lost appeal due to its senility" [1] without much evidence to declare the newer concept superior.

(2) The second exemplar is provided by *moving the emphasis* given to studies of particular problems as a result of economic, political and/or strategic influences, thus causing a shift of scientific interest or preferences from one sub-discipline to another. Such "roller-coaster" patterns in geologic studies can be exemplified by the interests in coal, uranium, oil-sands, tar-shales, and many other commodities. To be even more specific: in the oil industry, the interest in limestone versus sandstone reservoir petrology (e.g., diagenesis) has had its periodic ups and downs.

With this roller-coaster pattern in developing or broadening of concepts, hypotheses and theories, with the discovery of new techniques, as well as the periodic shift of geologic interests, one can also notice a parallel roller-coaster, but progressive, evolvement of nomenclatures and classifications.

The discussions in Peters' (1978) excellent book have reiterated my believe that the present summary chapter is a necessity. Peters' arguments are directly applicable to the theme of classification, nomenclatures, and communication in general: already in his "introduction" he pointed out the need of definitions to clarify situations, maintaining that "Sometimes . . . (it) is the heroic person who rescues a floundering discussion; sometimes it is the hairsplitting pedant who derails an entire train of developing idea. . . . , the ubiquitous 'definer' thrives on terms that should have been agreed on at the onset of a discussion — or a book. One way to avoid derailment is to set ground rules, a short list of key-word meanings that can be the shades in semantics."

The "hairsplitting pedant" (one could use a less derogative-sounding expression) (well, as a student said to his professor: "I don't mind you splitting a hair, but I don't believe in having it quartered for me!") is usually a super-logical individual who has seen too many technical and non-technical deliberations (whole committee meetings, mind you) disintegrate into nothingness precisely because abstract and concrete terms were used differently (both purposively and inadvertently, it should be stressed) by the various individuals. It is here also, where the "Standard of Appropriateness", mentioned earlier, must be employed.

A chapter such as this one, is not the place to discuss the purposive misuse of terms to support *un*professional aims — Peters (1978) has discussed this briefly in his book (e.g., pages 8, 285, 464, 491–492, 496, 506, 512, 515, 523, 530, 541–542, 549, 556–557, 558–560, 575). I call it "verbal/oral salting" or "*terminological or nomenclatural salting*" and it is to be strictly avoided. In this respect, the publication by Thouless (1971), referred to above, makes fruitful reading. The same applies to hiding behind scientific

---

[1] "Everywhere in the world is the 'Law of Diminishing Appeal' valid: the third glass of water is less tasty to the thirsty than the first."

L. Reiners (1963, p.187; freely translated)

slogans, double-talk or deliberate vaqueness, as well as to pretences of importance by overly utilizing jargon (thus creating a "specialist's" report that can "only be meaningfully interpreted to the uninitiated by the specialist").

What makes Peters' book a rarity is his separate chapter on "Communication", supported by references to communication in many of his other chapters. Examples: "The need for classification and the design of an appropriate system of classification depend heavily on communication ..." (p.133). While discussing exploration examples, Peters states (p.121) that "once the rock classification changed, the conceptual model had to change", shown by the Mount Isa exploration history as well as by those of the Witwatersrand (Chapters 1 and 2, Vol. 7, of the present Handbook), Zambian/Rhodesian copper belt (Chapter 6, Vol. 6), Broken Hill (Chapter 6, Vol. 4), and numerous others. What Peters called "data screening" (p.460) includes the selection of the most appropriate terminologies for each step in the study. "Pattern recognition" possible after some work has been done may induce the changes from one type of nomenclature and classification to another more suitable one. To do this, personal attributes of experience, imagination and perception are required.

As to the use of computers, Peters opines (p.133) that it " ... is the most helpful and the most dangerous ... because it adheres to classifications much more faithfully than do geologists or managers. It recognizes what a geologist has told it to recognize, and it delivers only the data asked for. No interpretation — no exceptions!" Peters (p.285) also stated that "one man's message is another man's noise"; noise being the irrelevant part of the signal or of the data, and message is the part we can actually use, namely the information. The same applies to terminologies and classifications, especially to genetic classificatory models — they assist us in recognizing the "message" and sort out the "noise".

In "Elements of Style" Strunk and White (1972) declared that the style of a writer, and equally of a researcher, reflects his attitude, degree of motivation, capabilities, experiences, as well as the reliability of the data collected, methods utilized, interpretations and conclusions. Part of this "style" is the type of nomenclatures and classifications selected (often preferentially). We have to realize that personal aptitudes and tastes are operative in the planning of a project, its implementation and execution. Let us consider one very simple, practical example, namely the preparation of a geologic map legend (note: a "classification" of sorts). Examining some published versions, it becomes obvious that a hybrid of descriptive and genetic/interpretive divisions have been used: e.g., environmental (talus, alluvial, wadi for Quaternary deposits; and marine, geosynclinal for older rocks), processes-genetic (turbidite, fluvial; nueé ardente, water-laid tuff; fore-reef talus), compositional (arkose, greywacke; crystal tuff, rhyolite tuff; granite), grain-size (sandstones, fine pyroclastic), and so on. Various combinations of the above have been utilized, e.g., composition + grain-size + environmental — but even on the same map consistency is too often absent. The critical evaluation of something as simple as a map legend already supports my contention that we have not wholly solved our terminological/classificatory enigmas. As Peters has stated (p.559), "the geologist may need to provide a certain amount of education as well as evaluation"!

Those who insist that "exploration uses science, but it is not science, since it aims are fundamentally different" because "science strives for understanding, whereas exploration strives solely for discovery — with or without understanding — by whatever means" (Northern Miner, 1978), may take exception to the proposal that our terminologies and classification should be refined still further. I maintain, however, that the more "scientific" data and understanding a geologist has, the more capable he will be to make the best "guesstimate" where to put the expensive drill hole. I have seen too many instances where the lack of basic "scientific" comprehension has not only resulted in the selection of the wrong drill-site, but also has consequented the delay in deciding to terminate the drilling because of not recognizing "academic" criteria that would have compelled the abandoning of the earlier-chosen locality, i.e., more "logical" planning would have been the result.

Many earth scientists still adhere to the difference between "practical/applied" and "academic" fields of endeavour as if they were abstract antipoles, although it has become quite obvious that for philosophical discussions (only!) one could consider them as two end-members along a spectrum with all gradational/transitional cases in between. (Is there a theory or technique that is, say, 80% applied and 20% academic? Problem to be left for esoteric discussions!) Although an exploration geologist has to win his crust of bread by being "practical", the new trend of putting more meaning and personal satisfaction into one's daily professional performances can be easily accomplished by stressing pride in both "extrinsic" and "intrinsic" duties. At least some "intrinsic" perfunctory performances should be permitted to be undertaken by a geologist; e.g., work that shows no immediate "extrinsic" or "practical" value, except that it adds a little to the large, possibly long-term, project. These routine intrinsic duties could bring a sense of accomplishment if they are done with a highly developed curiosity and an enjoyment in intellectual achievement. Anyone interested in such "philosophical", but practical aspects should consult Willoughby's (1975) proposal in his *"Stimulating Creativity"*. An ore petrologist— economic geologist may be involved in work that is purely profit-oriented, but if he is the slightest bit intellectually stimulated (and what is education?), he wishes to study problems merely "because they are there"; this in the vein of the following two quotations: "The scientist does not study nature because it is useful; he studies it because he delights in it, and he delights in it because it is beautiful" (Henri Poincaré), and "Scientific discovery and scientific knowledge have been achieved only by those who have gone in pursuit of them without any practical purpose whatsoever in view" (Max Planck, in "Where is Science Going?", 1932). Definitely, a well-balanced compromise has to be found! As to terminologies and classifications, which are part of our scientific communicational methodologies, we are far from perfection and we should have the courage to prepare an occasional "Encyclopaedia of Our Geological Ignorance" (cf. Duncan and Weston-Smith, 1977) to guide our extrinsic and intrinsic pursuits for betterment.

ACKNOWLEDGEMENTS

This contribution was started in 1971 at the Laurentian University, Sudbury, Ontario, Canada, when the writer was Associate Professor of Geology and held research grants received from the Canadian National Research Council and the Geological Survey of Canada. Their assistance is gratefully acknowledged. The chapter underwent up to its completion several stages of periodic updating and expansion.

The project of reviewing and summarizing pertinent ideas on the terminology and classification plus related topics could not have materialized without the direct and indirect contributions of innumerable individuals who have made their experience available in different languages in widely dispersed publications. Many authors have over the past twenty years generously responded to hundreds of reprint requests, which have enabled the present writer to reduce the time envolved in preparing this and similar syntheses.

Although creative syntheses in science may lead to new ideas and point the way to unique avenues of future research, the credit goes largely to the earlier workers who have furnished the fundamental data. As Samuel Smiles has pointed out: "Human knowledge is but an accumulation of small facts made by successive generations of men — the little bits of knowledge and experience carefully treasured up by them growing at length into a mightly pyramid."

REFERENCES

Adler, E., 1970a. Praktische Tektonik, No. 1: Definition, Aufgaben, Arbeitsmethodik und Gliederung. *Zentralbl. Geol. Paläontol.*, Teil 1, 4: 601–621 (35–55).

Adler, E., 1970b. Tektonische Geologie, photogrammetrische Datenerhebung und elektronische Datenbearbeitung. *Glückauf-Forschungsh.*, 31(6): 318–332 (1–15).

Amstutz, G.C., 1959a. Syngenetic zoning in ore deposits. *Proc. Geol. Assoc. Can.*, 11: 95–113.

Amstutz, G.C., 1959b. Syngenese und Epigenese in Petrographie und Lagerstättenkunde. *Schweiz. Mineral. Petrogr. Mitt.*, 39: 1–84.

Amstutz, G.C., 1960. Some basic concepts and thoughts on the space—time-analysis of rocks and mineral deposits in orogenic belts. *Geol. Rundsch.*, 50: 165–189.

Amstutz, G.C., 1964. Introduction. In: G.C. Amstutz (Editor), *Sedimentology and Ore Genesis.* Elsevier, Amsterdam, pp.1–7.

Amstutz, G.C., 1965. Some comments on the genesis of ores. In: *Problems of Post-Magmatic Ore Deposition, 2.* Geol. Surv. Czechoslovakia, Praque, pp.147–150.

Amstutz, G.C., 1972. Observational criteria for the classification of Mississippi Valley–Bleiberg–Silesia type of deposits. In: *2nd. Int. Symp. on the Mineral Deposits of the Alps,* Ljubljana, pp.207–215.

Amstutz, G.C. (Editor), 1974. *Spilites and Spilitic Rocks*. Springer, Heidelberg, 492 pp.

Amstutz, G.C. and Bubenicek, L., 1967. Diagenesis in sedimentary mineral deposits. In: G. Larsen and G.V. Chilingar (Editors), *Diagenesis in Sediments*. Elsevier, Amsterdam, pp.417–475.

Association of Professional Engineers of the Province of Ontario, 1972. *Performance Standards for Professional Engineers Advising on and Reporting on Mineral Properties*. Association of Professional Engineers of the Province of Ontario, Toronto, Ont., 16 pp.

Australian Institute of Mining and Metallurgy, 1972. *Report by Joint Committee on Ore Reserves*. Australasian Institute of Mining and Metallurgy, Parville, Vic., 5 pp. (Supplement to the Aust. I.M.M. Bull., No. 355, July, 1972.)

Barnes, F.Q. and Ruznicka, V., 1972. A genetic classification of uranium deposits. In: *24th Int. Geol. Congr.*, Montreal, Sect. 4, Mineral Deposits, pp.159–166.

Bataman, A.M., 1950. *Economic Mineral Deposits*. Wiley, New York, N.Y., 916 pp.

Bates, R.L., 1960. *Geology of the Industrial Rocks and Minerals*. Harper and Row, New York, N.Y.

Baumann, L. and Tischendorf, G., 1976. *Metallogenie und Mineragonie*. VEB Deutscher Verlag für Grundstoffindustrie, Leipzig, 458 pp.

Beales, F.W. and Jackson, S.A., 1968. Pine Point – a stratigraphic approach. *Can. Min. Metall. Bull.*, 1968: 1–12.

Beerbaum, B., 1977. Die Genese der marin-sedimentären Phosphatlagerstätte von Al Hasa (westliches Zentraljordanien). *Geol. Jahrb., Reihe D*, 24: 3–55.

Beuss, A.A. and Grigorian, S.V., 1977. *Geochemical Exploration Methods for Mineral Deposits*. Applied Publishing Ltd., Wilmette, Ill., 287 pp.

Beveridge, W.I.B., 1957. *The Art of Scientific Investigation*. Vintage Book, V-129, 239 pp.

Beyer, W., 1971. Grundlagen einer Theorie der Erzbilding. *Z. Angew. Geol.*, 17: 526–530.

Bilibin, Yu.A., 1968. *Metallogenic Provinces and Metallogenic Epochs*. Geol. Bull., Queens College Press, Flushing, N.Y. 35 pp.

Bloom, B.S., 1956. *Taxonomy of Educational Objectives*. Longman, London.

Bogdanov, Y.V. and Kutyrev, E.I., 1973. Classification of stratified copper and lead–zinc deposits and the regularities of their distribution. In: G.C. Amstutz and A.J. Bernard (Editors), *Ores in Sediments*. Springer, Heidelberg, pp.58–63.

Bono, E. de, 1967. *The Use of Lateral Thinking*. Penguin-Pelican Books, Middlesex, 141 pp.

Booth, J.K.B., 1974. The influence of the natural abundance of metals on exploration. *Can. Min. Metall., Trans.*, 74: 92–97.

Borchert, H., 1951. Die Zonengliederung der Mineralparagenesen in der Erdkunde. *Geol. Rundsch.*, 39: 81–94.

Borchert, H., 1957. Der initiale Magmatismus und die zugehörigen Lagerstätten. *Neues Jahrb. Mineral. Abh.*, 91: 541–572.

Borchert, H., 1959. Zur Entstehung der Kieslagerstätten des japanischen Aussenbogens. *Neues Jahrb. Mineral., Monatsh.*, 1959: 49–54.

Borchert, H., 1960a. Genesis of marine sedimentary iron ores. *Trans. Inst. Min. Metall.*, 69: 261–279.

Borchert, H., 1960b. Geosynklinale Lagerstätten, was dazu gehört und was nicht dazu gehört, sowie deren Beziehungen zu Geotektonik und Magmatismus. *Freiberg. Forschungsh. C*, 79: 7–61.

Borchert, H., 1961. Zusammenhänge zwischen Lagerstättenbildung, Magmatismus und Geotektonik. *Geol. Rundsch.*, 50: 131–165.

Borchert, H., 1962. Chemismus und Petrologie der Erdschalen sowie die Entstehung und Ausgestaltung der wichtigsten Diskontinuitäten der Erdkruste. *Neues Jahrb. Mineral. Monatsh.*, 718: 143–163.

Borchert, H., 1965. Formation of marine sedimentary iron ores. In: J.P. Riley and G. Skirrow, (Editors), *Chemical Oceanography*. 2. Academic Press, New York, N.Y., pp.159–204.

Borchert, H., 1968. Der Wert gesteins- und lagerstättengenetischer Forschung für die Geologie und Rohstoffnutzung. *Ber. Dtsch. Ges. Geol. Wiss., B. Mineral. Lagerstättenforsch.*, 13(1): 65–116.

Borchert, H., 1978. *Lernblätter zur Geochemie und Lagerstättenkunde der mineralischen Rohstoffe*. Glückauf, Essen, 119 pp.

Borchert, H. and Tröger, E., 1950. Zur Gliederung der Erdkruste nach geophysikalischen und petrologischen Gesichtspunkten. *Gerlands Beitr. Geophys.*, 62: 101–126.

Bostick, N.H., 1970. Electronic data processing applied to uranium resource prediction and exploration. *Soc. Min. Eng., AIME Trans.*, 247: 4–9.

Botvinkina, L.N., 1973. Genetic classification of volcanic-sedimentary rocks and some aspects of their facies analysis. *Lithol., Miner. Resour. (USSR)*, 7(2): 147–160.

Bowman, H.N. and Stevens, B.P.J., 1978. The philosophy, design and applications of the New South Wales geological survey's metallogenic mapping programs. *J. Geol. Soc. Aust.*, 25(3): 121–140.

Boyle, R.W., 1968. The geochemistry of silver and its deposits. *Geol. Surv. Can. Bull.*, 160: 264 pp.

Boyle, R.W., 1974a. Elemental association in mineral deposits and indicator elements of interest in geochemical prospecting (revised). *Geol. Surv. Can. Pap.*, 74-45: 40 pp.

Boyle, R.W., 1974b. The use of major elemental ratios in detailed geochemical prospecting utilizing primary halos. *J. Geochem. Explor.*, 3: 345–369.

Boyle, R.W., 1976. Mineralization processes in Archean greenstone and sedimentary belts. *Geol. Surv. Can. Pap.*, 75-15: 45 pp.

Brown, W., 1974. *Organization.* Penguin Books, Middlesex, 493 pp.

Bundesanstalt, Hannover, 1974. *Untersuchungen über Angebot und Nachfrage mineralogischer Rohstoffe, V. Zink.* Bundesanstalt für Bodenforschung, Hannover, Germany, 259 pp. (See also Nos. I–IV on lead, copper, aluminum and fluorite, respectively.)

Čada, M. and Novák, J.K., 1974. Spatial distribution of greisen types at the Cínovec-South tin deposit. In: M. Štemprok (Editor), *Metallization Associated with Acid Magmatism*, 1: 383–388.

Callahan, W.H., 1964. Paleophysiographic premises for prospecting for stratabound base metal mineral deposits in carbonate rocks. In: *Proc. CENTO Symposium, Mining, Geology and Base Metals.*, Turkey, Sept. 14–18: pp.191–248.

Carey, S.W., 1962. Scale of geotectonic phenomena. *J. Geol. Soc. India*, 3: 97–105.

Cargill, S.M., Meyer, R.F., Picklyk, D.D. and Urquidi, F., 1977. Summary of resource assessment methods resulting from the International Geological Correlation Program Project 98. *J. Int. Assoc. Math. Geol.*, 9(3): 211–220.

Chilingar, G.V., Bissell, H.J. and Wolf, K.H., 1967. Diagenesis of Carbonate Rock. In: G. Larsen and G.V. Chilingar (Editors), *Diagenesis in Sediments.* Elsevier, Amsterdam, pp.179–322.

Cissarz, A., 1951. *Einführung in die Lagerstättenlehre.* Schweizerbart, Stuttgart, 170 pp.

Cissarz, A., 1964. Die Stellung der Lagerstätten im geologischen Bildungsablauf. *Geol. Jahrb.*, 82: 75–98.

Cissarz, A., 1965. *Einführung in die allgemeine und systematische Lagerstättenlehre.* Schweizerbart, Stuttgart, 2nd ed., 228 pp.

Clark, G.B. and Lewis, R.S., 1964. *Elements of Mining.* Wiley, New York, N.Y., 3rd ed., 768 pp.

Conybeare, C.E.B., 1979. *Lithostratigraphic Analysis of Sedimentary Basins.* Academic Press, New York, 384 pp.

Corder, J.W., 1978. *Handbook of Current English.* Scott, Foresman, and Co., Glenview, Ill., 494 pp.

Crain, H.M. (Editor), 1950. Economics of the mineral industry. *Q. Colo. Sch. Mines*, 45(1A): 47 pp.

Dahlkamp, F.J., 1978. Classification of uranium deposits. *Miner. Deposita*, 13: 83–104.

Derry, D.R., 1970. Geochemistry – the link between ore genesis and exploration. *Can. Inst. Min. Metall. Bull.*, 63: 655–658.

Dickinson, W.R., 1970. Interpreting detrital modes of graywackes and arkoses. *J. Sediment. Petrol.*, 40: 695–707.

Dimroth, E., 1968. Sedimentary textures, diagenesis and sedimentary environment of certain Precambrian ironstones. *Neues Jahrb. Geol. Paläontol. Abh.*, 130: 247–274.

Dixon, B., 1973. *What is Science For?* Penguin-Pelican Books, Harmondsworth, Middlesex, 284 pp.

Duncan, R. and Weston-Smith, M., 1977. *The Encyclopaedia of Ignorance.* Pergamon, Frankfurt, 433 pp.

Dutch, R.A., 1962. *Roget's Thesaurus of English Words and Phrases.* Longmans, London, 1309 pp.

Elliston, J., 1966. The genesis of the Peko orebody. *Australas. Inst. Min. Metall. Proc.*, 218: 9–17.

Erickson, R.L., 1973. Crustal abundances of elements, and mineral reserves and resources. *U.S. Geol. Surv. Prof. Pap.*, 820.

Fersman, A.E., 1952. Les Pegmatites, 1. Les Pegmatites Granitiques. (French translation of 1st ed., 1931.) Louvain/Bruxelles, 671 pp.

Folinsbee, R.E., 1977. World's view – from Alph to Zipf. *Geol. Soc. Am. Bull.,* 88(7): 897–907.

Folk, R.L., 1959. Practical petrographic classification of limestones. *Bull. Am. Assoc. Pet. Geol.,* 43: 1–38.

Folk, R.L., 1974. *Petrology of Sedimentary Rocks.* Hemphill, Austin, Texas, 182 pp.

Friedman, G.M., 1978. The Golden Age of the geoscientist. *Science,* 201: p.4352.

Frietsch, R., 1978. On the magmatic origin of iron ores of the Kiruna type. *Econ. Geol.,* 73: 478–485.

Füchtbauer, H. and Müller, G., 1970. *Sedimente und Sedimentgesteine.* Schweizerbart, Stuttgart, 726 pp.

Gabelman, J.W., 1973. Growth of a concept of metallotectonics. *U.S. Geol. Surv. Private Circulation,* 25 pp.

Garrels, R.M. and McKenzie, F.T., 1971. *Evolution of Sedimentary Rocks.* W.W. Norton and Co., 397 pp.

Garrett, R.G., 1977. Exploration geochemistry in resource appraisal. *J. Int. Assoc. Math. Geol.,* 9(3): 245–258.

Gary, M., McAfee, R. Jr. and Wolf, C.L. (Editors), 1974. *Glossary of Geology.* American Geological Institute, Washington, D.C., 857 pp.

Geotimes, 1974. BUMines, Survey revise definitions of mineral terms. *Geotimes,* Sept. 1974, pp.18–19.

Gilmour, P., 1962. Notes on a non-genetic classification of copper deposits. *Econ. Geol.,* 57: 450–455.

Gilmour, P., 1971. Stratabound massive pyritic sulfide deposits – a review. *Econ. Geol.,* 66: 1239–1249.

Ginsburg, I.I., 1963. *Grundlagen und Verfahren geochemischer Sucharbeiten auf Lagerstätten der Buntmetalle und seltener Metalle.* Akademie-Verlag, Berlin. (Transl. from the Russian.)

Goldschmidt, V.M., 1954. *Geochemistry.* Oxford, 730 pp.

Gomez, P., 1978. Organic problem-solving in public administration: a systems methodology. In: J.W. Sutherland (Editor), *Management Handbook for Public Administrators.* Van Nostrand/Reinhold, New York, pp.40–67.

Goranson, R.W., 1936. Silicate–water systems. The solubility of water in albite melt. *Trans. Am. Geophys. Union,* 17: 257–259.

Govett, G.J.S., 1977. World mineral supplies – the role of exploration geochemistry. *J. Geochem. Explor.,* 8: 3–16.

Govett, G.J.S. and Govett, M.H., 1972. Mineral resource supplies and the limits of economic growth. *Earth-Sci Rev.,* 8: 275–290.

Gowers, E., Sir, 1975. *Fowler's Modern English Usage.* Oxford Clarendon Press, Oxford, 2nd ed., 725 pp.

Gross, G.A., 1965. Geology of iron deposits in Canada. Vol. I: General Geology and Evaluation of Iron Deposits. *Geol. Surv. Can., Econ. Geol. Rep.* No. 22, 181 pp.

Halbouty, M.T., King, R.E., Klemme, H.D., Dott, R.H. and Meyerhoff, A.A., 1970. Factors affecting formation of giant oil and gas fields, and basin classification, Parts 1 and 2. In: *Geology and Giant Petroleum Fields – A Symposium. Am. Assoc. Pet. Geol. Mem.,* 14: 502–528; 528–555.

Hallberg, R.O., 1972. Sedimentary sulfide mineral formation – an energy circuit system approach. *Miner. Deposita,* 7: 189–201.

Hayakama, S.I., 1962. *The Use and Misuse of Language.* Fawcett, Greenwich, Conn., 240 pp.

Hubaux, A., 1973. A new geological tool – the data. *Earth-Sci. Rev.,* 9(2): 159–196.

Jancovič, S., 1974. Genesis of ore deposits as a basis for prospecting. In: *Problems of Ore Deposits,* 4th IAGOD Symp., Varna, Vol. 2, pp.347–355.

Jaques, E., 1965. Speculation concerning level of capacity. In: W. Brown and E. Jacques (Editors), *Glacier Project Papers.* Heinemann, London.

Joralemon, P., 1975. The ore finders. *Min. Eng.,* 27(12): 32–35.

Kirchmayer, M., 1961. Untersuchungsbereiche in der Strukturgeologie. *Neues Jahrb. Geol. Paläontol. Monatsh.,* 3: 151–155.

Korolev, A.V. and Shekhtman, P.A., 1968. Classification of postmagmatic ore fields. *Int. Geol. Rev.*, 4: 908–915.

Kosygin, Yu.A., 1970. Methodological problems of the systems investigations in geology. *Acad. Sci USSR Geotectonics* (Engl.), No. 2: 75–79.

Kreiter, V.M., 1968. *Geological Prospecting and Exploration.* MIR, Moscow. 383 pp. (Translated from the Russian.)

Krumbein, W.C. and Graybill, F., 1965. *An Introduction to Statistical Models in Geology.* McGraw-Hill, New York, N.Y., 475 pp.

Kunitz, W., 1929. Die Mischungsreihen in der Turmalingruppe und die genetischen Beziehungen zwischen Turmalin und Glimmern, *Chem. Erde*, 4: 208–225.

Kuzvart, M. and Böhmer, M., 1978. *Prospecting and Exploration of Mineral Deposits.* Elsevier, Amsterdam, 431 pp.

Laffitte, P., 1967. In: *Genesis of Stratiform Lead–Zinc–Barium–Fluorite Deposits (Mississippi Valley Type Deposits) – A symposium. Econ. Geol. Mem.,* 3.

Lambert, R.St.J. and McKerrow, W.S., 1978. Comparison of the San Andreas fault system with the Highland Boundary fault-system of Scotland. *Geol. Mag.*, 115: 367–372.

Lange, H., 1965. Zur Genese der Metåbasite im sächsischen Erzgebirge. *Freiberger Forschungsh. C,* 177.

Lasky, S.G., 1950. Mineral-resource appraisal by the U.S. Geological Survey. *Q. Colo. Sch. Mines,* 45(1A): 1–28.

Laznicka, P., 1978. Concept of exotic ore deposits deserves more attention. *North. Miner* (Canada), March 1978: C7–C8.

Leech, G., 1974. *Semantics.* Penguin Books, Middlesex, 386 pp.

Lehman, M.M., 1977. Human thought and action as an ingredient of system behaviour. In: R. Duncan and M. Weston-Smith (Editors), *The Encyclopaedia of Ignorance, 2.* Pergamon, Frankfurt, pp.347–354.

Lewis, N., 1974. *Word Power Made Easy.* Simon and Schuster, New York, N.Y., 412 pp.

Little, H.W., 1970. Distribution of types of uranium deposits and favourable environments for uranium exploration. In: *Uranium Exploration Geology,* Proc. Panel Vienna, April 13–17, 1970. Int. Atomic Energy Agency, Vienna, pp.35–48.

Lovering, T.G., 1963. Epigenetic, diplogenetic, syngenetic and lithogene deposits. *Econ. Geol.,* 58: 315–331.

Lovering, T.G. and McCarthy, J.H. Jr. (Editors), 1978. Conceptual models in exploration geochemistry. *J. Geochem. Explor.,* 9: 113–276.

Mason, B., 1958. *Principles of Geochemistry* Wiley, New York, N.Y., 2nd ed.

McKelvey, V.E., 1973. Mineral estimates and public policy. *U.S. Geol. Surv. Prof. Pap.,* 820: 19 pp.

Meine, K.-K., 1979. Thematic mapping: present and future capabilities. Dept. Int. Econ. World Cartography, XV: 1–17, Social Affairs, U.N. Publ., ST-ESA-SER, L-15 (Sales No. E.78. I.14), New York, N.Y.

Middle East, 1979. Synthesis = progress – early Arab style. *Middle East,* Jan. '79, pp.95–96.

Mills, J.W. and Eyrich, H.T., 1966. The role of unconformities in the localization of epigenetic mineral deposits in the United States and Canada. *Econ. Geol.,* 61: 1232–1257.

Naboko, S.I., 1959. Volcanic exhalations and their reaction products. *Studies of the Laboratory for Volcanology, 6.* AN SSSR, Moscow. (In Russian.)

Nasseef, A.O. and Gass, I.G., 1977. Granitic and metamorphic rocks of the Taif area, western Saudi Arabia. *Geol. Soc. Am. Bull.,* 88: 1721–1730.

Nicolini, P., 1970. *Gîtologie des Concentration Minérales Stratiformes.* Gauthier-Villars, Paris, 792 pp.

Niggli, P., 1937. *Das Magma und seine Produkte.* Leipzig, 379 pp.

Niggli, P., 1952. *Rocks and Mineral Deposits.* Freeman, San Francisco, Calif.

Northern Miner, 1978a. Mineral deposit research – whither now? *Northern Miner* (Canada), Oct. 19, 1978: p.B17.

Northern Miner, 1978b. The art of exploration. Good scientists find orebodies. *Northern Miner* (Canada), Nov. 9, 1978. p.5.

O'Keefe, J.A., 1978. The tektite problem. *Sci. Am.,* 239: 98–107.

Ovchinnikov, L.N., Shliapnikov, D.S. and Shoor, A.S., 1964. Mobilization and transportation of matter in the endogenic ore-formation. In: *22nd Int. Geol. Congr.,* Pt. V, New Delhi, pp. 596–622.

Page, N.J. and Creasy, S.C., 1975. Ore grade, metal production, and energy. *U.S. Geol. Surv., J. Res.,* 3(1): 9–13.

Pälchen, W. and Tischendorf, G., 1974. Some special problems of petrology and geochemistry in the Erzgebirge, G.D.R. In: M. Štemprok (Editor), *Metallization Associated with Acid Magmatism,* vol. 1, pp.206–209.

Pälchen, W. and Tischendorf, G., 1978. Some special problems of petrology and geochemistry of the Variscan granites of the Erzgebirge, G.D.R. In: M. Štembrok, L. Burnal and G. Tischendorf (Editors), *Metallization Associated with Acid Magmatism,* vol. 3, pp.257–266 (Int. Geol. Correl. Program, Geol. Surv., Praque.)

Palliser, B., 1957. *Admirable Discourses.* Univ. of Illinois Press, Urbana, Ill., pp.148, 186. (Transl. by Al. La Rocque.)

Parák, T., 1975. Kiruna iron ores are not "intrusive-magmatic ores of the Kiruna-type". *Econ. Geol.,* 70: 1242–1258.

Park, G., 1963. A model for concept learning. In: Proc. 10th Int. Congr. on Electronics, Rome.

Pélissonnier, H., 1962. Classifications métallogéniques: problèmes et essais de synthèse. *Chronique des Mines d'Outre-Mer et de la Recherche Minière,* Nos. 306 and 307: 1–28.

Pélissonnier, H. and Michel, H., 1972. Les dimensions des gisements de cuivre du monde. *Bur. Rech. Géol. Min. France,* No. 57: 400 pp.

Pereira, J., 1963. Reflections on ore genesis and exploration. *Min. Mag.,* 108: 9–22.

Peters, W.C., 1978. *Exploration and Mining Geology.* Wiley, New York, N.Y., 696 pp.

Petrascheck, W.E., 1968. Die Entstehung der Erzlagerstätten. In: H. Murawski (Editor), *Vom Erdkern · bis zur Magnetosphäre.* Umschau Verlag, Frankfurt/Main, pp.191–204.

Pettijohn, F.J., Potter, P.E. and Siever, R., 1972. *Sand and Sandstone.* Springer, New York, N.Y., 618 pp.

Reiners, L., 1963. *Stilfibel (Primer of Writing Style).* Deutscher Taschenbuch Verlag, Nördlingen, 268 pp.

Reiners, L., 1976. *Stilkunst. Ein lehrbuch der Prosa (Art of Style – a textbook of prose).* C.H. Beck, Munich, 784 pp.

Rickard, M.J. and Scheibner, E., 1976. The philosophical basis and terminology for tectonic nomenclature. In: *25th Int. Geol. Congr.,* symp. 103. 3: 16–29, Sydney, Australia.

Ridge, J.D., 1972. Annotated bibliographies of mineral deposits in the Western Hemisphere. *Geol. Soc. Am. Mem.,* 131: 675 pp,

Ridge, J.D., 1973. Volcanic exhalations and ore deposition in the vicinity of the sea floor. *Miner. Deposita,* 8: 332–348.

Roscoe, S.M., 1965. The importance of ore deposit classifications to the exploration geologist. *Geol. Soc. Can. Pap.,* 65–6: 101–107.

Rösler, H.J. and Lange, H., 1972. *Geochemical Tables.* Elsevier, Amsterdam, 568 pp.

Routhier, P., 1963. *Les Gisements Métalliferès.* Geologie et Principès de Recherche, Vols. 1 and 2. Masson et Cie., Paris, 1282 pp.

Routhier, P., 1969a. *Essai critique sur les méthodes de la géologie (de l'object à la genèse).* Masson, Paris, 202 pp.

Routhier, P., 1969b. Surtois principes généraux de la métallogenie et de la recherche minérale. *Miner. Deposita,* 4: 213–218.

Routhier, P. and Delfour, J., 1974. Lithology of massive sulfide deposits associated with acid volcanism and its application to the search for such deposits in the Precambrian of Saudi Arabia. *B.R.G.M. and D.G.M.R. Publ.* 75 JED 10, 201 pp.

Rowlands, N.J. and Sampey, D., 1976. Zipf's Law – an aid to resource inventory prediction in partially explored areas. Symp. 116, *25th Int. Geol. Congr.,* Sydney, Australia (preprint).

Rundkvist, D.V., 1978. Classifications of mineralizations related to acid intrusive magmatism. Stud. Geol. Univ. Salamance, 14: 149–160.

Ruznicka, V., 1971. Geological comparison between East European and Canadian uranium deposits. *Geol. Soc. Can. Pap.*, 70–48: 196 pp.

Saager, R., 1967. Drei Typen von Kieslagerstätten im Mofjell-Gebiet, Nordland, und ein neuer Vorschlag zur Gliederung der Kaledonischen Kieslager Norwegens. *Nor. Geol. Tidsskr.*, 47 (pt. 4): 333–358.

Saksela, M., 1957. Die Entstehung des Outokumpu-Erze im Lichte der tektonisch-metamorphen Stoffmobilisierung. *Neues Jahrb. Mineral. Abh.*, 91: 278–302.

Sawkins, F.J. and Rye, D.M., 1974. Relationship of Homestake-type gold deposits to iron-rich Precambrian sedimentary rocks. *Inst. Min. Metall., Trans. Sect. B,* 83: B56–B59 (see discussions, 84: B37–B38, 1975).

Schneiderhöhn, H., 1941. *Lehrbuch der Erzlagerstättenkunde.* Gustav Fischer, Jena, 858 pp.

Schneiderhöhn, H., 1952. Genetische Lagerstättengliederung auf geotektonischer Grundlage. *Neues Jahrb. Mineral., Monatsh.,* 1952: 47–89.

Schneiderhöhn, H., 1958. *Die Erzlagerstätten der Frühkristallisation, 1.* Gustav Fischer, Stuttgart, 315 pp.

Schneiderhöhn, H., 1961. *Die Pegmatite, 2.* Gustav Fischer, Stuttgart, 720 pp.

Schneiderhöhn, H., 1962. *Erzlagerstätten – Kurzvorlesungen.* Gustav Fischer, Stuttgart, 371 pp.

Schneiderhöhn, H. and Borchert, H., 1956. Zonale Gliederung der Lagerstätten. *Neues Jahrb. Mineral. Monatsh.,* 6/7: 136–161.

Schüller, A., 1961. Die Druck-, Temperatur- und Energie-Felder der Metamorphose. *Neues Jahrb. Mineral. Abh.,* 96: 250–290.

Seyhan, I., 1971. Die Entstehung des vulkanischen Kaolins und das Andesit-Problem. *Bull. Mineral. Res. Explor.,* Inst. Turkey (Foreign Edition), April '71, No. 76: 117–129.

Sheldon, R.P., 1978. Replenishing non-renewable mineral resources – a parodox. *Va. Miner. Mag.,* 24(2): 10–16.

Sillitoe, R.H., 1973. The tops and bottoms of porphyry copper deposits. *Econ. Geol.,* 68: 799–815.

Skinner, B.J. and Barton, P.B., Jr., 1973. Genesis of Mineral Deposits. *Annu. Rev. Earth Planet. Sci.,* 1: 183–211.

Smirnov, V.I., 1972. Convergence of sulfide deposits. *Int. Geol. Rev.,* 7: 978–983.

Smirnov, V.I., 1976. *Geology of Mineral Deposits.* MIR, Moscow, 520 pp. (Translated from the Russian.)

Smirnov, V.I., 1977. *Ore deposits of the USSR, 1, 2 and 3.* Pitman, London. (Translated from the Russian.)

Snyder, F.G., 1967. Criteria for origin of stratiform ore bodies with application to southeast Missouri. In: J.S. Brown (Editor), *Genesis of Stratiform Lead–Zinc–Barium–Fluorite Deposits in Carbonate Rocks (Mississippi Valley Type Deposits) – A Symposium. Econ. Geol. Mem.,* 3: 1–13.

Stanton, R.L., 1972. *Ore Petrology.* McGraw-Hill, New York, N.Y. 713 pp.

Štemprok, M., 1977. The source of tin, tungsten and molybdenum of primary ore deposits. In: M. Štemprok, L. Burnol and G. Tischendorf (Editor), *Metallization Associated with Acid Magmatism,* 2, pp.127–168.

Štemprok, M., 1978a; Classification criteria of tin, tungsten and molybdenum deposits. *Stud. Geol. Univ. Salamanca,* 14: 119–143.

Štemprok, M., 1978b. A suggestion for the classification of tin, tungsten and molybdenum deposits associated with plutonic rocks, *Stud. Geol. Univ. Salamanca,* 14: 161–169.

Štemprok, M., Burnol, L. and Tischendorf, G., 1977–1978. *Metallization Associated with Magmatism, 1, 2 and 3.* Int. Geol. Correlation Program, Geol. Surv. Praque: 410 pp, 166 pp, 446 pp.

Stevenson, W.G., 1969. A classification of mineral prospects. *Western Miner,* July, 1969.

Stille, H., 1939. Zur Frage der Herkunft der Magmen. *Abh. Preuss. Akad. Wiss. Math.-nat. Kl.* No. 19, Berlin 1940.

Stille, H., 1950. Der "subsequent" Magmatismus. *Abh. Geotektonik,* No. 3, Akademie-Verlag, Berlin, 1950.

336

Strakhov, N.M., 1967, 1969, 1970. *Principles of Lithogenesis, 1, 2 and 3*. Consultants Bureau, New York, N.Y., 245 pp, 609 pp, 577 pp.

Strunk, W. and White, E.B., 1972. *The Elements of Style*. MacMillan, New York, N.Y., 78 pp.

Sullivan, C.J., 1957. The classification of metalliferous provinces and deposits. *Can. Min. Metall. Bull.,* 50(546): 599–601.

Sullivan, C.J., 1970. Relative discovery potential of the principal economic metals. *Bull. Can. Inst. Min. Metall.,* July '70: 773–784.

Taylor, R.G., 1978. A classification of tin provinces. *Stud. Geol. Univ. Salamanca,* 14: 171–181.

Taupitz, K.-C., 1954. Über Sedimentation, Diagenese, Metamorphose und Magmatismus und die Entstehung der Erzlagerstätten. *Chemie Erde,* 17: 104–164.

Thomas, L.J., 1973. *An Introduction to Mining – Exploration, Feasibility, Extraction, Rock Mechanics*. Hicks, Smith and Sons, Sydney, Wiley, New York, N.Y., 421 pp.

Thornton, J.B., 1961. The place of hypotheses in the teaching of chemistry. In: *Approach To Chemistry,* The Univ. New South Wales, Sydney, Australia, pp.1–13.

Thouless, R.H., 1971. *Straight and Crooked Thinking*. Pan Piper Books Ltd., London, 189 pp.

Tischendorf, G., 1968. Das System der metallogenetischen Faktoren und Indikatoren bei der Prognose und Suche endogener Zinnlagerstätten. *Z. Angew. Geol.,* 14: 393–405.

Tischendorf, G., 1977. Geochemical and petrographic characteristics of silicic magmatic rocks associated with rare-element mineralization. In: M. Štemprok, L. Burnol and G. Tischendorf (Editors), *Metallization Associated with Acid Magmatism, 2*: 41–98.

Tischendorf, G., Schust, F. and Lange, H., 1978. Relation between granites and tin deposits in the Erzgebirge, G.D.R. In: M. Štemprok, L. Burnol and G. Tischendorf (Editors), *Metallization Associated with Acid Magmatism, 3*: 123–138. Intern. Geol. Correl. Program, Geol. Surv., Prague.

Titus, H.H., 1953. *Living Issues in Philosophy*. American Book Co., New York, N.Y., 500 pp.

Ullman, S., 1951. *Words and Their Use*. Man and Society Series, Frederick Muller Ltd., London, 108 pp.

United States Geological Survey, 1975. Mineral resources perspectives 1975. *U.S. Geol. Surv. Prof. Pap.,* 940: 24 pp.

USSR Ministry Naval Defense, 1974. *Atlas Okeanov-World ocean atlas – The Pacific Ocean,* Sections 1 to 7. Collet's Denington Estate, Wellingborough, Northants, England.

Valjaško, M.G., 1958. Die wichtigsten geochemischen Parameter für die Bildung der Kalisalzlagerstätten. *Freiberger Forschungsh. A,* 123: 197–235.

Varlamoff, N., 1974–1978. Classification and spatio-temporal distribution of tin and associated mineral deposits. In: M. Štemprok (Editor), *Metallization Associated with Acid Magmatism, 1*: 137–144. (*See also vol. 3* (1978): 139–156.)

Weeks, L.G., 1961. Origin, migration and occurrence of petroleum. In: G.B. Moody (Editor), *Petroleum Exploration Handbook*. McGraw-Hill, New York, N.Y. pp.5.1–5.50.

Weemann, C.E., 1963. Tectonic patterns at different levels. *Geol. Soc. S. Afr. Annexure* 66: 1–78.

Whitten, E.H.T., 1966. *Structural Geology of Folded Rocks*. Rand McNally and Co., Chicago, Ill., 678 pp.

Wilke, A., 1952. *Die Erzgänge von St. Andreasberg im Rahmen des Mittelharz-Ganggebietes*. Hannover.

Willoughby, R.A., 1975. Stimulating creativity – the employee's responsibility. *Eng. Digest,* 21(8): 23–25.

Wilson, R.C., Picard, M.D. and Harp, E.L., 1974. Geologic cycle of mars: a comparison with the earth and moon. *Geology,* 2(3): 121–124.

Wolf, K.H., 1973. Conceptual models, Part I. Examples in sedimentary petrology, environmental and stratigraphic reconstructions, and soil, reef, chemical and placer sedimentary ore deposits. *Sediment. Geol.,* 9: 153–194.

Wolf, K.H., 1976a. Introduction. In: K.H. Wolf (Editor), *Handbook of Stratabound and Stratiform Ore Deposits, 1*. Elsevier, Amsterdam, pp.1–9.

Wolf, K.H., 1976b. Conceptual models in geology. In: K.H. Wolf (Editor), *Handbook of Strata-bound and Stratiform Ore Deposits, 1*. Elsevier, Amsterdam, pp.11–78.

Wolf, K.H., 1976c. Ore genesis influenced by compaction. In: G.V. Chilingar and K.H. Wolf (Editors), *Compaction of Coarse-grained Sediments, 2*. Elsevier, Amsterdam, pp.475–676.

Wolf, K.H., 1980. Mineral exploration – a comparative book review of: Kuzvart, M. and Böhmer, M., 1978. *Prospecting and Exploration of Mineral Deposists*. Elsevier, Amsterdam, 431 pp.; and Peters, W.C., 1978. *Exploration and Mining Geology*. Wiley, New York, 696 pp. *Earth-Science Rev.*, 15(4): 404–409.

Wolf, K.H. and Chilingar, G.V., 1976. Diagenesis of sandstones and compaction. In: G.V. Chilingar and K.H. Wolf (Editors), *Compaction of Coarse-Grained Sediments, 2*. Elsevier, Amsterdam, pp,69–444.

Ziegler, V., 1974. Essai de classification metallotectonique des gisements d'uranium. In: *Formation of Uranium Ore Deposits*. Proc. Symp. Athens, Int. Atomic Energy Assoc., Vienna, pp.661–678.

Zimmermann, R.A. and Amstutz, G.C., 1964. Small-scale sedimentary features in the Arkansas barite district. In: G.C. Amstutz (Editor), *Sedimentology and Ore Genesis*. Elsevier, Amsterdam, pp.157–163.

Zipf, G.K., 1949. *Human Behaviour and the Principle of Least Effort*. Addison-Wesley/Hafner, New York, N.Y., 573 pp.

Zitzmann, A., 1977. The distribution of iron ore deposits. In: A. Zitzmann (Editor), *The Iron Ore Deposits of Europe and Adjacent Areas*. Bundesanstalt für Geowissenschaften und Rohstoffe, Hannover, pp.37–68.

Zwartendyk, J., 1972. What is "mineral endowment" and how should we measure it? *Mineral Bull. MR 126*, Dep. Energy, Mines, Resources, Ottawa, Canada, 17 pp.

Chapter 2

EARLY EARTH EVOLUTION AND METALLOGENY

I.B. LAMBERT and D.I. GROVES

INTRODUCTION

Concepts of Earth evolution and metallogeny are continually changing as new data from a variety of disciplines make significant impacts on geological thought.

This dissertation, which is divided into five main sections, endeavours to present a balanced view of the current state of knowledge of these matters. It outlines constraints and identifies assumptions that have been made in arriving at various hypotheses for the early evolution of the lithosphere, atmosphere, hydrosphere and biosphere. In so doing, it presents a number of conclusions concerning early Earth history and metallogeny which have not been emphasized previously.

The first section reviews evidence concerning the origin of the Earth, approximately 4.6 Gyr BP [1], and its evolution prior to formation of the oldest recognized crust, some 3.9 Gyr BP. The second section assesses geological evolution in the Archaean, taking account of relevant conclusions concerning earlier Earth history, but detailed discussion of the important types of Archaean mineralization is reserved for the third section. The fourth section covers the Archaean–Proterozoic transition, with its distinctly different lithotectonic features and styles of mineralization; this transition began as early as ca. 3 Gyr BP in some terrains and is taken here as ending with the development of an oxygenous atmosphere some 2 Gyr BP. The final section presents a succinct summary of evolutionary trends after this major punctuation point in Earth history.

PRE-ARCHAEAN EARTH HISTORY

Origin of the Earth

Several different general models have been proposed for the origin of the Earth. During the 1960's in particular, it was widely held that the Earth accreted from chon-

---

[1] Gyr BP = giga (10^9) years before the present.

dritic materials containing intimate mixtures of silicate and metal particles which had formed earlier in the solar nebula (e.g. Urey, 1962; Elsasser, 1963; Birch, 1965). These models predicted that the Earth accreted cold (average temperature <1000°C) from relatively homogeneous, well-mixed materials, and that the core did not separate until a considerable period after accretion, when heating by long-lived radioactive elements led to melting of the metal phase.

A number of more recent hypotheses (e.g. Anders, 1968; Clark et al., 1972; Murthy, 1976) invoked secular changes in the composition of materials accreting to the Earth from initial high-temperature, metal-rich condensates to later, lower-temperature condensates.

In contrast, Ringwood (1969) proposed accretion of materials that were compositionally similar to type I carbonaceous chondritic meteorites, with metallic Fe forming by reaction of carbon in the accreting materials and oxidized iron. This model envisaged that the Earth formed "inside out", with a cool, oxidized nucleus grading into a metal-rich, devolatilized outer zone, and that gravitational instability eventually led to rapid over-turn and core segregation.

However, recent critical assessments by Ringwood (1975, 1977a, b) showed that it is difficult to reconcile each of the general models outlined above with all of the known constraints on the formation of the Earth. He considers that the main factors, which have to be accounted for in any model, are the following:

(1) Compared with cosmic abundances, the Earth is depleted in elements more volatile than Si, in degrees proportional to the volatilities of these elements in the solar nebula.

(2) The Earth's magnetic field, and other evidence, indicates that the Earth's core is composed predominantly of metallic Fe, but density considerations indicate that it must also contain a significant amount of a light element(s).

(3) Samples of upper mantle rocks which have found their way to the surface (e.g. as portions of ophiolite complexes or as fragments in kimberlites and alkali basalts) are peridotitic and most are depleted in elements enriched in the crust.

(4) There is considerable evidence for local heterogeneity of the mantle (Bailey et al., 1978). However, two main lines of evidence imply that it was essentially homogeneous peridotite when it formed and that *large-scale* homogeneity has persisted to the present time despite the dynamic behaviour of the mantle. Firstly, the fact that the abundant Archaean low-Mg tholeiites have similar compositions to the common Phanerozoic oceanic basalts strongly suggests that, throughout geological time, such low-potash basaltic rocks have been derived from ultramafic mantle with fairly constant and restricted abundance ranges for Si, Mg, Fe, Mn, Ni, Co, Zn and other crystallographically compatible elements. Secondly, the major seismic discontinuities in the mantle do not support the presence of compositionally distinct zones, as these discontinuities are adequately explained by phase changes within a largely homogeneous peridotitic composition. As the elements in the mantle display extremely diverse behaviour in the solar neb-

ula, these lines of evidence imply that there was an efficient homogenization process during, or before formation of the Earth and, also, that the mantle was not extensively melted and differentiated during accretion or core formation. The long-term dynamic behaviour of the mantle has apparently resulted in cycles of downsinking of depleted upper mantle and ascent of undepleted mantle.

(5) The abundances of Ni, Co, Au, and other siderophile elements are much higher than would be expected if the mantle had ever equilibrated with a metallic Fe phase which subsequently segregated into the core. Also, the mantle is much more oxidized than would be expected if it accreted in equilibrium with metallic Fe.

Ringwood (1977a, b) concluded that the only model which satisfies all of these constraints in an uncontrived manner is one involving accretion of dust, planetesimals and larger bodies in the ecliptic plane of a "cocoon" nebula around the proto-sun, on a time scale of 10^6 to 10^8 years. Solids which spiralled into the ecliptic plane and subsequently accreted into the planets would have condensed over a wide range of temperatures. As a result, they would have had a wide diversity of compositions, ranging from highly reduced, devolatilized, silicate-iron planetesimals close to the proto-sun, to oxidized, volatile-rich planetesimals resembling carbonaceous chondrites in the much cooler outer regions of the nebula. The depletion of volatile elements in the Earth essentially reflects incorporation of a major proportion of high-temperature condensates.

During the early stages of accretion, significant amounts of volatile components from the low-temperature condensates would be trapped. However, once the mass of the growing nucleus exceeded about 10% of that of the present Earth, planetesimals hitting the surface would have sufficiently high velocities to cause strong transient heating with evaporation of volatiles, as well as total fragmentation with consequent mixing and homogenization of diverse compositions. As surface temperatures almost certainly exceeded the boiling point of water, a water-rich atmosphere would have existed and only the largest projectiles could have passed through this at high velocities without large proportionate loss of mass by vapourization. The most volatile compounds, for example H_2, H_2O, CH_4, CO_2, N_2, NH_3, some H_2S and inert gases, would have remained in the atmosphere, whilst less volatile components like Pb, Tl, B, and some S and FeS would have largely recondensed to be incorporated into the solid Earth. The water vapour in this atmosphere would have reacted with infalling metallic Fe to form FeO and H_2.

The presence of FeO results in marked lowering of the melting curve of metallic Fe, and could have lowered it well below the solidus of mantle silicates (Fig. 1). Thus, the molten iron core with oxygen as the light component could have segregated without equilibrating with the largely unmelted peridotitic mantle. Once the temperature in the Earth's interior had risen sufficiently to permit metal segregation (probably $\sim 1500°C$ in the upper mantle), the additional heat liberated during rapid core formation would not have caused extensive melting of the silicate fraction, because strong convection would rapidly lead to the establishment of an adiabatic gradient.

The formation of the core would have caused a significant fractionation of Pb from

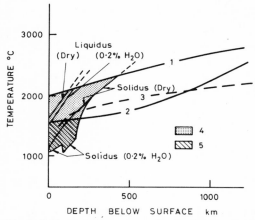

Fig. 1. Estimated temperature distributions in the early Earth in relation to the solidus (beginning of melting) and liquidus (total melting) for peridotitic mantle (adapted from Ringwood, 1975, 1977a). *1* = Maximum temperature distribution post core-formation (adiabat). *2* = Post-accretion temperature distribution; this approximates minimum possible temperature distribution following core formation. *3* = Temperature distribution in early Earth assumed by Shaw (1972, 1976), after Hanks and Anderson (1969). *4* = Zone of melting after catastrophic core formation. *5* = Zone of melting for core formation at time scale of ca. 10^6 years. Melting curve for pure iron is approximately mid-way between mantle solidus and liquidus.

U, with a relatively substantial proportion of Pb entering the molten Fe component. Therefore, the ~4.5 Gyr Pb isotopic "age" of the Earth probably refers to the time of core formation, when the present overall Pb/U ratio of the upper mantle/crust system was established (Oversby and Ringwood, 1971). A minimum age of ~4.4 Gyr for core formation is indicated independently by the oldest ages determined for lunar materials (Tera and Wasserberg, 1974; Lugmair et al., 1975), *if* the Moon was derived from the Earth soon after core formation [as proposed by Ringwood and Kesson (1977) and Delano and Ringwood (1978)]. Estimates of the time interval between completion of accretion of the Earth and completion of core formation are less than a few hundred million years for a wide variety of initial assumptions (Oversby and Ringwood, 1971; Vollmer, 1977; Ringwood, 1977b).

Development of the primitive crust

It is likely that heating, during accretion/core segregation caused melting in the outer Earth. Cooling would have occurred by major convection within the Earth, coupled with surface radiation, and the first crust would have formed when temperatures fell sufficiently for solidification of the melt. As yet, signs of this crust have not been recognized in the geological record, so its nature can only be deduced from theoretical considerations. The only limits on the time of its formation are that it was after accretion and core segregation, but before the time of formation of the oldest dated crustal segments, some 3.9 Gyr BP.

The composition and thickness of the primordial crust would have been controlled primarily by the composition of the mantle and the extent to which it was melted. A secondary influence is likely to have been the effects of bombardment by giant projectiles analogous to those which hit the Moon before ~3.9 Gyr (e.g. Green, 1972; Goodwin, 1976).

Several papers dealing with the origin of the Earth (e.g. Clark et al., 1972; Levin, 1972; Wetherill, 1972; Murthy, 1976) extend their models to briefly consider the early crust and many papers dealing with the evolution of the Archaean crust include short sections on the author's prejudices about the pre-Archaean period. The main publications dealing specifically with the composition of the primitive crust are by Shaw (1972, 1976), who considered the matter in the light of partition coefficients for selected lithophile elements. He concluded that the initial crust consisted of a few kilometres of anorthositic rocks overlying 10—15 km of granitic rock, and that this graded down to a thick zone of gabbroic rocks and eclogite. However, a major assumption in his model is that the mantle passed through a stage when it was largely or wholly molten and the primitive crust fractionated from the resultant very large volume of melt. This is unlikely because differentiation processes in the molten mantle would have led to marked chemical zoning as it solidified, which is contrary to evidence cited above. In fact, the maximum temperatures Shaw assumed (after Hanks and Anderson, 1969), are only sufficient to have caused partial melting of the outer ~200 km of peridotitic mantle (cf. Fig. 1, and discussion below).

Here, we draw on recent experimental petrological findings to assess the extent of melting and the volume of sialic rocks that are likely to have fractionated into the primordial crust for two extreme temperature distributions. Ringwood (1975, 1977a, b) considered the thermal effects of core segregation over periods varying from ~10^3 to ~10^6 years. He concluded that if core segregation took ~10^6 years, most of the core-forming energy would have been radiated away by a thin layer of magma at the surface, and the net temperature distribution in the mantle would have been little changed from the postaccretional temperature distribution which he estimated to be ~1500°C in the outer Earth (Fig. 1). At the opposite extreme, catastrophic core formation (10^3 to 10^4 years) could have elevated temperatures to ~2000°C near the surface and 3500°C at the mantle—core boundary (Fig. 1). In considering the products of melting and fractionation, it is necessary to take into account the following conclusions from experimental studies (for reviews, see Wyllie, 1971; Nicholls and Ringwood, 1972; Ringwood, 1975; Green, 1976a, b):

(1) Partial melting of peridotite under a wide range of conditions yields basaltic magmas; ultramafic melts can be produced under conditions of very high geothermal gradients.

(2) Silicic magmas ($>60\%$ SiO_2) cannot be the direct melting product of peridotite, or the fractionation product of peridotitic liquids, under any p_{H_2O} conditions, but they can form by two-stage melting with the first product being basic rocks.

(3) Magmas of intermediate (basaltic andesite) composition can form directly from peridotite by wet melting only at pressures $\lesssim 10$ kbar, which correspond to depths $\lesssim 30$ km in the Earth. Mysen and Boettcher (1972) reported andesitic glasses in experiments with hydrous peridotite at 20 kbar, but Nicholls and Ringwood (1974) argue that these were quenched from the vapour phase, and are much more silicic than the melts produced under these conditions.

(4) Fractionation of basic melts at $\lesssim 30$ km can give rise to silicic residual liquids, but in very minor proportions (as evidenced in large layered mafic intrusions and in igneous rocks produced in oceanic environments).

(5) Andesine and more calcic plagioclase feldspars are only stable at depths $\lesssim 30$ km.

These observations, coupled with the melting curves for mantle peridotite (Fig. 1), mean that the minimum estimated temperatures of Ringwood (op. cit.) would have yielded basaltic magmas by partial melting to depths of ~ 120 km for dry conditions. Water is likely to have been present in amounts $\lesssim 0.5\%$, and this could have extended the depth of partial melting to ~ 200 km. On the other hand, the maximum temperature distribution would have generated a more ultramafic overall melt composition as a result of complete melting to depths of ~ 120 km, and partial melting from there to ~ 400 km, under wet or dry conditions (Fig. 1).

The main products of fractional crystallization of the melts produced under both low and high extremes of temperature would have been ultramafic and mafic rocks, with the former crystallizing earlier and, in general, accumulating beneath the latter. The constraints on the formation of granitic and anorthositic rocks from a peridotitic source imply that the volume of sialic rocks in the first-formed crust must have been considerably less than in the present crust, even for maximum temperature estimates; *if* the primordial granitic rocks had concentrated into a world-wide layer in the crust it is very unlikely that such a layer could have exceeded a thickness of 2—3 km. Mafic rocks would have been produced in much greater abundances by the initial melting event and could have formed a crust extending to depths of roughly 100 km for Ringwoods's maximum estimated temperatures, and roughly 30 km for his lowest estimated temperatures (Fig. 2).

Although no direct evidence has yet been recognized, it is likely that mega-impact events influenced the evolution of the outer regions of the Earth. Some of the projectiles could have been significantly larger than those which bombarded the Moon, commensurate with the greater gravitational attraction of the Earth. They would have caused extensive disruption and brecciation of the primitive crust, and major fracturing well into the mantle. Pressure release beneath the giant impact craters should have caused melting and diapiric uprise of mantle materials. This would have generated lavas and intrusives of predominantly ultramafic and mafic compositions, but associated partial melting of the abundant mafic rocks and fractional crystallization of mafic melts should have given rise

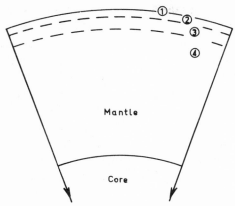

Fig. 2. Schematic section through outer part of the Earth following core segregation and crystallization of primordial crust. *1* = Possible sialic layer; *if* all sialic material fractionated into an outer crust,
this is unlikely to have been more than 3 km thick. *2* = Simatic crust, from ~30 km to ~100 km thick
(for minimum and maximum temperature distributions, respectively). *3* = Depleted peridotitic mantle,
~90 to ~300 km thick (for minimum and maximum temperature distributions, respectively). *4* =
Undepleted peridotitic mantle (temperatures were not high enough to cause melting).

to additional sialic and anorthositic components in the crust. They probably would also
have caused intense "degassing", with associated accumulation of "igneous" exhalative,
chemically precipitated sediments. Thus, the immediate effects of impacting on the
primitive crust should have been the production of a complex mixture of ultramafic,
mafic, felsic and anorthositic igneous rocks, plus exhalative sediments and components
from the projectiles.

On a more speculative note, Goodwin (1976) suggested that major impact sites
could have been loci for subsequent generation of sialic crustal nuclei.

Development of the early atmosphere and hydrosphere

The atmosphere and hydrosphere are considered together as they developed and
evolved as interacting entities. The problems associated with their formation have been
discussed in numerous publications including Rubey (1951), H. Brown (1952), Holland
(1972, 1976), Cloud (e.g. 1974, 1976), Fanale (1971) and J.C.G. Walker (1976a, b). In
this section, salient points that have been made by these authors are outlined, but discussion of the evolution of the atmosphere—hydrosphere system as it relates to the deposition of Archaean bedded barite and banded-iron-formation (BIF) is reserved until later.

The extreme depletions of non-radiogenic inert gases relative to cosmic abundances
indicate that only a negligible portion of the atmosphere could have been directly supplied from primordial gases of approximately solar composition. The mechanism of Earth
formation must have militated against retention of substances that existed at that time in
the gaseous state. A number of arguments, including radiogenic argon abundances and

mass balance considerations, indicate that the present atmosphere—hydrosphere system developed largely by subsequent additions of juvenile components from the Earth's interior (degassing), with superposed evolutionary modifications by a variety of physical, chemical and biological processes. There is some debate as to whether degassing was effectively completed at a very early stage of Earth history (e.g. Fanale, 1971; J.C.G. Walker, 1976a, b), or whether it has been a continuous or recurrent process which is still occurring (e.g. Rubey, 1951). The answer is probably closer to the first of these viewpoints: it seems logical that there would have been a relatively high rate of degassing associated with initial melting and crust-forming processes, and that degassing would have continued after this at rates decreasing in general sympathy with the degree of igneous activity. Taylor (e.g. 1977) concluded from $\delta^{18}O-\delta D$ studies on igneous rocks that there has been very little degassing of juvenile water in the past 2.5—3 Gyr.

A constraint on the timing of degassing, and on the development of surface temperatures low enough for condensation of water, is that the oldest known volcano-sedimentary sequence provides evidence of at least local subaqueous environments ~3.9 Gyr BP.

There are differences of opinion concerning the composition of the early atmosphere. A highly reducing atmosphere is favoured by many biologists. This is largely an outcome of the classic Miller—Urey synthesis of amino acids, the building blocks of life, by spark discharge in an atmosphere of NH_3 (later experiments used N_2), CH_4, H_2 and H_2O (S.L. Miller, 1953). However, the composition of volcanic gases and fluid inclusions in mantle materials, supported by geochemical arguments based on the oxidation state of the Earth's mantle, imply a less-reduced atmosphere containing H_2O, CO_2, with lesser CO, N_2, H_2S, SO_2, HCl, and little, if any NH_3 and CH_4. It appears that amino acids could also have been synthesized from this atmosphere: Abelson (1966) and Hubbard et al. (1971) achieved similar results to S.L. Miller (op. cit.) using CO, H_2 and H_2O. The lack of success with abiotic syntheses of amino acids in the presence of free oxygen militates against O_2 as a stable component of the early atmosphere. In this regard, it is noteworthy that there is no juvenile source of O_2 in the Earth, so that its presence in the atmosphere must be a result of secondary processes. As enlarged on later, microbial photosynthesis was probably the major process involved (e.g. Cloud, 1974, 1976; Muir, 1979); the amount of O_2 produced by photodissociation of H_2O vapour is likely to have been very minor according to Schidlowski (e.g. 1976), but Towe (1978) regards the evidence as equivocal.

The rare gases, N_2 and CO would, as at present, have been strongly partitioned into the early atmosphere, whilst H_2O and HCl would have concentrated in the early hydrosphere. By virtue of their low molecular weights, He and free H_2 would have escaped back into the solar nebula. Sulphur species would have existed in the atmosphere, in solution in the hydrosphere, and they would have precipitated as sulphide, sulphite(?) and sulphate minerals. CO_2 would have been concentrated in the atmosphere, but significant amounts would have been dissolved in the hydrosphere and precipitated as carbonate minerals in the lithosphere. Carbonated waters falling on dry land would have reacted with the silicate rocks to produce waters enriched in elements, such as Na, K, Mg, Ca and Si, which flowed into the oceans/seas. However, the contents of these elements may have built up slowly as there is no evidence for extensive areas of land throughout much of the Archaean. Furthermore, whilst there are early Archaean sulphate deposits (cf. discussion of barite in section on Archaean mineralization), no evidence exists for major halite deposits prior to the ~2 Gyr old evaporitic sequences in the Great Slave Lake area, Canada (Badham and Stanworth, 1977). This implies a very slow build-up of NaCl-concentrations in the Archaean hydrosphere, a factor which would have restricted the leaching ability of surface waters circulating into the lithosphere. The pH of the present ocean is buffered by silicates (Sillén, 1967) and possibly also by the carbonate system, but it is likely that it took some time for such pH-buffers to become established. However, it is not clear whether the primitive hydrosphere was slightly acid because of dissolved CO_2, HCl, H_2S etc., or, as maintained by Abelson (1966), it was alkaline as a result of interaction of outgassing volatiles with the lithosphere. The presence of carbonate and chert sediments in the 3.8 Gyr old Isua sequence indicates that there were at least local basins of neutral to slightly alkaline waters at this time.

Development of the first life forms

Microorganisms were probably largely instrumental in the generation of an oxygenous atmosphere (e.g. Cloud, 1974) and they are likely to have played important roles in the formation of various types of mineral deposits (e.g. Trudinger et al., 1972; Trudinger, 1976, I.B. Lambert, 1977). Therefore, it is pertinent to briefly review speculations as to how and when the first life forms developed.

Major discussions of the origin of life were published by Haldane (1929) and Oparin (1957). More recent ideas have been reviewed by Orgel (1974) and Sylvester-Bradley (1976).

It has already been mentioned that mixtures of compounds, which occur in contemporary biochemistry can be chemically synthesized from mixtures of components in the Earth's primitive atmosphere/hydrosphere. Recent research has also focussed attention on a possible extra-terrestrial source for the molecules which ultimately gave rise to life forms (e.g. Orgel, 1974; Hoyle and Wickramasinghe, 1977). This research has documented a wide range of "source materials" (e.g. ammonia, formaldehyde, hydrogen cyanide, alcohols, methane) in molecular clouds in collapsing nebulae, and amino acids have been isolated from carbonaceous chondrite meteorites (but these amino acids are not optically active, as are those in living matter). Hoyle and Wickramasinghe (op. cit.) believe the low-molecular-weight organic compounds and radicals in interstellar clouds are the result of ultraviolet and cosmic-ray degradation of much more complex structures. They suggest that these complex molecules exist widely in considerable quantities within the interstellar medium, and that they could have been incorporated in the outer Earth during the final stages of accretion. This is hard to reconcile with the high temperatures envisaged during formation of the Earth's primordial crust, although subsequent influx of organic molecules cannot be ruled out.

The conditions under which prebiotic compounds evolved into living systems are, of course, extremely speculative, but three general environments in which this could have occurred have been proposed. Haldane's (1929) paper gave impetus to the hypothesis of a marine "hot dilute soup" environment, in which solar energy powered the evolutionary processes whereby relatively small organic molecules gave rise to autotrophic life forms (i.e. microorganisms capable of self-nourishment using inorganic compounds). Orgel (1974) envisaged that small organic molecules were polymerized and life ultimately evolved in desert environments where hot, dry conditions alternated with cold, damp conditions. Most recently, Sylvester-Bradley (1976) has argued for life evolving as a result of violently oscillating conditions in active volcanic areas, where terrestrial energy (heat), concentrated solutions, immiscible polymers and prebiological synthesis of nutrients gave rise to heterotrophic life forms (i.e. microorganisms dependent for food on organic nutrients).

Abiotic synthesis of living organisms may have been rapid and may have occurred locally at many different places on Earth, and there could have been multiple origins, evolutions and extinctions of primitive living forms on the early Earth. There are no recognizable micro-fossils and no isotopic evidence for photosynthesis in the oldest recognized (~3.9 Gyr) metasediments near Isua (Oehler and Smith, 1977), although it is

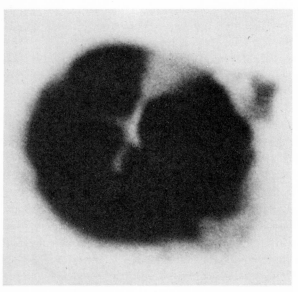

Fig. 3. Photomicrograph showing microfossil in ~3.45 Gyr old chert, North Pole, Pilbara Block, Western Australia: tetragonal tetrad enclosed in an almost transparent membrane. Width of field approximately 10 μm. Kindly provided by J.S.R. Dunlop.

likely that these would have been destroyed during metamorphism. The oldest putative microfossils so far reported (Fig. 3) occur in ~3.45 Gyr old cherts in Western Australia (Dunlop et al., 1978) and southern Africa (e.g. Schopf, 1975; M.D. Muir and Grant, 1976).

Interim summary

(1) The least contrived model of Earth accretion/core segregation predicts melting of homogeneous peridotitic mantle to a maximum depth of ca. 400 km and formation of a predominantly simatic primordial crust, between 30 and 100 km thick, as this melt underwent fractional crystallization. Major impact events probably occurred before 3.9 Gyr BP.

(2) The primordial atmosphere—hydrosphere system developed largely as a result of degassing of the Earth. The atmosphere probably consisted mainly of H_2O, CO_2, with lesser CO, N_2, H_2S, SO_2, HCl and possibly some NH_3 and CH_4. There was no juvenile source of O_2. The hydrosphere likely had a low NaCl content, but degassing probably enriched it in iron, silica and carbonate species. Abiotic synthesis of living organisms apparently occurred before 3.45 Gyr BP, the age of the oldest putative microfossils.

ARCHAEAN CRUSTAL EVOLUTION

An overview of the Archaean Era

The Archaean is regarded here as commencing with formation of the oldest recognized crust and culminating with stabilization of granitoid-rich cratons, some 2.5 Gyr ago in most terrains. However, the upper boundary of the Archaean is diachronous; for example, epicratonic sedimentary and volcanic rocks as old as ~3 Gyr, but of distinctly Proterozoic aspect, occur in the eastern Kaapvaal craton of southern Africa and these will be discussed in the section on the Archaean—Proterozoic transition.

Archaean rocks outcrop in little more than 5% of the Earth's land area, but they are believed to underlie most Proterozoic sequences, and some areas of Phanerozoic rocks. They have been found in the shields of every continent (Fig. 4), and a recent review of the published literature on many of these terrains has been given by Windley (1977).

Archaean shields have been divided (e.g. Windley, 1977) into low-grade granitoid [1]-greenstone terrains, containing most of the economic mineralization, and high-grade gneiss-supracrustal complexes. The former comprise volcano-sedimentary (greenstone) sequences in arcuate to linear belts surrounding granitoid intrusives which are commonly

---

[1] The terms "granitoids" and "granitic rocks" are used synonymously here to describe granites (sensu lato) plus associated quartzofeldspathic gneisses.

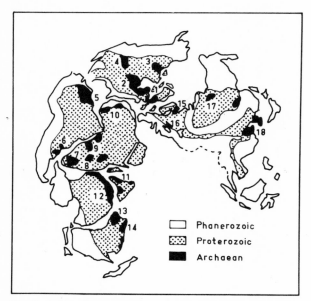

Fig. 4. A Permian pre-drift reconstruction of the continents (Pangea) showing the distribution of Archaean cratons, Proterozoic rocks and Phanerozoic rocks (adapted from Hurley and Rand, 1969; Windley, 1977). *1* = North Atlantic craton; *2* = Superior Province; *3* = Slave Province; *4* = Wyoming Province; *5* = Guyanan craton; *6* = Guapore (Brazilian) craton; *7* = Kaapvaal craton; *8* = Rhodesian craton; *9* = central African craton; *10* = West African craton; *11* = Dharwar (Indian) craton; *12* = Antarctic craton (extrapolated between known Archaean terrains); *13* = Yilgarn Block; *14* = Pilbara Block; *15* = Baltic shield; *16* = Ukrainian shield; *17* = Siberian (Aldan) craton; *18* = North China craton.

dome-shaped (Fig. 5). In addition to relatively massive bodies, the granitoids contain areas of quartzofeldspathic gneisses that are generally similar to those of high-grade terrains (Fig. 5). The greenstone sequences are characterized by abundant basalts, plus variable proportions of ultramafic, andesitic and felsic volcanics, chemical and clastic sedimentary rocks and felsic to ultramafic intrusives. They have commonly suffered low- to medium-grade, low-pressure metamorphism, and most ultramafic rocks are extensively serpentinized or carbonated.

The high-grade gneiss complexes comprise mainly amphibolite to granulite facies (low- to high-pressure) quartzofeldspathic gneisses and migmatites, with enclaves of metavolcanic and sedimentary rocks (supracrustals), and younger intrusive granitoids. Doleritic dyke swarms and layered intrusives characterized by ultramafic, mafic and anorthositic phases are features of both high- and low-grade terrains; there is commonly evidence for more than one age of emplacement of these intrusives in individual cratonic units.

It is clear that both sialic and simatic components were present in the oldest-known terrain, which developed between ~3.9 and 3.7 Gyr BP. However, existing geochronological information indicates that the bulk of the Earth's sialic crust was not in existence at

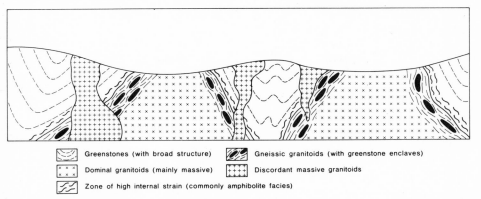

Fig. 5. Schematic section through an Archaean granitoid—greenstone terrain showing relationships between major components; adapted from Hickman (1975).

this time; sialic components were added to the crust throughout the Archaean, interspersed with accumulation of greenstone sequences.

Table I summarizes general features evident from the Archaean record and compares these with features seen in Proterozoic sequences.

Oldest-known terrain

It is pertinent to briefly review the general lithological features of the oldest segment of crust so far documented, which is in west Greenland (e.g. McGregor, 1973; Allart, 1976; cf. discussion under "Radiometric age determination" below); this is broadly similar to the east Labrador terrain (e.g. Collerson et al., 1976), also in the North Atlantic craton.

The Amitsoq gneisses, as well as younger quartzofeldspathic gneisses in west Greenland, are largely the product of deformation and metamorphism of intrusive tonalitic to granodioritic rocks (e.g. McGregor, 1973). This caused elongation of feldspar megacrysts and quartz grains, metamorphic segregation, local anatexis and folding. The best-banded gneisses, which at first sight might be thought to have the simplest structure are in fact zones of ductile shear.

The Amitsoq gneisses intrude the slightly older, arcuate belt of Isua supracrustals, which is metamorphosed at least to middle amphibolite facies. The main rock types in the Isua sequence are iron-rich, tholeiitic amphibolites (metabasalts and dykes); siliceous and carbonate-bearing biotite—muscovite schists which contain large numbers of stretched fragments or boudinaged layers of metarhyolite; talc schists with relics of dunite; biotite—garnet schists with local staurolite and graphite; quartzites; banded iron formation; calc-silicates; ankeritic marbles; and conglomerate with elongated quartzite pebbles (Allart, 1976).

TABLE I

Comparison of major features of Archaean and Proterozoic terrains

	Archaean	Proterozoic
Regional setting	Greenstone belts and extensive granitoid complexes. Several ages of greenstone and granitoid formation result in variable rock relationships.	Extensive epicratonic sequences plus intercratonic mobile belts.
Main structures and metamorphism	Synclinal or nappe-like greenstone belts; strike-slip faults common; mainly low to moderate metamorphism (low pressure). Complex structure and moderate to high grade metamorphism (low to high pressure) in gneissic terrains.	Epicratonic sequences little deformed or metamorphosed; mobile belts strongly folded, faulted and metamorphosed up to granulite facies (low to high pressure).
Volcanic rocks	Ultramafic–mafic komatiites, low-K tholeiites, Na-rich dacites and rhyolites. Andesites locally important.	Tholeiitic basalts, predominantly normal-K; acid volcanics mainly K-rich. Andesites locally abundant. Minor komatiite-like rocks.
Intrusive rocks	In some terrains, early granitoids mainly tonalite, trondjhemite and granodiorite, whereas late granitoids mainly granodiorite, adamellite and granite (s.s.); temporal overlap between subvolcanic ultramafic, mafic and felsic intrusives in greenstone belts. Dolerite dyke swarms.	Large, layered mafic–ultramafic intrusives, dolerite dyke swarms. Granodiorite, adamellite, granite (including rapikivi types). Subvolcanic felsic porphyries. Na-rich granitoids uncommon.
Sedimentary rocks	Chert, BIF, conglomerate, volcaniclastic greywacke and siltstone; minor carbonate. Quartz-rich sandstone, aluminous shale are uncommon, except in some late Archaean sequences.	Quartz-rich sandstone (quartzite) arkose, polymictic conglomerate, carbonate (mainly dolomite), chert, BIF, shale, greywacke.
Biological activity	Photosynthesis probably established before 3.5 Gyr; putative microfossils from ~3.5 Gyr; sporadic stromatolite occurrences. Bacterial sulphate reduction not of widespread significance.	Proliferation of microorganisms: common stromatolites, sulphate reducing bacteria. First eucaryotic microorganisms and megascopic algae.
Important mineralization	Ni–Cu, Cr and asbestos in ultramafic volcanics and intrusive; Fe and Au in BIF; Au in altered greenstones; Cu–Zn (–Pb) in mafic–felsic volcanics. Minor Ba in sediments and Cu–Mo in felsic porphyries.	Cr, Ni–Cu, Pt in large layered mafic–ultramafic intrusives: Fe in BIF; Au–U in quartz-rich clastics; U associated with unconformities; Pb–Zn–Ag in shales; Cu in "red bed" sandstone–dolomitic–shale sequences.

Radiometric age determination

Radiometric dates on Archaean rocks (cf. recent review by O'Nions and Pankhurst, 1978) are predominantly in the range 3.0–2.6 Gyr, with crust older than 3.4 Gyr so far being authenticated only from restricted localities in some of the shields. However, with the current state of interest in finding the "oldest" rocks, there is little doubt that more occurrences of early Archaean crust will be recognized in the near future.

It is necessary to include a note of caution here to the effect that published age determinations which are based on limited numbers of samples, or samples chosen without good geological control, are of doubtful significance. Geochronological laboratories are now adopting the attitude that the reliability of their results is a function of careful and adequate sampling, coupled with the utilization of two or more independent age determination techniques. This has been done for West Greenland, but in the following discussion of other Archaean terrains, it must be borne in mind that many of the published ages do not fully satisfy these criteria for reliability. Therefore, in many cases, it is not possible to make meaningful global correlations, or to erect evolutionary models, solely on the basis of currently available age data.

Whilst it is not possible to lay down hard-and-fast rules concerning radiometric dating, the following observations apply, in general terms, to the techniques that have been applied to Archaean rocks:

(1) U—Pb zircon dating is best suited to felsic igneous and metamorphic rocks, commonly yielding older ages than the Rb—Sr technique; caution is needed in interpreting results for zoned zircons which may reflect ages of source rocks, emplacement and/or metamorphism.

(2) Nd—Sm dating has recently evolved as a technique which can also be used for satisfactorily dating felsic rocks and which is particularly suitable for dating mafic to ultramafic rocks; few studies have been conducted to date.

(3) Rb—Sr and Pb—Pb total rock and mineral ages are best suited to felsic rocks; they commonly yield disturbed isochrons reflecting post-formational events and these have to be interpreted with care.

Resetting of radiometric ages is discussed further in the section on diachronous growth of sialic crust, below.

Early Archaean terrains. Very old K—Ar ages, up to an impossible 6 Gyr, have been reported from the Baltic Shield; these are now considered to reflect excess Ar in the metamorphic rocks, and the terrain is probably ~3.3 Gyr old (e.g. Tugarinov and Bibicova, 1973; Bowes, 1976). The oldest, well-authenticated ages have been obtained in the high-grade terrain of southwestern Greenland. Here, the Isua supracrustal sequence has yielded ages of 3.75–3.82 Gyr (Baadsgaard, 1976; Moorbath et al., 1977; Michard-Vitrac et al., 1977; Hamilton et al., 1978). The predominant rock type in this area, the Amitsoq gneiss, has yielded ages from 3.65 to 3.75 Gyr (Moorbath et al., 1977). It is noteworthy that Rb—Sr, Pb—Pb (whole rock), U—Pb (zircon) and Nd—Sm (whole rock), methods have yielded reasonably similar ages. Whilst these ages may be reflecting the severe metamorphism of these rocks, the initial $^{87}Sr/^{86}Sr$ ratios are low, implying that the primary ages could not have been more than ~3.9 Gyr (e.g. Moorbath, 1977; O'Nions and Pankhurst, 1978; cf. discussion in section on "Diachronous growth of sialic crust").

The Uivak and Hebron gneisses of eastern Labrador, which are correlatable with the Amitsoq gneiss of Greenland are ≥3.6 Gyr old (Hurst et al., 1975; Collerson, in prep.). A Pb—Pb age of ~3.4 Gyr was published for the Vesteralen gneisses of northwestern Norway by P.N. Taylor (1975), but recent work by Jacobsen and Wasserberg (1978) suggests these rocks were originally laid down ~2.6 Gyr ago.

In Africa, the Mashaba gneisses of Rhodesia have been dated at ~3.6 Gyr and associated enclaves of Sebakwian greenstones are at least as old (Moorbath et al., 1976; Wilson et al., 1978), in contrast to the more abundant Bulawayan greenstones which are 2.5—2.7 Gyr old (Hawkesworth et al., 1975). In the Kaapvaal craton of southern Africa, the age of the Lower Onverwacht greenstone sequence is >3.5 Gyr (Jahn and Shih, 1974) and dates of up to ~3.4 Gyr have been obtained from the nearby Ancient Gneiss complex (R.D. Davies and Allsopp, 1976). The Limpopo Belt separating the Rhodesian and Kaapvaal cratons contains a small strip of probable basement gneisses containing dolerite dykes recently dated at ~3.6 Gyr by Barton et al. (1977), who also refer to an unpublished ~3.8-Gyr isochron on the gneisses.

In North America, Goldich and Hedge (1974) claimed that the Montevideo Gneiss of Minnesota, which intrudes a complex of gabbro and dioritic gneisses is ~3.8 Gyr old; however, this has been challenged by Farhat and Wetherill (1975) and, although the terrain may contain rocks as old as ~3.8 Gyr, this has not yet been unambiguously established.

In Australia, the little metamorphosed or deformed greenstones of the Warrawoona Group in the Pilbara region of Western Australia are also very old; here, a felsic volcanic unit in the Duffer Formation has yielded a zircon age of 3.45 Gyr (Pidgeon, 1978a) and this unit is underlain by the Talga Talga subgroup, which is up to ~8 km thick (Hickman and Lipple, 1975). Granitoids in the Pilbara region have yielded zircon ages of 3.2—3.45 Gyr (Pidgeon, 1978b).

Younger Archaean terrains. No rocks older than ~3.2 Gyr have been reported from the Yilgarn Block (Western Australia; Turek and Compston, 1971; Arriens, 1971; Oversby, 1975; Roddick et al., 1976; Cooper et al., 1978), the Superior and Slave Provinces (Canada; Baragar and McGlynn, 1976) or the Scottish shield (Moorbath et al., 1975; Hamilton et al., 1979), where the great bulk of the radiometric (mainly Rb—Sr, Pb—Pb) ages are between 2.6 and 3.0 Gyr. McCulloch and Wasserberg (1978) obtained a Sm—Nd model age of ~3.1 Gyr for the source rocks of a volcaniclastic greywacke sample from a greenstone belt in the Yilgarn Block.

Diachronous growth of sialic crust

Relatively few granitoids in Archaean terrains exhibit high initial $^{87}Sr/^{86}Sr$ ratios (and other isotopic ratios), indicative of reworking of much older sialic crust. The normal interpretation of these low initial ratios is that the parent rocks could not have been greatly older than the measured ages (e.g. Pidgeon and Hopgood, 1975; Moorbath, 1977; Moorbath and Taylor, 1978; O'Nions and Pankhurst, 1978), implying diachronous generation of the sialic crust.

A less-favoured interpretation is that these low initial ratios are the result of resetting of the isotopic systems on huge scales during reworking of older sial, either by preferential loss of radiogenic daughter products or by isotopic exchange with mantle Sr, Pb and Nd. Three possible mechanisms for achieving this resetting have been proposed and critical appraisals of these follow:

(1) Armstrong (1968) proposed that the bulk compositions and volume of the mantle crust and ocean have been constant for at least the last 2.5 Gyr, during which time there has been mixing of crust and mantle (with consequent Sr and Pb isotopic exchange) in continental margin cordillera and island arcs. However, extension of this model back through the Archaean to account for resetting of ancient ages is difficult. For instance, whereas the surface rocks are eroded into trenches and cycled through the mantle, it is the deep crustal rocks which have to be isotopically reset in reworked basement hypotheses. Also, evidence to be reviewed in the section on Archaean global tectonics militates against the existence of plate tectonic regimes like those envisaged in the Armstrong model.

(2) The second general model invokes preferential loss of radiogenic daughter products during granulite facies metamorphism (e.g. Heier, 1964; Arriens and I.B. Lambert, 1969). This is based largely on the observations that anhydrous granulite facies rocks have low abundances of Rb, U, Th and heavy REE elements, presumably because there were concentrated into ascending aqueous fluids and granitic melts (e.g. Heier, 1964; I.B. Lambert and Heier, 1967; R.St.J. Lambert et al., 1976; Tarney, 1976); It is argued that radiogenic daughter products were residing metastably in lattice sites originally occupied by these depleted elements (e.g. ^{87}Sr in ^{87}Rb sites), and might have migrated out of the granulites with these elements before isotopic homogenization could occur. In this manner, it is theoretically possible that isotopic evidence for a long crustal pre-history could be lost during high-grade metamorphism. However, no direct evidence has yet been found for preferential loss of daughter isotopes in a manner that gives rise to good regional-scale "isochrons".

(3) The final model postulates modification of the Sr, Pb and Nd isotope ratios of the deep crust by exchange with mantle Sr, Pb and Nd in outgassing fluids (e.g. Collerson and Fryer, 1978). Two problems with this process are the enormous scale and pervasive nature of this exchange (e.g. Pidgeon and Hopgood, 1975), and the uncertainty as to whether outgassing fluids were enriched in mantle-derived Sr, Pb and Nd.

The very fact that *some* late Archaean granitoids *do* preserve isotopic ratios indicative of reworking of much older crust is a major factor militating against the extensive resetting of isotopic systems in ancient sialic crust. It is contrived to call on such resetting in some instances but not in others.

It is therefore concluded that most Archaean granitoids formed directly from mantle-derived simatic crust, or from sialic crust that had not been in existence for long periods prior to reworking.

Relationships between major components of Archaean terrains

In attempting to assess associations between major components of Archaean terrains, it is important to understand two types of relationships. The first is between domal granitoids and adjacent greenstone sequences (e.g. Ancient Gneiss complex/Onverwacht Group; Pilbara batholiths/Warrawoona Group). The second relationship is that between extensive areas of high-grade gneisses and low-grade granitoid—greenstone terrains; the high-grade terrains may be:

(1) discrete entities (e.g. North Atlantic craton);

(2) marginal belts of individual cratons [e.g. the West Yilgarn (Wheatbelt) gneiss domain] or intracratonic belts flanked by low-grade granitoid—greenstone terrains (e.g. English River Gneiss of the Superior Province); or

(3) mobile belts between very extensive, lower-grade, cratonic regions (e.g. Limpopo Belt between Kaapvaal and Rhodesian cratons.

The relationships between domal granitoids and greenstones will be discussed in more detail under petrogenesis of Archaean igneous rocks. Basically, there are two schools of thought: one is that the granitoids represent diapirs of sialic basement emplaced into younger greenstones, and the other is that granitoid magmas were generated by partial melting of underlying, predominantly simatic crust. Neither of these models can be ruled out on the basis of existing geochronological data, which indicate generally similar ages (that are not necessarily accurate measurements of ages of *formation*) for spatially associated granitoids and greenstones.

The relationship between high- and low-grade terrains is also contentious, except for type (3) terrains where structural, metamorphic and lithological transitions have been reported with adjacent low-grade terrains (e.g. Coward et al., 1976; Shackleton, 1976). In other cases, "boundaries" between low- and high-grade terrains cannot be interpreted in the field, being either poorly exposed (e.g. West Yilgarn gneiss domain/ Eastern Goldfields in Western Australia), or major tectonic zones (e.g. English River Gneiss/Uchi Belt in the Superior Province). In such cases, relationships have to be assessed indirectly and this is most readily done by comparing the overall lithological features of supracrustal sequences in each type of terrain, as discussed below.

The type (1) high-grade Isua supracrustals, it was shown earlier, comprise mainly metamorphosed mafic rocks with some dunites, talc schists, carbonate-bearing siliceous schists, ankeritic carbonates, quartzites and banded iron formations. The schistose rocks could well be altered tuffites, with carbonate, silica and iron components being introduced into the hydrosphere by hydrothermal activity (Allart, 1976). In this model, the quartz-rich metasediments could have formed from cherts rather than quartz-sandstones, and the ankeritic lenses need not be equivalent to normal marine carbonates in younger sequences. As enlarged on later, cherts, BIF and carbonates occur in low-grade, early Archaean greenstone sequences. On this evidence alone, therefore, it is possible that terrains of the West Greenland type are high-grade equivalents of early greenstone—granitoid belts.

Marginal cratonic and intracratonic high-grade terrains [type (2)] are not known before ~3 Gyr BP. Metamorphosed supracrustal sequences in such terrains in India, southwest Australia and elsewhere have higher metasediment to metavolcanic ratios than are typical of basal parts of greenstone sequences, and they contain abundant quartzites (some with probable cross-bedding) and/or marbles. These features led Windley (1977) to conclude that the precursors of these high-grade supracrustals formed in epicontinental platform/shelf sequences dissimilar to the depositional environment of most penecontemporaneous greenstone belts: a similar conclusion was reached by Gee (1979), who considered the West Yilgarn Gneiss terrain formed before the granitoid—greenstone belts of the Eastern Goldfields Province.

These conclusions based on lithological comparisons are not unequivocal and further resolution of relationships between low-grade granitoid—greenstone terrains and high-grade complexes [types (1) and (2)] must await detailed geochronological studies which reliably establish the ages of formation of all relevant rock types.

Deformation and metamorphism of greenstone belts

Before discussing the stratigraphic, lithologic and petrogenetic aspects of greenstone belts, it is important to briefly consider their deformational and metamorphic histories. These provide important constraints on the certainty of stratigraphic reconstructions and the degree of confidence that can be placed on geochemical data from the metavolcanic rocks.

On a macroscopic scale, the greenstone belts commonly occur as synclinal keels flanked by intrusive granitoids, with or without subordinate gneissic rocks. Most mesoscopic folds are upright with variable development of subvertical foliation which is normally subparallel to the elongation of the greenstone belts. Structural studies, following the pioneering work by Ramsay (1963), have shown that there are commonly two or more phases of folding with steeply inclined axial surfaces and shallowly to steeply plunging axes. One or more phases may be linked to diapiric granitoid uprise (see for example Fig. 6). Strike faults appear to be very common in areas of good exposure and low-angle reverse faults may be locally developed. Deformation is heterogeneous with high strain commonly concentrated in metasedimentary layers within volcanic sequences. Mesoscopic to macroscopic recumbent (nappe-like) folds are undetected in most structural studies, but some authors (e.g. Stowe, 1974; Coward et al., 1976; Archibald et al., 1978) present evidence for their existence early in the deformational sequence. This is further discussed in the section on Archaean global tectonics. Granitoid—greenstone ter-

358

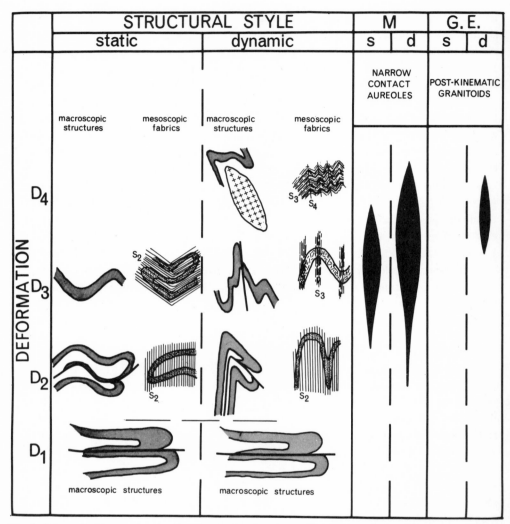

STRUCTURAL STYLE				M		G. E.	
static		dynamic		s	d	s	d

Fig. 6. Schematic diagram showing the structural and metamorphic sequence deduced from part of the Yilgarn Block, Western Australia by Archibald et al. (1978). Note that recumbent folds ascribed to D1 in this instance may be absent from other terrains.

rains are further complicated by regional-scale shear zones (often carbonated and/or mineralized), mylonite zones (commonly subparallel to greenstone belt elongation) and major cross-cutting fracture zones (many of which are occupied by early Proterozoic dykes).

Metamorphic patterns also appear to be relatively consistent, characterically increasing from sub-greenschist in the central areas to high amphibolite facies towards the intrusive granitoid contacts (e.g. Anhaeusser, 1971a; Binns et al., 1976; Glikson, 1976a; Jolly,

1978). The metamorphism is commonly of the low-pressure type, with low internal strain towards the central parts of greenstone belts resulting in good preservation of structures and textures in these areas. In contrast, zones adjacent to granitoid contacts are commonly areas of high strain in which original features of the lithologies and even previous deformational fabrics are poorly preserved. In most areas, peak metamorphism post-dates early deformation phases. These relationships suggest a fundamental link between granitoid uprise and metamorphism, but the metamorphic pattern is not the result of contact metamorphism in the normally accepted sense as contact aureoles around some post-kinematic granitoids clearly overprint the regional metamorphism (e.g. Binns et al., 1976).

An important aspect of the metamorphism in the Yilgarn Block emphasized by Binns et al. (1976) is the distribution of relict igneous minerals in volcanic rocks at varying metamorphic grades. For example, in komatiitic and tholeiitic volcanics in subgreen-schist facies environments, igneous clinopyroxene is commonly present, whereas olivine is serpentinized and plagioclase is altered. On the other hand, in the same lithologies at amphibolite grade, igneous olivine and plagioclase are commonly preserved, while clino-pyroxene is converted to amphibole. This indicates that the higher-grade rocks did not experience complete alteration during sea-floor and/or early burial metamorphism, and suggests the possibility that water circulation was restricted in these sequences prior to the onset of deformation and granitoid uprise. It is important to determine whether such relationships exist in other greenstone terrains.

Due to the common occurrence of low-strain metamorphic domains in greenstone belts, the original nature of the lithologies is clear in many cases. Igneous and sedimentary terminology are thus used to describe these rocks in the following sections, but it must always be borne in mind that all rocks are metamorphosed to some degree. Terms such as amphibolite, schist, gneiss, etc. are used to indicate lithologies that have experienced high-strain metamorphism.

Temporal variation in greenstone sequences

An overview. The existence of at least two major, temporally distinct greenstone sequences has been suggested in many Archaean terrains. In reviewing the evidence from several continents, Glikson (e.g. 1976b, 1977a) contended that lower (older) greenstones can be distinguished from upper (younger) greenstones because the latter are not intruded by the large domal granitoid complexes, they have unconformities or disconformities (e.g. chert beds of regional extent) at their base, and they tend to have lower proportions of ultramafic volcanics and higher proportions of calc-alkaline volcanics and volcaniclastic sedimentary rocks than the lower greenstones. Most of these relationships have been described from the Swaziland sequence of South Africa (R.P. Viljoen and M.J. Viljoen, 1969; Anhaeusser, 1971b); this thick sequence apparently accumulated in the period ~3.5 to ~3.1 Gyr BP, implying that the postulated major break in volcanicity could not have exceeded a few hundred million years, and that at least the upper parts of the sequence formed after the adjacent ~3.4 Gyr old Ancient Gneiss complex.

It has been argued, however, that there is no clear evidence for temporally distinct greenstone successions in some cratons, for example the Yilgarn Block (Archibald et al., 1978), and the Superior Province (Goodwin, 1977). As discussed above, geochronological studies in these terrains have not produced greenstone ages greater than ~2.9 Gyr, but there is current dispute about the "primary" ages of the stratigraphically oldest greenstones. Thus, the case for the existence of distinct greenstone sequences in these terrains rests mainly with the available geological evidence which, in many cases, is equivocal. Alternative interpretations to those of Glikson (op. cit.) are outlined and discussed briefly below.

(1) Unconformities within greenstone belts may be of local significance and they need not reflect long time breaks; there are probably many examples of this within greenstone successions. Major breaks in volcanism cannot be proven unless units beneath the unconformities have been subjected to more phases of metamorphism/deformation. This information is not available for most terrains where upper and lower greenstone sequences have been proposed (including the Swaziland sequence), but N.J. Archibald and M.E. Barley (pers. comm., 1979) believe there are no structural/metamorphic breaks within the dominantly volcanic sequences of the Yilgarn and eastern Pilbara greenstone belts, respectively.

(2) Disconformities may likewise only reflect short lulls in volcanism. This almost certainly applies to many localized cherty sediments. Regionally developed chert beds may reflect significant breaks in volcanic activity if persistent marked changes in lithological patterns are associated with them. In most cases, the detailed studies required to establish this have not been carried out.

(3) Diapiric granitoid domes could have stopped ascending at a particular lithological or structural level, well beneath the surface in greenstone successions which accumulated without major breaks in volcanic activity. In most cases, no criteria exist which can resolve the differences in interpretation of granitoid level.

It has also been suggested that, where there are greenstone sequences of different ages, these do not always exhibit secular lithological trends. For example, Bickle and Nisbet (1977) and Wilson et al. (1978) do not see any fundamental lithological differences between the approximately 3.5, 2.9 and 2.7 Gyr old greenstones in part of the Rhodesian craton. This matter is discussed further in the next section.

Cyclicity of greenstones has been proposed at various scales from that of "upper" and "lower" greenstones, down to multiple ultramafic—mafic to felsic cycles within these sequences (e.g. Anhaeusser, 1971b). However, felsic volcanics commonly appear to have formed at local centres and, therefore, may have accumulated beside, and interfingered with, contemporaneous ultramafic—mafic volcanics (as illustrated in the interpretive section through the eastern Pilbara Block, Fig. 7). This must be borne in mind when examining apparent volcanic "cycles", when making correlations on lithological grounds and when estimating thickness of greenstone sequences.

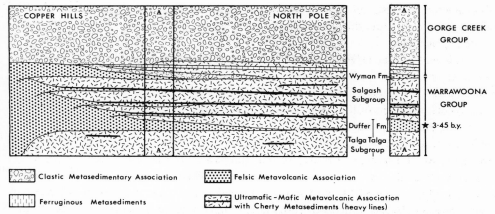

Clastic Metasedimentary Association

Felsic Metavolcanic Association

Ferruginous Metasediments

Ultramafic-Mafic Metavolcanic Association
with Cherty Metasediments (heavy lines)

Fig. 7. Schematic section through part of the eastern Pilbara Block, Western Australia, showing relationships between felsic and mafic—ultramafic intrusive sequences of the Warrowoona Group and relations of this Group to the predominantly sedimentary Gorge Creek Group. The section at A—A is similar to that of the Barberton Mountain Land according to Anhaeusser (1971a, b); from Barley et al. (1979).

Generalizations concerning lithological features. Volcanics normally dominate in the lower parts of sequences, whereas clastic sedimentary rocks commonly dominate the upper parts. Ultramafic rocks are normally most abundant at lower stratigraphic levels, the ratio of tholeiites to komatiites increases upwards in many cases, and calc-alkaline rocks are most common in upper sequences. However, it must be stressed that exceptions to the above generalizations do occur.

In areas of low-grade metamorphism and low internal strain, lavas typically have well-preserved relict igneous textures. Although a compositional continuum exists between most volcanic rock types in greenstone belts, it is usually possible to distinguish lavas with komatiitic, tholeiitic and calc-alkaline affinities (e.g. Jolly, 1978; Hawkesworth and O'Nions, 1977).

Komatiitic volcanics include komatiitic peridotite units with distinctive upper, spinifex-textured A zones (Plate I, A) and lower cumulate B zones (e.g. Pyke et al., 1973) that typically form sequences with thinner, less magnesian flows becoming more abundant upwards. These are commonly overlain by komatiitic basalts typically with distinctive skeletal clinopyroxene textures (Plate I, C), that also form thick sequences of massive, pillowed or differentiated flows (e.g. Arndt et al., 1977) without associated peridotites. Cherty sediments and tholeiitic basalts are commonly intercalated between flow units. In some places, the komatiite sequences are complicated by the occurrence of thin units of less magnesian composition, and small subspherical blebs (ocelli) of similar composition (Plate I, B) can also occur within basaltic rocks in the komatiitic sequences (e.g. Ferguson and Currie, 1972); similar structures occur in some Fe-rich tholeiites (e.g.

Gelinas et al., 1976). Differentiated sills and layered intrusions of probable komatiitic affinity also occur, and thick dunitic to peridotitic pods of uncertain origin are present in some sequences.

Tholeiitic basalts typically have ophitic to sub-ophitic textures (Plate I, D) and they occur as massive flows, pillowed flows commonly with elongate branching pillows or flow lobes (Fig. 8), or thick differentiated flows (e.g. Arndt et al., 1977). Differentiated sills and layered intrusions of tholeiitic parentage also occur.

Calc-alkaline rocks are generally less abundant than komatiitic or tholeiitic volcanics, but are locally abundant in some belts (e.g. Abitibi Belt of Superior Province, western part of Rhodesian Craton, western and eastern Pilbara Block; see Fig. 7). They form both laterally extensive sequences and more localized volcanic centres. The sequences are normally dominated by pyroclastic or volcaniclastic rocks, and contain intercalated mafic or even ultramafic flows or tuffs. Dacites appear to be the dominant felsic lithology. The proportion of andesites is highly variable, but is generally subordinate to felsic volcanics, except in calc-alkaline centres in the Abitibi belt where it has been estimated that they constitute 20–30% of the volcanic rocks (Baragar and Goodwin, 1969; Goodwin, 1978). However, it is noteworthy that some "andesites" in the Superior Province formed by silicification of basic volcanics (e.g. MacGeehan, 1978).

Geochemical features of Archaean igneous rocks

Volcanics. The literature dealing with the geochemistry of Archaean volcanics has been reviewed by Glikson (1971), Condie (1975), Windley (1977), Gunn (1976) and Hallberg (1972), with the latter two cautioning that some of the chemical features are the result of alteration. Fig. 9 summarizes the variation and overlap of volcanic suites in terms of $MgO-CaO-Al_2O_3$.

Komatiites, distinguished on the basis of their highly magnesian character and distinctive quench (spinifex) textures, have widely varying chemical compositions (e.g. R.P. Viljoen and M.J. Viljoen, 1969; Nesbitt and Sun, 1976; Sun and Nesbitt, 1977), but they are normally characterized by higher Mg/Al ratios and lower Ti contents than associated tholeiites. The komatiitic basalts commonly have La/Sm ratios <1 and flat heavy REE patterns, whereas periodotitic volcanics normally have slightly enriched light

PLATE I

Photomicrographs illustrating structures and textures of Archaean ultramafic and mafic flows.
A. Spinifex texture defined by altered olivine blades from A-zone of komatiitic ultramafic flow, Miriam, Yilgarn Block, Western Australia.
B. Ocelli in tholeiitic basalt, Kelly Greenstone Belt, east Pilbara, Western Australia.
C. Quench-texture defined by clinopyroxene crystals in komatiite basalt. Kelly greenstone Belt. Note skeletal overgrowths on pseudomorphs after subhedral olivine microphenocrysts.
D. Sub-ophitic texture in tholeiitic basalt, Kelly Greenstone Belt.

Photomicrographs kindly provided by M.E. Barley.

PLATE I

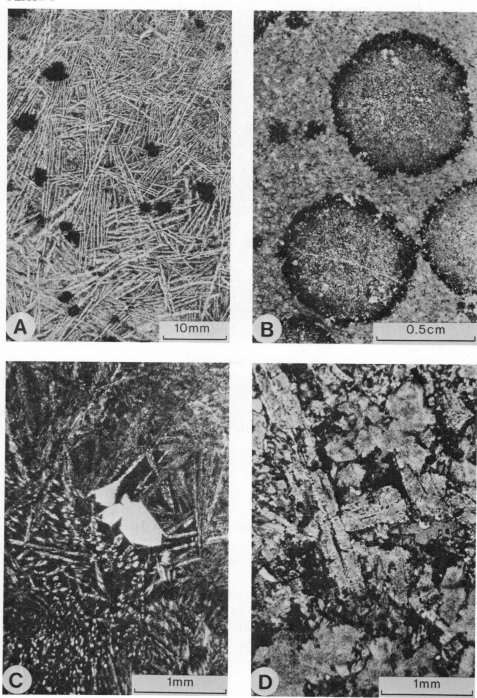

Fig. 8. Pillowed metabasalts showing elongate pillows or flowlobes and excellent preservation of volcanic structures; Salgash Subgroup, Kelly Greenstone Belt, Pilbara Block. Photograph courtesy of M.E. Barley.

REE patterns. The ultramafic komatiites from the "type" sequence in South Africa are important exceptions in that they display depleted heavy REE patterns; they are also characterized by higher Ca/Al ratios than komatiites from most other greenstone belts. It is particularly difficult to identify rocks of the komatiite suite at low MgO contents (< ca. 8% MgO), where they overlap with tholeiites in terms of major element composition (Fig. 9). Locally, Fe-rich tholeiites can be distinguished from komatiitic basalts on the basis of lower $Al_2O_3/[FeO/(FeO + MgO)]$ ratios, but most tholeiites cannot be distinguished in this manner (Groves and Hudson, 1981). The presence of distinctive skeletal textures (Plate I, C) may be an important criterion for distinguishing komatiitic basalts in view of the variable geochemical features, although such textures do occur in some modern tholeiites.

Archaean tholeiites typically have low contents of K, Rb, Ba and other large-ion lithophile elements (LIL). Whilst they basically resemble modern oceanic tholeiites, the Archaean examples commonly have higher abundances of transition metals (V, Cr, Ni and Co), higher Fe/Mg and lower Al and Ti (Glikson, 1971); in addition, they often have slightly enriched light REE patterns which contrast with the depleted light REE patterns of most ocean-floor basalts. There is some evidence that basalts associated with felsic volcanics at high stratigraphic levels in greenstone belts tend to resemble modern island-arc tholeiites in their contents of K, Rb, Sr and Ba (Glikson, 1971; Condie, 1975), but these elements are notorious for their secondary mobility.

Archaean andesites exhibit erratic levels of alkalies and alumina and, on average, are low K and Sr types (Gunn, 1976). The dacites and rhyolites also exhibit variable chemical features, but they are predominantly high Na/K porphyries; high-K rhyolites do occur, although Gunn (op. cit.) ascribes these to secondary illite formation. Compared with modern calc-alkaline volcanics, these Archaean examples commonly have lower Al_2O_3, higher Ni and Cr contents, and heavy REE depletions.

Granitoids. Recent reviews of Archaean granitoids include Lister (1973), Windley (1977) and Glikson (1979).

Many granitoid terrains are not well exposed and, as outcrops are dominated by rocks that are most resistant to weathering and erosion, it is difficult to assess proportions of various lithologies. Thus, in general, plagioclase-rich granitoids are less resistant than

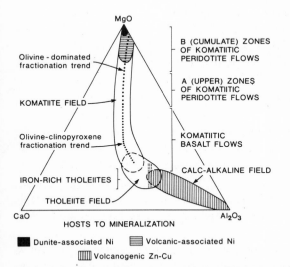

Fig. 9. $CaO-MgO-Al_2O_3$ diagram showing fields of major Archaean volcanic suites, subdivisions of ultramafic rocks and their fractionation trends, and host rocks for stratabound Archaean base metal mineralization.

K-feldspar-rich granitoids, and gneisses are less resistant than massive granitoids. Observations in relatively well-exposed regions suggest that certain generalizations can be made. High-grade terrains comprise predominantly tonalitic and trondhjemitic gneisses ("grey gneisses"); granodioritic to granitic (s.s.) gneisses ("pink gneisses") are locally important. In low-grade terrains there are large composite granitoid domes comprising tonalites, trondhjemites and granodiorites which are broadly conformable within gneisses of similar to more mafic compositions. Adamellites and granites (s.s.) occur in widely varying proportions in low-grade terrains; in some cases they are obviously younger, discordant bodies, but, in poorly-exposed areas in the Yilgarn Block, intrusives and gneisses of granodioritic to adamellitic compositions are the dominant rocks outcropping (Archibald et al., 1978).

The bulk of the rocks in the early formed granitoid domes (particularly in southern Africa) have Na/K ratios >1, initial $^{87}Sr/^{86}Sr$ ratios <0.705 and strongly sloping REE patterns with marked depletions in heavy REE. Plots of SiO_2 against other component oxides (Harker diagrams) for many of the published analyses from individual gneiss terrains and granitoid batholiths show straight-line relationships similar to those in younger intrusive granitoid batholiths (White and Chappell, 1977). In several cratons (e.g. Kaapvaal, Rhodesian, Pilbara), the youngest postkinematic granitoid intrusives are characterized by Na/K <1, variable initial $^{87}Sr/^{86}Sr$ ratios which commonly exceed 0.705, negative Eu anomalies, and no consistent heavy REE depletions; the compositions of Pilbara examples, associated with tin mineralization are given in Table II.

Petrogenesis of Archaean igneous rocks

An overview. Archaean geothermal gradients must have been higher than in the present Earth because of residual heat from accretion/core formation, heat produced by short-lived radioisotopes and, possibly, the effects of giant impacting, (e.g. Hart et al., 1970; Fyfe, 1973; Glikson and I.B. Lambert, 1976; R.St.J. Lambert, 1976, 1978; Collerson and Fryer, 1978). This would have been an important factor in Archaean petrogenesis, causing much more widespread igneous activity and higher degrees of melting than is characteristic of younger periods. However, there is evidence against extremely high geothermal gradients post $\sim$3.8 Gyr: for example, Glikson and I.B. Lambert (1976) argued that the low metamorphic grades of thick sequences of early Archaean greenstones (e.g. in the Pilbara and Kaapvaal cratons) suggest average regional gradients less than $\sim$40°C (Fig. 10).

Geochemical, petrographic, experimental and field data are satisfied by the following general petrogenetic processes, which are thought to have been recurrent during the Archaean:

(1) Generation of mafic—ultramafic volcanics of greenstone sequences by mantle melting, and andesitic to rhyolitic volcanics by partial melting of predominantly simatic crust or fractionation of basic melts.

(2) Generation of tonalites, trondhjemites and granodiorites by partial melting of

TABLE II

Comparison of composition of post-tectonic Archaean granitoids associated with tin mineralization in Western Australia (Blockley, 1973) with the average composition (with standard deviation) of mainly Phanerozoic specialized granitoids associated with tin mineralization (Tischendorf, 1977).

	1	2	3	4	(S.D.)
Percentage					
SiO_2	74.9	73.5	71.8	73.38	1.39
TiO_2	0.1	0.2	0.4	0.16	0.10
Al_2O_3	13.4	13.3	13.7	13.97	1.07
Fe_2O_3	0.4	0.6	1.3	0.80	0.47
FeO	0.8	1.2	1.2	1.10	0.47
MnO	0.05	0.07	0.04	0.045	0.040
MgO	0.1	0.3	0.5	0.47	0.56
CaO	0.8	1.1	1.4	0.75	0.41
Na_2O	4.1	3.9	3.5	3.20	0.61
K_2O	4.6	4.6	5.0	4.69	0.68
ppm					
F	1050	2730	2640	3700	1500
Li	190	50	315	400	200
Sn	10	10	2	40	20
Rb	520	490	405	580	200
Sr	40	45	130	–	–

1 Average of 11 analyses, Moolyella Granite, Pilbara Block (Blockley, 1973).
2 Average of 11 analyses, Cooglegong Granite, Pilbara Block (Blockley, 1973).
3 Granitoid stock near Poona tin deposits, Murchison Province (Blockley, 1973).
4 Specialized granitoids associated with tin mineralization (Tischendorf, 1977).

predominantly simatic primitive crust or ancient greenstone lithologies.

(3) Generation of more potassic igneous rocks by anatexis of relatively Na-rich granitoids, metavolcanics and metasedimentary rocks, or more complete melting of earlier granodiorites/adamellites.

(4) Generation of high-grade quartzofeldspathic gneiss terrains by metamorphism and deformation of intrusive granitoids.

(5) Eventual downsinking into the mantle of dense, anhydrous, basic to ultrabasic residual rocks formed during anatexis.

The existence of extensive sialic crust prior to formation of the early Archaean greenstone sequences is still advocated by a major proportion of geologists. However, considerations of the origin of the Earth and of the significance of radiometric age studies (presented above) suggest that the amount of sialic material present in the earliest Archaean was minor relative to that which was generated subsequently. Furthermore, field and geochronological evidence indicate that granitic gneisses in the oldest known crustal segment (in the North Atlantic craton) were emplaced into older greenstone-like, supracrustal sequences. The clear field and geochronological evidence that some ≤3.2 Gyr old greenstones accumulated on and/or beside slightly older sialic crust does not bear on the

368

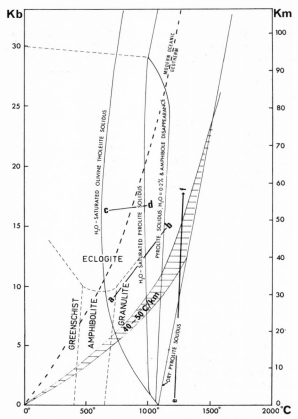

Fig. 10. Pressure–temperature diagram showing relations between a maximum estimated Archaean geothermal gradient of 40–50°C/km and solidus curves for mantle peridotite (pyrolite; Green, 1973) and olivine tholeiite (I.B. Lambert and Wyllie, 1972). The dry solidus for olivine tholeiite is very close to that for pyrolite with 0.2% H_2O up to ~23 kbar. Metamorphic facies boundaries are after F.J. Turner (1968). Line $a–b$ signifies incoming of garnet in anhydrous olivine tholeiite (Green and Ring-wood, 1967). $c–d$ approximates the incoming of garnet in the same composition for water saturated conditions (I.B. Lambert and Wyllie, 1972), $c–f$ = anhydrous tonalite liquidus (Stern et al., 1975); the water-saturated tonalite liquidus is close below the water-saturated pyrolite solidus in the melting region). Adapted from Glikson and Lambert (1976).

existence of extensive sialic basement in earliest Archaean times. This matter of the nature of the basement to the ancient greenstones will be discussed further in the section enlarging on granitoid genesis.

Volcanics. Pillowed lavas (Fig. 8), quench textures (Plate I, A and C), and structures and textures of intercalated sedimentary rocks attest to submarine deposition of the green-stone sequences. Furthermore, the considerable thicknesses of these sequences imply progressive or intermittent subsidence during accumulation.

The highly magnesian nature of quench-textured ultramafic volcanics led Green (1972) to conclude that they formed by very high degrees (up to ~80%) of melting of mantle peridotite, with rapid upwelling of little-fractionated magma, and extrusion temperatures of 1600°C or above (Fig. 10). This would require melting of superheated mantle rising from considerable depths as diapirs, plumes and/or convection cells; Green (op. cit.) favoured impact-triggered mantle diapirism. However, an alternative model involving smaller degrees of partial melting of residuum left after previous partial melting of mantle peridotite is suggested by considerations of melt viscosities and settling velocities of residual crystals (Arndt, 1977), and by geochemical features of the peridotitic komatiites, particularly light REE depletions (Arth et al., 1977; Bickle et al., 1976; Sun and Nesbitt, 1978).

Arndt (op. cit.) envisaged that about 20% melting in rising mantle diapirs produced olivine tholeiite magmas which separated from the diapirs. The diapirs continued to rise, melting at a reduced rate until a second batch of magma separated, giving rise to mafic komatiitic magmas. Continuation of the ascent and melting process would eventually enrich the crystalline residue sufficiently in refractory components that these would yield peridotitic komatiitic magmas by relatively low degrees of melting (cf. Fig. 22).

A modification of this general model which deserves consideration is based on our earlier conclusion that melting associated with core segregation generated a predominantly mafic outer zone underlain by a peridotitic residuum zone (Fig. 2). Diapris rising from this postulated depleted peridotitic zone should have been capable of yielding komatiitic magmas with less protracted sequential melting than in the Arndt model. If this were the mechanism of komatiite production, such a depleted peridotite zone would have to have existed in the upper mantle through the Archaean, probably requiring that its density was lowered by minor proportions of entrained melt.

The wide variations in the compositions of komatiitic and other lavas could be reflecting varying degrees of melting prior to separation of crystalline residue, polybaric assimilation of crystalline fractions and/or high-level fractionation processes.

Alkaline and high-alumina basalts occur in the Superior Province (e.g. Goodwin, 1977), but overall there are few such basalts in Archaean greenstone sequences. Tholeiites form by higher degrees of melting of peridotite, and their abundance in Archaean greenstone sequences suggests that they represent the minimal amount of melt (~20%) that commonly segregated and ascended during mantle melting under the prevailing high geothermal-gradient conditions.

The andesites could have formed by partial melting of mafic rocks, fractionation of mafic magmas at depths ≤30 km or partial melting of "wet" peridotitic rocks at similarly shallow depths. The original geochemical characteristics of the andesites are not yet well-enough established to distinguish between these processes.

Experimental data outlined earlier preclude generation of Na-rich felsic volcanics by direct melting of mantle peridotite; these dacites and rhyolites could have been generated by partial melting of amphibole and/or garnet-bearing primordial simatic crust or deep

370

levels of earlier-formed greenstone sequences (e.g. Glikson, 1972, 1976a; Arth and Hanson, 1972; Condie, 1975), or fractionation of basaltic magmas (e.g. Jolly, 1978).

Granitoids. The bulk of the quartzofeldspathic "grey gneisses" in high-grade terrains evidently formed by metamorphism and deformation of intrusive granitoids, during and after their emplacement. All such terrains contain large to small inclusions of metamorphosed supracrustal rocks which are intruded by granitoids. Many of the bands and irregular masses of adamellite and granite (s.s.) compositions ("pink gneisses) in these terrains were probably generated by partial melting of "grey gneisses". In general, paragneisses appear to become more abundant with decreasing age.

However, the genesis of the granitoid domes at higher crustal levels is disputed, with two fundamentally different models being proposed (Fig. 11). One is that the granitoids represent basement which was diapirically intruded into younger greenstones (e.g. Windley and Bridgwater, 1971; Hunter, 1974; Hickman, 1975; Archibald et al., 1978), and the other is that the gneisses represent complexly deformed zones formed during emplacement of anatectic granitoid magmas into older greenstone sequences (Anhaeusser, 1973; Anhaeusser and Robb, 1978; Glikson, 1979).

The granitic basement model receives impetus from structural studies which suggest

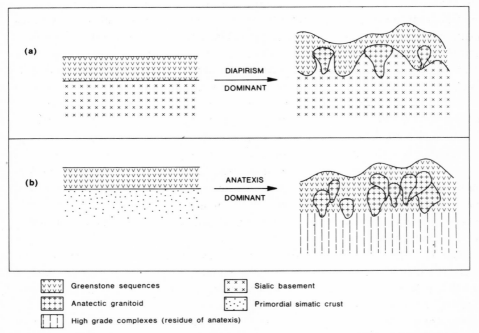

Fig. 11. Schematic diagram showing different interpretations of formation of domal granitoids in early Archaean terrains.

that the deformation history of some gneissic rocks is more complex than that of adjacent greenstone belts (e.g. Hunter, op. cit.; Archibald et al., op. cit.). These workers conclude that such gneisses represent basement which was diapirically emplaced into younger greenstone sequences with the intrusive granitoids being derived from the gneisses by solid-state and melting processes. Production of the intrusive granitic rocks in these domes by melting of the geochemically similar gneisses requires that either:

(1) the melting was virtually complete; i.e. that temperatures were $\geq 950°C$ for tonalitic compositions (Fig. 10) or $\geq 850°C$ for granodioritic compositions, depending on p_{H_2O}, or

(2) there was partial melting, requiring temperatures $\geq 700°C$ for tonalitic or granodioritic compositions, and the residuum did not separate from the melt.

Such temperatures require that the sialic basement ascended from depths of at least 20 km, or that the geothermal gradient was significantly above $40°C/km$. In this model, the structurally coherent mafic—ultramafic inclusions in the postulated basement gneisses would have to be relics of earlier greenstone sequences in the predominantly sialic basement.

In contrast, proponents of the second model interpret the mafic—ultramafic inclusions in the gneisses as being rafted off the basal parts of older greenstone sequences during granitoid emplacement, noting that lines of inclusions can be traced along strike into greenstone belts and that there are lithological similarities between such inclusions and associated greenstones (e.g. Anhaeusser and Robb, 1978). In this model, the structural complexities of the gneisses in granitoid—greenstone terrains are interpreted as reflecting multi-stage granitoid emplacement and shearing, but this remains to be substantiated by detailed studies. The model proposes that granitic melts formed by anatexis of primordial, predominantly simatic crust and/or deep levels of greenstone sequences, and that gneisses formed in zones of marginal solidification (high crystal/liquid ratios) during ascent; liquid "squeezed out" of these largely crystallized marginal zones could have generated the aplite—pegmatite veins characteristic of the domal granitoid complexes. The presence of amphibole and/or garnet as a residual phase in the source rocks could explain the depleted heavy REE patterns of the granitoids (Arth and Hanson, 1972, 1975).

Migmatitic rocks are commonly associated with the mafic—ultramafic inclusions in the gneisses. They could have formed by extensive invasion and partial assimilation of the inclusions by granitic magmas, metamorphic differentiation processes related to severe chemical gradients at granitoid—mafic rock contacts, and/or partial melting of inclusions.

The generally discordant, post-tectonic adamellites and granites (s.s.) commonly occur along greenstone margins, between older plutons, and in other zones of weakness (Fig. 5; e.g. Hickman, 1975). Experimental data, field relations and geochemical features (including initial $^{87}Sr/^{86}Sr$ ratios) are all compatible with derivation of these relatively potash-rich granitoids mainly from older, more plagioclase-rich granitoids (e.g. Glikson, 1979).

Archaean sedimentary rocks

Very interesting insights into early Archaean sedimentary environments have come from recent investigations of the early Archaean greenstone sequences in the Pilbara Block (Lipple, 1975; Dunlop, 1978; Barley, 1978; Barley et al., 1979), and the Swaziland System (Lowe and Knauth, 1977; Heinrichs and Reimer, 1977). These studies have documented abundant cherty sedimentary rocks which preserve structures and textures indicative of widespread shallow-water environments, with probable local evaporitic conditions. At least some of these metasediments apparently formed largely by silicification of carbonates and volcaniclastic sediments in large, volcanically active, but otherwise stable basins with subdued marginal relief. It has been noted above that the sedimentary environments of the early Archaean supracrustal sequences in high-grade granite-gneiss terrains (e.g. Isua) are not markedly dissimilar to those in the Pilbara and Barberton greenstone belts.

At higher stratigraphic levels in these terrains (e.g. Fig. 7) and in upper parts of greenstone sequences elsewhere (Anhaeusser, 1971b; Glikson, 1971; Veizer, 1976), there are thick sequences of immature clastic metasediments. These are mainly volcaniclastic arenites (greywackes), siltstones and conglomerates which appear to have accumulated mainly as marginal fluvial, alluvial fan or submarine fan deposits in high-energy, shallow- to relatively deep-water environments (e.g. C.C. Turner and R.G. Walker, 1973; R.G. Walker and Pettijohn, 1971; Henderson, 1972). Oxide-type BIF, jaspilite, sulphidic sediments and, less commonly, Fe-carbonate sediments, can also be important components of younger Archaean greenstone belts. The oxide facies BIF can form layers of several kilometres strike length and are widely considered to be of volcano-exhalative origin. In comparison, the sulphidic sediments are commonly laterally discontinuous, with strike lengths of less than 1 km and with wide fluctuations in thickness. They contain considerable quantities of Al_2O_3, MgO and/or CaO and alkalies (Groves et al., 1978; Sangster, 1978) and thus are not simply sulphidic equivalents of oxide-type BIF, but represent sulphide-rich tuffaceous and exhalative sediments related to local volcanism. They are normally developed between successive flows in basaltic, ultramafic and/or felsic volcanic sequences.

The formation of monotonous, plane-laminated oxide-type BIF implies deposition below wave-base with most authors favouring relatively shallow-water environments (<100 m), although R.G. Walker (1976) has questioned whether there is evidence for nearshore deposition. The plane-laminated sulphidic sediments presumably also formed below wave-base, but there is no diagnostic evidence for very deep-water deposition. Sulphur activity could have been more important than water depth (cf. Goodwin, 1973b) in controlling the distribution of oxide- and sulphide-rich sediments. BIF is discussed further in the sections on Archaean mineralization, early Proterozoic mineralization and development of an oxygenous atmosphere.

Veizer and Compston (1976) used the Sr isotopic compositions of carbonates as

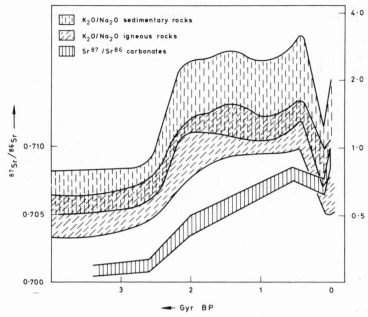

Fig. 12. Variations through geological time in the average K_2O/Na_2O ratios of igneous sedimentary rocks (after Engel et al., 1974) and the $^{87}Sr/^{86}Sr$ ratios of carbonates (after Veizer and Compston, 1976). Note increases in these ratios at around 2.5 Gyr, reflecting the increased proportion of high K_2O-Rb rocks in the upper crust in the late Archaean.

an index of crustal evolution; these reflect contemporaneous seawater composition and, thereby, the general nature of the crust exposed to weathering. They found a strong increase in $^{87}Sr/^{86}Sr$ ratios in the period 2.5–2.1 Gyr, which is in accord with evidence (above) for a secular variation during the Archaean and pre-Archaean from a crust dominated by mafic and high-Sr, (Na + Ca) granitic rocks to one in which relatively Rb- and K-rich rocks became more abundant in the upper parts. This is illustrated in Fig. 12, together with the composition of Na/K ratios of sedimentary and igneous rocks through geological time (after Engel et al., 1974).

A number of workers have reported secular changes in the oxygen isotope compositions of cherts through geological time, with Precambrian cherts having lower ^{18}O contents than younger examples (e.g. Perry, 1967; Perry and Tan, 1972; Knauth and Epstein, 1976; Knauth and Lowe, 1978). This has been interpreted by Perry (op. cit.) in terms of changes of the isotopic composition of seawater in line with evolutionary trends in the Earth, but Knauth and his coworkers (op. cit.) argue that the existing data are best satisfied by progressively declining mean climatic temperatures from ~70°C at ~3.4 Gyr, to present-day values.

Biological activity in the Archaean

The oldest probable microfossils so far reported are spheroidal, unicellular micro-structures, commonly 5—10 μm across (Fig. 3). These occur in the ~3.45 Gyr old North Pole chert—barite sequence, in the east Pilbara region, Western Australia (Dunlop, 1978; Dunlop et al., 1978); this sequence also contains possible stromatolites (R. Buick and J. Dunlop, pers. comm., 1979). There is dispute concerning the biogenicity of microstructures in ~3.5 Gyr old cherts in the Lower Onverwacht Group of the Swaziland System, South Africa, but others in the overlying ~3.2 Gyr old Fig Tree Group exhibit the currently accepted criteria for biogenic origin (Schopf, 1975; M.D. Muir and Grant, 1976).

Relatively few Archaean microfossil localities have been documented to date, although new examples have been reported with increasing frequency over the past few years (cf. Schopf, op. cit.). Many more microfossil localities are known in Proterozoic sedimentary rocks (e.g. Schopf, op. cit.), paralleling the much higher incidence of stromatolites in this era. It is amazing the extent to which recent micropalaeontological studies have pushed back the time of origin of life on Earth, which only fifteen years or so ago was widely accepted to have been not long before the advent of the first macro-fossils, near the base of the Cambrian.

There are no morphological or other characteristics of the Archaean microfossils which indicate their metabolic processes, but there is some indirect evidence for photosynthesis. Isotopic and geological studies of the occurrences of Archaean bedded barite deposits suggest that these formed by surficial oxidation of reduced exhalative sulphur,

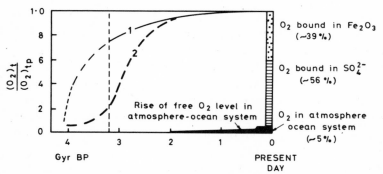

Fig. 13. Evolution of terrestrial oxygen budget with oxygen reservoirs expressed as fractions of the present reservoir (adapted from Schidlowski, 1976). Column on right indicates partitioning of the present budget between "bound" and "free" oxygen reservoirs; note that molecular oxygen, comprises only 5% of the total photosynthetic oxygen. The filled-in area at the base shows tentative rise of free oxygen in the atmosphere—ocean system as inferred from palaeontological and geological evidence. Curve *1* is Schidlowski's estimate which implies some 80% of the present oxygen reservoir existed by ~3.3 Gyr BP; curve *2* indicates the more gradual oxygen build up, favoured by the present authors because of evidence for limited proportions of oxygen reservoirs (e.g. carbonate, sulphate and oxide species) in the early Archaean hydrosphere—lithosphere.

most likely involving primitive blue-green algae or photosynthetic sulphur bacteria (Perry et al., 1971; I.B. Lambert et al., 1978; see discussion of barite in Archaean mineralization section). Furthermore, carbon isotopic data for Archaean carbonates and organic carbon led Schidlowski and co-workers (e.g. 1975, 1976, 1977, 1978) to conclude that the total terrestrial reservoir of photosynthetic oxygen should already have amounted to ~80% of the present reservoir by ~3.3 Gyr ago (Fig. 13; note that this does not necessarily mean the atmosphere was oxygenated at this early stage). However, the possibility of a more gradual build-up of the oxygen reservoir (line 2, Fig. 13) is suggested by the evidence preserved in the geological record concerning major oxygen reservoirs in the lithosphere/hydrosphere: it appears that sulphates, oxides and carbonates were less abundant in the early Archaean than in younger periods. The slower build-up of oxygen would probably require that some factor(s) inhibited proliferation of oxygen producing microorganisms in the early Archaean; Cloud (e.g. 1973; 1976) proposed that one such factor might have been the absence of internal oxygen-mediating enzymes.

On the other hand, sulphur isotopic studies of Archaean sulphidic sedimentary rocks suggest that bacterial sulphate reduction was not of widespread significance in this era (Donnelly et al., 1978; I.B. Lambert, 1978).

Thickness of the Archaean crust

Early Archaean sialic continental nuclei are commonly envisaged as thin and mobile. The characteristic mineral assemblages in the oldest high-grade terrains are in accord with this in that they could have formed without deep burial. However, there is evidence that some late-Archaean high-grade rocks formed at considerable depths (Condie, 1973; Engel et al., 1974; Wells, 1976; O'Hara, 1977; Newton, 1977); mineral assemblages developed locally in a number of $\lesssim 2.8$ Gyr old granulite facies terrains suggest burial depths of 20–30 km, with the maximum estimated depths being ~50 km ($T \simeq 1250°C$) for the Scourie gneisses (O'Hara, 1977).

Other features which have been used to estimate the thickness of the Archaean crust, but which are of dubious significance are: (1) the estimated stratigraphic thicknesses of exposed sections of greenstone sequences — most estimates fall between 8–17 km, and (2) geochemical indices (K, Rb, Sr contents relative to SiO_2 levels), which are based on the highly contentious assumption that Archaean volcanic and plutonic rocks formed above subduction zones analogous to those in post-Mesozoic times — they yield estimates of crustal thickness between 12 and >35 km (Condie, 1973).

The fact that Archaean, moderate- to high-pressure granulite facies terrains now exposed at the surface are underlain by 35–40 km of crust does not necessarily indicate that the crust in these regions was originally much thicker. It is more likely that the granulite terrains are fragments from deep in the crust that have been tectonically emplaced at higher levels.

Archaean global tectonics

The best insight into Archaean tectonic regimes should come from low-grade grani-toid—greenstone systems. However, whilst there has been much theorizing on the subject, critical evidence has so far proved elusive, partly because of the vagaries of preservation of ancient environments in the rock record. Thus, there is far from a consensus on the general nature of Archaean tectonic processes (cf. discussion and references in the recent paper by Groves et al., 1978).

From the evidence cited earlier, it appears that the best documented early supra-crustal sequences ($\geqslant$3.4 Gyr) were deposited in slowly subsiding, shallow-water basins, with low marginal relief. This stable behaviour could be reflecting the overall homoge-neous nature of the early crust. In contrast, many younger Archaean greenstone sequences appear to have been deposited in higher energy, shallow- to deep-water troughs (pseudo-eugeosynclines), the tectonic axes of which display a remarkable parallelism on Gondwanaland reconstructions (e.g. Dearnley, 1966); these troughs could have been mul-tiple rift systems which did not evolve into oceans as we know them today. Tectonic instability may have been a consequence of the development of a more heterogeneous crust following diachronous addition of more sialic components.

It is commonly considered that greenstone belts are essentially downwarped syn-forms, containing numerous high-angle strike faults. This implies that the structure was largely controlled by vertical movements, involving (1) downbuckling of the greenstones into unstable primitive crust (e.g. MacGregor, 1951; Anhaeusser, 1971a), possibly facili-tated by numerous small convection cells in the hot Archaean mantle (e.g. Fyfe, 1973), and/or (2) ascent of granitoid diapirs (e.g. Glikson, 1972; Anhaeusser, 1973). The domi-nantly domal pattern of early granitoid—greenstone terrains may be reflecting a relatively constant crustal thickness with no major heterogeneities that could influence diapiric uprise of the granitoids. On the other hand, the more linear nature of many younger greenstone belts may reflect their formation in elongate troughs (rifts), the margins of which influenced granitoid uprise (cf. Archibald and Bettenay, 1977).

More complicated situations have been proposed for some greenstone belts. For example, horizontal nappe-like structures have been invoked at several localities in south-ern Africa (e.g. Stowe, 1974; Coward et al., 1976; compare with Ramsay, 1963), and in part of the Yilgarn (Fig. 6; Archibald et al., 1978; compare with Platt et al., 1978). The existence of nappes suggests large-scale horizontal tectonics of the type associated with mountain building in the modern global tectonic regime, but this raises several problems when applied to Archaean environments:

(1) The classic Phanerozoic nappe structures are not known in predominantly vol-canic sequences; they formed within largely sedimentary sequences as a result of major crustal shortening/thrusting.

(2) Molasse troughs and other important features of Phanerozoic mountain belts are absent from Archaean greenstone sequences.

(3) There is evidence, to be discussed in the remainder of this section, which militates against the existence of Phanerozoic-type global tectonics in the Archaean.

It is obvious that more detailed studies of these Archaean "nappes" are required, particularly to assess whether they could have formed in tectonic environments somewhat different from modern plate regimes, possibly as a result of emplacement of large granitoid domes.

A uniformitarian view of plate tectonics has been advocated by many geologists (e.g. Burke et al., 1976; Goodwin, 1973a; Talbot, 1973; Tarney et al., 1976; Windley, 1977). The main evidence cited is the broad petrochemical similarities between some Archaean metavolcanics and modern volcanics. This evidence is very much equivocal, however, as demonstrated by the conflicting interpretations of greenstone regimes given by the above-mentioned authors. O'Nions and Pankhurst (1978) emphasize the problems in linking magma types to specific modern tectonic regimes, and Nesbitt and Sun (1976) considered, on the basis of their detailed geochemical studies of Archaean komatiitic and tholeiitic rocks, that modern equivalents to greenstone sequences are lacking. Groves et al. (1978) concluded that the Archaean greenstone belts were not formed in environments analogous to modern marginal seas, or other plate tectonic settings; they pointed to the fact that intrusive granitoids formed penecontemporaneously over distances of roughly 700 km across the strike of greenstone belts in the eastern Yilgarn Block, but do not exhibit clear spatial—chemical variations of the type formed across modern plate boundaries.

Other lines of evidence which militate against the uniformitarian viewpoint of global tectonics are as follows:

(1) Whilst Archaean greenstones have certain broad petrochemical similarities to modern volcanics, they lack certain rock associations characteristic of Phanerozoic plate margin environments, namely ophiolite complexes with strong tectonic fabrics, sheeted dyke complexes, eclogites and glaucophane schists. Furthermore, there are very few instances of petrotectonic assemblages which suggest plate boundary conditions of the Phanerozoic type in rock sequences between 2.5 and 1.0 Gyr (e.g. Condie, 1976).

(2) Palaeomagnetic data for the Proterozoic suggest a single "supercontinent", or several closely spaced continental blocks (Piper, 1976; McElhinney and McWilliams, 1977; cf. Fig. 14). These data appear to preclude plate tectonic models involving convergence of previously widely separated (>1000 km) cratons; they do not rule out opening and closing of small intercratonic oceans with return of the continental nuclei to their same relative positions. This postulated "supercontinent" is not greatly different from the later Pangaea (Fig. 4), which probably formed after minor redistribution of sialic blocks (mainly the northern continents) in the late Proterozoic and Palaeozoic.

(3) Crustal province boundaries, structural grain and zones in which specific lithologies are concentrated (e.g. anorthosites, rapikivi granites, granulites and BIF) can be matched across many Precambrian belts and present continental margins (e.g. Hurley and Rand, 1969; Goodwin, 1973a; Piper, 1976; Condie, 1976). This would be very

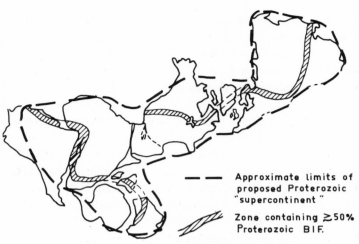

Fig. 14. Reconstruction of Proterozoic "supercontinent" (adapted from Piper, 1976). Zone of a BIF concentration is after Goodwin (1973a).

unlikely if there had been significant independent drift between continents or cratons through the Proterozoic.

The logical conclusion from these lines of evidence is that, for most of the Precambrian, the continental masses were in close proximity or welded into a single supercontinent, presumably surrounded by simatic, oceanic crust, and that any horizontal motions associated with the generation of the relatively high metamorphosed mobile belts between cratonic blocks must have been on a limited scale. The broad petrochemical similarities between Archaean and Phanerozoic (excluding continental-type) volcanics are explicable in terms of the development of generally similar melting conditions in different tectonic situations during Earth history. Thus, the prime controls on magma composition, given a fixed source composition, are T, p_{H_2O}, p_{total} in the source regions, plus fractionation history during ascent — it is not difficult to envisage similar conditions occurring in more than one tectonic regime.

If, as proposed by Ringwood (e.g. 1975), subduction of oceanic lithosphere is primed by density increase as basaltic rocks transform to eclogite, an absence of Phanerozoic-type plate tectonics in the Precambrian can possibly be explained in terms of high geothermal gradients which normally precluded eclogite formation (Green et al., 1975; Baer, 1977; cf. Fig. 10).

A note of caution should be inserted here concerning the size of the Earth in the Precambrian. It is normally assumed that the Earth has not expanded to any significant degree, although Carey (e.g. 1975, 1976) suggested post-Palaeozoic expansion. Glikson (1976) noted that a number of features of the Precambrian geological record —notably the rarity of evidence for Proterozoic oceanic lithosphere — warrant serious consideration of the possibility that, until ~1 Gyr ago, the Earth was significantly smaller than its present size. He envisaged that continental crust occupied most of the Earth's surface through the Proterozoic. This is a thought-provoking hypothesis which should be borne in mind in considerations of the Precambrian rock record, but it is impossible to discuss it in any more detail at the present state of knowledge; it is noteworthy that McElhinney et al. (1978) raise serious objections to Mesozoic expansion.

Interim summary

(1) The oldest crust presently recognized is in the North Atlantic craton and this formed ~3.9 Gyr BP. It comprises mainly tonalitic and trondhjemitic gneisses intruding older, metamorphosed volcano-sedimentary sequences (supracrustals).

(2) Sialic components were added to the crust throughout the Archaean; tonalitic and trondhjemitic granitoids were probably derived from anatexis of simatic crust, and more potassic from melting involving earlier-formed granitoids. There is also evidence for diachronous accumulation of Archaean greenstone sequences; ultramafic and mafic volcanics formed by extensive upper mantle melting and more silicic volcanics by fractionation of mafic magmas or crustal melting. Different relationships pertain between greenstones and intrusive granitoids, and between low- and high-grade Archaean terrains, at different stages of crustal evolution.

(3) With some exceptions, basal parts of greenstone sequences tend to have greater proportions of komatiitic peridotites and basalts than upper parts, which commonly have relatively high proportions of low-potassium tholeiites, calc-alkaline volcanics and volcaniclastic sedimentary rocks. High geothermal gradients are indicated by the widespread komatiites, but they need not have been more than a factor of two greater than in the present Earth.

(4) The oldest greenstones contain laminated cherty sediments and some probable evaporites; they formed in more stable environments than did upper greenstone sequences which contain relatively high-energy sediments, and which accumulated in elongated basins suggestive of rift zones. This change to less stable conditions may be reflecting the structural response to increasing heterogeneity in the crust as more sialic rocks were intruded into it. There is no conclusive evidence for extensive terrestrial volcanism in the Archaean, nor for modern-style plate tectonics.

ARCHAEAN MINERALIZATION

An overview

General reviews of Archaean mineralization have been presented by Boyle (1976) and Watson (1973, 1976). In addition, there have been several reviews of specific terrains (e.g. SouthAfrica — Anhaeusser, 1976a; Canada — Hutchinson et al., 1971; Goodwin,

1971; India — Radhakrishna, 1976), and reviews of specific types of deposits that are particularly important in Archaean terrains (e.g. volcanogenic Cu—Zn—Sangster and Scott, 1976; volcanic-associated Ni—Cu — Groves and Hudson, 1981). Here we emphasize the development of mineralization as an inherent part of the evolution of Archaean terrains as set out above, and attempt to define the most important parameters that control the distribution of ore deposits.

Archaean mineralization is largely a function of the high levels of igneous and exhalative activity during this era. The main metals represented are those enriched in ultramafic and mafic rocks (viz. Fe, Ni, Cr, Pt, Cu, Zn, Au), reflecting the importance of such rocks in the mineralised sequences and provenance regions. The deposits can be divided into four major groups:

(1) Strata-bound deposits associated with felsic to mafic volcanics: Fe, Au, Mn, Cu—Zn, Cu—Mo, Ba.

(2) Strata-bound and disseminated deposits in ultramafic/mafic flows and intrusives: Ni—Cu—(Pt), Cr.

(3) Deposits associated with granitoids and pegmatites: Sn, W, Li, Ta, Be.

(4) Deposits related to metamorphism and alteration of greenstones: Au, As, Hg, Sb, W, industrial minerals.

In addition, the stabilized Archaean cratons are sites for the development of other younger types of ore deposits. This aspect has been emphasized by Watson (1976). Examples include sedimentary Au—U ores, kimberlite-associated diamonds, carbonatite-associated mineralization of various types, large layered intrusives, and deposits related to Cainozoic weathering (e.g. bauxite, pisolitic iron ores, lateritic Ni (+Co), calcrete U, alluvial Sn—Ta and heavy-mineral beach sands).

Before dealing systematically with Archaean mineralization it is instructive to first summarize several important general features:

(1) Most economic deposits occur in the granitoid—greenstone terrains, whereas the higher-grade gneiss terrains are relatively poorly mineralized. In many cases, this probably reflects different environments of formation of supracrustals in these terrains, but another factor may be lower proportions of "eugeosynclinal-type" volcano-sedimentary sequences and high-level granitoids in deeply eroded high-grade terrains. High grades of metamorphism, *per se*, cannot explain the rarity of base-metal deposits in the gneiss terrains as, for example, some of the Western Australian Ni—Cu deposits occur in high amphibolite facies terrains (e.g. Barrett et al., 1977), and the huge Pb—Zn deposits at Broken Hill in New South Wales have survived granulite facies metamorphism (e.g. Binns, 1964; Both and Rutland, 1976).

(2) The most widespread types of mineralization in Archaean terrains are BIF-associated iron ores, BIF-associated gold ores, epigenetic gold (± copper) ores, minor chromite deposits and pegmatite-associated deposits.

(3) Certain deposits are concentrated in one or perhaps two Archaean cratons and are rare in the others. Examples include the large proportion of known volcanogenic Cu—

Zn deposits in the Superior Province, Canada, the abundance of volcanic- and dunite-associated Ni—Cu deposits of the Yilgarn Block, Western Australia and to a lesser extent Rhodesia, and the virtual absence of economic Archaean chromite deposits outside Rhodesia.

(4) The major deposits tend to be concentrated in relatively small areas of the cratons.

Points (3) and (4) may reflect true distributions, or they may be functions of limited exposure and intensity of exploration.

(5) Certain major base-metal deposit types are unimportant or absent in Archaean terrains. There are no "shale-hosted" Pb—Zn ores (McArthur, Mount Isa, Sullivan, Broken Hill, Rammelsberg types); "carbonate-hosted" Pb—Zn ores (Mississippi Valley, Alpine types); sedimentary Cu ores ("red bed", Copperbelt, Kupferschiefer types); Cyprus type Cu ores; U ores (the fluvial—deltaic U—Au ores of South Africa, Canada, and probably also Brazil, formed in the transition from the Archaean to the Proterozoic). Archaean porphyry (s.l.) Cu—Mo, Sn, W, Sb and Hg deposits are mainly small, low-grade and less common than in younger periods, and most significant Archaean volcanogenic massive sulphide deposits are Pb- and sulphate-poor.

Deposits associated with felsic to mafic volcanics

Banded iron formation. Recent publications dealing with the features and genesis of BIF include Gross (1967), Stanton (1972), the special issue of *Economic Geology* edited by James and Sims (1973), the proceedings of the Kiev Symposium (Anonymous, 1973) and Boyle (1976). BIF are protore for rich iron deposits which formed largely by structurally controlled leaching of silica, and some have significant gold concentrations.

Archaean BIF are characteristically spatially associated with felsic to mafic volcanic rocks. Although the BIF are most common in the "eugeosynclinal" type younger greenstone sequences, they are important in the earliest Archaean Isua Supracrustals (the oldest known significant mineralization), and they also occur in early Archaean greenstone sequences in the Pilbara and Kaapvaal cratons. Following Canadian terminology (Gross, 1967; cf. Wolf, 1981, Chapter 1, Vol. 8, fig.52 and table LXX, in this Handbook), these volcanic-associated BIF are referred to here as Algoma type, and the more abundant early Proterozoic BIF, which commonly do not have a close association with volcanics, are distinguished as Superior type (cf. discussion in section on Archaean—Proterozoic transition). The Algoma-type deposits are most important in Ontario (Michipicoten and other districts); they supply roughly one-quarter of Canada's iron. Other major Algoma-type deposits include those at Mount Goldsworthy in the Pilbara Block of Western Australia and the Greater Krivoy Rog district of the Ukranian shield of the USSR. Whilst Algoma-type BIF are mainly of Archaean age, there are younger examples such as the Ordovician New Brunswick deposits, Canada.

Archaean BIF are characteristically smaller than the major Proterozoic examples.

The former range from several hundred metres to less than one metre thick, and rarely extend more than a few kilometres along strike, although en echelon lenses can occur over tens of kilometres strike length. The iron content of BIF is commonly in the range of 25–40% Fe, but many ore-grade enriched zones contain 60–65% Fe.

Many Archaean BIF show a general zoning, with iron sulphide, carbonate and silicate types closer to volcanic centres than the predominant oxide facies. Whereas the BIF are mainly thin-banded, the carbonate facies is characterically relatively massive. Goodwin (e.g. 1973b) has proposed a conceptual model in which the changes from sulphide (mainly pyrite + pyrrhotite), through carbonate (mainly siderite + ankerite), to oxide (mainly magnetite ± hematite) types are reflecting decreasing water depths in the depositional environments. However, decreasing sulphur activity with distance from outlets of hydrothermal springs is probably a more important control.

There can be little doubt that the formation of Archaean BIF is related to the high levels of igneous and exhalative activity characterizing this era. Fluids enriched in Fe, Si, Au, etc., could have been derived from degassing of cooling lavas, release of aqueous phases from ascending magmas, or circulation of surface waters within sequences of fragmental and fractured volcanics. The latter two mechanisms are feasible in sequences containing highly permeable felsic volcaniclastic rocks, which commonly exhibit discrete zones of strong alteration. However, such evidence for discrete exhalative conduits is rare in mafic-dominated sequences (e.g. Boyle, 1976), suggesting that degassing of lavas and lava–seawater interactions may have been more important than convective water circulation. Whatever the fluid source, precipitation of iron formations would have been induced by changing physicochemical conditions after the hydrothermal fluids debouched into the hydrosphere.

Isotope studies of sulphide minerals from Archaean BIF and sulphide-rich, tuffaceous sediments (Fig. 15; Goodwin et al., 1976; Donnelly et al., 1978; Fripp et al., 1979; Schidlowski, 1979; Seccombe et al., 1977) indicate that the great bulk of the $\delta^{34}S$ values are in the "magmatic" range (0 ± 4‰; see Fig. 15). These imply magmatic emanations, dissolution of sulphides from greenstone sequences, or the presence of reduced sulphur species of deep-seated origin in surface waters circulating in the volcanic pile. However, Goodwin et al. (1976) called on bacterial sulphate reduction in the Michipicoten iron formation, Canada, on the basis of a minority of values outside the "magmatic" range (Fig. 15); these are the oldest rocks (ca. 2.7 Gyr) in which it has been claimed that there is isotopic evidence for significant biological sulphate reduction.

Cloud (e.g. 1973, 1976) believes that the oxide facies BIF formed because Fe^{2+} in the ancient hydrosphere was a "sink" for oxygen produced by microbial photosynthesis. It is noteworthy that the main primary oxide mineral in most Archaean BIF appears to have been magnetite which will form in the absence of free oxygen, under low sulphur and carbonate activity conditions. Thus, there is no evidence that photosynthetic processes were already established during deposition of the ∼3.9 Gyr old Isua BIF, which contains magnetite to the exclusion of other oxide minerals (Allart, 1976); admittedly

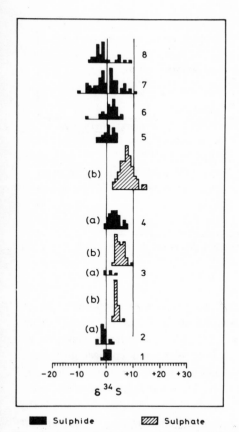

Sulphide   Sulphate

Fig. 15 Sulphur-isotope compositions of Archaean sedimentary sulphides and sulphates, arranged in *general* order of decreasing age from bottom to top (adapted from Schidlowski, 1979). These are concentrated in the range for magmatic derivation, in contrast to sedimentary sulphur in most younger rocks. *1* = Isua BIF, Greenland (Schidlowski, 1979); *2* = North Pole area, Pilbara Block, Western Australia (I.B. Lambert et al., 1978); *3* = Barberton area, Swaziland System, South Africa (Perry et al., 1971; Vinogradov et al., 1976; I.B. Lambert et al., 1978); *4* = Aldan Shield, Siberia, lower and upper suites (Vinogradov et al., 1976); *5* = Rhodesian BIF, older and younger sequences (Fripp et al., 1979); *6* = Yilgarn Block, Western Australia (Donnelly et al., 1978); *7* and *8* = Michipicoten and Woman River BIF, Superior Province, Canada (Goodwin et al., 1976).

hematite could have been present originally but destroyed during metamorphism. The occurrence of some primary hematite in BIF < 3.5 Gyr old implies the presence of locally oxygenated waters, probably as a result of microbial photosynthesis, but the low Eh necessary for coexistence of magnetite, hematite and water militates against the existence of an oxygenated atmosphere. Oxidized hydrosphere could have existed in a world-wide surficial zone, or locally in seas with high levels of photosynthetic activity (Fig. 16).

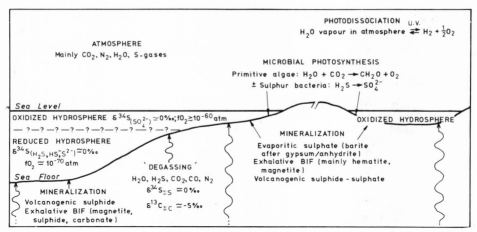

Fig. 16. Schematic representation of preferred geobiochemical model for the Archaean. Refer to text for discussion.

Gold in banded iron formation. Gold mineralization associated with arsenopyrite in sulphide- and carbonate-facies iron formation occurs in Rhodesia (Fripp, 1976a, b), Canada (Hutchinson, et al., 1971) and Australia (Woodall, 1975). The Rhodesian examples are the best documented, and although they occur in both older and younger greenstone sequences as defined by structural and isotopic studies, there is a regional-scale stratigraphic control, with mineralization being best developed in lower sequences. Most deposits contain both stratiform and vein mineralization and it is thought that the latter is the result of remobilization of original syngenetic/diagenetic mineralization by subsequent metamorphic and hydrothermal processes (e.g. Sawkins and Rye, 1974; Boyle, 1976; Fripp, 1976a, b). Many authors consider the gold was transported as chloride complexes in the Fe-rich hydrothermal fluids which formed the BIF, but Seward (1979) reasoned that thio-complexes are more likely.

Recent studies of inter-flow sulphidic metasediments within dominantly komatiitic sequences in Western Australia show that many of them are significantly enriched in Au, although not of ore grade (Bavinton and Keays, 1978; Chapman and Groves, 1979).

Manganese. It appears that Archaean manganese oxide deposits are rare. Most Algoma-type iron formations contain <0.5% MnO (Boyle, 1976). Manganese deposits of Precambrian age are scattered throughout India, mostly associated with phyllites, schists, gneisses, marbles and cherty quartzites. According to Roy (1966, 1976), some of them formed by volcanic processes in the Archaean, and these are much smaller than the sedimentary deposits of Proterozoic and younger ages. There has been extensive modification of the deposits by later enrichment processes.

The rarity of manganese deposits in Archaean sequences may be attributed to the dominance of soluble Mn^{2+} in most depositional basins because of generally low f_{O_2} conditions. Even in modern oxygenated environments, oxidation of Mn^{2+} is very sluggish, (e.g. Seyfried and Bischoff, 1977), leading to distinctive zoning of Fe/Mn rates in many metalliferous sea-floor deposits (e.g. Fryer and Hutchinson, 1976).

Volcanogenic base-metal sulphide deposits. In Archaean shields, the only major volcanogenic base-metal deposits are massive Cu–Zn sulphide ores associated with felsic volcanic centres in younger greenstone sequences (Hutchinson, 1973; Boyle, 1976; Sangster and Scott, 1976). These are most important in the Abitibi Belt of the Superior Province, Canada, where their age is not generally known with any accuracy, but galena Pb-isotope model ages clustering around 2.9 Gyr are the best guides (Sangster and Scott, 1976). Massive Cu–Zn deposits also occur in the adjacent greenstone belts of the Superior Province, and minor occurrences are recorded from the Slave Province (Canada), southern Africa (Anhaeusser, 1976a), the Pilbara Block (Reynolds et al., 1975), and two as yet uneconomic deposits were recently discovered in the Yilgarn Block (Golden Grove and Teutonic Bore). These deposits are variably metamorphosed and deformed, but the deposits in the famous Noranda field of the Abitibi Belt display many well-preserved primary features.

There are no Cyprus-type, ophiolite-associated copper deposits in Archaean terrains despite the common predominance of mafic–ultramafic volcanic sequences with certain gross similarities to ophiolite sequences. This may be a function of the absence in the Archaean of mid-ocean ridge environments, similar to those in which the Cyprus-type deposits are thought to have formed in the Phanerozoic. These are zones in which multiple tensional fractures enable extensive water circulation, sea-floor metamorphism and metal leaching (e.g. Chapman and Spooner, 1977). A lack of convective water circulation in the Archaean other than in areas of explosive volcanism is borne out by the lack of discrete alteration zones in mafic sequences and the petrographic evidence for nonprogressive metamorphism in greenstone belts in the Yilgarn Block (Binns et al., 1976), although it is not clear if this is an Archaean-wide feature. Another factor may have been that surface waters circulating within the greenstones had low chloride concentrations relative to modern seawater (cf. earlier section on development of hydrosphere). The Archaean mineralization most similar to the Cyprus-type deposits is pyritic and pyrrhotitic cherty units interbedded in mafic–ultramafic sequences, but these generally have only minor Cu enrichments, although some are Ni-, Zn- or Au-rich (e.g. Groves and Hudson, 1981). Appel (1979) describes the oldest known strata-bound copper sulphides in iron formations and basaltic tuffs from the pre 3.76 Gyr Isua sequence of West Greenland. Evidence is presented that these formed by direct submarine exhalation rather than by leaching of volcanic sequences by convective water circulation.

Significant Cu–Zn deposits occur in clusters around felsic volcanic centres, normally of calc-alkaline affinities in the Superior Province (e.g. Descarreaux, 1973). Most are closely associated with felsic agglomerates and coarse tuffs (proximal type), although some are more immediately associated with fine tuffs and volcaniclastic sediments (distal type). They are also associated with pyritic sediments (sulphide exhalites of Ridler and Shilts, 1974), and there are no known cases of major Cu–Zn deposits in zones dominated by oxide-type iron formations (Sangster, 1978). A submarine environment is indicated by associated pillow basalts, greywackes and cherty sediments. Economic deposits are commonly restricted to a particular volcanic cycle in a given region (e.g. Sharpe, 1968; Spence and De Rosen-Spence, 1975); whilst mineralization can also occur in younger volcanics, this is usually subeconomic and characterized by higher Zn/Cu ratios. The main features of the Archaean volcanogenic Cu–Zn deposits are summarized schematically in Fig. 17.

At a detailed scale, there are considerable variations between individual massive sulphide deposits. However, in general terms, each mine consists of a number of strata-bound, massive lenses. Sangster (1977) noted that 88% of the Canadian deposits contain <10% combined Cu + Pb + Zn, the most likely combined metal grade is in the 5–6% range, and Pb grades are commonly negligible. He also emphasized that Zn shows the widest range of grades and the Zn-rich deposits have a higher average size; those with Zn/(Zn + Cu) > 0.5 average 6.2 million tonnes compared with 4.2 million tonnes for deposits with lower Zn/(Zn + Cu) ratios. The largest massive sulphide deposit so far discovered is Kidd Creek in the Abitibi Belt (e.g. R.R. Walker and Mannard, 1974), which as a total size of over 120 million tonnes at roughly 9% Zn, 1.5% Cu, 0.3% Pb and 4.2

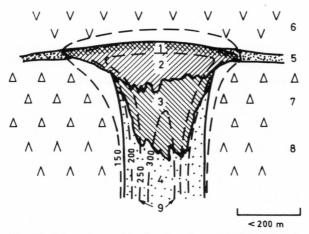

Fig. 17. Schematic section showing characteristic general features of Archaean volcanogenic Cu–Zu sulphide deposits. *1* = banded pyrite, sphalerite (± galena) ore; *2* = massive chalcopyrite, pyrite ore; *3* = stringer chalcopyrite, pyrrhotite, pyrite ± magnetite ore; *4* = alteration pipe; *5* = pyritic chert bed; *6* = hanging-wall volcanics; *7* = felsic volcanics, including agglomerates, coarse lithic tuffs and quartz porphyries (altered); *8* = footwall andesites and basalt (altered); *9* = hypothetical temperature during formation of mineralization from hydrothermal fluids ascending to the sea floor (adapted from Large, 1977).

oz/tonne Ag. At the opposite extreme, many mines contain very small massive sulphide lenses, commonly a few hundred thousand tonnes, with maximum dimensions less than ~150 m.

The upper contacts of the massive sulphide lenses are usually sharp, but not necessarily planar. Laterally, many of the massive sulphide bodies grade rapidly (~0.1 to 2.5 m) into cherty, pyritic tuff beds. The central and lower portions of the massive sulphide lenses commonly have a brecciated, blocky appearance. Towards the tops and lateral margins, there is commonly some crude banding, mainly defined by alternating sphalerite- and pyrite-rich layers, load casts, flame-structures, graded-bedding and angular fragments of ore in the base of overlying flows. Downwards, the proximal massive sulphides grade through semi-massive ore into stringer mineralization (veinlets typically ~2 to ~30 cm wide), which can persist for hundreds of metres into the footwall volcanics. Less commonly, the massive sulphides are underlain by a thin zone of stringer disseminated mineralization which grades into barren footwall rocks within 1 m or so.

Distinct mineralogical and chemical zoning is evident within most deposits. The massive ores are relatively Cu-rich at the centre and grade out to Zn-rich. Their mineralogy is essentially pyrite, pyrrhotite, chalcopyrite and sphalerite, with minor magnetite and traces of galena, arsenopyrite, mackinawite and native silver; silicate gangue minerals include quartz, chlorite, biotite and amphibole. The footwall stringer mineralization has high Cu/Zn ratios and consists of pyrrhotite and chalcopyrite with lesser pyrite, and sphalerite, with or without magnetite. There is commonly a thin Zn-rich outer stringer zone containing pyrite, sphalerite and minor chalcopyrite.

Footwall alteration is normally quite marked in pipe-like zones surrounding the stringer mineralization. The main alteration minerals are chlorite and sericite, but there is also some silicification. Biotite, cordierite, anthophyllite, garnet and amphibole are less common alteration minerals, which presumably formed during later thermal metamorphism. Alteration pipes in the footwall rocks have been observed to change direction suddenly; those beneath Millenbach and Amulet in the Noranda field appear to be heading towards a trondhjemitic stock several kilometres to the west (Flavrian Lake Granite). Beyond the obvious alteration pipes, there is extensive addition of Fe and Mg, depletion of Na and Ca and variable addition or depletion of Si, in the footwall volcanics (e.g. E.M. Cameron, 1975). Hydrothermal alteration of the hanging-wall rocks is weak or not observed.

The features of the massive sulphide deposits are in accord with their formation from hydrothermal solutions on the sea floor and within "feeder zones" in the footwall volcanics. Explosive felsic volcanism evidently played a major role in fracturing the pile and opening up channel ways for rapid ascent of the ore solutions, in contrast with the situation at mafic–ultramafic eruptive sites. The sulphides probably formed by precipitation at and below the seafloor, in reverse order of solubility along sharp temperature-composition gradients within the feeder zones (Fig. 17; Large, 1977).

Existing information is equivocal on whether the metals were leached from underlying rocks or released during fractionation of subvolcanic magmas. Base-metal analyses of the volcanics generally show normal to high Cu and Zn contents. However, this is not conclusive evidence against leaching of metals by heated seawater circulating in the volcanic pile, as it is possible that (1) the volcanics originally had higher metal contents, or (2) the waters circulated mainly along faults, fissures and volcanic breccia zones, with leaching being restricted largely to the vicinities of such channel ways. In the latter regard, it is important that there is a common association between mineralization and faults. Viable alternative models are that leaching was concentrated within and immediately around hot, subvolcanic intrusives, or that the bulk of the metals were supplied by magmatic fluids released from fractionating magmas. An involvement of subvolcanic granitoids is supported for the deposits in the Noranda field that have alteration pipes extending towards such a body, which contains low-grade disseminated chalcopyrite.

$\delta^{34}S$ values of sulphides from the massive deposits are close to 0‰, strongly suggesting sulphur of magmatic derivation. This again could have been derived directly from fractionating magmas, leached from earlier-formed greenstones or present as dissolved sulphide of deep-seated origin in surface waters circulating in the volcanics. There is no evidence for incorporation of isotopically heavy seawater sulphate, as there is in Phanerozoic volcanogenic deposits (cf. Sangster, 1976).

An attractive genetic model involves explosive release of metal- and sulphur-enriched fluids from ascending magmas as fluid pressures exceed load pressures, followed by mixing of these fluids with seawater circulating in the volcanic pile and rapid ascent of the resultant mixed brines to the surface. This model represents an extension, to submarine volcanic trough environments, of concepts of the genesis of porphyry Cu–Mo deposits in geanticlinal zones [cf. discussion of porphyry (s.l.) Cu–Mo deposits, next section].

The Archaean volcanogenic deposits differ in a number of respects from most younger volcanogenic deposits associated with felsic volcanics. For instance, there is generally greater Mg-metasomatism associated with the Archaean mineralization, probably reflecting leaching of the abundant andesites and basalts and/or ultramafics beneath the deposits. Secondly, Sangster (1972) and Hutchinson (1973) emphasized the trend within calc-alkaline-associated volcanogenic ores from largely Cu–Zn–Au–Ag deposits without sulphates in the Archaean, to Cu–Pb–Zn–Ag deposits, commonly with sulphates, in Phanerozoic terrains. This has been attributed (e.g. Hutchinson, 1973; Sangster and Scott, 1976; Solomon, 1976) to the low Pb content of either the underlying volcanic pile and/or mantle source in the Archaean, and to the lack of oxidized sulphur species in Archaean waters. The Broken Hill deposit in New South Wales, which has a metamorphic age of ~1.7 Gyr, is the oldest known Pb-rich ore deposit and it appears that upward concentration of Pb in the lithosphere was a prerequisite for formation of major deposits of this metal. It is noteworthy, however, that some of the oldest-known, but minor, volcanogenic deposits in the Pilbara Block (e.g. the ca. 3.5 Gyr Big Stubby deposit – Reynolds et al., 1975; Sangster and Brook, 1977) are rich in Pb and Ag and contain significant barite.

The occurrence of the barite is particularly significant in that these deposits were formed in a volcanic terrain containing shallow-water oxidized environments (as deduced from the nearby North Pole barite occurrences) that could represent a source of SO_4^{2-} in the ore fluid (Fig. 16). The contrast between the Pilbara occurrences and the Canadian deposits further emphasizes that there were important differences between depositional basins in Archaean terrains. The rarity and limited economic significance of the sulphate-bearing volcanogenic deposits in the most-primitive, shallow-water greenstone basins suggests that the latter were not normally suitable repositories for volcanogenic Cu—Zn mineralization.

Most Phanerozoic volcanogenic deposits occur in tectonic situations interpreted as being related to subduction or spreading zones (e.g. Sawkins, 1972; Sillitoe, 1972; Mitchell and Garson, 1976), and modern calc-alkaline rocks are currently virtually restricted to the former tectonic setting. Despite this, there is no definitive evidence that the Archaean volcanogenic Cu—Zn deposits were related to subduction processes; although the geochemistry of Archaean andesite—rhyolite sequences is commonly broadly similar to modern suites, both the intercalated volcanic associations and sedimentary sequences contrast with modern analogues. It is likely that the factors necessary for volcanogenic ore formation existed in different tectonic environments at different stages of crustal evolution. Apart from the presence of metal- and sulphur-rich fluids, the main factors required for their formation appear to be heat sources, submarine depositional basins and permeability for rapid ascent of metalliferous fluids. Volcanism was obviously the heat source, and explosive eruptions during felsic volcanism were probably largely instrumental in generating the permeability.

Porphyry (s.l.) copper—molybdenum. Low-grade Cu—Mo mineralization occurs in some Archaean felsic volcanic centres, commonly associated with sub-volcanic, porphyritic granitoids. Several instances of this general type of mineralization have been described from the Superior and Slave Provinces, Canada (e.g. Kirkham, 1972; Boyle, 1976; Findlay, 1976; Findlay and Ayres, 1977; J.F. Davies and Lutha, 1978), and similar mineralization is known, but not yet described, from the Pilbara Block, Western Australia.

Most such mineralization occurs in disseminations, stockworks, shears and quartz veins in zones of widespread hydrothermal alteration, and it is rarely of economic grade. The only major deposit to be mined is the Parmour (formerly McIntyre) Mine, Ontario, where grades of 0.8% Cu and 0.03% Mo occur in a lens of altered felsic schist (Davies and Lutha, op. cit.); disseminated chalcopyrite—bornite—molybdenite mineralization is associated with pyrite, and adjacent to gold-bearing quartz veins.

This Archaean mineralization exhibits broad similarities in form, mineralogy and alteration to post-Palaeozoic porphyry Cu—Mo deposits, although some mineralization and alteration may post-date early deformation and metamorphism (e.g. Davies and Lutha, op. cit.). The scarcity of this mineralization in the Archaean is probably real, and not a function of exposure. Most of the younger porphyry Cu—Mo deposits are considered to have formed in geanticlinal zones (at active continental margins or in island arcs — e.g. Hollister, 1975; Sillitoe, 1976). Whilst this renders it feasible that the rapidly decreasing importance of porphyry deposits in sequences older than Mesozoic is reflecting the relative ease with which mountain belts are destroyed by erosion, it does not appear that such an explanation can be extrapolated back as far as the Archaean because there is no firm evidence for extensive geanticlinal volcanic belts in this era. It seems more likely that the predominantly submarine nature of the Archaean volcanic belts favoured formation of volcanogenic massive sulphide deposits rather than porphyry-type mineralization. It was proposed above that rapid ascent of felsic intrusives into such submarine volcanic belts led to the explosive release of metalliferous fluids which, together with heated "seawater" in the volcanic pile, migrated through fractured and fragmented volcanics to the sea floor.

Barite. Strata-bound mineralization is rare in Archaean volcanic—sedimentary sequences that show firm evidence for very shallow-water deposition; bedded barite is the only important type of deposit apart from BIF.

Strata-bound barite is preserved in low-strain, low-metamorphic-grade parts of the Swaziland Supergroup of South Africa (e.g. Perry et al., 1971; Heinrichs and Reimer, 1977), the east Pilbara Block, Western Australia (e.g. Hickman, 1973; Dunlop, 1978) and the Peninsular Gneissic Complex, India (Radhakrishna and Vasudev, 1977). It is interesting that all of the sequences containing barite have ages greater than 3.0 Gyr, with some as old as ~3.5 Gyr (e.g. Jahn and Shih, 1974; Sangster and Brook, 1977; Radhakrishna and Vasudev, 1977; Pidgeon, 1978a).

The largest Archaean barite deposits are at North Pole in the Pilbara region, where barite beds lie within a ~30 m thick sequence consisting of laminated chert, silicified arenite and conglomerate, between slightly metamorphosed mafic and ultramafic volcanics (Hickman, 1973; Dunlop, 1978). Stratigraphic correlations suggest the North Pole sequence is of similar age to felsic volcanics of the nearby Duffer Formation, which has been dated at 3.45 Gyr (Pidgeon, 1978a; zircon age). Various structures and textures suggest that the bedded barite and chert replaced an original evaporitic gypsum—carbonate sequence (Dunlop, op. cit.). The barite beds are composed of bottom-nucleated crystal groups, up to 20 cm in length, radiating across bedding (Fig. 18). These contrast markedly with barite from both volcanic-exhalative and primary sedimentary deposits (Shawe et al., 1969; Igarishi et al., 1974), but they strongly resemble "cavoli", cabbage- or cauliflower-like gypsum in evaporite deposits (e.g. Richter-Bernberg, 1973; Schreiber and Kinsman, 1975; Schreiber et al., 1976). Some North Pole barite crystals have swallow-tail twins and where it has been possible to measure interfacial angles on undeformed single crystals, these have indicated crystal faces typical of gypsum but not recorded for barite (Fig. 18). That chert replaced gypsiferous carbonate sediments at North Pole is indicated by relic carbonate which is partly replaced by silica, stylotitized dolomite rhombs, rhombic voids, and clear cryptocrystalline pseudomorphs after single or twinned

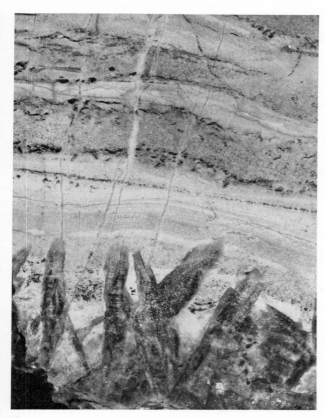

Fig. 18. Section of stratiform barite–chert horizon from North Pole showing large barite pseudo-morphs after gypsum, and overlying laminated (possibly stromatolitic) chert. Universal stage measurements confirm that undeformed barite crystals have gypsum interfacial angles. Photograph courtesy of J.S.R. Dunlop.

gypsum crystals (Fig. 18). Peritidal deposition of chert precursors is implied by laminated (possibly stromatolitic) bedding, graded- and cross-bedding, scour-and-fill structures, intraclast and edgewise conglomerates, ooid grainstones, ripple marks and probable desiccation features (Dunlop, 1978).

Synsedimentary formation of the precursor gypsum beds is indicated by eroded crystal tops, and sediment drapes and fills between crystals (Fig. 18); intraformational conglomerates containing clasts of bedded barite attest to early lithification of precursor gypsum, if not early baritization. The origin of the Ba^{2+} is uncertain, but it may have been derived from exhalations related to felsic volcanism, or from diagenetic-burial metamorphic reactions.

The Barberton barite occurrences exhibit generally similar features, leading Dunlop and Groves (1978) to suggest they also formed by diagenetic replacement of gypsum in an evaporite sequence; Lowe and Knauth (1977) recorded possible evaporite pseudomorphs elsewhere in the Swaziland supergroup.

The barite occurrences are important as they indicate the presence of at least local sources of sulphate in Archaean basins. Isotopic data (Perry et al., 1971; I.B. Lambert, 1978; I.B. Lambert et al., 1978; see Fig. 15) indicate that the sulphate was derived by rapid and virtually complete surficial oxidation of juvenile sulphur ($\delta^{34}S \simeq 0\%$) in restricted basins. Significantly, sulphur- isotope data imply that the minor sulphides associated with barite were not derived by bacterial reduction of SO_4^{2-}, but apparently from minor unoxidized H_2S in the depositional basin (I.B. Lambert et al., 1978). The evidence for precursor evaporative gypsum necessitates high concentrations of SO_4^{2-} in the basin waters, but not necessarily the presence of free oxygen in the hydrosphere nor the stable existence of oxygen in the contemporaneous atmosphere. The oxidation of reduced sulphur species of photosynthetic activity of sulphur bacteria or blue-green algae is in accord with the observed presence of microfossils both at North Pole (Dunlop et al., 1978) and at Barberton (M.D. Muir and Grant, 1976), and with the carbon-isotopic evidence for photosynthesis before 3.3 Gyr ago (e.g. Schidlowski et al., 1975; Schidlowski, 1976). Photodissociation of water vapour in the atmosphere by U.V. radiation may have led to incorporation of some oxygen into the hydrosphere, but Schidlowski (1976) argues that this is unlikely to have been a significant process.

In summary, the bedded barite deposits formed in the early Archaean as a result of (1) the existence of stable, shallow-water, evaporative environments; (2) the influx of juvenile sulphurous exhalations, reflecting generally high levels of degassing and volcanism; and (3) surficial oxidation, probably involving microbial photosynthesis, which gave rise to at least locally high sulphate/sulphide ratios in the hydrosphere (Fig. 16).

Deposits in ultramafic–mafic flows and intrusives

Volcanic-associated nickel–iron–copper. The major economic interest in ultramafic–mafic sequences over the past decade or so has centred on recently recognized classes of Ni–Cu (± platinoid) deposits associated with the ultramafic members of the komatiite suite. There are two major types of deposits; those intimately associated with komatiitic peridotite flow sequences, and those associated with dunitic pods of uncertain origin. Of lesser importance are deposits in stratiform intrusions.

It is appropriate to present a relatively detailed discussion of these nickel sulphide deposits as, by virtue of their recent recognition, they are not generally well-known to the international audience.

Volcanic-associated Ni–Fe–Cu sulphide deposits (Groves and Hudson, 1981) occur in groups within greenstone sequences for which radiometric dates > 3 Gyr have not yet been reported. These sequences contain sulphide-rich, tuffaceous and exhalative sediments, but rarely oxide-facies iron formation. Their geographic distribution is different from that of the volcanogenic Cu–Zn deposits. They are best represented in the

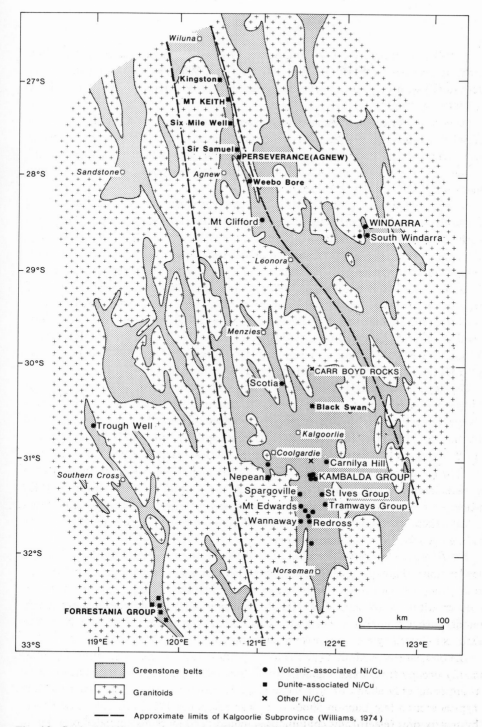

Fig. 19. Geological sketch map of the Eastern Goldfields Province of the Yilgarn Block, Western Australia, showing the distribution of volcanic-associated and dunite associated Ni—Fe—Cu sulphide deposits in relation to greenstone belts and the Kalgoorlie Subprovince.

Yilgarn Block (e.g. Kambalda, Windarra, Scotia, Nepean, Mt. Edwards, Spargoville, Redross, Wannaway), less well developed in Rhodesia (e.g. Shangani, Damba-Inyati, Trojan, Hunters Road) with only minor deposits in the Superior Province of Canada (e.g. Langmuir, Texmont, Alexo).

In Western Australia, the major deposits all occur within, or adjacent to, a single structural province (Fig. 19), the Kalgoorlie subprovince of I.R. Williams (1974), which probably represents an important zone of rifting within the greenstone basins (e.g. Archibald et al., 1978). Deposits in Canada and Rhodesia may also occur within rifting environments (e.g. Goodwin, 1977; Wilson, in prep.).

All known deposits have suffered regional metamorphism and deformation with the economically most important deposits in Western Australia being confined to upper greenschist and amphibolite facies metamorphic domains, commonly of high strain (e.g. Barrett et al., 1977). The smaller Canadian deposits and lower tenor Rhodesian deposits commonly occur in greenschist (e.g. Williams, 1979) or even subgreenschist facies environments (e.g. J.E. Muir and Comba, 1979). The Kambalda deposits that occur in an upper greenschist—lower amphibolite transition facies, low-strain environment may be taken as the type examples as they are the best explored, economically the most significant and the best documented (e.g. Woodall and Travis, 1969; Ewers and Hudson, 1972; Keele and Nickel, 1974; Ross and Hopkins, 1975; Barrett et al., 1977; Bavinton and Keays, 1978; Ross and Keays, 1979; Marston and Kay, in press).

The deposits show an intimate relationship with sequences of komatiitic volcanics that form part of a mixed tholeiite—komatiite sequence or a new cycle of komatiitic volcanism. Mineralization appears to be stratigraphically controlled at the regional and local scale (e.g. Pyke and Middleton, 1971; Gemuts and Theron, 1975; Wilson, in prep.; D.A.C. Williams, 1979). Sulphide ores occur at or towards the base of thick (generally ca. 50 m), highly magnesian (>35% MgO volatile-free) komatiitic peridotite units that in most cases represent the basal (less commonly the second or third) flows in thick sequences of progressively thinner and less magnesian peridotitic and pyroxenitic flows (Fig. 20). These have upper, spinifex-textured zones (Plate I, A) and lower cumulate zones; their geochemistry is summarized in Fig. 9.

The flows are thought to be the result of fissure eruption (e.g. Fig. 21). Although no feeder zones have yet been recorded for mineralized ultramafic units in Western Australia, they are known in Rhodesia and in one instance (Shangani), mineralization may occur within a volcanic rock (D.A.C. Williams, 1979). Sulphidic metasediments commonly occupy interflow positions (Figs. 20, 21) and the ultramafic komatiite sequence is followed by a mixed tholeiitic and komatiitic basalt suite.

The ores, largely pyrrhotite—pentlandite—pyrite—chalcopyrite—magnetite/chromite, commonly occupy trough-like embayments in footwall rocks from which interflow sediments are commonly absent or attenuated (Fig. 21). There is considerable variation, but in a typical section (e.g Lunnon Shoot — Ross and Hopkins, 1975) massive ores are overlain by matrix ores (electrically continuous sulphides) and more disseminated ores; inter-

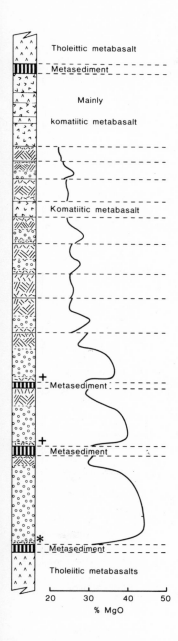

Fig. 20. Schematic section showing a hypothetical komatiitic ultramafic flow sequence with basal interflow metasediments and major (∗) and relatively minor (+) ores. Note that, in the real situation, there are commonly many more flow units, particularly thin (<5 m) units. From Groves and Hudson (1981).

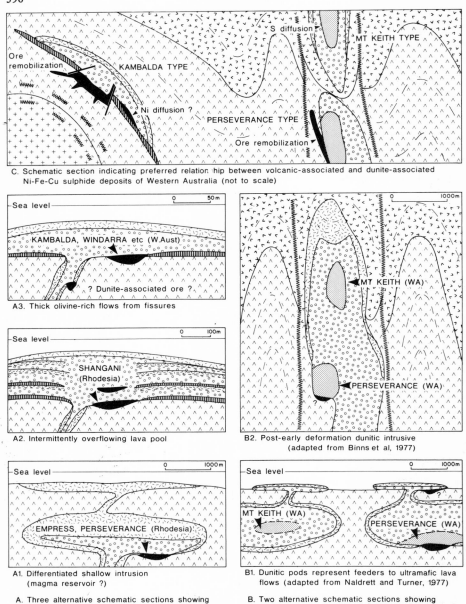

C. Schematic section indicating preferred relationship between volcanic-associated and dunite-associated Ni-Fe-Cu sulphide deposits of Western Australia (not to scale)

A3. Thick olivine-rich flows from fissures

A2. Intermittently overflowing lava pool

B2. Post-early deformation dunitic intrusive (adapted from Binns et al, 1977)

A1. Differentiated shallow intrusion (magma reservoir ?)

B1. Dunitic pods represent feeders to ultramafic lava flows (adapted from Naldrett and Turner, 1977)

A. Three alternative schematic sections showing formation of volcanic-associated Ni-Fe-Cu sulphide ores (adapted from Naldrett, 1973)

B. Two alternative schematic sections showing formation of dunite-associated Ni-Fe-Cu sulphide ores

Granitoid

Upper part of greenstone sequence (mafic-felsic association)

Lower part of greenstone sequence (mafic-ultramafic association)

Peridotite/pyroxenite (gabbro)

Dunite/peridotite

Sulphidic metasediments

Disseminated ore

Massive/matrix ore

Shear zone

Fault

Fig. 21. Schematic diagram showing a number of possible relationships for generation of volcanic-associated and dunite-associated Ni—Fe—Cu sulphide deposits. Refer to text for discussion.

pillow sulphides may be developed in underlying metabasalts. Individual ore shoots are generally small (<5 million tonnes ore) but high-grade (2—4% bulk Ni), although larger (up to 20 million tonnes ore) deposits of lower grade (≤1% Ni) do occur in Rhodesia. Geochemical parameters of ores are generally consistent. Ni/Cu ratios are normally in the range 10—15 (e.g. Barrett et al., 1977). S/Se ratios and $\delta^{34}S$ values are commonly in the normal magmatic range (e.g. Seccombe et al., 1977; Donnelly et al., 1978). Precious-metal tenors are significantly high (e.g. Ross and Keays, 1979) and there are distinctive chondritic precious-metal abundance patterns (e.g. Naldrett et al., 1979). There may be systematic variations through the ore zone with Cu and Pd enriched and Ir depleted in matrix ore relative to massive ore.

The ore bodies are typically situated around granitoid-cored domes and their orientation is structurally controlled; the ores have been remobilized to varying degrees. In some environments, ores have been upgraded or their compositions have been altered during metamorphism, particularly the disseminated deposits and, locally massive ores have formed, but there is no definitive evidence that the majority of massive ores formed by metamorphic processes.

Most features of the ores are consistent with derivation from oxy-sulphide liquid droplets carried by phenocryst-rich komatiitic ultrabasic melts that were extruded on the sea floor and concentrated in depressions (Fig. 21, models A2 and A3). Various genetic models are discussed by Naldrett (1973), Ross and Hopkins (1975), Groves et al. (1979), Usselman et al. (1979), and Marston and Kay (in press). Gravity separation of olivine phenocrysts and immiscible oxy-sulphide liquid droplets during eruption, with differential flow between silicate liquid, olivine-rich ultrabasic mush and sulphide liquid due to density and viscosity contrasts, were probably the critical processes in ore formation.

Naldrett (1973) suggested that Archaean ultramafic komatiitic magmas contained significant sulphides as a result of sulphide melting in the Archaean upper mantle below about 100 km (Fig. 22). It is significant in this respect that many samples of the first melting fraction of Archaean mantle (i.e. basaltic rocks) are sulphur-saturated (e.g. Naldrett et al., 1978), that most Ni—Cu sulphide deposits formed during the Archaean, and that younger deposits commonly exhibit evidence that the sulphur necessary to form ores was derived via crustal contamination (e.g. Kovalenker et al., 1975). The possibility that some sulphur from associated metasediments was incorporated into Archaean Ni—Fe—Cu deposits cannot be overlooked in view of the similar S/Se ratios and $\delta^{34}S$ values in the ores and metasediments (e.g. Groves et al., 1979).

Several features, including variable Fe/Ni ratios of ores, calculated high phenocryst contents of mineralized ultramafic flows and problems of entraining high-density oxy-sulphide droplets in low-viscosity ultrabasic liquids, all suggest that the sulphide-rich ultrabasic mushes differentiated within, and were erupted from, high-level komatiitic magma chambers. Whether these are now represented by large layered komatiitic sills or dunitic (i.e. residual) pods (e.g. Fig. 21, model A1 or B1) is unclear.

Lusk (1976) pointed out the similarity of many features of the volcanic-associated

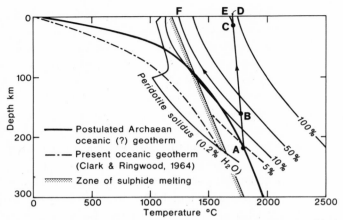

Fig. 22. Depth—temperature diagram showing possible relationship between modern and Archaean oceanic geotherms and melting of komatiitic magmas and mantle sulphides (after Naldrett and Cabri, 1976). A mantle diapir (sulphur-rich) generated at *A* rises to *B* at which stage ca. 25% melting occurs and basaltic liquid separates and rises along a non-adiabatic path *B—F*. The residue of partial melting rises from *B* to *C* resulting in a ca. 30% further melting and separation of komatiitic magma that rises along path *C—E*.

Ni—Cu ores with volcanogenic Cu—Zn deposits of felsic volcanic association and suggested exhalative precipitation of the Ni—Cu. While there appears overwhelming evidence against this as a general process for their formation (e.g. Groves et al., 1979), there are some deposits that are intimately associated with sulphidic interflow metasediments (e.g. Windarra breccia ore and F Shoot — Seccombe et al., 1977; R.J. Marston, pers. comm., 1979: Sherlock Bay, Pilbara Block — L.J. Miller and Smith, 1975), which may represent examples of volcanic-exhalative Ni—Cu deposits. Most sulphidic metasediments in ultramafic sequences are relatively Ni-poor with Zn > Cu > Ni (e.g. Groves et al., 1978).

Dunite-associated nickel—iron—copper. Dunite-associated Ni—Fe—Cu sulphide deposits represent an even more recently recognized type of deposit in granitoid—greenstone terrains than do the volcanic-associated deposits of these metals. Although examples occur elsewhere (e.g. Dumont—Eckstrand, 1975), economically important deposits of this type associated with greenstone sequences are only known from the Yilgarn Block of Western Australia. The distribution is thus broadly similar to the volcanic-associated Ni—Fe—Cu deposits, but in detail the two types of deposit are spatially separated (Fig. 19). Whereas the volcanic-associated deposits are most abundant in the southern part of the Kalgoorlie subprovince, the dunite-associated deposits are concentrated along its northeast margin [Mt. Keith—Perseverance (Agnew) line] and to the southwest of the subprovince (Forrestania line); the linearity of the groups of deposits is evident in Fig. 19 and contrasts with the more clustered patterns of the volcanic-associated deposits. The host dunites have consistently high magnesium contents (>50% MgO), as shown in Fig. 9.

Dunite-associated Ni—Cu deposits in Western Australia are perhaps best exemplified by those occurring along the northeastern margin of the Kalgoorlie subprovince (Fig. 19). These deposits and associated dunites have been described by Martin and Allchurch (1975), A.R. Turner and Ranford (1975), Burt and Sheppy (1975), Nickel et al. (1977), Binns et al. (1977), Naldrett and A.R. Turner (1977) and Groves and Keays (1979). There are no published data on the Forrestania deposits apart from some information in Binns et al. (1977), but a small discrete deposit (Black Swan) from the Kalgoorlie area is described by Groves et al. (1974) and Keays et al. (in press).

The deposits in the northeast Kalgoorlie subprovince are associated with a series of discontinuous, elongate, steeply dipping metamorphosed dunite pods roughly coincident with a north-northwest-trending lineament for ca. 150 km, from south of Perseverance to north of Wiluna (Fig. 19). The dunite bodies are normally concordant with country rocks, but are locally discordant. Marginal shearing and boudinage of the bodies during subsequent tectonism has obscured original relationships with country rocks in most cases. In contrast to the volcanic-associated deposits, the mineralized dunites are not specifically associated with ultramafic flow sequences, but country rock sequences include felsic metavolcanics and volcaniclastic rocks together with mafic and ultramafic rocks. This is true also for Black Swan and Forrestania and in the latter case there is also an association with oxide facies BIF (cf. volcanic-associated ores). The dunitic pods do not show the asymmetric olivine distribution nor the spinifex-textured zones shown by komatiitic peridotite flows.

There has been some controversy concerning the timing of emplacement of the dunitic bodies and their relationship to komatiitic peridotite flow sequence and associated mineralization. Some authors (e.g. Burt and Sheppy, 1975; Binns et al., 1977) envisage that the dunitic bodies were emplaced as steeply dipping subconcordant bodies along major fracture zones, presumably following initial deformation of the greenstone sequences, whereas others (e.g. Naldrett and A.R. Turner, 1977) suggest that they may represent sill-like feeder chambers for overlying komatiitic volcanics; these alternatives, which are not necessarily mutually exclusive, are represented schematically in Fig. 21, models B1 and B2. Further investigation is required to resolve these relationships, but it appears very unlikely that the dunitic bodies were important sources of sulphide-rich ultrabasic flows due to the marked spatial separation of economically important volcanic-associated and dunite-associated deposits.

The dunite-associated deposits, occur in a wide range of metamorphic environments, with the Mt. Keith—Perseverance line cutting across the regional metamorphic isograds (e.g. Binns et al., 1976). Unpublished data (University of Western Australia — mainly R.J. Gunthorpe) suggest that there is a *broad* geographic variation in the degree of fractionation of the dunitic bodies and in the nature of contained mineralization that tends to parallel the variation in metamorphic grade and degree of internal strain in enclosing rock sequences. In very low-grade metamorphic environments (e.g. Wiluna), the dunites show marginal fractionation to orthopyroxenite, norite and possibly even sodic granophyre and contain no significant mineralization; in intermediate-grade environments of low strain (e.g. Mt. Keith, Black Swan) the dunites have a varied development of marginal orthopyroxenite and have high disseminated massive sulphide ratios; in the higher-grade, higher-strain environments (e.g. Perseverance, most Forrestania occurrences), the dunites normally show only minor, if any, marginal fractionation and contain both disseminated and massive sulphides. Binns et al. (1977) suggested that the above relationships could be simply explained in terms of increasing fractionation of a single or related series of ultramafic intrusions with increasing crustal level, but evidence is equivocal and the possibility of several periods of emplacement cannot be ruled out.

The centrally disposed disseminated deposits in greenschist facies or amphibolite/greenschist transition environments are typically large, low-grade deposits: for example Mt. Keith contains ca. 290 million tonnes of 0.6% Ni. The ores typically occur in serpentinite cores surrounded or irregularly cut by talc-carbonate zones. The mineralogy of the disseminated sulphides is strongly controlled by Fe-related redox mechanisms related to serpentinization and talc—carbonation reactions, varying from strongly reduced assemblages such as heazlewoodite—awaruite through pyrrhotite—pentlandite—magnetite to strongly oxidized pyrite—millerite—magnetite assemblages (e.g. Eckstrand, 1975). Some dunite cores have escaped hydrous alteration and contain disseminated pentlandite and chromite.

In higher-grade, higher-strain environments, transitions from disseminated through matrix to massive ores commonly occur on one margin (e.g. Perseverance), or both margins of the dunite bodies (e.g. some Forrestania ores) which are normally retrogressively serpentinized, metamorphic olivine—talc rocks or relict recrystallized dunites. Massive ores may be strongly remobilized into the country rocks (e.g. Fig. 21C) and, in combination with more disseminated ore, may be very large (e.g. ca. 45 million tonnes of ca. 2% Ni in *three* main shoots at Perseverance) compared with volcanic-associated deposits (e.g. ca. 30 million tonnes of ca. 3.2% Ni in *fourteen* main deposits at Kambalda—St. Ives). The ore mineralogy is normally pyrrhotite—pentlandite—pyrite—chalcopyrite—magnetite. Ni/Cu ratios are generally in the range 20—60 and S/Se ratios are also generally higher than for volcanic-associated ores, with some massive ores from Perseverance having very high ratios (e.g. Binns et al., 1977; Groves and Keays, 1979). The ores have significant platinoid contents (Keays and Davison, 1976).

As in the case of the volcanic-associated ores, there is overwhelming evidence for the early existence of magmatic sulphides in the dunitic pods (e.g. Binns et al., 1977; Groves and Keays, 1979), although there is little detailed evidence bearing on the source of sulphur; Donnelly et al. (1978) reported small negative $\delta^{34}S$ values for a very limited sample population. However, regional metamorphism appears to have played an important role in controlling the final nature of sulphide ores in the dunitic hosts. During deformation, the previously relatively impermeable greenstone sequences became fractured and deformed allowing ingress of carbonated solutions of juvenile or metamorphic derivation towards the cores of dunitic intrusions. Resultant reactions strongly modified original disseminated magmatic sulphides and apparently introduced externally derived sulphur, which reacted with Fe released during alteration processes to produce additional sulphides without significantly increasing nickel tenor (e.g. Binns et al., 1977; Groves and Keays, 1979; Keays et al., in press). In higher-grade, higher-strain areas metamorphically derived or original magmatic massive sulphides were concentrated in dilatant zones during penetrative deformation. The relative roles of magmatic and metamorphic processes require further investigation before hypotheses can be erected to explain the centrally disposed disseminated ores and marginally disposed massive plus disseminated ores, although a combination of flow and gravity segregation of dense oxy-sulphide liquid is likely. Apparently, undepleted Ni contents of the mineralized dunites argue against in situ formation of oxy-sulphide liquids (cf. Duke and Naldrett, 1978) and suggest emplacement of crystal mushes carrying oxy-sulphide droplets. If this is correct, centrally disposed *magmatic* mineralization is difficult to explain in terms of model B1 (Fig. 21), whereas marginal *magmatic* massive sulphides are difficult to explain in terms of model B2 (Fig. 21), unless a "footwall" trap as shown in the figure is postulated (cf. Martin and Allchurch, 1975). These points illustrate the problems in defining precise genetic models without greater knowledge of the role of metamorphism and deformation.

Thompson and Ungava-type nickel-copper. The Ni—Cu mineralization occurring in the Thompson and Ungava Belts along the contact of the Superior and Churchill Provinces of northern Canada shows many similarities to the dunite-associated deposits of Western Australia. However, the age of the Thompson hosting sequences and ultramafic intrusives is still debatable (e.g. Peredery, 1979) because of possible crustal reworking along this contact, and the Ungava Belt is clearly not a greenstone terrain in the normally accepted sense, but has a distinctive Proterozoic aspect.

The Thompson Belt nickel-sulphide deposits are associated with regionally concordant, metamorphosed peridotitic to dunitic intrusives in a complex elongate tectonic zone (Peredery and Coats, 1978; Peredery, 1979).

The ultramafic bodies are interpreted to be differentiated ultramafic sills intruded into a complex sequence comprising dominantly high-grade paragneisses and basic amphibolites, with elongate belts of deformed metasedimentary and metavolcanic sequences that have been variably metamorphosed up to amphibolite grade. The metasedimentary rocks include arenaceous, pelitic and calcareous rocks together with iron formations, and the metavolcanic rocks include massive and pillowed basaltic flows and possible ultramafic flows. Importantly, some highly magnesian metabasalts exhibit quench textures and have geochemical characteristics normally attributed to the komatiite suite (e.g. Naldrett and Cabri, 1976), but no definitive textures (e.g. spinifex textures) have been observed in ultramafic rocks.

Isotopic dating (e.g. Cranstone and Turek, 1976) indicates a major metamorphic event at ca. 1.7 Gyr, but some gneisses record an earlier event at ca. 2.8 Gyr. There is considerable uncertainty whether the metasedimentary—metavolcanic sequences (plus mineralized ultramafic intrusives) are Proterozoic in age, representing an infolded cover sequence on Archaean gneiss basement, or whether the whole sequence is a variably reworked Archaean sequence. Thus, the age of the Ni—Cu deposits is in doubt. However, in terms of their association with komatiitic volcanics and intrusives their high Ni/Cu ratios (ca. 15) and chondritic precious-metal distributions (Naldrett et al., 1979), they are very similar to the Archaean group of komatiite-associated Ni—Cu deposits.

The ores are typically composed of pyrrhotite and pentlandite with lesser, variable chalcopyrite, cubanite, pyrite, magnetite and various alteration products. They vary in character from disseminated deposits in serpentinite (e.g. Moak, Mystery), vein networks in serpentinite (Ospwagen), matrix sulphides in serpentine breccia (Pipe 2), fault-controlled sheet-like bodies (N. Mystery), to massive, breccia or stringer mineralization on contacts with country rocks or within country rocks (e.g. Birchtree, Thompson). There is a strong deformational and metamorphic imprint with considerable separation of massive ores from host ultramafic units, but this aspect has been little studied.

The ores are again generally regarded as metamorphosed magmatic ores formed by gravity settling of immiscible oxy-sulphide liquids from ultrabasic magmas. Detailed studies of the Pipe deposits by Naldrett et al. (1979) indicate that they are depleted in Ni, Cu, Pt and Pd relative to most komatiite-associated deposits and they suggest an origin involving localized assimilation of barren country-rock sulphides.

The Ungava Belt is similar in many respects to the Thompson Belt. It occupies a similar tectonic position, comprises a similar sequence of shallow- to deep-water clastic metasediments and carbonates and similar pillowed metabasalts or thick differentiated flows that commonly show skeletal textures or are variolitic (or ocelli-rich). These flows are probably related to the komatiite suite (J. Scoates, pers. comm., 1978); their geochemistry is consistent with this interpretation (e.g. Naldrett and Cabri, 1976). They are, however, in generally lower metamorphic grade and lower-strain environments than the Thompson Belt, so that original structures and textures are better preserved. The sequences appear to be Proterozoic (ca. 1.7—1.6 Gyr) in age and to represent filling of ensialic basins in an incipient rift zone around the margin of the Superior Province: importantly no felsic volcanics are present (J. Scoates, op. cit.).

A number of mafic and/or ultramafic sills intrude the sequence and the Ni—Cu deposits are associated with the ultramafic sills or ultramafic layers of multiple differentiated sills. Sulphide deposits may be disseminated within serpentinites, concentrated in basal depressions as massive ores (e.g. Kilburn et al., 1969), or occur as breccia ores in country rocks adjacent to basal contacts of sills. Only limited data exist on the composition of the ores but Ni/Cu ratios are commonly between 3 and 5 (e.g. Naldrett, 1973) and are, therefore, more Cu-rich than other komatiite-associated ores described above. The deposits are presumably also magmatic sulphide concentrations.

Stratiform intrusive-associated nickel—copper. These deposits represent an economically less important group in Archaean terrains than the volcanic- and dunite-associated mineralizations. Nevertheless, examples of Ni—Cu sulphides occurring as disseminated to massive lenses within ultramafic layers or cross-cutting pegmatoids of Archaean stratiform intrusions are known from Western Australia (Carr Boyd, Mt. Sholl), Canada (Lynn Lake, Dumbarton), Rhodesia (Empress, Perseverance?) and USSR (Kola Peninsula). These deposits are generally more Cu-rich (Ni/Cu ~ 1.5—3) than other Archaean Ni/Cu concentrations. The host intrusions probably represent subvolcanic equivalents of basaltic extrusives (e.g. Hallberg and Williams, 1972) and are much smaller than the subsequent, post-cratonic intrusions (e.g. Bushveld, Great Dyke).

The problems of distinguishing tholeiitic from komatiitic basalts over an overlapping range of MgO contents (e.g. Fig. 9) have been discussed above. The problem is accentuated for layered bodies which are commonly "beheaded" and where undisputed chill zones are seldom present, so that their parentage is generally unclear. Although some mineralized intrusions may have a komatiitic affinity (e.g. Mt. Sholl, Empress, Perseverance), others have been grouped in the tholeiitic suite (e.g. Lynn Lake, Dumbarton, Naldrett, 1973; Kola Peninsula, Smirnov, 1976), and a detailed study of the Carr Boyd Complex by Purvis et al. (1972) suggests that the parent magma was a high-alumina tholeiite. Thus, although most Archaean Ni/Cu deposits are of komatiitic affinity, this is not necessarily an exclusive relationship.

The possibility of reworking of Archaean ores during Proterozoic tectonism is well demonstrated by the Ni—Cu sulphide deposits at Pikwe-Selebi in the Limpopo Mobile Belt. Here, Archaean Ni—Fe—Cu sulphide deposits associated with amphibolites of troctolitic composition have been reworked during tectonic-metamorphic events (2.7—2.0 Gyr) in the evolution of the belt (Wakefield, 1976).

Chromite. Chromitite layers in ultramafic-dominated Archaean intrusions have been known for a considerable time. Sub-economic chromitite layers occur in anorthositic intrusives in ancient gneiss terrains such as the Fiskenaesset areas of southwest Greenland (e.g. Windley, 1973) and also in anorthosites in reworked Archaean terrains exemplified by the Limpopo Mobile Belt (e.g. Messina Formation, Anhaeusser, 1976a). Magmatic layering is commonly well preserved despite deformation and metamorphism.

However, the economically most important deposits occur in ultramafic intrusive bodies and/or ultramafic flows within granitoid—greenstone terrains, and in large layered concentrations intruded into stabilized Archaean cratons. Both Anhaeusser (1976a) and Watson (1976) have pointed out the remarkable chromite province in southern Africa were both types of deposit, ranging in age from ca. 3.5 to ca 2.0 Gyr, combine to produce the largest concentrations of this mineral known on Earth. Deposits in greenstone sequences are briefly described below and those in layered intrusions formed at the Archaean—Proterozoic transition are discussed later.

The oldest economic chromite deposits are those of the Selukwe area of Rhodesia, which are ca. 3.5 Gyr in age (e.g. Wilson et al., 1978). The chromite occurs as cumulate sheets within differentiated ultramafic to mafic (komatiitic?) sills, emplaced into tholeiitic and komatiitic metabasalts and ultramafic flows, and unconformably overlain by conglomerate containing gneissic tonalite and granodiorite and massive adamellite clasts. The mineralized sequence is thought to form part of a large nappe-like structure. The thickest development of chromite occurs in downwarps in the basal zone and magmatic structures are preserved in the chromitite layer despite deformation.

Chromite lenses also occur in the Mashaba Complex within serpentinized dunites and peridotites that may also be ca. 3.5 Gyr old (Wilson, in prep.), although a younger age was previously suggested (e.g. Anhaeusser, 1976a).

Although chromite appears as an early cumulate phase in younger Archaean ultramafic intrusions and flows in Rhodesia, there are no economic concentrations. A similar situation exists in other granitoid—greenstone belts where there appear to be no major chromite deposits in ultramafic flows and only minor mineralizations in ultramafic sills.

The significance of the major chromite province in southern Africa is discussed in the later section on stratiform intrusives in the early Proterozoic.

Deposits associated with granitoids

Pegmatite-associated tin, tantalum, lithium, beryllium, tungsten, molybdenum. Mineralization associated with granitoid emplacement represents a relatively minor facet of Archaean metallogeny. There are few deposits within the granitoids themselves, mineralization generally being associated with pegmatities that contain Sn—Ta (less commonly W—Mo), non-metallic resources (Li, Be, Cs), industrial minerals (mica, feldspar, corundum) and emeralds. Cassiterite and tantalite may be concentrated in simple pegmatites or in complex pegmatites containing lepidolite, petalite, spodumene, pollucite, beryl and other Li-, Be- and REE-bearing minerals. Emeralds are commonly found in metasomatic reaction zones around pegmatite contacts with greenstone lithologies.

Mineralized pegmatites are generally associated with post-kinematic, potassic granitoids (mainly adamellites) that have geochemical characteristics similar to those of specialized granitoids in younger terrains, although incompatible elements in Archaean granitoids are commonly slightly lower and Sn contents considerably lower (Table II). Most deposits occur close to granitoid—greenstone contacts or within greenstone enclaves in granitoid terrains. Although there are few data, most important mineralization appears to be associated with younger (<3.0, and generally <2.8 Gyr) granitoids (e.g. Bikita, Rhodesia — Anhauesser, 1976a; Pilbara Block, Western Australia — De Laeter and Blockley, 1972).

The style of Archaean mineralization and Sn—Ta—Li—Be association contrasts sharply with the more common quartz veins, greisens, sulphide-bearing lodes, skarns or other replacement bodies of Sn—W ($\pm$ Mo—Cu) association that typify younger terrains. The concentration of Sn and W mineralization in Archaean terrains is also much lower than that of Phanerozoic terrains (e.g. Pereira and Dixon, 1965; see Fig. 26).

Although the precise source zones of tin mineralization are still largely unknown (e.g. Stemprok, 1977), the high initial $^{87}Sr/^{86}Sr$ ratios of Archaean and later Sn-associated granitoids suggest that they formed by anatexis of pre-existing crustal rocks: fractional crystallization may have been another factor in generating very high Sr isotope ratios (e.g. McCarthy and Cawthorn, in press). The low Sn contents of the Archaean granitoids and the low concentration of Archaean Sn mineralization may thus be partly explained by lower Sn contents of source rocks that were either directly derived from the mantle or contained a relative paucity of clastic sediments in which repeated reworking of granitic detritus including heavy minerals had occurred. Repeated crustal reworking may be necessary to generate sufficiently high initial Sn contents for further concentration by fractional crystallization (e.g. Groves and McCarthy, 1978). Other factors affecting Sn concentration in late phases may be early removal of Sn from granitic magma (e.g. Sn may concentrate in sphene (Hesp, 1971), which is a common accessory mineral in Yilgarn granitoids), low initial volatile contents of magmas suppressing the separation of a Sn-bearing fluid phase, and/or low degree of interaction with meteoric waters in the dominantly volcanic sequences.

The last factors may also in part explain the paucity of hydrothermal deposits and alteration associated with Archaean granitoids, although a relatively deeper level of granitoid emplacement may also have promoted pegmatite formation (e.g. Hesp and Varlamoff, 1977) and inhibited complete separation of a discrete aqueous fluid phase. In this respect, the present level of erosion should be taken into account as most Phanerozoic deposits occur in the roof zones of high-level plutons which are not normally preserved in Archaean granitoid—greenstone terrains. At Mulgine Hill Yilgarn Block, for example, metamorphosed greisens and W and Mo mineralization do occur in the preserved roof zone of a two-mica granitoid pluton, and Boyle (1976) records small skarn deposits containing scheelite at some granitoid contacts.

Deposits related to metamorphism and alteration of greenstone sequences

Epigenetic gold—quartz veins. Unlike many types of Archaean mineralization, gold deposits occur in most cratons. They are most extensively developed and economically important in granitoid—greenstone terrains, and relatively few deposits are known from high-grade gneiss complexes.

The deposits occur in discordant quartz veins, normally 1—5 m thick and up to 2 km long, or more rarely quartz stockworks that are related to macroscopic fold structures or regional fracture zones. The lodes comprise groups of subparallel, commonly en echelon veins, often in several sets. Despite their discordant nature, the lodes in many areas tend to be localized within one or more distinctive stratigraphic units (e.g. Golden Mile Dolerite, Kalgoorlie; Woodall, 1975: Barberton area, South Africa; Anhaeusser, 1976b), or over a restricted stratigraphic interval (e.g. Gatooma area, Rhodesia; Fripp, 1976b).

Different authors emphasize different important ore-host rock associations. For example, Woodall (1975) notes the association of most Yilgarn gold deposits with mafic volcanics, Goodwin (1965) indicates an association with felsic volcanics in the Birch-Uchi Lakes area, Canada, and R.P. Viljoen et al. (1970) and Pyke (1975) suggest an association with mafic—ultramafic volcanics around Barberton, South Africa, and in the Timmins area, Canada, respectively; the Yelloknife deposits, Canada, are associated with sheared and altered mafic to felsic volcanics (e.g. Boyle, 1976). As noted above, various types of cherty metasediments or BIF can also be host to discordant Au lodes, as can greywacke—slate sequences (e.g. Boyle, 1976). Quartz—Au lodes only rarely occur in granitoids (e.g. less than 2% of lodes in the Yilgarn Block; Woodall, 1975).

The lodes are generally dominated by quartz, accompanied by lesser amounts of carbonate and commonly several percent sulphides. Pyrite, arsenopyrite and/or pyrrhotite are common sulphides with lesser amounts of sphalerite, galena, chalcopyrite, tetrahedrite, stibnite, bismuthinite and scheelite. Gold may occur as coarse-grained free gold, minute inclusions in sulphides, or tellurides. Mineralization is normally accompanied by wall-rock alteration, which may be confined to within a few centimetres of the lode or may extend over tens of metres. Silicification, carbonation, sericitization and/or propylitization are all common types of alteration.

The regional distribution of quartz—Au lodes appear to be broadly controlled by tectonic units within cratons and by granitoid distribution as well as regional metamorphic grade within granitoid—greenstone terrains: the last two are inter-related. On the broadest, craton-wide scale, gold appears to be concentrated in the internal parts of cratons (e.g. Fig. 23). There are few deposits in flanking high-grade "mobile belts", although the latter include reworked Archaean sequences, and deposits are rare in high-grade gneiss terrains. On a finer scale, the Au deposits in some cratons may be concentrated around the periphery of granitoid domes, but several kilometres inside the greenstone contact (e.g. Barberton area; Anhaeusser, 1976b; see Fig. 24), granitoids with this relationship vary from tonalites to adamellites. In areas where metamorphic grade has been established (e.g. Yilgarn Block), most large deposits (e.g. Kalgoorlie deposits) occur in low-metamorphic-grade, generally greenschist facies, environments within the greenstone belts.

Many deposits are younger than the major deformational events affecting the greenstone belts, and many post-date early metamorphic recrystallization as metamorphic assemblages are altered by mineralizing fluids. In the Golden Mile, Kalgoorlie, Woodall (1975) emphasizes the probable contemporaneity of Au mineralization with emplacement of felsic intrusives, which are themselves metamorphosed but show only minor deformation. In these cases, it is likely that gold deposition occurred during the waning stages of deformation and metamorphism affecting the greenstone belts. In other cases (e.g. Norseman, Western Australia; Woodall, 1975: Barberton; Ramsay, 1963), the Au—quartz veins, although discordant to the greenstone sequences, have been metamorphosed and suffered at least some of the deformational events recorded in surrounding rocks. It thus appears that mineralization occurred at various times thoughout green-

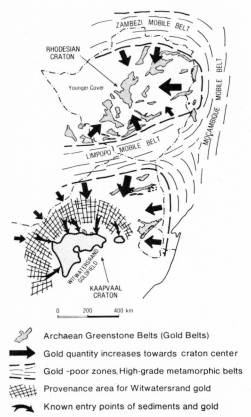

Archaean Greenstone Belts (Gold Belts)

Gold quantity increases towards craton center

Gold -poor zones. High-grade metamorphic belts

Provenance area for Witwatersrand gold

Known entry points of sediments and gold

Fig. 23. Schematic map of the southern African shield showing the distribution of gold mineralization (after Anhaeusser, 1976a, b). The high-grade mobile belts are virtually devoid of gold, as are greenstone belts immediately flanking them. The provenance area of the Witwatersrand Goldfield is now extensively covered by younger sequences, including the Bushveld complex.

stone evolution, but that in general it was post-early deformation and pre-late metamorphism: further data are required to fully substantiate this observation.

There are few Pb-isotope studies of sulphides from gold deposits. Saager and Köppel (1976) reported two-stage Pb in sulphides from Barberton Au mineralizations, and their data indicate an age for the lower Onverwacht Group between 3.8 and 3.45 Gyr, in agreement with the Rb/Sr ages of Jahn and Shih (1974).

The widespread occurrence of Au deposits in Archaean greenstone belts suggests a fundamental association between mineralization and greenstone belt evolution. The lack of a consistent association with specific granitoids or greenstone belt lithologies, the late-stage development of quartz—Au lodes, and their preferential occurrence in sub-amphibolite facies metamorphic domains, suggest a relationship to the late thermal history of granitoid—greenstone terrains. Several authors have proposed that Au was released from

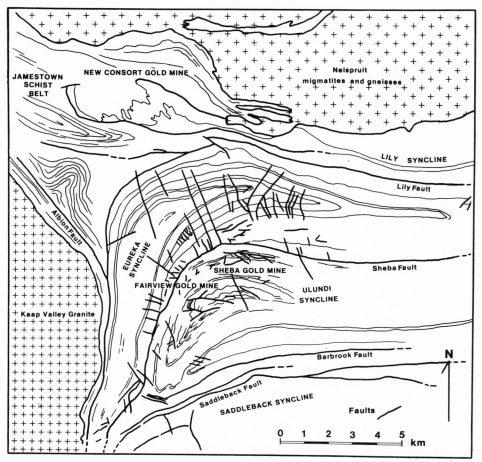

Fig. 24. Simplified structural map of the main gold region of the Barberton Mountain Land showing the restriction of gold-rich fractures to particular parts of the stratigraphy, and to lower-grade metamorphosed zones at some distance from intrusive granitoids (after Anhaeusser, 1976b).

the dominantly volcanic pile during prograde metamorphism and penecontemporaneous granitoid emplacement and migrated down the metamorphic gradient to be deposited in structurally prepared sites at temperatures below about 500°C (e.g. R.P. Viljoen et al., 1970; Henley, 1973; Anhaeusser, 1976a; Fripp, 1976b; Watson, 1976; Boyle, 1976; Karvinen, 1978). Supporting evidence for such a model is provided by trace-element patterns in pyrites from Barberton deposits (Saager and Köppel, 1976), and by calculations of SiO_2 leaching from host rocks at Kalgoorlie (e.g. Bartram and McCall, 1971).

Gold was transported as either, or both, chloro- or thio-complexes, and in either case theoretical deposition temperatures (i.e. significant decreases in solubility of gold) are around 450–500°C (e.g. Seward, 1973), in good agreement with prograde metamorphic temperatures reached at most mineralization sites.

Numerous authors have suggested ultramafic/mafic lithologies that tend to occur lower in the greenstone pile may represent an important source of the Au (e.g. R.P. Viljoen et al., 1970; Boyle, 1971; Anhaeusser, 1976b), with some authors (e.g. R.P. Viljoen et al., 1970) suggesting that these rocks may be Au-enriched. Available data on Archaean volcanic lithologies (e.g. Anhaeusser et al., 1975; Keays and Davison, 1976; Kwong and Crocket, 1978; Bavinton and Keays, 1978) suggest that these rocks do not generally have high Au contents; most values are in the range 1–2 ppb. This does not exclude these rocks as potential sources of Au, because Keays and Scott (1976) have demonstrated that mineralogical siting of the Au may be far more important than overall Au concentration: Au deposited on loosely bound sites (e.g. deuteric alteration products, grain boundaries, glass) and sulphides is more available to circulating fluids than that fixed in minerals, such as chromite or olivine. Keays and Scott (1976) further show that basalts with high MgO, Cr, Ni and Cu, common characteristics of Archaean basalts, may constitute a more favourable type of source rock than other basalts.

Gold-enriched cherty sediments (e.g. Fripp, 1976a; Bavinton and Keays, 1978) may also represent potential sources of Au for later circulating metamorphic solutions. Background Au values for these sediments are commonly in the range 5–10 ppb (D.G. Chapman and Groves, 1979) and average ore grades of ca. 11 ppm Au are recorded (e.g. Fripp, 1976a). Thus, although the sediments are volumetrically negligible compared with mafic–ultramafic rocks, they are relatively enriched in Au by ca. 5000–10,000 times. There have been no systematic studies of Au release from these sediments during metamorphism. Preliminary data from Western Australia (D.G. Chapman and Groves, 1979) suggest that there is no significant difference in background Au content between sediments from varying metamorphic grades, although it may be important that no metasediments from mid- to high-amphibolite facies environments contain anomalously high Au contents. Small losses, equivalent to the precision of Au analyses, may be all that are required to produce significant ore fluids (e.g. Gottfried et al., 1972), so that Au sources may be difficult to define.

Potentially, the most relevant studies relating to Au release during alteration and metamorphism of greenstone belts are those involving alteration of sulphide-bearing lithologies, as sulphide minerals commonly host the major proportion of Au in rocks. Keays and Kirkland (1972) for example, presented evidence to suggest that Au in the Wood's Point dyke swarm in Victoria, Australia, was derived from Cu-sulphides in adjacent ultramafic rocks. Keays et al. (in press) demonstrated the strong possibility that measurable Au was released during talc-carbonation of sulphide-bearing serpentinized dunites at the Archaean dunite-associated, Ni–Fe–Cu sulphide deposit at Black Swan, Western Australia (Fig. 19), where the metamorphic grade is greenschist–amphibolite transition. It is important that the resultant Au-bearing solutions also contained S and CO_2, major components of the discordant Au lodes. As there are thick sequences of talc-carbonated ultramafics, commonly containing disseminated sulphides, in Archaean terrains, these must represent an important potential source of Au. A cautionary note is, however, provided by studies of the Mt. Keith mineralization (Figs. 20, 21), which demonstrates no Au-loss from serpentinites during talc-carbonation (Groves and Keays, 1979); metamorphic grade at Mt. Keith is slightly lower than at Black Swan. Probable important factors in determining whether Au release occurs or not include degree of oxidation of alteration fluids (Black Swan was a highly oxidizing environment — Groves et al., 1974; Keays et al., in press) and temperature of alteration which controls Au-solubility.

A schematic diagram showing a possible model for Archaean Au mineralization is shown in Fig. 25. Available data are clearly in favour of an origin involving hydrothermal leaching of Au from Archaean volcanic sequences during metamorphism, but further data are required on Au distribution in Archaean lithologies, including disseminated sulphide ores, on conditions of deposition of Archaean Au deposits (e.g. fluid inclusion, stable isotope and wall-rock alteration studies) and on transport of Au from experimental studies (e.g. Seward, 1979). Donnelly et al. (1978) reported a reconnaissance sulphur-isotope study of the Kalgoorlie Au mineralization in which they demonstrate that sulphides in quartz-Au lodes are ^{32}S enriched relative to those in the host rocks. They interpret this to indicate an external source for the S, and the presence of oxidized ore fluids containing some SO_4^{2-}, a conclusion supported by the occurrence of minor barite and anhydrite in the lodes. This supports the indications above that oxidizing fluids were important in leaching gold from ultramafic sequences.

Antimony, arsenic, tungsten and mercury. Antimony, arsenic and tungsten commonly occur in sub-economic concentrations associated with Archaean quartz—Au deposits, and their origin is presumably related to that of gold. For example, arsenopyrite and stibnite, together with numerous sulphosalts are widespread in the Yellowknife belt of Canada, and are common accessory minerals in gold deposits in Western Australia (e.g. Woodall, 1975) and southern Africa (e.g. Anhaeusser, 1976a); cinnabar also occurs in the latter area. Scheelite, and less commonly wolframite, are also widespread minor phases in quartz—Au veins; even in scheelite-rich zones the WO_3 content is generally below 1%.

The only large antimony and mercury deposits of Archaean age occur in the Murchison greenstone belt in the Kaapvaal craton. Here, a number of antimony ores occur as lenses in sheared and carbonated intermediate to felsic volcanics over a strike length of roughly 50 km along the so-called "Antimony Line". These deposits, consisting of stibnite, antimony sulphosalts and oxides together with gold, pyrite, arsenopyrite and

---

Fig. 25. Conceptual model of formation of gold concentrations in greenstone sequences. Gold is initially concentrated during volcanism, subsequently remobilized during metamorphism and ultimately deposited in discordant quartz veins, commonly in sub-amphibolite grade domains. This model has been adapted from Fripp (1976b). Note that volcanic exhalations in *A* are shown, for convenience, as emanating from discrete zones in the mafic—ultramafic volcanics, although it is likely that lava degassing (including lava—seawater interaction) was an important factor in exhalite formation where there is no evidence for permeability and hydrothermal alteration in the volcanics.

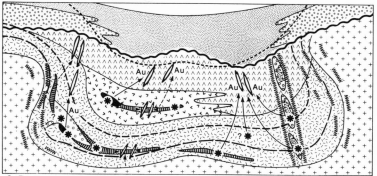

C. Emplacement of sulphide-bearing dunite dykes and granitoids, regional metamorphism and deformation, release of Au from sulphide-bearing lithologies, migration of Au (+ CO_2, S ±Sb, As) down metamorphic gradient to structural sites

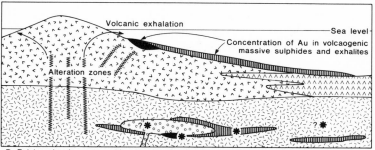

Volcanic exhalation

Sea level

Concentration of Au in volcaogenic massive sulphides and exhalites

Alteration zones

B. Felsic volcanic centre with associated exhalation and formation of Au-bearing volcanogenic ores and exhalites

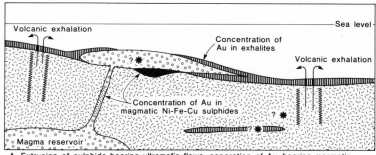

Volcanic exhalation

Sea level

Concentration of Au in exhalites

Volcanic exhalation

Concentration of Au in magmatic Ni-Fe-Cu sulphides

Magma reservoir

A. Extrusion of sulphide-bearing ultramafic flows, separation of Au-bearing magmatic Ni-Cu-Fe sulphides, concommitent exhalation and formation of Au-bearing sulphidic tuffaceous rocks and exhalites (these may also form during basaltic volcanism)

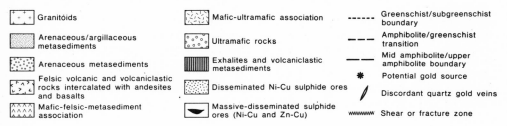

Granitoids

Arenaceous/argillaceous metasediments

Arenaceous metasediments

Felsic volcanic and volcaniclastic rocks intercalated with andesites and basalts

Mafic-felsic-metasediment association

Mafic-ultramafic association

Ultramafic rocks

Exhalites and volcaniclastic metasediments

Disseminated Ni-Cu sulphide ores

Massive-disseminated sulphide ores (Ni-Cu and Zn-Cu)

Greenschist/subgreenschist boundary

Amphibolite/greenschist transition

Mid amphibolite/upper amphibolite boundary

Potential gold source

Discordant quartz gold veins

Shear or fracture zone

cinnabar, are considered to be hydrothermally (metamorphically?) remobilized volcano-genic deposits related to felsic volcanism (e.g. Anhaeusser, 1976a). Considering that economic deposits of antimony and mercury are mainly of late Phanerozoic age, it is unexpected that the Archaean Consolidated Murchison Mine is amongst the largest-known antimony deposits in the world and is adjacent to the major Monarch cinnabar deposit.

Industrial minerals. Various industrial minerals occurring in greenstone belts are derived by alteration and/or metamorphism of greenstone lithologies. Ultramafic rocks are the most important host rocks for such mineralization which includes asbestos, magnesite, talc and ornamental stones (e.g. verdite at Barberton; Anhaeusser, 1976a).

Asbestos deposits are widespread in Archaean granitoid—greenstone terrains, occurring in layered stratiform highly magnesian intrusions, probably of komatiitic affinity. According to Anhaeusser (1976b), serpentinized dunites, peridotites and harzburgites represent the best hosts for asbestos development in southern Africa. Faulting, fracturing and folding all appear to play a role in the development of suitable conditions for formation of cross-fibre asbestos under dilational conditions and slip-fibre asbestos in shear zones.

Commercial talc deposits occur within ultramafic rocks around granitoid contacts, in fault zones, or within specific sections of fold structures. Magnesite commonly occurs in strongly fractured zones of primary alteration which have been modified by subsequent meteoric fluids.

Other industrial minerals include kyanite and corundum, the latter being produced on a large scale in Rhodesia. The corundum normally occurs in lenses within aluminous mica schists, close to granitoid—greenstone contacts.

Interim summary

Archaean mineralization is concentrated in the low-grade granitoid—greenstone terrains and it is largely reflecting the high incidence of igneous and exhalative activity, and the prevalence of submarine environments. The most important types of mineralization and their probable modes of formation are as follows:

(1) *Banded iron formation (Algoma type)*: These are widespread in greenstone belts of all cratons, but relatively few of them contain economic concentrations of Fe; some have high concentrations of Au. They occur in association with volcanics of felsic to ultramafic compositions and evidently formed as chemical precipitates on the sea floor from Fe- and Si-bearing fluids derived from ascending magmas, degassing of lavas, lava—seawater interactions, and/or circulation of surface waters into permeable zones in the volcanics. Sulphide-facies BIF probably reflect higher sulphur fugacities in the vicinities of volcanic centres. Hematite-bearing BIF require *relatively* oxidized, probably shallow-water conditions; these could have formed as a result of microbial photosynthesis, although photolysis cannot be entirely dismissed as a significant process. There is no

necessity for a contemporaneous oxygenated atmosphere, nor for oxidized conditions throughout the whole hydrosphere. Isotopic data suggest that bacterial sulphate reduction could have formed at least some of the sulphide in ca. 2.7 Gyr old sulphide-facies BIF in Canada; these are the oldest rocks so far reported in which there is evidence for this process.

(2) *Volcanogenic sulphide deposits*: These are most important in Canada and they occur at submarine calc-alkaline volcanic centres which are generally best developed in younger greenstone sequences. The ores characteristically have varying Cu/Zn ratios, low Pb and no sulphate. The main exceptions are small Pb-, Ag- and barite-rich deposits in both younger and older greenstones in Western Australia. The lack of Archaean Cyprus-type Cu deposits is probably reflecting the absence of environments analogous to those at Phanerozoic mid-ocean ridges, and a consequent lack of convective water circulation into hot mafic and ultramafic volcanics. This contrasts with the situation at centres of explosive Archaean calc-alkaline volcanism, where permeabilities were much higher and there is abundant evidence for hydrothermal alteration. Magmatic fluids could have been the main source of metals and sulphur in the volcanogenic Cu—Zn sulphide deposits, presumably mixing with circulating surface waters as they ascended to the sea floor to form the Cu—Zn sulphide lenses. Alternatively, surface waters could have leached ore components from the volcanic pile. The rarity of subaerial volcanic belts offers an explanation of the paucity of Archaean porphyry—Cu deposits. The low Pb contents of Archaean mineralization imply that several cycles of crustal enrichment were needed before economic concentrations of this metal formed.

(3) *Barite*: Bedded, evaporitic sulphate deposits formed in shallow-water, early Archaean sedimentary environments and were diagenetically baritized. They imply the oxidation of exhalative sulphur, possibly by photosynthetic microorganisms, in local shallow seas, or in a widespread, thin zone at the top of the hydrosphere. The rarity of barite in Archaean base-metal sulphide deposits supports deposition of the latter in relatively deep basins below any oxidized layer.

(4) *Nickel—copper deposits (± platinoids), and chromite deposits*: These are ultramafic-hosted magmatic deposits. Greenstones dated at <3 Gyr, mainly in Western Australia, contain Ni—Cu sulphide deposits in peridotite flows and dunite pods; these evidently formed from immiscible oxysulphide liquids in ultrabasic magmas. Some older terrains, notably the Rhodesian craton, have major Cr deposits formed by settling of chromite in differentiated ultramafic to mafic intrusives. Each type of mineralization is related to highly magnesian (komatiitic) magmatism which typified the Archaean but became less and less prolific into the Proterozoic; this is a major factor in the relative importance of Ni—Cu and Cr deposits in the Archaean and early Proterozoic. The deposits exihibit a strong tectonic control: volcanic-associated Ni—Cu ores formed in rift environments within greenstone basins typified by thick sequences of komatiitic peridotite flows and discontinuous sulphidic metasediments; dunite-associated deposits formed along major lineaments in bounding zones; and chromite-bearing layered intrusives probably formed in

association with major fractures in relatively stable crust.

(5) *Gold–quartz veins*: These are widespread in most Archaean cratons, occurring in association with a wide variety of rock types in lower and upper greenstone sequences. Most large deposits are in sub-amphibolite regimes, well removed from granitoid contacts. It appears that mineralization occurred at various times throughout greenstone evolution, but that it was normally post-early deformation and pre-late metamorphism. It appears likely that Au was released during prograde metamorphism and penecontemporaneous granitoid emplacement, and that it migrated down the metamorphic gradient, as thio- or chlor-complexes in metamorphic fluids, to be deposited in structurally prepared sites at temperatures $\leqslant 500°C$.

ARCHAEAN–PROTEROZOIC TRANSITION

An overview

The Archaean culminated with widespread anatectic-metamorphic events. These resulted in emplacement of abundant adamellites and granodiorites in the upper crust and final stabilization of sialic crustal blocks in a single "super-continent", or a small number of closely spaced continental nuclei (Fig. 14). Subsequent widespread erosion and weathering of the Archaean shields led to accumulation of quartz-rich clastic sedimentary units in shallow-water, epicratonic sequences which also contain carbonates, cherts and mafic to felsic volcanics, and which are intruded locally by potassic granitoids, dolerite dyke swarms and layered ultramafic–mafic intrusives. These sequences are preserved both as little disturbed cover successions and as relatively metamorphosed and deformed complexes in mobile belts. From the comparison of features of Archaean and Proterozoic terrains in Table I, it is clear that these epicratonal sequences are of Proterozoic aspect, even though some of them accumulated in the period ~3.0 to 2.5 Gyr (the Zuluan Wedge of Cloud, 1976), when typical Archaean granitoid–greenstone stabilization events were occurring in other regions. Thus, the lithostratigraphic Archaean–Proterozoic boundary is diachronous, and the 2.5 Gyr B.P. age commonly accepted for this boundary, whilst realistic in many terrains, is certainly not meaningful in all regions.

It has been proposed that continental crust remained largely below sea level until late Proterozoic times (Hargraves, 1976), but this seems difficult to reconcile with the geological record. Numerous features in Archaean greenstone belts (e.g. greywackes, conglomerates, probable evaporites) indicate at least local emergence (e.g. R.G. Walker, 1976), but there is no firm evidence for extensive land masses. However, the presence of voluminous clastic sediment and major erosional unconformities in early and middle Proterozoic sequences implies widespread emergence. The important mineralization in early Proterozoic sequences reflects the evolution of a thick, relatively stable crust.

Pre-2.5 Gyr sequences with Proterozoic aspect

The oldest-known rocks with Proterozoic aspect are in the Pongola System which formed locally on the stabilized eastern portion of the Kaapvaal craton some 3 Gyr ago (Burger and Coertze, 1973; Anhaeusser and Button, 1976; Cloud, 1976). The Pongola succession contains a higher proportion of basalt than most Proterozoic sequences, but it differs from typical Archaean greenstone belts in having extensive shallow-water quartzites, cryptalgal and stromatolitic carbonates; it also contains local, thin BIF units. More metamorphosed sequences of similar age occur in the Limpopo Mobile Belt (Messina Group) and in the Namaquland—Natal Mobile Belt (Kheis Group), which are to the north and south, respectively, of the Pongola Basin.

The next youngest sequence on the Kaapvaal craton is the localized Dominion Reef Group, comprising continental clastics and volcanics which are considered to be ~2.8 Gyr old (Haughton, 1969). More widespread is the overlying Witwatersrand Group (Haughton, 1964; Pretorius, 1976), an almost 14 km thick sequence which accumulated between 2.5 and 2.75 Gyr BP; it grades up from a probable marine deltaic assemblage of shale, quartzite and minor conglomerate, iron formation and mafic volcanics, to an economically more important sequence comprising mainly continental and marginal marine coarse clastics. It is noteworthy that the oldest putative tillites occur in the Witwatersrand System (Wiebols, 1955). Next in the stratigraphic succession are the thick and extensive, ~2.5 Gyr old Ventersdorp lavas and subordinate clastic sedimentary rocks, which in part conformably overlie the Witwatersrand System and in part unconformably overlie Archaean basement.

The Great Dyke of Rhodesia and the Usushwana Complex are major mafic intrusives emplaced in Southern Africa in the 2.9 to 2.5 Gyr interval.

At the present time, these sequences in southern Africa are the only well-documented examples of rocks of Proterozoic aspect formed pre-2.5 Gyr BP. However, it is likely that more detailed studies in other terrains will delineate analogous sequences of similar ages in some other regions. Recently, R. Pidgeon (oral communication, 1978) obtained a 2.76-Gyr age on zircons from a porphyry intrusive in the Fortescue Group, which unconformably overlies Archaean rocks of the Pilbara craton, and which is a typical Proterozoic sequence comprising basalts, pyroclastics, quartz-rich sandstones, arkoses, conglomerates, shales, jaspilites and stromatolitic carbonates. Also, the large layered Stillwater ultramafic—mafic complex of Montana was apparently emplaced in stabilized Archaean crust around 2.7 Gyr B.P.

Early Proterozoic mineralization

The main types of mineralization in early Proterozoic rocks are:

(1) Uraninite and gold in quartz-rich conglomerates and associated rocks; these are restricted to early Proterozoic sequences.

(2) Pitchblende in breccia and shear zones beneath major unconformities.

(3) Hematite, goethite and magnetite in extensive Superior-type BIF; these are restricted to the Lower Proterozoic.

(4) Manganese oxides, often associated with carbonates.

(5) Chromite in large layered intrusives.

It is noteworthy that massive base-metal sulphide deposits are relatively uncommon in the Proterozoic, although significant examples occur in the Flin Flon, Canada (e.g. Sangster and Scott, 1976), Jerome, U.S.A. (e.g. Sangster and Scott, 1976), and Skellefte, Sweden (e.g. Rickard and Zweifel, 1975) areas.

Conglomerate gold and uranium deposits

The most important examples of conglomerate Au—U deposits are the Witwatersrand-type fields in South Africa (e.g. Pretorius, 1976; Anhaeusser, 1976b). These occur at many different levels in the stratigraphic column, mainly in the Witwatersrand Group, but also in the underlying Pongola System. The main economic minerals are native gold, uraninite and thucolite (carbonaceous matter with Th and U; probably fossilized algal mats). The host rocks are commonly pyritic, quartz-rich conglomerates and arenites, but shales and silty dolomites are also mineralized. These rocks formed in fluvial fans or fan-deltas in downwarps between basement domes. The source of gold and the abundant pyrite in these deposits is considered to have been the >3.25 Gyr greenstone belts (Fig. 23), while the uraninite presumably came from granitoids (including pegmatites) intruding the greenstones. In addition to considerable evidence for detrital gold, uraninite, pyrite and other heavy minerals, it appears that some of the metals were also in solution. Critical factors in ore formation appear to have been continual uplift of source areas along the basin edge, repeated reworking of fans, and widespread development of algal mats. Gold, uraninite and other heavy metals were concentrated on unconformities separating transgressive sequences from later regressive phases. In addition, uranium and, to some extent, gold were concentrated in algal mats, presumably as a result of reducing conditions, adsorption and perhaps, microbial catalyzation of metal precipitation. Enigmatic structures resembling fungi or branching algae have been separated from thucolite (e.g. Hallbauer, 1975), although such eucaryotic organisms are incompatible with the supposedly prevailing procaryotic stage of microbial evolution. A possible solution to the enigma comes from experiments reported by Cloud (1976) which produced similar structures completely abiologically.

Broadly similar uranium mineralization occurs in pyritic, quartzose, conglomeratic sequences in the basal Proterozoic Huronian Supergroup, Canada (Elliot Lake region, Ontario; Montgomery Lake—Padlei region, Northwest Territories; Sakami Lake area, Quebec — Roscoe, 1973, Robertson, 1974). The Jacobina Series and probable early Proterozoic equivalents at various localities around the San Francisco Archaean craton in Brazil also contain significant conglomeratic U—Au mineralization (Robertson, 1974).

Detrital uraninite, gold (and pyrite) deposits of economic grade are not found in younger sequences. The significance of this will be considered further in the section dealing with evolution of an oxygenous atmosphere.

Unconformity-related uranium deposits. Uranium occurs in breccia and shear zones in Lower Proterozoic sedimentary sequences, most importantly in the Northern Territory of Australia (where the so-called Alligator River-type deposits constitute ~20% of the western world's known uranium reserves) and in midwest Canada. These deposits are described in International Atomic Energy Agency (1974), Kimberley (1978), the Special Issue of Economic Geology [vol. 73(8), 1978] and the International Uranium Symposium on the Pine Creek Geosyncline (1979). They are sometimes referred to as vein deposits, but this is not appropriate as many of the deposits are essentially conformable. The main primary uranium mineral is pitchblende (a variety of uraninite with negligible Th and REE), and common gangue minerals are chlorite and hematite.

These deposits show a general relationship to major unconformity surfaces at the top of Lower Proterozoic sequences, and mineralization tends to decrease with depth beneath these surfaces. Whether the unconformities reflect weathering surfaces for supergene enrichment, hydrologic barriers or windows to the basement, or whether they are of no importance at all, is still disputed.

In most cases, the mineralization clearly formed after regional metamorphism, which culminated around 1.8 Gyr in both Australia and Canada. Pitchblende "ages" for the deposits are variable (Köppel, 1968; Hills and Richards, 1972; Knipping, 1974), with maximum values near 1.8 Gyr, but most values being significantly younger than this. Fluid-inclusion studies on associated quartz and carbonates yield variable temperature estimates from around ambient values to >200°C (e.g. Pagel, 1975; Langford, 1978; Ypma and Fuzikawa, 1979).

The source of the uranium in these deposits was almost certainly uraniferous Archaean basement (cf. Heier and Rhodes, 1966 and Heier and I.B. Lambert, 1978, for radioactive element contents of northern Australian granitoids), and/or Lower Proterozoic metasediments derived from such basement. However, there is active debate concerning the relative importance of surficial weathering and subsurface "hydrothermal" processes in liberating the uranium and concentrating it in the ores:

(1) The surficial model proposes that uranium was liberated during weathering and transported in oxidizing waters into permeable zones (e.g. breccia zones formed by solution collapse or tectonic processes; joints; faults), where any of a number of possible changes of physiochemical conditions caused its precipitation. Subsequent remobilization involving subsurface "hydrothermal" fluids is seen as a secondary process which caused minor reworking of the deposits.

(2) In contrast, the "hydrothermal" model envisages that the bulk of the uranium was emplaced beneath the surface from intrastratal brines which were heated either by the normal geothermal gradient, by igneous activity and/or by regional metamorphism.

These subsurface brines migrated into zones of high permeability, and uranium precipitated in response to changing physiochemical conditions. In this model, supergene processes are regarded as late-stage over-prints which caused some modification of the original ore deposits.

A reasonable assessment of existing evidence would appear to be that both surficial and subsurface processes were involved in the generation of many unconformity-related uranium deposits, but the one process need not have been dominant in all cases.

Superior-type banded iron formation. Volcano-sedimentary sequences containing voluminous BIF of the Superior type were deposited either above unoxidized basal Proterozoic sequences, or unconformably on Archaean basement. The Superior-type BIF are reviewed in Gross (1967), James and Sims (1973), Anonymous (1973), Dimroth (1976), and Eichler (1976). As noted by Goodwin (1973a), they tend to fall within narrow continuous belts in the Proterozoic "supercontinent" (Fig. 14); the significance of these belts is uncertain, but they may have been major epicratonic rifts or downwarps. Available age data are not able to closely define depositional ages for many Superior-type BIF, but they most probably formed in the period 2.5 to 2 Gyr B.P., with only rare examples possibly younger than 1.8 Gyr.

In general, Proterozoic BIF are much larger than Algoma-type BIF, with individual members persisting along strike for up to several hundred kilometers. There are commonly several BIF members at different stratigraphic levels and major examples can exceed 100 m in thickness. They are characteristically associated with shales, sandstones and carbonates in shelf sequences, and do not generally have close spatial relationships with igneous rocks.

In most instances, hematite and magnetite appear to have been the dominant primary iron minerals; carbonates and silicates are generally less abundant and sulphides rare. Economic concentrations of iron are the result of secondary weathering processes, possibly at several different periods from the Proterozoic to the Cainozoic; they are accompanied by thinning of the BIF members, and are often associated with faults and folds which allowed access of waters to remove Si and enrich Fe (e.g. Trendall and Blockley, 1970).

Numerous hypotheses have been proposed, but there is still far from consensus on the genesis of the Superior-type BIF. However, it is unlikely that they could have formed directly from volcanic exhalations in the manner proposed for the Algoma-type BIF, which are generally smaller and more intimately associated with volcanic rocks. It seems feasible that there was a long-term build up of dissolved Fe and Si contents in deep unoxidized waters, mainly from degassing and sea-floor alteration reactions. This would have been facilitated by the absence of microorganisms capable of removing Si from solution, and by insufficient supply of sulphide ions to precipitate all of the Fe^{2+}. Such a deep hydrosphere could have upwelled into shallow-water, stable shelf basins, where photosynthetic microorganisms were flourishing, and where mild evaporation may have accelerated BIF precipitation. Whether or not iron bacteria catalysed precipitation of the iron minerals is a contentious point (e.g. Trudinger, 1976). Restricted influx of detrital materials is presumably reflecting low relief of adjacent land and, coupled with an absence of recognizable residual aluminous lateritic deposits in the geological record, is widely accepted as evidence against derivation of the bulk of the Fe and Si from terrestrial sources.

The origin of the rhythmic microbands in the BIF is an intriguing problem. These have been explained in terms of (1) evaporitic varves or rhythmites (Trendall, 1973), (2) annual variations in temperature in lakes (Eugster, 1969), and episodic growth of phyto-planktonic microbiota, coupled with glaciation which caused temperature gradients in the hydrosphere and upwelling of Fe-rich bottom waters (Cloud, 1973).

The virtual absence of major Superior-type BIF after ~2 Gyr BP has been explained by Cloud (op. cit.) in terms of early Proterozoic proliferation of algae with consequent increase of net oxygen production and eventual decrease of soluble iron in the hydrosphere to vanishingly low levels (cf. discussion of development of an oxygenated atmosphere). The overall paucity of sulphide-facies BIF in these Superior-type BIF can be explained by the absence, or restricted activity, of sulphate-reducing bacteria in the early Proterozoic depositional basins.

Manganese deposits. Significant manganese mineralization in pre-2 Gyr Proterozoic sequences is commonly associated with carbonates. The main examples are in the Transvaal and Loskop Systems of South Africa (De Villiers, 1960), the Amapa series and Minas Gerais succession of Brazil (Scarpelli, 1973) and the Aravalli Group of India (Roy, 1966, 1976). Small deposits, enriched by Proterozoic and/or Tertiary weathering occur towards the eastern end of the Hamersley and Bangemall basins of the Pilbara Block, Western Australia (Blockley, 1975); these appear similar to the South African manganese oxide concentrations which probably formed in favourable structures during weathering cycles which released and oxidized manganous ions from carbonate lattices. Whilst the Indian and Brazilian deposits are regarded as synsedimentary manganese accumulations, they have been enriched during younger oxidative weathering, and their original mineralogy is obscured.

Eriksson and Truswell (1978) point to distinct iron and manganese enrichments in the Lower Proterozoic Gamagara Shales of South Africa, which they relate to a geochemical facies relationship, reflecting the more oxidizing conditions required for manganese precipitation.

Chromite in layered intrusives. Large, layered mafic—ultramafic intrusives are features of the early Proterozoic and persist into the Middle Proterozoic. They represent major introductions of mantle-derived basic magma to stabilized upper crust where they underwent fractional crystallization, which led to major economic mineralization in some cases. They appear to have ascended in mega-fracture systems which formed in the newly stabilized crust, commonly oblique to the major trends of greenstone belts.

The oldest dated intrusion of this type is the ~2.9 Gyr old Usushwana complex in South Africa (Hunter, 1970) which consists essentially of gabbro and granophyre; it contains only minor concentrations of Cr, Ni and Cu. The ~2.75 Gyr old Stillwater complex of Montana (Jackson, 1969) and the ~2.55 Gyr old Great Dyke of Rhodesia (Bichan, 1970) both contain major chromite accumulations in their ultramafic fractions.

Important as the latter two intrusions are, they are overshadowed by the younger Bushveld complex of South Africa (e.g. Hunter, 1976) and the Sudbury irruptive of Canada (e.g. Naldrett et al., 1970). The immense 2 Gyr old Bushveld intrusive contains significant Ti–Fe, V and Ni concentrations in addition to major deposits of Cr and platinoids. It occurs along strike from the Great Dyke and probably reflects reactivation of the same major fracture system. The Sudbury irruptive is ~1.7 Gyr old and contains the world's greatest Ni deposits, plus important Cu, Co, Au and platinoid concentrations. There is widespread shock metamorphism in pre-irruptive rocks; whilst this is commonly interpreted as indicating that the igneous activity was triggered by meteorite impact, the possibility of a cryptoexplosive origin for the Sudbury structure cannot be ruled out.

Anhaeusser (1976a) and Watson (1976) suggested the possibility that the important chromium province of southern Africa, involving the Great Dyke, Bushveld and also Selukwe (discussed earlier under Archaean mineralization), is reflecting the long-term existence of a Cr-rich mantle under this region. If this is so, magmas formed by partial melting of this mantle should all be Cr-rich. However, there is no evidence of this in komatiitic volcanics in Rhodesia (e.g. Bickle et al., 1976), nor in the composition of the parent magma of the Great Dyke as deduced from chill phases of satellite dykes (0.11% Cr_2O_3, 19% MgO; I.D.M. Robertson and Van Breemen, 1970). Furthermore, a recent estimate of the composition of the Bushveld parent magma by E.N. Cameron (1978) indicated ≳0.1% Cr_2O_3 and ~13% MgO; this is more magnesian than tholeiitic magmas, but its Cr/Mg ratio is not necessarily indicative of Cr-rich source region in the mantle. The factors which could have been most important in formation of chromite layers are:

(1) tectonic conditions favourable for formations of large layered Mg-rich intrusives rather than extrusives, and

(2) precipitation of Cr_2O_3 from these intrusives as a result of reaction with sialic country rocks (Irvine, 1975), blending of picritic tholeiite liquid with relatively siliceous liquid from fractionation of earlier liquid of the same type (Irvine, 1977), or shifts in total pressure (Cameron, 1978).

Development of an oxygenated atmosphere

Abiotic syntheses of living organisms apparently required anoxic conditions and, although there is evidence for photosynthetic activity in the early Archaean, oxygen would only have become a stable component of the atmosphere after saturation of oxygen "sinks" (Fe^{2+}, H_2S, CH_4, etc.) in the hydrosphere–atmosphere system.

Cloud (1973, 1974, 1976) argued that the atmosphere was not oxygenated prior to about 2 Gyr BP. He suggested that (1) primitive algae required Fe^{2+} as an oxygen acceptor, a factor which inhibited their proliferation, and that advanced algae with internal oxygen-mediating enzymes did not evolve until the Lower Proterozoic; (2) the resultant rapid proliferation of algae, accompanied by a marked increase in net oxygen production, led to precipitation of the abundant Lower Proterozoic BIF from Fe^{2+} which had built up in previously reducing parts of the hydrosphere; and (3) once the seas were stripped of their Fe^{2+}, oxygen entered the atmosphere, an event preserved in the sedimentary record as the rapidly declining importance of Superior-type BIF and the formation of the oldest primary terrestrial red beds. As further evidence supporting an oxygen-free atmosphere prior to ~2 Gyr BP, Cloud cites an absence of oxidative weathering in the Archaean, and the lack of detrital uraninite-pyrite deposits after the Lower Proterozoic.

However, whilst Cloud offers a viable interpretation of the geological record, his arguments are not unequivocal and a number of workers have advocated much earlier development of an oxygenated atmosphere. For example, it has been argued that (1) synthesis of life forms probably occurred in locally reducing pools and these could have existed beneath an oxygenous atmosphere (e.g. M.D. Muir, 1979); (2) red beds and conglomerate clasts with oxidized margins are known from Archaean sequences (e.g. Dimroth and Kimberly, 1976), and red beds form diagenetically and do not necessarily reflect atmospheric composition (e.g. M.D. Muir, 1979); and (3) detrital uraninite and pyrite occur in modern sediments and thus do not indicate an absence of oxygen in the atmosphere (e.g. Simpson and Bowles, 1979). These arguments will now be discussed in turn:

(1) Given that life could well have evolved in reducing pools, this in no way supports the existence of an oxygenated atmosphere; in fact it is hard to envisage the stable existence of reducing pools beneath an oxidizing atmosphere before microbial processes became established.

(2) Cloud (op. cit.) maintains that Archaean "red-beds" formed by later oxidation of portions of originally reducing sediments, and the same could be the case for oxidized margins of conglomerate clasts. It is also possible that the oxidation was a primary feature formed in local, shallow-water, oxidizing environments which existed in the Archaean in the absence of an oxygenated atmosphere (cf. Fig. 16).

(3) The occurrence of relatively minor amounts of detrital uraninite and sulphide minerals in modern sediments may not be relevant to the formation of economic concentrations of these minerals in the early Archaean by repeated physical reworking in nearshore environments. The persistence of such readily oxidizable minerals during this reworking would obviously be more likely if the atmosphere was not oxygenated.

On balance, it appears that existing evidence is best interpreted in terms of the absence of an oxygenous atmosphere during formation of both the vast early Proterozoic BIF and the important Witwatersrand-type uraninite—pyrite concentrations. The rapid decline in importance of such deposits, and the incoming of extensive red beds are also most readily explained in terms of complete saturation of oxygen sinks in the hydrosphere-atmosphere system around 2 Gyr BP, with consequence development of a stable oxygenated atmosphere. As discussed by Cloud (1976) and Eriksson and Truswell (1978), the overlap of red bed and BIF deposition between 2.0 and 1.8 Gyr BP is not difficult to reconcile during this period of atmospheric transition.

Interim summary

(1) The Archaean—Proterozoic transition was characterized by the accumulation of widespread, epicratonic, shallow-water to subaerial sequences; these comprise mainly quartz-rich clastics, carbonates (some with stromatolites), cherts and mafic to felsic volcanics. It reflects extensive stripping of Archaean crust, which commenced as early as 3 Gyr BP in southern Africa, and the evolution of a thick stable continental crust.

(2) The main types of mineralization in early Proterozoic rocks are distinctly different from those of the Archaean:

(a) *Gold and uraninite deposits*: These occur in quartz- and pyrite-rich clastic sedimentary sequences. The source of the Au and pyrite was probably the quartz—Au mineralization in nearby, uplifted, greenstone belts, whilst uraninite may have come from granitoids (pegmatites). There is evidence that gold, uraninite and pyrite were transported as detrital grains, but it also appears Au and U were transported to some extent in solution. Critical factors in ore formation appear to have been continual uplift of source regions, repeated reworking of fluvial fan or fan delta sediments and the development of algal mats which concentrated metals from solution.

(b) *Unconformity-related uranium deposits*: These occur in breccia or shear zones in rocks beneath Lower to Middle Proterozoic unconformities. They are predominantly pitchblende deposits some of which exhibit evidence for both low- and moderate-temperature processes. Uranium appears to have been derived from Archaean basement and early Proterozoic cover sequences, via surficial waters and/or subsurface brines.

(c) *Banded iron formation (Superior type)*: These are of very widespread occurrence and they provide the bulk of the world's Fe. They are commonly larger than Archaean BIF and they frequently do not have close spatial relationships with volcanic rocks. It is feasible that these precipitated as a result of the general build-up of dissolved ferrous iron and silica in deep, unoxidized hydrosphere as a result of an exhalative activity and sea-floor alteration reactions. BIF would have precipitated as such waters upwelled onto stable shelf platforms where photosynthetic microorganisms were flourishing; mild evaporation may have facilitated BIF precipitation. Superior-type BIF are absent, or very rare, after ~1.8 Gyr, and primary red beds became increasingly abundant after this time.

(d) *Manganese deposits*: Many of these are associated with carbonates and other shallow-water sedimentary rocks; some are geochemical facies variants of iron deposits. They probably also reflect the long-term Mn build-up in the deep hydrosphere, and the movement of such water to oxidized shelf-type environments where conditions were suitable for Mn precipitation.

(e) *Chromite in layered intrusives*: Chromite accumulations in ultramafic fractions of large, Mg-rich (komatiitic?) layered intrusives are most important in southern Africa, where there are also major Archaean chromite deposits. The origin of this major Cr province may not lie in anomalously high Cr/Mg ratios in the underlying mantle, but rather in suitable tectonic conditions for formation of intrusives rather than flows, and favourable physicochemical conditions for chromite precipitation and accumulation. Other minerals concentrated in large, layered Proterozoic intrusives include platinoids, and Ni—Cu.

(3) There is no compelling evidence for the evolution of an oxygenated atmosphere before ~2 Gyr BP.

SUMMARY OF SUBSEQUENT EVOLUTIONARY TRENDS

Detailed discussion of evolutionary trends in the remainder of the Proterozoic and the Phanerozoic is beyond the scope of this chapter. The main trends are evident from Fig. 26 and from the brief outline which follows.

A number of important types of mineralization evolved after the early Proterozoic. These include diamonds in kimberlite pipes (particularly in southern Africa and Siberia) and various commodities in carbonatites (mainly P, REE, Nb, F and Ba less commonly Cu, Fe and U), which are again most important in Southern Africa. Both kimberlites and carbonatites are concentrated in stabilized Precambrian cratons, where they appear to have formed in association with large-scale uplift and distension, or rifting (Ferguson, 1976a, b). As summarized by Watson (1976), the earliest known kimberlites and carbonatites formed shortly before 2 Gyr BP and intrusives of different ages can occur within a single province; however, most of the known kimberlites and a high proportion of carbonatites formed during the late Phanerozoic.

A number of deposit types in predominantly sedimentary successions also evolved after the early Proterozoic. These include:

(1) Large stratiform Pb—Zn—Ag deposits in variably carbonaceous and pyritic shales (e.g. I.B. Lambert, 1976). These are important in the Middle Proterozoic and the early Phanerozoic and it is likely that at least part of their sulphide was generated biogenically. Metals may have been derived from the sediments and/or from igneous activity.

(2) Stratabound Pb—Zn deposits in carbonates (e.g. Brown, 1967). The first significant examples appeared in the Middle Proterozoic, but the most important ores occur through the Phanerozoic. Bacterial sulphate reduction and abiological sulphate reduction of sulphate have both been called upon as sulphide sources in such deposits, with the metals probably introduced by highly saline formation waters ascending from underlying sedimentary units.

(3) Stratabound Cu and/or U deposits associated with red beds, carbonates, evaporites and reduced strata (e.g. Fleischer et al., 1976; Jung and Knitzschke, 1976; G.E. Smith, 1976; A.C. Brown, 1978). The first major examples occur in the Upper Proterozoic, and other important deposits of this type occur in the Middle and Upper Phanerozoic. In

Gyr BP	4	3	2	1	0

PRE-ARCHAEAN	ARCHAEAN	PROTEROZOIC	PHAN-IEROZOIC

GEOBIOCHEMICAL MILESTONES

△ 1 △ 2 △ 3 △△△ 4 5 6 △ 7 △ 8 9

EVOLUTION OF CONTINENTAL CRUST

- Development of primordial, mainly simatic crust
- Meteorite impacts
- Growth of sialic continental crust
- Stable continental blocks, platform cover sequences, mobile belts
- Evidence for "modern-style" plate tectonics ←?

SEDIMENTARY ROCKS AND ASSOCIATED MINERALIZATION

- Cherts
- Dolomites
- Limestones (predominantly $CaCO_3$)
- Greywackes
- Shales and siltstones
- Carbonaceous sediments
- Arkoses, oligomictic conglomerates orthoquartzites
- Oxidized terrestrial sedimentary rocks (red beds)
- Phosphorites
- Halite (casts in Proterozoic)
- Sulphates (Ca sulphates largely pseudomorphed in Precambrian)
- Algoma type BIF deposits
- Superior type BIF deposits
- Manganese deposits
- Uraninite, gold, pyrite placer deposits
- Other uranium deposits
- Sedimentary copper deposits
- Sandstone type lead deposits
- Shale-hosted lead-zinc deposits
- Carbonate-hosted lead-zinc deposits
- Coal deposits
- Oil and gas fields
- Residual deposits

OTHER TYPES OF MINERALIZATION

- Volcanogenic massive sulphide deposits (copper, zinc ± lead)
- Gold deposits (all types)
- Nickel sulphide deposits (± copper, platinoids)
- Chromite deposits
- Porphyry (s. l.) copper deposits (± molybdenum, gold)
- Tin deposits
- Tungsten deposits
- Pegmatite-associated deposits
- Mercury deposits
- Antimony deposits
- Carbonatite-associated deposits
- Kimberlitic diamond pipes

many of these deposits, there is evidence for replacement of biogenic pyrite. Potential metal sources include the host sedimentary sequence and mineralized basement; Cu can also be released during alterations of mafic rocks.

(4) Phosphorites, which first became important in the uppermost Proterozoic and are important in lower Cambrian strata (e.g. Cook, 1976).

(5) Halite deposits, which are confined to the Phanerozoic (e.g. Veizer, 1976). However, there are abundant casts after halite (often associated with pseudomorphs after gypsum—anhydrite) in some Lower, Middle and Upper Proterozoic sequences.

(6) Oil and gas deposits (e.g. Levorsen, 1954). These are abundant in Phanerozoic sedimentary basins, and the oldest known economic oil deposit is probably of uppermost Proterozoic age (Murray, 1964).

(7) Coal deposits. These formed only after land plants became established in the middle Phanerozoic, although local concentrations of "coaly" matter (referred to as shungite, graphite, anthracite and thucolite) extend back as far as the Archaean—Proterozoic transition (e.g. M.D. Muir, 1979).

Many types of Phanerozoic mineralization associated with igneous rocks have been related to particular plate-tectonic environments (e.g. Table III, Fig. 27). Whilst some primitive form of plate tectonics is possible in the Archaean, it was shown above that there are difficulties in a uniformitarian extrapolation of modern global tectonics back further than the Upper Proterozoic. Nevertheless, several types of mineralization related to specific plate environments in the Phanerozoic are also represented in the Archaean. For example, the important massive sulphide mineralization in Archaean felsic volcanic-rich sequences is similar in several important respects to the subduction-related volcanogenic massive sulphide deposits of Phanerozoic age. Furthermore, the main deposits of Sn, W, Sb and Hg are also related to Phanerozoic convergent plate boundaries, but deposits of these metals are also present in the Archaean, albeit on a relatively minor scale; of these metals, Sn is the only one occurring in significant concentrations in the Proterozoic. From these observations, it appears unlikely that precise tectonic environment is the critical factor in the genesis of many types of mineralization. Exceptions to this generalization are the ophiolite-associated Cu—pyrite deposits (Cyprus type) and podiform Cr—Pt deposits (Alpine type). These are restricted to the Phanerozoic and their appearance can be related to the onset of ocean-floor spreading processes which caused thrusting (obduction) of slivers of oceanic lithosphere (ophiolites) onto continental crust.

Fig. 26. Diagram summarizing milestones in geobiochemical evolution, general evolutionary trends for the crust, sedimentary rocks and mineralization. Based on information from Cloud (1976), Laznika (1973), M.D. Muir (1979), Pereira and Dixon (1965), Veizer (1976), Watson (1973, 1976) and the authors' personal assessments. *1* = oldest putative microfossils; *2* = earliest isotopic evidence for biological sulphate reduction; *3* = development of oxygenous atmosphere; *4* = oldest eukaryotic microorganisms; *5* = oldest megascopic algae; *6* = oldest animal burrows; *7* = oldest animal body fossils; *8* = oldest animals with skeletons; *9* = oldest vascular land plants. Thick line signifies main period of formation, thin line signifies occurrence in moderate abundances; dashed line signifies sporadic occurrences.

TABLE III

Suggested plate tectonic settings for formation of various types of Phanerozoic mineralization

(From Mitchell and Garson, 1976)

Plate tectonic setting		Mineralization Type	For-mation	Em-place-ment	Ex-posure	Example Locality	Age
Ocean floor spreading related	Intercontinental Hot Spots and Rift Zones	Tin-fluorite–niobium	✓			Nigerian tin fields	Jurassic
		Carbonatite mineralization (Nb, Ce, P, Sr, Ba)	✓		✓	East African rift	Jurassic to present
		Benue-type lead deposits	✓			Benue trough, Nigeria	Cretaceous (?)
		Sullivan-type marine sulphides	✓		✓	Sullivan mine, Br. Columbia	Proterozoic (?)
	Intercontinental Rift Zones	Metal-rich muds (Zn, Cu)	✓			Red Sea deeps	Quaternary
		Mississippi Valley lead-zinc-barite	✓		✓	Red Sea coast, Saudi-Arabia	Cenozoic
	Oceanic Rises and Ocean Floor	Cyprus-type copper-lead-zinc massive sulphides	✓			None exposed	
		Podiform chromite	✓			None exposed	
		Nickel and platinum sulphides	✓			None exposed	
		Manganese nodules	✓		✓	Pacific Ocean floor	Quaternary
Subduction related	Island Arc Magmatic Belts	Porphyry copper-gold	✓		✓	Bougainville, Solomon Islands, Philippines	Late Cenozoic
		Mercury	✓		✓		Tertiary
		Kuroko-type zinc-copper-lead	✓		✓	Kosaka, Honshu; Vanua Levu,	Miocene
		Auriferous quartz veins	✓		✓	Hauraki Peninsula, Fiji; N. Zealand	Late Tertiary
		Gold tellurides and auriferous sulphides	✓			Vatukoula, Fiji	Early Tertiary / Pliocene
		Besshi-type massive sulphides	✓		✓	Besshi, Japan	Early Mesozoic ?
		Native sulphur-pyrite	✓		✓	Japan	Quaternary
	Andean Type Magmatic Belts	Porphyry copper–molybdenum	✓		✓	Braden, Chile	
		Tin-tungsten-fluorite	✓		✓	Eastern Cordillera, Peru	Pliocene
	Back Arc	Tin-tungsten-fluorite	✓		✓	South China	Late Mesozoic
	Continental Margin	Antimony	✓	✓	✓	Eastern Burma	Late Mesozoic (?)

	Deposits	Epithermal gold-silver veins	Basin and Range Province	Age
Outer Arcs		Epithermal gold-silver veins	?Chin Hills, Burma	Tertiary
		Auriferous quartz veins	Coast Ranges, California	Eocene
		Mercury		Late Mesozoic?
		Ocean rise and ocean floor or back arc basin deposits	None exposed	
Collision related — Continental Collision Magmatic Belts		Tin-tungsten-fluorite	Cornwall, Erzgebirge	Early Permian
		Iron-titanium in anorthosites		
		Native silver nickel cobalt arsenide	Cornwall, Erzgebirge	Early Permian
		Gemstone deposits?	Pakistan and Burma	Tertiary?
Continental Collision Tectonic Belts		Porphyry copper	Coed-y-Brenin, Wales	Palaeozoic
		Mercury		
		Kuroko-type	Umm Samiuki, Egypt	Proterozoic
			Buchans, Newfoundland	Palaeozoic
Island arc magmatic belt deposits		Auriferous quartz veins		
		Besshi-type sulphides		
		Iron-titanium in anorthosites	Grenville Province, Canada	Proterozoic
Obducted Ophiolites		Cyprus-type sulphides	Cyprus? Betts' Cove, Newfoundland	Cyprus–Late Mesozoic / Betts' Cove–Lower Palaeozoic
Ocean ridge and floor and back arc basin deposits		Podiform chromite	Philippines	Late Mesozoic to Early Tertiary
		Nickel and plat. sulphides	Philippines	
		?Manganese nodules	Semail nappe, Oman	Late Mesozoic
Interior of Underthrusing Continents		Irish-type base metal deposits	Navan, Silvermines, Ireland	Early Carboniferous
		Stratabound uranium	Molasse facies sediments, Himalayas	Tertiary

(continued)

TABLE III (continued)

Plate tectonic setting	For-mation	Em-place-ment	Ex-posure	Mineralization Type	Example Locality	Age
Post-collision Magmatic Belts	✓		✓	Uranium-rich alkaline rocks	Rome igneous province	Quaternary
Transform faults	✓		✓	Metal-rich muds	Red Sea Deeps	Quaternary
	✓		✓	Hydrothermal base metal deposits (Zn–Pb–Ba–Sr)	Red Sea Coast	Miocene
Oceanic Transform Faults	✓			Copper and nickel sulphides in ultramafic rocks	St John's Island, Red Sea Gabbro Akarem, Egypt	Proterozoic (?)
Continental Transform Fractures	✓		✓	Carbonatite (Nb, P, Ce, Ba) mineralization	Angola	Mesozoic/Cenozoic
	✓		✓	Kimberlite diamonds	Angola	Mesozoic/Cenozoic
	✓		✓	Porphyry copper	N. America, Philippines	Mesozoic/Cenozoic

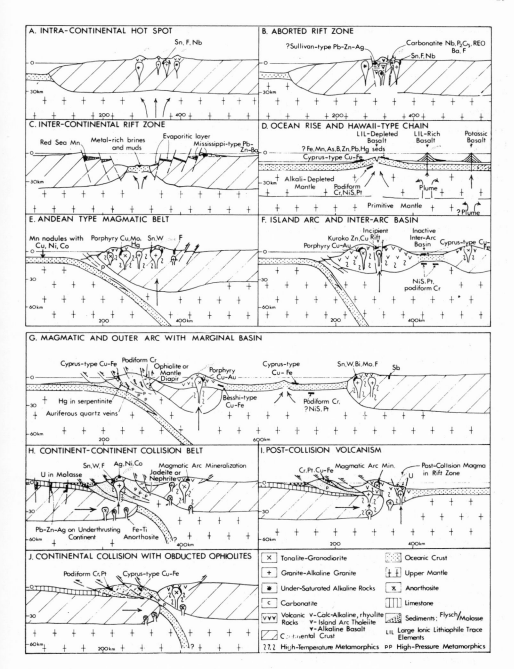

Fig. 27. Schematic sections showing relationships of various types of Phanerozoic mineralization to plate margins and related tectonic environments (from Mitchell and Garson, 1976).

ACKNOWLEDGEMENTS

We are grateful to M.E. Barley, J.S.R. Dunlop, D.R. Hudson, A.Y. Glikson and R.J. Marston for reading the manuscript and making constructive comments. D.I.G. had beneficial discussions with M.J. Bickle, and much of the section on Ni—Cu deposits has evolved from his joint studies with F.M. Barrett, R.A. Binns and R.J. Gunthorpe. Drafting was carried out by C. Steel and the CSIRO drawing office, North Ryde. The Baas Becking Laboratory is supported by the Australian Mineral Industries Research Association, Limited, the Bureau of Mineral Resources and the Commonwealth Scientific and Industrial Research Organization.

This represents a contribution to the International Geological Correlation Programme Projects No. 91 — Metallogeny of the Precambrian and No. 157 — Early Organic Evolution and Mineral and Energy Resources.

REFERENCES

Abelson, P.H., 1966. Chemical events on the primitive Earth. *Natl. Acad. Sci. Proc. U.S.A.*, 54: 1490–1497.

Allart, J.H., 1976. The pre-3760 m.y. old supracrustal rocks of the Isua area, central west Greenland, and the associated occurrence of quartz-banded ironstone. In: B.F. Windley (Editor), *The Early History of the Earth*. Wiley, London, pp. 177–190.

Anders, E., 1968. Chemical processes in the early solar system as inferred from meteorites. *Acc. Chem. Res.*, 1: 289–298.

Anhaeusser, C.R., 1971a. The Barberton Mountain Land, South Africa — a guide to the understanding of the Archaean geology of Western Australia. *Geol. Soc. Aust., Spec. Publ.*, 3: 103–120.

Anhaeusser, C.R., 1971b. Cyclic volcanicity and sedimentation in the evolutionary development of Archaean greenstone belts of shield areas. *Geol. Soc. Aust., Spec. Publ.*, 3: 57–70.

Anhaeusser, C.R., 1973. The evolution of the early Precambrian crust of southern Africa. *Philos. Trans. R. Soc. London, Ser. A*, 273: 359–388.

Anhaeusser, C.R., 1976a. Archaean metallogeny in southern Africa. *Econ. Geol.*, 71: 16–43.

Anhaeusser, C.R., 1976b. The nature and distribution of Archaean gold mineralization in southern Africa. *Miner. Sci. Eng.*, 8: 46–83.

Anhaeusser, C.R. and Button, A., 1976. A review of southern African stratiform ore deposits — their position in time and space. In: K.H. Wolf (Editor), *Handbook of Strata-Bound and Stratiform Ore Deposits, 5*. Elsevier, Amsterdam, pp. 257–319.

Anhaeusser, C.R. and Robb, L.J., 1978. Regional and detailed field and geochemical studies of Archaean trondhjemitic gneisses, migmatites and greenstone xenoliths in the southern part of the Barberton Mountain Land, South Africa. *Proc. 1978 Archaean Geochemistry Conf.*, Canada, pp. 322–325.

Anhaeusser, C.R., Mason, R., Viljoen, M.R. and Viljoen, R.P., 1969. A reappraisal of some aspects of Precambrian shield geology. *Geol. Soc. Am. Bull.*, 80: 2175–2200.

Anhaeusser, C.R., Fritze, K., Fyfe, W.S. and Gill, R.C.O., 1975. Gold in "primitive" Archaean volcanics. *Chem. Geol.*, 16: 129–135.

Anonymous (Editor), 1973. *Genesis of Precambrian Iron and Manganese Deposits. Proc. of Kiev Symposium, 1970.* UNESCO, Paris, 382 pp.

Appel, P.W.V., 1979. Stratabound copper sulfides in a banded iron-formation and in basaltic tuffs in the early Precambrian Isua supracrustal belt, West Greenland, *Econ. Geol.*, 74: 45–52.

Archibald, N.J. and Bettenay, L.F., 1977. Indirect evidence for tectonic reactivation of a pre-greenstone sialic basement in Western Australia. *Earth Planet. Sci. Lett.,* 33: 370–378.

Archibald, N.J., Bettenay, L.F., Binns, R.A., Groves, D.I. and Gunthorpe, R.J., 1978. The evolution of Archaean greenstone terrains, Eastern Goldfields Province, Western Australia. *Precambrian Res.,* 6: 103–131.

Armstrong, R.L., 1968. A model for the evolution of Sr and Pb isotopes in a dynamic Earth. *Rev. Geophys.,* 6: 175–199.

Arndt, N.T., 1977. Ultrabasic magmas and high-degree melting of the mantle. *Contrib. Mineral. Petrol.,* 64: 205–222.

Arndt, N.T., Naldrett, A.J. and Pyke, D.R., 1977. Komatiitic and iron-rich tholeiite lavas of Munro Township, northeast Ontario. *J. Petrol.,* 18: 319–369.

Arriens, P.A., 1971. The Archaean geochronology of Australia. *Geol. Soc. Aust., Spec. Publ.,* 3: 11–24.

Arriens, P.A. and Lambert, I.B., 1969. On the age and strontium isotopic geochemistry of granulite facies rocks from the Fraser Range, Western Australia, and the Musgrave Ranges, Central Australia. *Geol. Soc. Aust., Spec. Publ.,* 2: 377–388.

Arth, J.G. and Hanson, G.N., 1972. Quartz diorites derived by partial melting of eclogite or amphibolite at mantle depths. *Contrib. Mineral. Petrol.,* 37: 161–174.

Arth, J.G. and Hanson, G.N., 1975. Geochemistry and origin of the early Precambrian crust of northeastern Minnesota. *Geochim. Cosmochim. Acta.,* 39: 325–362.

Arth, J.G., Arndt, N.T. and Naldrett, A.J., 1977. Genesis of Archaean komatiites – trace element evidence from Munro Township. *Geology,* 5: 590–594.

Baadsgaard, H., 1976. Further U–Pb dates on zircons from the early Precambrian rocks of the Godthaabsfjord area, west Greenland. *Earth Planet. Sci. Lett.,* 33: 261–267.

Badham, J.P.N. and Stanworth, C.W., 1977. Evaporites from the Lower Proterozoic of the east arm, Great Slave Lake. *Nature,* 268: 516–517.

Baer, A.J., 1977. Speculations on the evolution of the lithosphere. *Precambrian Res.,* 5: 249–260.

Bailey, D.K., Tarney, J. and Dunham, K. (Organizers), 1978. *The Evidence for Chemical Heterogeneity in the Earth's Mantle.* R. Society, London, Abstract Volume.

Baragar, W.R.A. and Goodwin, A.M., 1969. Andesites and Archaean volcanism of the Canadian shield. In: A.R. McBirney (Editor), *Andesite Symposium. Oreg. Dep. Geol. Mineral., 2nd Bull.,* 65: 121–142.

Baragar, W.R.A. and McGlynn, J.C., 1976. Early Archaean basement in Canadian Shield: A review of the evidence. *Geol. Surv. Can., Pap.,* 76–14.

Barley, M.E., 1978. Shallow-water sedimentation during deposition of the Archaean Warrawoona Group, Eastern Pilbara Block, Western Australia. In: J.E. Glover and D.I. Groves (Editors), *Archaean Cherty Metasediments: their Sedimentology, Micropalaeontology, Biogeochemistry and Significance to Mineralization, 2.* Dep. Geol. and Extension Service, Univ. West. Aust., Perth, pp. 22–29.

Barley, M.E., Dunlop, J.S.R., Glover, J.E. and Groves, D.I., 1979. Sedimentary evidence for a distinctive Archaean shallow-water volcanic-sedimentary facies, eastern Pilbara Block, Western Australia. *Earth Planet. Sci. Lett.,* 43: 74–84.

Barrett, F.M., Binns, R.A., Groves, D.I., Marston, R.J. and McQueen, K.G., 1977. Structural history and metamorphic modification of Archaean volcanic-type nickel deposits, Yilgarn Block, Western Australia. *Econ. Geol.,* 72: 1195–1223.

Barton, J.M., Fripp, R.E.P. and Ryan, B., 1977. Rb/Sr ages and geological setting of ancient dykes in the Sand River area, Limpopo Mobile Belt, South Africa. *Nature,* 267: 487–490.

Bartram, G.D. and McCall, G.J.H., 1971. Wall-rock alteration associated with auriferous lodes in the Golden Mile, Kalgoorlie. *Geol. Soc. Aust., Spec. Publ.,* 3: 191–199.

Bavinton, O.A. and Keays, R.R., 1978. Precious metal values from interflow sedimentary rocks from the komatiite sequence at Kambalda, Western Australia. *Geochim. Cosmochim. Acta,* 42: 1151–1163.

Bichan, R., 1970. The evolution and structural setting of the Great Dyke, Rhodesia. In: T.N. Clifford

432

and I.G. Gass (Editors), *African Magmatism and Tectonics.* Oliver and Boyd, London, pp. 51–71.

Bickle, M.J. and Nisbet, E.G., 1977. Earliest Precambrian ultramafic-mafic volcanic rocks: ancient oceanic crust or relic terrestrial maria: Comment – the chickens or the egg. *Geology,* 5: 68.

Bickle, M.J., Hawkesworth, C.J., Martin, A., Nisbet, E.G. and O'Nions, R.K., 1976. Mantle compositions derived from the chemistry of ultramafic lavas. *Nature,* 263: 577–580.

Bickle, M.J., Ford, C.E. and Nisbet, E.G., 1977. The petrogenesis of peridotitic komatiites: evidence from high-pressure melting experiments. *Earth Planet. Sci. Lett.,* 37: 97–106.

Binns, R.A., 1964. Zones of progressive metamorphism in the Willyama Complex, Broken Hill district, N.S.W. *J. Geol. Soc. Aust.,* 11: 280–330.

Binns, R.A. and Groves, D.I., 1976. Iron–nickel partition in metamorphosed olivine–sulfide assemblages from Perseverance, Western Australia. *Am. Mineral.,* 61: 782–787.

Binns, R.A., Gunthorpe, R.J. and Groves, D.I., 1976. Metamorphic patterns and development of greenstone belts in the eastern Yilgarn Block, Western Australia. In: B.F. Windley (Editor), *The Early History of the Earth.* Wiley, London, pp. 303–316.

Binns, R.A., Groves, D.I. and Gunthorpe, R.J., 1977. Nickel sulfides in Archaean ultramafic rocks of Western Australia. In: A.V. Sidorenko (Editor), *Correlation of the Precambrian, 2.* Nauka, Moscow, pp. 349–380.

Birch, F., 1965. Speculations on the Earth's thermal history. *Geol. Soc. Am. Bull.,* 76: 133–154.

Blockley, J.G., 1973. Geology of Western Australian tin deposits. *Australas. Inst. Min. Metall., West. Aust. Conf.,* 2: 131–140.

Blockley, J.G., 1975. Pilbara manganese province, W.A. In: C.L. Knight (Editor), *Economic Geology of Australia and Papua-New Guinea, 1. Metals.* Australas. Inst. Min. Metall., Melbourne, pp. 1019–1020.

Both, R.A. and Rutland, R.W.R., 1976. The problem of identifying and interpreting stratiform ore bodies in highly metamorphosed terrains: the Broken Hill example. In: K.H. Wolf (Editor), *Handbook of Strata-Bound and Stratiform Ore Deposits, 4.* Elsevier, Amsterdam, pp. 261–325.

Bowes, D.R., 1976. Archaean crustal history in the Baltic shield. In: B.F. Windley (Editor), *The Early History of the Earth.* Wiley, London, pp. 481–488.

Boyle, R.W., 1971. The geology, geochemistry and origin of the gold deposits of the Yellowknife district. *Can. Geol. Surv. Mem.,* 310.

Boyle, R.W., 1976. Mineralization processes in Archaean greenstone and sedimentary belts. *Geol. Surv. Can., Pap.* 75-15.

Brown, A.C., 1978. Stratiform copper deposits – evidence for their post-sedimentary origin. *Miner. Sci. Eng.,* 10: 172–181.

Brown, H., 1952. Rare gases and the formation of the Earth's atmosphere. In: G. Kuiper (Editor), *Atmospheres of the Earth and Planets.* Univ. Chicago Press, 2nd ed., pp. 258–266.

Brown, J.S. (Editor), 1967. Genesis of stratiform lead–zinc–barite–fluorite deposits. *Econ. Geol. Monogr.,* 3.

Burger, A.J. and Coertze, F.J., 1973. Radiometric age measurements on rocks from southern Africa to the end of 1971. *A.A. Geol. Surv. Bull.,* 58: 46 pp.

Burke, K., Dewey, J.F. and Kidd, W.S.F., 1976. Dominance of horizontal movements, arcs and micro-continental collisions during the later permobile regime. In: B.F. Windley (Editor), *The Early History of the Earth.* Wiley, London, pp. 113–130.

Burt, D.R.L. and Sheppy, N.R., 1975. Mt. Keith nickel sulphide deposits. In: C.L. Knight (Editor), *Economic Geology of Australia and Papua-New Guinea, 1. Metals.* Australas. Inst. Min. Metall., Melbourne, pp. 159–168.

Cameron, E.M., 1975. Geochemical methods of exploration for massive sulphide mineralization in the Canadian shield. In: I.L. Elliott (Editor), *Proc. 5th Int. Geochem. Explor. Symp., Vancouver, 1974,* pp. 21–49.

Cameron, E.N., 1978. The lower zone of the eastern Bushveld Complex in the Olifants River Trough. *J. Petrol.,* 19: 437–462.

Carey, S.W., 1975. The Expanding Earth – an essay review. *Earth-Sci. Rev.,* 11: 105–143.

Carey, S.W., 1976. *The Expanding Earth.* Elsevier, Amsterdam.

Chapman, D.G. and Groves, D.I., 1979. Preliminary study of the distribution of gold in Archaean interflow sulphidic metasediments, Yilgarn Block, Western Australia. In: J.E. Glover and D.I. Groves (Editors), *Gold Mineralization.* Dep. Geol. and Extension Service, Univ. West. Aust., 3: 76–88.

Chapman, H.J. and Spooner, E.T.C., 1977. [87]Sr enrichment of ophiolitic sulphide deposits in Cyprus confirms ore formation by circulating seawater. *Earth Planet. Sci. Lett.,* 35: 71–78.

Clark, S.P., Turekian, K. and Grossman, L., 1972. Model for early history of the Earth. In: E.C. Robertson (Editor), *The Nature of the Solid Earth.* McGraw-Hill, New York, N.Y., pp. 3–18.

Cloud, P., 1973. Paleoecological significance of the banded iron-formation. *Econ. Geol.,* 68: 1135–1143.

Cloud, P., 1974. Atmosphere, development of. *Encyclopaedia Britannica.* 15th ed., pp. 313–319.

Cloud, P., 1976. Major features of crustal evolution. *Geol. Soc. S.Afr., Spec. Publ.,* 79(annex): 32 pp.

Collerson, K.D. and Fryer, B.J., 1978. The role of fluids in the formation and subsequent development of early continental crust. *Contrib. Mineral. Petrol.,* 67: 151–169.

Collerson, K.D., Jessan, C.W. and Bridgwater, D., 1976. Crustal development of the Archaean gneiss complex: Eastern Labrador. In: B.F. Windley (Editor), *The Early History of the Earth.* Wiley, New York, N.Y., pp. 237–256.

Condie, K.C., 1973. Archaean magmatism and crustal thickening. *Geol. Soc. Am. Bull.,* 84: 2981–2992.

Condie, K.C., 1975. A mantle plume model for the origin of Archaean greenstone belts based on trace element distributions. *Nature,* 258: 413–414.

Condie, K.C., 1976. *Plate Tectonics and Crustal Evolution.* Pergamon, New York, N.Y., 288 pp.

Condie, K.C. and Harrison, N.M., 1976. Geochemistry of the Archaean Bullawayan Group, Midlands greenstone belt, Rhodesia. *Precambrian Res.,* 3: 253–271.

Cook, P.J., 1976. Sedimentary phosphate deposits. In: K.H. Wolf (Editor), *Handbook of Strata-Bound and Stratiform Ore Deposits,* 7. Elsevier, Amsterdam, pp 505–536.

Cooper, J.A., Nesbitt, R.W., Platt, J.P. and Mortimer, G.E., 1978. Crustal development in the Agnew region, Western Australia, as shown by Rb/Sr isotopic and geochemical studies. *Precambrian Res.,* 7: 31–60.

Coward, M.P., B.C. Lintern and Wright, L.I., 1976. The pre-cleavage deformation of the sediments and gneisses of the northern part of the Limpopo Belt. In: B.F. Windley (Editor), *The Early History of the Earth.* Wiley, London, pp. 323–330.

Cranstone, D.A. and Turek, A., 1976. Geological and geochronological relationships of the Thompson nickel belt, Manitoba. *Can. J. Earth Sci.,* 13: 1058–1069.

Davies, J.F. and Luhta, L.E., 1978. An Archean "Porphyry-type" disseminated copper deposit, Timmins, Ontario. *Econ. Geol.,* 73(3): 383–396.

Davies, R.D. and Allsopp, H.L., 1976. Strontium isotopic evidence relating to the evolution of the lower Precambrian granitic crust in Swaziland. *Geology,* 4: 553–556.

Dearnley, R., 1966. Orogenic fold belts and a hypothesis of Earth evolution. In: L. Ahrens, F. Press, S.K. Runcorn and H.C. Urey (Editors), *Physics and Chemistry of the Earth, 8.* Pergamon, Oxford, pp. 1–114.

De Laeter, J.R. and Blockley, J.G., 1972. Granite ages within the Archaean Pilbara Block, Western Australia. *J. Geol. Soc. Aust.,* 19: 363–370.

Delano, J.W. and Ringwood, A.E., 1978. Siderophile elements in the lunar highlands: nature of the indigenous component and implications for the origin of the moon. *Proc. 9th Lunar Sci. Conf.*

Descarreaux, J., 1973. A petrochemical study of the Abitibi volcanic belt and its bearing on the occurrences of massive sulphide ores. *Can. Min. Metall. Bull.,* 66(2): 61–69.

de Villiers, J., 1960. The manganese deposits of the Union of South Africa. *Geol. Surv. S.Afr. Handb.,* 2: 280 pp.

434

Dimroth, E., 1976. Aspects of the sedimentary petrology of cherty iron formation. In: K.H. Wolf (Editor), *Handbook of Strata-Bound and Stratiform Ore Deposits, 7.* Elsevier, Amsterdam, pp. 203–247.

Dimroth, E. and Kimberley, M.M., 1976. Precambrian atmospheric oxygen: evidence in the sedimentary distributions of carbon, sulphur, uranium and iron. *Can. J. Earth Sci.,* 13: 1161–1185.

Donnelly, T.H., Lambert, I.B., Oehler, D.Z., Hallberg, J.A., Hudson, D.R., Smith, J.W., Bavinton, O.A. and Golding, L., 1978. A reconnaissance study of stable isotope ratios in Archaean rocks from the Yilgarn Block, Western Australia. *J. Geol. Soc. Aust.,* 24: 409–420.

Duke, J.M. and Naldrett, A.J., 1978. A numerical model of the fractionation of olivine and molten sulfide from komatiite magma. *Earth Planet. Sci. Lett.,* 39: 255–266.

Dunlop, J.S.R., 1978. Shallow water sedimentation at North Pole, Pilbara Block, Western Australia. In: J.E. Glover and D.I. Groves (Editors), *Archaean Cherty Metasediments: their Sedimentology, Micropalaeontology, Biogeochemistry and Significance to Mineralization, 2.* Geol. Dep. and Extension Service, Univ. of West. Aust., Perth, pp. 30–38.

Dunlop, J.S.R. and Groves, D.I., 1978. Sedimentary barite of the Barberton Mountain Land: a brief review. In: J.E. Glover and D.I. Groves (Editors), *Archaen Cherty Metasediments: their Sedimentology, Micropalaeontology, Biogeochemistry and Significance to Mineralization, 2.* Dep. Geol. and Extension Service, Univ. West. Aust., Perth, pp 39–44.

Dunlop, J.S.R., Muir, M.D., Milne, V.A. and Groves, D.I., 1978. A new microfossil assemblage from the Archaean of Western Australia. *Nature,* 274: 676–678.

Eckstrand, O.R., 1975. The Dumont serpentinite: A model for control of nickeliferous opaque mineral assemblages by alteration reactions in ultramafic rocks. *Econ. Geol.,* 70: 183–201.

Eichler, J., 1976. Origin of the Precambrian banded iron-formations. In: K.H. Wolf (Editor), *Handbook of Strata-Bound and Stratiform Ore Deposits, 7.* Elsevier, Amsterdam, pp. 157–202.

Elsasser, W.M., 1963. Early history of the Earth. In: J. Geiss and E. Goldberg (Editors), *Earth Science and Meteorites.* North Holland, Amsterdam, pp. 1–30.

Engel, A.E.J., Itson, S.P., Engel, C.G., Stickney, D.M. and Cray, E.J., 1974. Crustal evolution and global tectonics: a petrogenic view. *Geol. Soc. Am. Bull.,* 85: 843–858.

Eriksson, K.A. and Truswell, J.F., 1978. Geological processes and atmospheric evolution in the Precambrian. In: D.H. Tarling (Editor), *Evolution of the Earth's Crust.* Academic Press, New York, N.Y., pp. 219–238.

Eugster, H.P., 1969. Inorganic cherts from the Magadi area, Kenya. *Contrib. Mineral. Petrol.,* 22: 1–31.

Ewers, W.E. and Hudson, D.R., 1972. An interpretative study of a nickel–iron sulfide ore intersection, Lunnon Shoot, Kambala, Western Australia. *Econ. Geol.,* 67: 1075–1092.

Fanale, F.P., 1971. A case for catastrophic early degassing of the Earth. *Chem. Geol.,* 272: 521–536.

Farhat, J.S. and Wetherill, G.W., 1975. Interpretation of apparent ages in Minnesota. *Nature,* 257: 721–723.

Ferguson, J., 1976a. Pay versus non-pay kimberlites. *B.M.R. J. Aust. Geol. Geophys.,* 1(3): 252.

Ferguson, J., 1976b. Economic mineralization associated with carbonatites. *B.M.R. J. Aust. Geol. Geophys.,* 1(3): 252.

Ferguson, J. and Currie, K.L., 1972. Silicate immiscibility of the ancient "basalts" of the Barberton Mountain Land Transvaal. *Nature Phys. Sci.,* 235: 86–89.

Findlay, D.J., 1976. Lang Lakes, an Archean porphyry copper-molybdenum deposit. *Centr. Precambrian Studies, Univ. Manitoba, 1975 Annu. Rep.,* pp. 103–107.

Findlay, D.J. and Ayres, L.D., 1977. Lang Lake – an early Precambrian porphyry copper–molybdenum in northwestern Ontario. *Rep. of Activities, Part B; Geol. Surv. Can., Pap.,* 77-1B: 25–28.

Fleischer, V.D., Garlick, W.G. and Haldane, R., 1976. Geology of the Zambian Copper belt. In: K.H. Wolf (Editor), *Handbook of Strata-Bound and Stratiform Ore Deposits, 6.* Elsevier, Amsterdam, pp. 223–352.

Fripp, R.E.P., 1976a. Stratabound gold deposits in Archaean banded iron-formation, Rhodesia. *Econ. Geol.,* 71: 58–75.

Fripp, R.E.P., 1976b. Gold metallogeny in the Archaean of Rhodesia. In: B.F. Windley (Editor), *The Early History of the Earth*. Wiley, London, pp. 455–466.

Fripp, R.E.P., Donnelly, T.H. and Lambert, I.B., 1979. Sulphur isotope studies for Archaean banded iron-formation, Rhodesia. *J. Geol. Soc. S.Afr., Spec. Publ.,* in press.

Fryer, B.J. and Hutchinson, R.W., 1976. Generation of metal deposits on the sea floor. *Can. J. Earth Sci.,* 13: 126–136.

Fyfe, W.S., 1973. The granulite facies, partial melting and Archaean crust. *Philos. Trans. R. Soc. London, Ser. A.,* 273: 457–462.

Gee, R.D., 1979. Tectonics of the Western Australian Shield, Tectonophysics, 58(3/4): 327–369.

Gelinas, L., Brooks, C. and Trzcienski Jr., W.E., 1976. Archaean variolites – quenched immiscible liquids. *Can. J. Earth Sci.,* 13: 210–230.

Gemuts, I. and Theron, A., 1975. The Archaean between Coolgardie and Norseman – stratigraphy and mineralization. In: C.L. Knight (Editor), *Economic Geology of Australia and Papua-New Guinea, 1. Metals*. Australas. Inst. Min. Metall., Melbourne, pp. 66–74.

Glikson, A.Y., 1971. Primitive Archaean element distribution patterns: chemical evidence and geotectonic significance. *Earth Planet. Sci. Lett.,* 12: 309–320.

Glikson, A.Y., 1972. Early Precambrian evidence of a primitive oceanic crust and island nuclei of sodic granite. *Geol. Soc. Am. Bull.,* 83: 3323–3344.

Glikson, A.Y., 1976a. Stratigraphy and evolution of primary and secondary greenstones: significance of data from shields of the southern hemisphere. In: B.F. Windley (Editor), *The Early History of the Earth*. Wiley, London, pp. 257–278.

Glikson, A.Y., 1976b. Earliest Precambrian ultramafic–mafic volcanic rocks: ancient oceanic crust or relic terrestrial maria. *Geology,* 4: 202–205.

Glikson, A.Y., 1977a. Earliest Precambrian ultramafic–mafic volcanic rocks: ancient oceanic crust or relic terrestrial maria: Reply – vestiges of a beginning. *Geology,* 5: 68–71.

Glikson, A.Y., 1977b. Evidence on the radius of the Precambrian Earth. *B.M.R.J. Aust. Geol. Geophys.,* 2: 229–232.

Glikson, A.Y., 1979. Early Precambrian tonalite-trondhjemite sialic nuclei. *Earth Sci. Rev.,* 15: 1–73.

Glikson, A.Y. and Lambert, I.B., 1973. Relations in space and time between major Precambrian shield units: an interpretation of Western Australian data. *Earth Planet. Sci. Lett.,* 20: 395–403.

Glikson, A.Y. and Lambert, I.B., 1976. Vertical zonation and petrogenesis of the early Precambrian crust in Western Australia. *Tectonophysics,* 30: 55–89.

Goldich, S.S. and Hedge, C.E., 1974. 3800 Myr granitic gneiss in south-Western Minnesota. *Nature,* 252: 467–468.

Goodwin, A.M., 1965. Volcanism and gold deposition in the Birch-Uchi Lakes area. *Trans. Can. Inst. Min. Metall.,* 68: 91–104.

Goodwin, A.M., 1971. Metallogenic patterns and evolution of the Canadian Shield. *Geol. Soc. Aust., Spec. Publ.,* 3: 157–174.

Goodwin, A.M., 1972. The Superior Province. In: A.Y. Price and R.J.W. Douglas (Editors), *Variations in Tectonic Styles in Canada. Geol. Assoc. Can., Spec. Paper,* 11: 527–564.

Goodwin, A.M., 1973a. Plate tectonics and evolution of Precambrian crust. In: D.H. Tarling and S.K. Runcorn (Editors), *Implications of Continental Drift to the Earth Sciences, 2*. Academic Press, London, pp. 1047–1069.

Goodwin, A.M., 1973b. Archaean iron-formations and tectonic basins of the Canadian shield. *Econ. Geol.,* 68: 915–933.

Goodwin, A.M., 1976. Giant impacting and the development of continental crust. In: B.F. Windley (Editor), *The Early History of the Earth*. Wiley, London, pp. 77–95.

Goodwin, A.M., 1977. Archaean basin–craton complexes and the growth of Precambrian shields. *Can. J. Earth Sci.,* 14: 2737–2759.

Goodwin, A.M., 1978. *Archaean Volcanism in Superior Province, Canadian shield. Geol. Assoc. Can., Spec. Pap.,* 16: 205–241.

Goodwin, A.M., Monster, J. and Thode, H.G., 1976. Carbon and sulphur isotope abundances in Archaean iron-formations and early Precambrian life. *Econ. Geol.,* 71: 870–892.

Gottfried, D., Rowe, J.J. and Tilling, R.I., 1972. Distribution of gold in igneous rocks. *U.S., Geol. Surv., Prof. Pap.,* 727.

Green, D.H., 1972. Archaean greenstone belts may include terrestrial equivalents of lunar maria. *Earth Planet. Sci. Lett.,* 15: 263–270.

Green, D.H., 1973. Experimental melting studies on a model upper-mantle composition of high pressure under water-saturated and water-undersaturated conditions. *Earth Planet. Sci. Lett.,* 19: 37–53.

Green, D.H., 1976a. Experimental petrology in Australia – a review. *Earth-Sci. Rev.,* 12: 99–138.

Green, D.H., 1976b. Experimental testing of "equilibrium" partial melting of peridotite under water saturated, high-pressure conditions. *Can. Mineral.,* 14: 255–268.

Green, D.H. and Ringwood, A.E., 1967. The genesis of basaltic magmas. *Contrib. Mineral. Petrol.,* 15: 103–190.

Green, D.H., Nicholls, I.A., Viljoen, M.J. and Viljoen, R.P., 1975. Experimental demonstration of the existence of peridotitic liquids in earliest Archaean magmatism. *Geology,* 3: 11–15.

Gross, G.A., 1967. Geology of iron deposits in Canada, vols. I, II and III. *Geol. Surv. Can., Econ. Geol. Rep.* No. 22.

Groves, D.I. and Hudson, D.R., 1981. Volcanic associated Ni–Fe–Cu sulphide deposits. In: K.H. Wolf (Editor), *Handbook of Strata-bound and Stratiform Ore deposits,* 9. Elsevier Amsterdam, pp. 305–410.

Groves, D.I. and Keays, R.R., 1979. Mobilization of ore-forming elements during alteration of dunites, Mt. Keith-Bethero, Western Australia. *Can. Mineral.,* 17: 373–390.

Groves, D.I. and McCarthy, T.S., 1978. Fractional crystallization and the origin of tin deposits in granitoids. *Miner. Deposita,* 13: 11–26.

Groves, D.I., Hudson, D.R. and Hack, T.B.C., 1974. Modification of iron–nickel sulfides during serpentinization and talc–carbonate alteration at Black Swan, Western Australia. *Econ. Geol.,* 69: 1265–1281.

Groves, D.I., Archibald, N.J., Bettenay, L.F. and Binns, R.A., 1978. Greenstone belts as ancient marginal basins or ensialic rift zones. *Nature,* 273: 460–461.

Groves, D.I., Barrett, F.M. and McQueen, K.G., 1979. The relative roles of magmatic segregation, volcanic exhalation and regional metamorphism in the generation of volcanic-associated nickel ores of Western Australia. *Can. Mineral.,* 17(2): 319–336.

Gunn, B.M., 1976. A comparison of modern and Archaean oceanic crust and island-arc petrochemistry. In: B.F. Windley (Editor), *The Early History of the Earth.* Wiley, London, pp. 389–404.

Haldane, J.B.S., 1929. The origin of life. Rationalist Annual (reprinted in Bernal, J.D., 1967. *The Origin of Life.* Weidenfeld and Nicholson, London, 345 pp.).

Hallbauer, D.K., 1975. The plant origin of the Witwatersrand carbon. *Miner. Sci. Eng.,* 7: 111–131.

Hallberg, J.A., 1972. Geochemistry of Archaean volcanic rocks in the Eastern Goldfields region of Western Australia. *J. Petrol.,* 13: 45–56.

Hallberg, J.A., 1978. Acid/intermediate volcanism in the Yilgarn Block, Western Australia. *Proc. 1978 Archaean Geochemistry Conf.,* Canada, pp. 349–352.

Hallberg, J.A. and Williams, D.A.C., 1972. Archaean mafic and ultramafic rock associations in the Eastern Goldfields Region, Western Australia. *Earth Planet. Sci. Lett.,* 15: 191–200.

Hamilton, P.J., O'Nions, R.K. Evensen, N.M., Bridgwater, D. and Allaart, J.H., 1978. Sm–Nd isotopic investigations of Isua supracrustals and implications for mantle evolution. *Nature,* 272: 41–43.

Hamilton, P.J., Carter, S.R., Evensen, N.M. O'Nions, R.K. and Tarney, J., 1979. Sm–Nd systematics of lewisian gneisses. Implications for the origins of granulites. *Nature.*

Hanks, T.C. and Anderson, D.L., 1969. The early thermal history of the Earth. *Phys. Earth Planet. Inter.,* 2: 19–29.

Hargraves, R.B., 1976. Precambrian geologic history. *Science,* 193: 363–371.

Hart, S.R., Brooks, C., Krogh, T.E., Davies, G.L. and Nava, D., 1970. Ancient and modern volcanic rocks: A trace element model. *Earth Planet. Sci. Lett.,* 10: 17–28.

Haughton, S.H. (Editor), 1964. *The Geology of Some Ore Deposits in Southern Africa,* 1. Geol. Soc. S. Africa, 625 pp.

Haughton, S.H., 1969. *Geological History of Southern Africa*. Geol. Soc. S. Africa, 535 pp.

Hawkesworth, C.J. and O'Nions, R.K., 1977. The petrogenesis of some Archaean volcanic rocks from Southern Africa. *J. Petrol.*, 18: 487–520.

Hawkesworth, C.J., Moorbath, S., O'Nions, R.K. and Wilson, J.F., 1975. Age relationships between greenstone belts and "granites" in the Rhodesian Archaean craton. *Earth Planet. Sci. Lett.*, 25: 251–262.

Heier, K.S., 1964. Rubidium/strontium and strontium-87/strontium-86 ratios in deep crustal material. *Nature*, 202: 477–478.

Heier, K.S. and Lambert, I.B., 1978. A compilation of potassium, uranium and thorium abundances and heat productivities of Australian rocks. *Tech. Rep., Res. Sch. Earth Sci., Aust. Natl. Univ.*

Heier, K.S. and Rhodes, J.M., 1966. Thorium, uranium and potassium concentrations in granites and gneisses of the Rum Jungle Complex, N.T., Australia. *Econ. Geol.*, 61: 563–571.

Heinrichs, T.K. and Reimer, T.O., 1977. A sedimentary barite deposit from the Archaean Fig Tree Group of the Barberton Mountain Land (South Africa). *Econ. Geol.*, 72: 1426–1441.

Henderson, J.B., 1972. Sedimentology of Archaean turbidites at Yellowknife, Northwest Territories, *Can. J. Earth. Sci.*, 9: 882.

Henley, R.W., 1973. Solubility of gold in hydrothermal chloride solutions. *Chem. Geol.*, 11: 73–87.

Hesp, W.R., 1971. Correlations between the tin content of granitic rocks and their chemical and mineralogical composition. *Can. Inst. Min. Metall. Spec. Publ.*, 11: 341–353.

Hesp, W.R. and Varlamoff, N., 1977. Temporal and spatial relations between the formation of acid magmatic rocks and tin deposits. In: M. Stemprok, L. Burnol and G. Tischendorf (Editors), *Metallization Associated with Acid Magmatism, 2.* pp. 23–40.

Hickman, A.H., 1973. The North Pole barite deposit, Pilbara Goldfield. *Geol. Surv. West Aust., Annu. Rep., 1972:* 57–60.

Hickman, A.H., 1975. Precambrian structural geology of part of the Pilbara region. *Geol. Surv. West. Aust., Annu. Rep., 1974:* 68–72.

Hickman, A.H. and Lipple, S.L., 1975. Explanatory notes on the Marble Bar 1 : 250,000 Geological Sheet, Western Australia. *West. Aust. Geol. Surv. Rec., 1974/20.*

Hills, J.H. and Richards, J.R., 1972. The age of uranium mineralization in Northern Australia. *Search*, 3: 382–385.

Holland, H.D., 1972. The geologic history of seawater – an attempt to solve the problem. *Geochim. Cosmochim. Acta*, 36: 637–651.

Holland, H.D., 1976. The evolution of seawater. In: B.F. Windley (Editor), *The Early History of the Earth*. Wiley, London, pp. 559–568.

Hollister, V.F., 1975. An appraisal of the nature and source of porphyry copper deposits. *Miner. Sci. Eng.*, 7: 225–233.

Hoyle, F. and Wickramasinghe, N.C., 1977. Origin and nature of carbonaceous material in the galaxy. *Nature*, 270: 701–703.

Hubbard, J.S., Hardy, J.P. and Horowitz, N.H., 1971. Photocatalytic production of organic compounds from CO and H_2O in a stimulated Martian atmosphere. *Proc. Natl. Acad. Sci. U.S.A.*, 68: 574–578.

Hunter, D.R., 1970. The Geology of the Usushwana complex in Swaziland. *Geol. Soc. S.Afr., Spec. Publ.*, 1: 645–660.

Hunter, D.R., 1974. Crustal development of the Kaapvaal craton, The Archaean. *Precambrian Res.*, 1: 259–294.

Hunter, D.R., 1976. Some enigmas of the Bushveld Complex. *Econ. Geol.*, 71: 229–248.

Hurley, P.M. and Rand, J.R., 1969. Pre-drift continental nuclei. *Science*, 164: 1229–1242.

Hurst, R.W., Bridgwater, D., Collerson, K.D. and Wetherill, G.W., 1975. 3600 m.y. Rb-Sr ages from the very early Archaean gneisses from Saglek Bay, Labrador. *Earth Planet. Sci. Lett.*, 27: 393–403.

Hutchinson, R.W., 1973. Volcanic sulphide deposits and their metallogenic significance. *Econ. Geol.*, 68: 1223–1246.

438

Hutchinson, R.W., Ridler, R.H. and Suffel, G.G., 1971. Metallogenic relationships in the Abitibi belt, Canada. A model for Archaean metallogeny. *Trans. Can. Inst. Min. Metall.*, 74: 106–115.

Igarishi, T., Okabe, K. and Yajima, J., 1974. Massive barite deposits in west Hokkaido. In: S. Isihara (Editor), *Geology of Kuroko Deposits. Mines Geol. Spec. Issue*, 6: 39–44.

International Uranium Symposium on the Pine Creek Geosyncline, 1979. *Extended Abstracts Volume.* Sydney, June 1979.

Irvine, T.N., 1975. Crystallization sequences in the Muskox intrusion and other layered intrusions, II. Origin of chromite layers and similar deposits of other magmatic ores. *Geochim. Cosmochim. Acta*, 39: 991–1020.

Irvine, T.N., 1977. Origin of chromite layers in the Muskox intrusion and other stratiform intrusions: a new interpretation. *Geology*, 5: 173–277.

Jackson, E.D., 1969. Chemical variation in coexisting chromite and olivine in chromitite zones of the Stillwater Complex. In: H.D.B. Wilson (Editor), *Magmatic Ore Deposits. Econ. Geol. Monogr.*, 4: 41–70.

Jacobsen, S.B. and Wasserberg, G.J., 1978. Interpretation of Nd, Sr and Pb isotope data from Archaean migmatites in Lofoten-Vesteralen, Norway. *Earth Planet. Sci. Lett.*, 41: 245–253.

Jahn, B.M. and Shih, C.Y., 1974. On the age of the Onverwacht Group, Swaziland Sequence, South Africa. *Geochim. Cosmochim. Acta*, 38: 611–627.

James, H.L. and Sims, P.K. (Editor), 1973. *Precambrian Iron-Formations of the World. Econ. Geol.*, 68: 913–1179.

Jolly, W.T., 1978. Relations between Archaean lavas and intrusive bodies of the Abitibi Greenstone Belt, Ontario-Quebec. In: W.R.A. Baragar, L.C. Coleman and J.M. Hall (Editors), *Volcanic Regimes in Canada. Geol. Assoc. Can., Spec. Pap.*, 16: 311–330.

Jung, W. and Knitzschke, G., 1976. Kupferschiefer in the German Democratic Republic (GDR) with special reference to the Kupferschiefer deposit in the southeastern Harzforeland. In: K.H. Wolf (Editor), *Handbook of Strata-Bound and Stratiform Ore Deposits, 6*. Elsevier, Amsterdam, pp. 353–406.

Karvinen, W.O., 1978. The Porcupine camp – A model for gold exploration in the Archaean. *Can. Min., Sept. 1978*: 48–53.

Keays, R.R. and Davison, R.M., 1976. Palladium, iridium, and gold in the ores and host rocks of nickel sulfide deposits in Western Australia. *Econ. Geol.*, 71: 1214–1228.

Keays, R.R. and Kirkland, M.C., 1972. Hydrothermal mobilization of gold from copper–nickel sulfides, and ore genesis at the Thomson River copper mine, Victoria, Australia. *Econ. Geol.*, 67: 1263–1275.

Keays, R.R. and Scott, R.B., 1976. Precious metals in ocean-ridge basalts: implications for basalts as source rocks for gold mineralization. *Econ. Geol.*, 71: 705–720.

Keays, R.R., Groves, D.I. and Davison, R.M., in press. Ore element remobilization during progressive alteration of sulphide-bearing ultramafic rocks at the Black Swan nickel deposit, Western Australia: results of a precious metal study. *Econ. Geol.*

Keele, R.A. and Nickel, E.H., 1974. The geology of a primary millerite-bearing sulfide assemblage and supergene lateration at the Otter Shoot, Kambalda, Western Australia. *Econ. Geol.*, 69(7): 1102–1117.

Kilburn, L.C., Wilson, H.D.B., Graham, A.R., Ogura, Y., Coats, C.J.A. and Scoates, R.F.J., 1969. Nickel sulfide ores related to ultrabasic intrusions in Canada. *Econ. Geol., Monogr.*, 4: 276–293.

Kimberley, M.M. (Editor), 1978. *Uranium Deposits: their Mineralogy and Origin.* Min. Assoc. Can., Short Course Handb., 3. Toronto, October, 1978.

Kirkham, R.V., 1972. Geology of copper and molybdenum deposits. *Rep. of Activities, Part A, Geol. Surv. Can., Pap.*, 72-1A: 82–87.

Knauth, L.P. and Epstein, S., 1976. Hydrogen and oxygen isotope ratios in nodular and bedded cherts. *Geochim. Cosmochim. Acta*, 40: 1095–1108.

Knauth, L.P. and Lowe, D.R., 1978. Oxygen isotope geochemistry of cherts from the Onverwacht group (3.4 billion years), Transvaal, South Africa, with implications for secular variations in the isotope compositions of cherts. *Earth Planet. Sci. Lett.,* 41: 209–222.

Knipping, H.D., 1974. The concepts of supergene versus hypogene emplacement of uranium at Rabbit Lake, Saskatchewan, Can. In: *Formation of Uranium Ore Deposits.* Proc. Symp., Athens, Int. Atomic Energy Agency, pp. 531–549.

Köppel, V., 1968. Age and history of the uranium mineralization of the Beaverlodge area, Saskatchewan. *Geol. Surv. Can., Pap.,* 67-31.

Kovalenker, V.A., Gladyshev, G.D. and Nasik, L.P., 1975. Isotopic composition of sulfide sulfur from deposits of Talnakh ore made in relation to their selenium content. *Int. Geol. Rev.,* 17: 725–733.

Kwong, Y.T.J. and Crocket, J.H., 1978. Background and anomalous gold in rocks of an Archaean greenstone assemblage, Kakagi Lake area, Northwestern Ontario. *Econ. Geol.,* 73: 50–63.

Lambert, I.B., 1971. The composition and evolution of the deep continental crust. *Geol. Soc. Aust., Spec. Publ.,* 3: 419–428.

Lambert, I.B., 1976. The McArthur zinc–lead–silver deposit: features, metallogenesis and comparisons with some other stratiform ores. In: K.H. Wolf (Editor), *Handbook of Strata-Bound and and Stratiform Ore Deposits, 6.* Elsevier, Amsterdam, pp. 535–585.

Lambert, I.B., 1977. Notes on exploration guides for lead–zinc ores, geochemical and geobiological evolution in the Precambrian, and massive sulphide deposits of the Noranda area. *CSIRO, Min. Res. Lab, Tech. Comm.,* 61.

Lambert, I.B., 1978. Sulphur-isotope investigations of Archaean mineralization and some implications concerning geobiochemical evolution. In: J.E. Glover and D.I. Groves (Editors), *Archaean Cherty Metasediments: their Sedimentology, Micropalaeontology, Biogeochemistry, and Significance to Mineralization, 2.* Geol. Dep. and Extension Service, Univ. West. Aust., Perth, pp. 45–56.

Lambert, I.B. and Heier, K.S., 1967. The vertical distribution of uranium, thorium and potassium in the continental crust. *Geochim. Cosmochim. Acta,* 31: 377–390.

Lambert, I.B. and Heier, K.S., 1968. Geochemical investigations of deep-seated rocks in the Australian shield. *Lithos,* 1: 30–53.

Lambert, I.B. and Wyllie, P.J., 1970. Melting in the deep crust and upper mantle and the nature of the low velocity layer. *Phys. Earth Planet. Inter.,* 3: 316–322.

Lambert, I.B. and Wyllie, P.J., 1972. Melting of gabbro (quartz eclogite) with excess water to 35 kilobars, with geological implications. *J. Geol.,* 80: 693–708.

Lambert, I.B., Donnelly, T.H., Dunlop, J.S.R. and Groves, D.I., 1978. Stable isotopic compositions of early Archaean sulphate deposits of probable evaporitic and volcanogenic origins. *Nature,* 276: 808–811.

Lambert, R.St.J., 1976. Archaean thermal regimes, crustal and upper mantle temperatures, and a progressive evolutionary model for the Earth. In: B.F. Windley (Editor), *The Early History of the Earth.* Wiley, London, pp. 363–373.

Lambert, R.St.J., 1978. The early thermal history of the Earth. *Proc. 1978 Archaean Geochemistry Conf.,* Canada, pp. 353–354.

Lambert, R.St.J., Chamberlain, V.E. and Holland, J.G., 1976. The geochemistry of Archaean rocks. In: B.F. Windley (Editor), *The Early History of the Earth.* Wiley, London, pp. 377–388.

Langford, F.F., 1978. Origin of the unconformity-type pitchblende deposits in the Athabasca basin of Saskatchewan. In: M.M. Kimberley (Editor), *Uranium Deposits: Their Mineralogy and Origin.* Min. Assoc. Can., Short Course Handb., 3: 485–500.

Large, R.R., 1977. Chemical evolution and zonation of massive sulphide deposits in volcanic terrains. *Econ. Geol.,* 72: 549–572.

Laznicka, P., 1973. Development of non-ferrous metal deposits in geological time. *Can. J. Earth Sci.,* 10: 18–25.

Levin, B.J., 1972. Origin of the Earth. *Tectonophysics,* 13: 7–29.

Levorsen, A.I., 1954. *Geology of Petroleum.* Freeman, San Francisco, Calif.

Lipple, S.L., 1975. Definitions of new and revised stratigraphic units of the eastern Pilbara region. *Annu. Rep. Geol. Surv. West. Aust.,* 1973: 58–63.

Lister, L.A. (Editor), 1973. *Symposium on Granites, Gneisses and Related Rocks. Salisbury. Geol. Soc. S.Afr., Spec. Publ.,* 3: 509.

Lowe, D.R. and Knauth, L.P., 1977. Sedimentology of the Onverwacht Group (3.4 billion years), Transvaal, South Africa, and its bearing on the characteristics and evolution of the early Earth. *J. Geol.,* 85: 699–723.

Lugmair, G.W. Scheinin, N.B. and Marti, K., 1975. Sm–Nd age and history of Apollo 17 basalt 75075: evidence for early differentiation of lunar exterior. *Proc. 6th Lunar Sci. Congr.,* pp. 1419–1429.

Lusk, J., 1976. A possible volcanic-exhalative origin for lenticular nickel sulfide deposits of volcanic association with special reference to those in Western Australia. *Can. J. Earth Sci.,* 13: 451–458.

MacGeehan, P.J., 1978. The geochemistry of altered volcanic rocks at Matagami, Quebec: a geothermal model for massive sulphide genesis. *Can. J. Earth Sci.,* 15: 551–570.

MacGregor, A.M., 1951. Some milestones in the Precambrian of southern Africa. *Proc. Geol. Soc. S. Afr.,* 54: 27–71.

Marston, R.J. and Kay, B.D., in press. The distribution and petrology of nickel ores at Otter-Juan Shoot complex in relation to ore genesis theories at Kambalda, Western Australia. *Econ. Geol.*

Martin, J.E. and Allchurch, P.D., 1975. Perseverance nickel deposit, Agnew. In: C.L. Knight (Editor), *Economic Geology of Australia and Papua-New Guinea, 1. Metals.* Australas. Inst. Min. Metall., Melbourne, pp. 149–155.

McCarthy, T.S. and Cawthorn, R.G., in press. Changes in initial $^{87}Sr/^{86}Sr$ ratio during protracted crystallization in igneous complexes. *J. Petrol.*

McCullogh, M.T. and Wasserberg, G.J., 1978. Penultimate provenances of crustal rocks. *Science,* 200: 1003–1011.

McElhinney, M.W. and McWilliams, M.O., 1977. Precambrian geodynamics: a palaeomagnetic view. *Tectonophysics,* 40: 137–159.

McElhinny, M.W., Taylor, S.R. and Stevenson, D.J., 1978. Limits to the expansion of Earth, Moon, Mars and Mercury and to changes in the gravitational constant. *Nature,* 271: 316–321.

McGregor, V.M., 1973. The early Precambrian gneisses of the Godthaab district, west Greenland. *Philos. Trans. R. Soc. London, Ser. A,* 273: 343–358.

Michard-Vitrac, A., Lancelot, J., Allegre, C.J. and Moorbath, S., 1977. U–Pb ages on single zircons from the early Precambrian rocks of west Greenland and the Minnesota River valley. *Earth Planet. Sci. Lett.,* 35: 449–454.

Miller, L.J. and Smith, M.E., 1975. Sherlock Bay nickel–copper. In: C.L. Knight (Editor), *Economic Geology of Australia and Papua-New Guinea, 1. Metals.* Australas. Inst. Min. Metall., Melbourne, pp. 168–174.

Miller, S.L., 1953. A production of amino acids under possible primitive earth conditions. *Science,* 117: 528–529.

Mitchell, A.H.G. and Garson, M.S., 1976. Mineralization at plate boundaries. *Miner. Sci. Eng.,* 2: 129–169.

Moorbath, S., 1977. Ages, isotopes and the evolution of the Precambrian crust. *Chem. Geol.,* 20: 151–187.

Moorbath, S. and Taylor, P., 1978. Lead isotope geochemistry of Archaean cratons, with special reference to Greenland. *Proc. 1978 Archaean Geochemistry Conf.,* Canada, pp. 357–359.

Moorbath, S., Powell, J.L. and Taylor, P.N., 1975. Isotopic evidence for the age and origin of the grey-gneiss complex of southern Outer Hebrides, north west Scotland. *J. Geol. Soc. London,* 131: 213–232.

Moorbath, S., Wilson, J.F. and Cotterill, P., 1976. Early Archaean age for the Sebakwian Group at Selukwe, Rhodesia. *Nature,* 264: 536–538.

Moorbath, S., Allaart, J.H., Bridgwater, D. and McGregor, V.R., 1977. Rb–Sr ages of early Archaean supracrustal rocks and Amitsoq gneisses at Isua. *Nature,* 270: 43–45.

Muir, J.E. and Comba, C.D.A., 1979. The Dundonald deposit: an example of volcanogenic nickel-sulphide mineralization. *Can. Mineral.,* 17(2).

Muir, M.D., 1979. Palaeontological evidence bearing on the evolution of the atmosphere and hydrosphere. In: P.R. Grant (Editor), *Proceedings of a Meeting: Evolution of the Atmosphere and Hydrosphere.*

Muir, M.D. and Grant, P.R., 1976. Micropalaeontological evidence from the Onverwacht Group, South Africa. In: B.F. Windley (Editor), *The Early History of the Earth.* Wiley, London, pp. 595–604.

Murray, G.E., 1964. Indigenous Precambrian petroleum. *Bull. Am. Assoc. Pet. Geol.,* 49: 3–31.

Murthy, V. Rama, 1976. Composition of the core and the early chemical history of the Earth. In: B.F. Windley (Editor), *The Early History of the Earth.* Wiley, London, pp. 21–32.

Mysen, B.O. and Boettcher, A.L., 1972. Melting in a hydrous mantle: phase relationships in periodotite – H_2O-CO_2 systems. *Program Annu. Meet. Geol. Soc. Am.,* p. 608.

Mysen, B.O. and Kushiro, I., 1974. A possible mantle origin for andesitic magmas: Discussion of a paper by Nicholls and Ringwood. *Earth Planet. Sci. Lett.,* 21: 221–229.

Naldrett, A.J., 1973. Nickel sulfide deposits – their classification and genesis, with special emphasis on deposits of volcanic association. *Trans. Can. Inst. Min. Metall.,* 76: 183–201.

Naldrett, A.J. and Cabri, L.J., 1976. Ultramafic and related rocks: their classification and genesis with special reference to the concentration of nickel sulphides and platinum-group elements. *Econ. Geol.,* 71: 1131–1158.

Naldrett, A.J. and Turner, A.R., 1977. The geology and petrogenesis of a greenstone belt and related nickel sulfide mineralization at Yakabindie, Western Australia. *Precambrian Res.,* 5: 43–103.

Naldrett, A.J., Bray, J.G., Gasparrini, E.L., Podolsky, T. and Rucklidge, J.C., 1970. Cryptic variation and the petrology of the Sudbury Nickel Irruptive. *Econ. Geol.,* 65: 122–155.

Naldrett, A.J., Goodwin, A.M., Fisher, T.L. and Ridler, R.H., 1978. The sulfur content of Archaean volcanic rocks and a comparison with ocean floor basalts. *Can. J. Earth. Sci.,* 15: 715–728.

Naldrett, A.J., Hoffman, E.L., Green, A.H., Chou, C.-L. and Alcock, R.A., 1979. The composition of Ni-sulfide ores with particular reference to their content of PGE and An. *Can. Mineral.,* 17(2).

Nesbitt, R.W. and Sun, S.S., 1976. Geochemistry of Archean spinifex-textured periotites and magnesian and low-magnesian tholeiites. *Earth Planet. Sci. Lett.,* 31: 433–453.

Newton, R.C., 1977. Experimental evidence for the operation of very high pressures in Archaean granulite metamorphism. *Archaean Geochemistry – IUGS – UNESCO – IGCP Symposium,* India. Abstracts, pp. 73–74.

Nicholls, I.A. and Ringwood, A.E., 1972. Effect of water on olivine stability in tholeiites and the production of silica-saturated magmas in the island-arc environment. *J. Geol.,* 81: 285–300.

Nicholls, I.A. and Ringwood, A.E., 1974. A possible mantle origin for andesitic magmas: Discussion. *Earth Planet. Sci. Lett.,* 21: 221–229.

Nickel, E.H., Allchurch, P.D., Mason, M.G. and Wilmshurst, J.R., 1977. Supergene alteration at the Perseverance nickel deposit, Agnew, Western Australia. *Econ. Geol.,* 72: 184–203.

Oehler, D.Z. and Smith, J.W., 1977. Isotopic composition of reduced and oxidized carbon in early Archaean rocks from Isua, Greenland. *Precambrian Res.,* 5: 221–228.

O'Hara, M.J., 1977. Thermal history of excavation of Archaean gneisses from the base of the continental crust. *J. Geol. Soc. London,* 134: 185–200.

O'Nions, R.K. and Pankhurst, R.J., 1978. Early Archaean rocks and geochemical evolution of the Earth's crust. *Earth Planet. Sci. Lett.,* 38: 211–236.

Oparin, A.I., 1957. *The Origin of Life on the Earth.* Oliver and Boyd, London, 3rd ed.

Orgel, L.E., 1974. The synthesis of life molecules. In: J.P. Wild (Editor), *In the Beginning . . . the Origin of Planets and Life.* Australian Academy of Science, Canberra, pp. 85–101.

Oversby, V.M., 1975. Lead isotope systematics and ages of Archaean acid intrusives in the Kalgoorlie-Norseman area, Western Australia. *Geochim. Cosmochim. Acta,* 39: 1107–1125.

Oversby, V.M. and Ringwood, A.E., 1971. Time of formation of the Earth's core. *Nature,* 234: 463–465.

Pagel, M., 1975. Microthermometry and chemical analysis of fluid inclusions from the Rabbit Lake uranium deposit, Saskatchewan, Canada (Abstr.). *Inst. Min. Metall., Trans.*, 86: B157.

Peredery, W.V., 1979. Relationship of the ultramafic amphibolites of the metavolcanites and serpentinites in the Thompson Belt, Manitoba. *Can. Mineral.*, 17(2).

Peredery, W.V. and Coats, C.J.A., 1978. *Guide Book for Nickel Sulfide Field Conference, 1978: Thompson Belt Geology.* Mineralogical Assoc., Canada.

Pereira, J. and Dixon, C.J., 1965. Evolutionary trends in ore deposition. *Inst. Min. Metall. Trans.*, 74: 505–527.

Perry Jr., E.C., 1967. The oxygen isotopic chemistry of ancient cherts. *Earth Planet. Sci. Lett.*, 3: 62–66.

Perry Jr., E.C. and Tan, F.C., 1972. Significance of oxygen and carbon isotope determinations in early Precambrian cherts and carbonate rocks of southern Africa. *Geol. Soc. Am. Bull.*, 83: 647–664.

Perry, E.C., Monster, J. and Reimer, T., 1971. Sulphur isotopes in Swaziland System barites and the evolution of the Earth's atmosphere. *Science*, 171: 1015–1016.

Pidgeon, R.T., 1978a. 3450 m.y.-old volcanics in the Archaean layered greenstone succession of the Pilbara Block, Western Australia. *Earth Planet. Sci. Lett.*, 37: 421–428.

Pidgeon, R.T., 1978b. Geochronological investigation of granite batholiths of the Archaean granite-greenstone terrain of the Pilbara Block, Western Australia. *Proc. 1978 Archaean Geochemistry Conf.*, Canada, pp. 360–362.

Pidgeon, R.T. and Hopgood, A.M., 1975. Geochronology of Archaean gneisses and tonalites from north of the Frederikshabs isblink, S.W. Greenland. *Geochim. Cosmochim. Acta*, 39: 1333–1346.

Piper, J.D.A., 1976. Palaeomagnetic evidence for a Proterozoic super-continent. *Philos. Trans. R. Soc. London*, 280: 469–490.

Platt, J.P., Allchurch, P.D. and Rutland, R.W.R., 1978. Archaean tectonics in the Agnew supracrustal belt, Western Australia. *Precambrian Res.*, 7: 3–30.

Pretorius, D.A., 1976. The nature of the Witwatersrand gold–uranium deposits. In: K.H. Wolf (Editor), *Handbook of Strata-Bound and Stratiform Ore Deposits, 7.* Elsevier, Amsterdam, pp. 28–88.

Purvis, A.C., Nesbitt, R.W. and Hallberg, J.A., 1972. The geology of part of the Carr Boyd Rocks Complex and its associated nickel mineralization, Western Australia. *Econ. Geol.*, 67(8): 1093–1113.

Pyke, D.R., 1975. On the relationship of gold mineralization and ultramafic rocks in the Timmins area, Ontario. *Ont., Div. Mines Misc. Rep.*, 62: 23 pp.

Pyke, D.R. and Middleton, R.S., 1971. Distribution and characteristics of the sulphide ores of the Timmins area. *Bull. Can. Inst. Min.*, 74: 157–168.

Pyke, D.R., Naldrett, A.J. and Eckstrand, O.R., 1973. Archaean ultramafic flows in Munro Township, Ontario. *Geol. Soc. Am. Bull.*, 84: 955–978.

Radhakrishna, B.P., 1976. Mineralization episodes in the Dharwar craton of peninsular India. *J. Geol. Soc. India*, 17: 79–88.

Radhakrishna, B.P. and Vasudev, V.N., 1977. The early Precambrian of the southern Indian Shield. *J. Geol. Soc. India*, 18: 525–554.

Ramsay, J.G., 1963. Structural investigations in the Barberton Mountain Land, eastern Transvaal. *Trans. Geol. Soc. S. Afr.*, 66: 353–398.

Reynolds, D.G., Brook, W.A., Marshall, A.E. and Allchurch, P.D., 1975. Volcanogenic copper–zinc deposits in the Pilbara and Yilgarn Archaean Blocks. In: C.L. Knight (Editor), *Economic Geology of Australia and Papua-New Guinea, 1. Metals.* Australas. Inst. Min. Metall., Melbourne, pp. 185–194.

Richter-Bernberg, G., 1973. In: W. Drooger (Editor), *Messinian Events in the Mediterranean.* North-Holland, Amsterdam, pp. 124–149.

Rickard, D.T. and Zweifel, H. 1975. Genesis of Precambrian sulphide ores, Skellefte District, Sweden. *Econ. Geol.*, 70: 255–274.

Ringwood, A.E., 1969. Composition and evolution of the upper mantle. In: *The Earth's Crust and Upper Mantle. Am. Geophys. Union, Geophys. Monogr.*, 13: 1–17.

Ringwood, A.E., 1975. *Composition and Petrology of the Earth's Mantle*. McGraw-Hill, New York, N.Y., 618 pp.

Ringwood, A.E., 1977a. Composition of the core and implications for origin of the Earth. *Geochem. J.*, 11: 111–135.

Ringwood, A.E., 1977b. Composition and origin of the Earth. *Res. Sch. Earth Sci., Aust. Natl. Univ. Publ.*, 1299: 65 pp.

Ringwood, A.E. and Kesson, S.E., 1977. Composition and origin of the moon. *Proc. 8th Lunar Sci. Conf.*, pp. 371–398.

Robertson, D.S., 1974. Basal Proterozoic units as fossil time markers and their use in uranium prospecting. In: IAEA, Vienna, *Formation of Uranium Ore Deposits*, pp. 495–512.

Robertson, I.D.M. and Van Breemen, O., 1970. The southern satellite dykes of the Great Dyke, Rhodesia. *Geol. Soc. S. Afr., Spec. Publ.*, 1: 621–644.

Roddick, J.C., Compston, W. and Durney, D.W., 1976. The radiometric age of the Mount Keith Granodiorite, a maximum age estimate for an Archaean greenstone sequence in the Yilgarn Block, Western Australia. *Precambrian Res.*, 3: 55–78.

Roscoe, S.M., 1973. The Huronian Supergroup, a paleo-Aphebian succession showing evidence of atmospheric evolution. *GAC-Spec. Pap.*, 12: 31–47.

Ross, J.R. and Hopkins, G.M.F., 1975. Kambalda nickel sulphide deposits. In: C.L. Knight (Editor), *Economic Geology of Australia and Papua-New Guinea, 1. Metals*. Australas. Inst. Min. Metall., Melbourne, pp. 100–121.

Ross, J.R. and Keays, R.R., 1979. Precious metals in volcanic-type nickel sulfide deposits in Western Australia, Part I: Relationship with the composition of the ores and their host rocks. *Can. Mineral.*, 17(2).

Roy, S., 1966. *Syngenetic Manganese Formations of India*. Jadavpur Univ., Calcutta, 219 pp.

Roy, S., 1976. Ancient manganese deposits. In: K.H. Wolf (Editor), *Handbook of Strata-Bound and Stratiform Ore Deposits, 7*. Elsevier, Amsterdam, pp. 395–476.

Rubey, W.W., 1951. Geologic history of sea water. *Geol. Soc. Am. Bull.*, 62: 1111–1147.

Saager, R. and Köppel, V., 1976. Lead isotopes and trace elements from sulphides of Archaean greenstone belts in southern Africa – A contribution to the knowledge of the oldest known mineralizations. *Econ. Geol.*, 71: 44–58.

Sangster, D.F., 1972. Precambrian volcanogenic massive sulphide deposits in Canada: a review. *Geol. Surv. Can., Pap.*, 72-22: 44.

Sangster, D.F., 1976. Sulphur and lead isotopes in strata-bound deposits. In: K.H. Wolf (Editor), *Handbook of Strata-bound and Stratiform Ore deposits, 2*. Elsevier, Amsterdam, pp. 219–266.

Sangster, D.F., 1977. Some grade and tonnage relationships among Canadian volcanogenic massive sulphide deposits. *Geol. Surv. Can., Pap.*, 77-1A: 5–12.

Sangster, D.F., 1978. Exhalites associated with Archaean volcanogenic massive sulphide deposits. In: J.E. Glover and D.I. Groves (Editors), *Archaean Cherty Metasediments: their Sedimentology, Micropalaeontology, Biogeochemistry and Significance to Mineralization, 2*. Dep. Geol. and Extension Service, Univ. West Aust., Perth, pp. 70–81.

Sangster, D.F. and Brook, W.A., 1977. Primitive lead in an Australian Zn-Pb-Ba deposit. *Nature*, 270: 423.

Sangster, D.F. and Scott, S.D., 1976. Precambrian stratabound massive Cu–Zn–Pb sulfide ores of North America. In: K.H. Wolf (Editor), *Handbook of Strata-Bound and Stratiform Ore Deposits, 6*. Elsevier, Amsterdam, pp. 129–222.

Sawkins, F.J., 1972. Sulphide ore deposits in relation to plate tectonics. *J. Geol.*, 80: 377–397.

Sawkins, F.J. and Rye, D.M., 1974. Relationship of Homestake-type gold deposits in iron-rich Precambrian sedimentary rocks. *Inst. Min. Metall., Trans.*, 83: B56–59.

Scarpelli, W., 1973. The Serra do Navio manganese deposit (Brazil). In: *The Genesis of Precambrian Iron and Manganese Deposits*. Proc. of Kiev Symposium, 1970: UNESCO, Paris, pp. 217–228.

Schidlowski, M., 1976. The Archaean atmosphere and evolution of the terrestrial oxygen budget. In: B.F. Windley (Editor), *The Early History of the Earth*. Wiley, London, pp. 525–534.

Schidlowski, M., 1979. Antiquity and evolutionary status of bacterial sulphate reduction: sulphur isotope evidence. In: C. Ponnamperuma (Editor), *Limits of Life* (Proc. 4th College Park Colloq. on Chemical Evolution, Univ. Maryland, October, 1978), (in press).

Schidlowski, M. and Eichmann, R., 1977. Evolution of the terrestrial oxygen budget. In: C. Ponnamperuma (Editor), *Chemical Evolution of the Early Precambrian.* Academic Press, New York, N.Y., pp. 87–99.

Schidlowski, M., Eichmann, R. and Junge, C.E., 1975. Precambrian sedimentary carbonates: carbon and oxygen isotope geochemistry and implications for the terrestrial oxygen budget. *Precambrian Res.,* 2: 1–69.

Schopf, J.W., 1975. Precambrian paleobiology: Problems and perspectives. *Annu. Rev. Earth Planet. Sci.,* 3: 213–249.

Schreiber, B.C. and Kinsman, D.J.J., 1975. New observations on the Pleistocene evaporites of Montallegro, Sicily, and a modern analogue. *J. Sediment. Petrol.,* 45: 469–479.

Schreiber, B.C., Friedman, G.M., Decima, A. and Schreiber, E., 1976. Depositional environments of Upper Miocene (Messinian) evaporite deposits of the Sicilian Basin. *Sedimentology,* 23: 729–760.

Seccombe, P.K., Groves, D.I., Binns, R.A. and Smith, J.W., 1977. A sulphur isotope study to test a genetic model for Fe–Ni sulphide mineralization at Mt. Windarra, Western Australia. In: B.W. Robinson (Editor), *Stable Isotopes in the Earth Sciences. DSIR Bull.,* 220: 187–200.

Seward, T.M., 1973. Thio complexes of gold and the transport of gold in hydrothermal ore solutions. *Geochim. Cosmochim. Acta,* 37: 379–400.

Seward, T.M., 1979. Hydrothermal transport and deposition of gold. In: J.E. Glover and D.I. Groves (Editors), *Gold Mineralization, 3.* Geol. Dep. and Extension Service, Univ. West Aust., Perth, pp. 45–55.

Seyfried, W. and Bischoff, J.L., 1977. Hydrothermal transport of heavy metals by sea water: the role of seawater/basalt ratio. *Earth Planet. Sci. Lett.,* 34: 71–77.

Shackleton, R.M., 1976. Shallow and deep-level exposures of the Archaean crust in India and Africa. In: Windley, B.F. (Editor), *The Early History of the Earth.* Wiley, London, pp. 317–322.

Sharpe, J.I., 1968. Geology and sulphide deposits of the Mattagami area. *Quebec Dep. Natl. Res. Geol. Rep.,* 137.

Shaw, D.M., 1972. The development of the early continental crust. Part 1: Use of trace element distribution coefficient models for the Protoarchaean crust. *Can J. Earth Sci.,* 9: 1577–1595.

Shaw, D.M., 1976. Development of the early continental crust. Part 2: Prearchaean, Protoarchaean and later eras. In: B.F. Windley (Editor), *The Early History of the Earth.* Wiley, London, pp. 33–54.

Shawe, D.R., Poole, F.G. and Brosst, D.A., 1969. Newly discovered bedded barite deposits in East Northumberland Canyon, Nye County, Nevada. *Econ. Geol.,* 64: 242–254.

Sillén, L.G., 1967. The ocean as a chemical system. *Science,* 156: 1189–1197.

Sillitoe, R.H., 1972. Formation of certain massive sulphide deposits at sites of sea-floor spreading. *Inst. Min. Metall., Trans., Sect. B,* 81: B141–148.

Sillitoe, R.H., 1976. Andean mineralization: a model for the metallogeny of convergent plate margins. In: D.F. Strong (Editor), *Metallogeny and Plate Tectonics. Geol. Assoc. Can., Spec. Pap.,* 14: 59–100.

Simpson, P. and Bowles, J., 1979. Detrital uraninite and pyrite, and evidence for a reducing atmosphere. In: P.R. Grant (Editor), *Proceedings of a Meeting: Evolution of the Atmosphere and Hydrosphere.*

Smirnov, V.I., 1976. *Ore Deposits of the USSR, II.* Pitman, London, pp. 3–79.

Smith, G.E., 1976. Sabkha and tidal-flat facies control of stratiform copper deposits in north Texas. In: K.H. Wolf (Editor), *Handbook of Strata-Bound and Stratiform Ore Deposits, 6.* Elsevier, Amsterdam, pp. 407–446.

Smith, J.V., 1976. Development of the Earth-Moon system with implications for the geology of the early Earth. In: B.F. Windley (Editor), *The Early History of the Earth.* Wiley, London, pp. 3–20.

Solomon, M., 1976. "Volcanic" massive sulphide deposits and their host rocks – a review and an explanation. In: K.H. Wolf (Editor), *Handbook of Strata-Bound and Stratiform Ore Deposits, 6*. Elsevier, Amsterdam, pp. 21–54.

Spence, C.D. and de Rosen-Spence, A.F., 1975. The place of sulphide mineralization in the volcanic sequence at Noranda, Quebec. *Econ. Geol.,* 70: 90–101.

Stanton, R.L., 1972. *Ore Petrology*. McGraw-Hill, New York, N.Y., 713 pp.

Stemprok, M., 1977. The source of tin, tungsten and molybdenum of primary ore deposits. In: M. Stemprok, L. Burnol and G. Tischendorf (Editors), *Metallization Associated with Acid Magmatism, 2*. pp. 127–166.

Stern, C.R., Huang, W.L. and Wyllie, P.J., 1975. Basalt–andesite–rhyolite–H_2O: crystallization intervals with excess H_2O and H_2O-undersaturated liquidus surfaces to 35 kilobars, with implications for magma genesis. *Earth Planet. Sci. Lett.,* 28: 189–196.

Stowe, C.W., 1974. Alpine-type structures in the Rhodesian basement complex at Selukwe. *J. Geol. Soc. London,* 130: 411–426.

Sun, S.-S. and Nesbitt, R.W., 1977. Chemical heterogeneity of the Archaean mantle, composition of the Earth and mantle evolution. *Earth Planet. Sci. Lett.,* 35: 429–448.

Sun, S.-S. and Nesbitt, R.W., 1978. Petrogenesis of Archaean ultra-basic and basic volcanics: evidence from rare earth elements. *Contrib. Mineral. Petrol.,* 65: 301–325.

Sylvester-Bradley, P.C., 1976. Evolutionary oscillation in prebiology: igneous activity and the origins of life. *Origins Life,* 7: 9–18.

Talbot, C.J., 1973. A plate tectonic model for the Archaean crust. *Philos. Trans. R. Soc. London, Ser. A,* 273: 413–428.

Tarney, J., 1976. Geochemistry of Archaean high-grade gneisses, with implications as to the origin and evolution of the Precambrian crust. In: B.F. Windley (Editor), *The Early History of the Earth*. Wiley, London, pp. 405–418.

Tarney, J., Dalziel, I.W.D., and de Wit, M.J., 1976. Marginal basin "Rocas Verdes" complex from south Chile: A model for Archaean greenstone belt formation. In: B.F. Windley (Editor), *The Early History of the Earth*. Wiley, London, pp. 131–146.

Taylor, H.P., 1977. Water/rock interactions and the origin of H_2O in granitic batholiths. *J. Geol. Soc. London,* 133: 509–558.

Taylor, P.N., 1975. An early Precambrian age for migmatitic gneisses from Vikan i Bo, Vesteralen, North Norway. *Earth Planet. Sci. Lett.,* 27: 35–42.

Tera, F. and Wasserberg, G.J., 1974. U–Th–Pb systematics in lunar rocks and inferences about lunar evolution and the age of the Moon. *Proc. 5th Lunar Sci. Congr.,* pp. 1571–1599.

Tischendorf, G., 1977. Geochemical and petrographic characteristics of silicic magmatic rocks associated with rare-element mineralization. In: M. Stemprok, L. Burnol and G. Tischendorf (Editors), *Metallization Associated with Acid Magmatism, 2*. pp. 41–98.

Towe, K.M., 1978. Early Precambrian oxygen: a case against photosynthesis. *Nature,* 274: 657–661.

Trendall, A.F., 1973. Iron formations of the Hamersley Group of Western Australia: type examples of varved Precambrian evaporites. In: *Genesis of Precambrian Iron and Manganese Deposits*. UNESCO, Paris, pp. 257–270.

Trendall, A.F. and Blockley, J.G., 1970. The iron formation of the Precambrian Hamersley Group, Western Australia. *Bull. Geol. Surv. West. Aust.,* 119: 366 pp.

Trudinger, P.A., 1976. Microbial processes in relation to ore genesis. In: K.H. Wolf (Editor), *Handbook of Strata-Bound and Stratiform Ore Deposits, 2*. Elsevier, Amsterdam, pp. 135–190.

Trudinger, P.A., Lambert, I.B. and Skyring, G.W., 1972. Biogenic sulphide ores: a feasibility study. *Econ. Geol.,* 67: 1114–1127.

Tugarinov, A.L. and Bibicova, E.V., 1973. Geochronology of the ancient Kola Formation. Abstr. *3rd Europ. Colloq. Geochronology*.

Turek, A. and Compston, W., 1971. Rb/Sr geochronology in the Kalgoorlie region. *Geol. Soc. Aust., Spec. Publ.,* 3: 72.

Turner, A.R. and Ranford, L.C., 1975. Nickel deposits of Mt. Sir Samuel, Western Australia. In: C.L. Knight (Editor), *Economic Geology of Australia and Papua-New Guinea, 1. Metals*. Australas. Inst. Min. Metall., Melbourne pp. 156–159.

Turner, C.C. and Walker, R.G., 1973. Sedimentology, stratigraphy and crustal evolution of the Archaean greenstone belt near Sioux Lookout, Ontario. *Can. J. Earth Sci.,* 10: 817—845.

Turner, F.J., 1968. *Metamorphic Petrology.* McGraw-Hill, New York, N.Y.

Urey, H.C., 1962. Evidence regarding the origin of the Earth. *Geochim. Cosmochim. Acta,* 26: 1—13.

Usselman, T.M., Hodge, D.S., Naldrett, A.J. and Campbell, I.H., 1979. Physical contraints on the localization of nickel sulfide ore in ultramafic lavas. *Can. Mineral.,* 17(2).

Veizer, J., 1976. Evolution of ores of sedimentary affiliation through geologic history: relations to the general tendencies in evolution of the crust, hydrosphere, atmosphere and biosphere. In: K.H. Wolf (Editor), *Handbook of Strata-Bound and Stratiform Ore Deposits, 3.* Elsevier, Amsterdam, pp. 1—41.

Veizer, J. and Compston, W., 1976. $^{87}Sr/^{86}Sr$ in Precambrian carbonates as an index of crustal evolution. *Geochim. Cosmochim. Acta,* 40: 905—914.

Viljoen, R.P. and Viljoen, M.J., 1969. Evidence for the composition of the primitive mantle and its products of partial melting, from the study of rocks of the Barberton Mountain Land, South Africa. *Geol. Soc. S. Afr., Spec. Publ.,* 2: 275—296.

Viljoen, R.P., Saager, R. and Viljoen, M.J., 1970. Some thoughts on the origin and processes responsible for the concentrations of gold in the early Precambrian of southern Africa. *Mineral. Deposita,* 5: 164—180.

Vinogradov, V.I., Reimer, T.O., Leites, A.M. and Smelon, S.B., 1976. The oldest sulfates in the Archaean formations of the South African and Aldan Shields, and the evolution of the Earth's oxygen atmosphere. *Lithol. Miner. Resour.,* 11: 407—420.

Vollmer, R., 1977. Terrestrial lead isotopic evolution and formation time of the Earth's core. *Nature,* 270: 144—147.

Wakefield, J., 1976. The structural and metamorphic evolution of the Pikwe Ni—Cu sulfide deposit, Selebi-Pikwe, Eastern Botswana. *Econ. Geol.,* 71(6): 988—1005.

Walker, J.C.G., 1976a. Implications for atmospheric evolution of the inhomogeneous accretion model of the origin of the Earth. In: B.W. Windley (Editor), *The Early History of the Earth.* Wiley, London, pp. 535—546.

Walker, J.C.G., 1976b. *Evolution of the Atmosphere.* Hafner, New York, N.Y.

Walker, R.G., 1976. A critical appraisal of Archaean basin—craton complexes. *Can. J. Earth Sci.,* 15: 1213—1218.

Walker, R.G. and Pettijohn, F.J., 1971. Archaean sedimentation: analysis of the Minnitaki Basin, northwestern Ontario, Canada. *Geol. Soc. Am. Bull.,* 82: 2099—2130.

Walker, R.R. and Mannard, G.W., 1974. Geology of the Kidd Creek Mine — A progress report. *Can. Min. Metall. Bull.,* 1974: 1—17.

Watson, J., 1973. Influence of crustal evolution on ore deposition. *Inst. Min. Metall., Trans., Sect. B,* 82: 107—114.

Watson, J., 1976. Mineralization in Archaean provinces. In: B.F. Windley (Editor), *The Early History of the Earth.* Wiley, London, pp. 443—454.

Wells, P.R.A., 1976. Late Archaean metamorphism in the Buksefjorden region, southwest Greenland. *Contrib. Mineral. Petrol.,* 56: 229—242.

Wetherill, G.W., 1972. The beginning of crustal evolution. *Tectonophysics,* 13.

White, A.J.R. and Chappell, B.W., 1977. Ultrametamorphism and granitoid genesis. *Tectonophysics,* 43: 7—22.

Wiebols, J.H., 1955. A suggested glacial origin for the Witwatersrand conglomerates. *Trans. Geol. Soc. S. Afr.,* 58: 367—382.

Williams, D.A.C., 1972. Archaean ultramafic, mafic and associated rocks, Mt. Monger, Western Australia. *J. Geol. Soc. Aust.,* 19: 163—188.

Williams, D.A.C., 1979. The Association of some nickel sulfide deposits with komatiitic volcanism in Rhodesia. *Can. Mineral.,* 17(2).

Williams, I.R., 1974. Structural subdivision of the Eastern Goldfields Province, Yilgarn Block. *W. Aust. Geol. Surv., Annu. Rep.,* 1973: 53—59.

Wilson, J.F., in pre. A preliminary reappraisal of the Rhodesian basement complex.

Wilson, J.F., Bickle, M.J., Hawkesworth, C.J., Martin, A., Nisbet, E.G. and Orpen, J.L., 1978. Granite–greenstone terrains of the Rhodesian Archaean craton. *Nature*, 271: 23–27.

Windley, B.F., 1973. Archaean anorthosites: a review with the Fiskenaesset complex, west Greenland, as a model for interpretation. *Geol. Soc. S. Afr., Spec. Publ.*, 3: 312–332.

Windley, B.F., 1977. *The Evolving Continents*. Wiley, London, 385 pp.

Windley, B.F. and Bridgwater, D., 1971. The evolution of Archaean low- and high-grade terrains. *Geol. Soc. Aust., Spec. Publ.*, 3: 33–46.

Woodall, R., 1975. Gold in the Precambrian shield of Western Australia. In: C.L. Knight (Editor), *Economic Geology of Australia and Papua-New Guinea. 1. Metals.* Australas Inst. Min. Metall., Melbourne, pp. 175–185.

Woodall, R. and Travis, G.A., 1969. The Kambalda nickel deposits, Western Australia. *Proc. 9th Commonw. Min. Metall. Congr., London*, 2: 517–533.

Wyllie, P.J., 1971. *The Dynamic Earth*. Wiley, New York, N.Y., 416 pp.

Ypma, P. and Fuzikawa, K., 1979. Fluid inclusion studies as a guide to the composition of uranium ore-forming solutions of the Nabarlek and Jabiluka deposits, N.T., Australia. In: *Int. Uranium Symp. on Pine Creek Geosyncline, Extended Abstracts*, Sydney, June, 1979.

NOTE ADDED IN PROOF

Recent papers by D.R. Lowe (*Nature*, 284: 441–443, 1980) and M.R. Walter et al. (*Nature*, 284: 443–445, 1980) have presented good evidence for stromatolites in the Warrawoona Group of the eastern Pilbara. The latter authors cited a personal communication by P.J. Hamilton that volcanics from low in the Warrawoona Group have yielded a Sm–Nd age of 3520 ± 60 Myr. H.D. Phlug and H. Jaeschke-Boyer (*Nature*, 280: 483–486, 1979) suggested that microstructures within >3.7 Gyr old metasediments of the Isua Group, western Greenland, are biogenic.

Chapter 3

THE CONCEPT OF ORE TYPES – SUMMARY, SUGGESTIONS AND A PRACTICAL TEST

PETER LAZNICKA

INTRODUCTION

Ore (or mineralization) type is part of our present geological methodology, and there has hardly been any recent conference or symposium volume dealing with metallogeny that would fail to mention examples of ore types. In the first seven volumes of this Handbook (Wolf, 1976a) reference to ore types is made in 47 out of 59 chapters, and in four chapters such reference appears in the title. Yet despite the present popularity, definitions of ore type are conspicuously missing in recent glossaries of geology, currently available in Western literature. This may indicate that ore type is considered to be one of the numerous informal, but convenient, terms in geology which most of us have been using without much concern as to whether or not it conforms with its defined meaning.

Ore or mineralization type (English); Erztypus or Lagerstättentypus (German); type de gítes, dépôts, or gisements (French); rudnyi tip (Russian); and tipo de yacimiento or deposito (Spanish), are easily recognizable in European languages and have a close meaning in all of them. In this review, the whole spectrum of ore types is considered and a restriction of coverage to stratiform and strata-bound deposits has not been considered proper.

The first use of the etymological grouping "type de gítes" in an influential paper is usually attributed to de Launay (1900, 1913), but many earlier authors gave examples of ore types although they did not refer to them as such. Among the earliest writers is Agricola (1556), who, for example, distinguished venae profundae, venae dilatatae, and venae cumulatae – in fact three morphological types of hydrothermal ore deposits. One of the most massive early typifications of local ores by Breithaupt (1849) listed numerous "vein formations" (Gangformazionen) complete with letter codes, based on persistent parageneses in the Freiberg (Saxony) ore district. Von Groddeck (1879) and Stelzner and Bergeat (1904–1906) were among the first to use type localities (Mansfeld-Cu; Schemnitz, now Banská Štiavnica – Au, Ag) to characterize distinct and repeated styles of mineralization. The book of Bain (1911) may have been the first one that contains reference to ore types in its title, but in style and contents it differs little from contemporary textbooks on ore deposits except for its coverage restricted to several American ore types, such as the Clinton-type (Fe); Lake Superior-type (Fe), as well as "Flats and Pitches of the Wisconsin lead and zinc district".

Following the introduction of ore type into the literature, the French school had added considerably to it (e.g. Blondel, 1951, 1955; Routhier, 1958). It provided a definition of "types de gisements", which "are based on mineralogical associations present in deposits and on petrographic relations among the deposits and certain eruptive or sedimentary rocks" (Raguin, 1961). The French, furthermore, coined the use of the word "typologie" for the branch of metallogeny dealing with the use and systematics of mineralization types (e.g. Bauchau, 1971).

The German-language schools (German, Austrian, Swiss, partly Czech) contributed many classic concepts to the development of economic geology and metallogeny, including numerous ore types of worldwide recognition (Kupferschiefer or Mansfeld-Cu; Lahn–Dill, Fe; Rammelsberg-Pb, Zn, Cu). Schneiderhöhn, in several editions of his textbooks, extensively subdivided the system of ore deposits into very detailed genetic and paragenetic categories and provided a type locality ("Typus") for many of them. This greatly encouraged the use of dual reference to narrow categories of ore deposits that could either be referred to by their longer, descriptive characteristic (e.g. "Kata- bis mesothermale Verdrängungslagerstätten mit silberreichen Blei–Zinkerzen"), or by corresponding type locality (Leadville). The third edition of *Erzlagerstätten* (Schneiderhöhn, 1955) lists 85 locality types – one of the most extensive listings of such types available (see Wolf's chapter in this volume).

In the U.S.S.R., the systematics of ore deposits has been the most formal of all the national schools discussed, and universally applied countrywide. The basic unit is "rudnaya formatsiya" – literally "ore formation", but better expressed as ore association. In some English texts, "rudnaya formatsiya" has been translated as "ore type" – for example in the paper of Domarev (1968; translated by the International Geology Review staff). The definition of "rudnaya formatsiya", for example by Konstantinov and Sirotinskaya (1974), reads "a group of deposits with corresponding, and compositionally persistent mineral associations, connected with repetitive geological conditions of all deposits". This indeed stands very close to the few available definitions of ore type in Western literature.

The Soviet "rudnye formatsii", however, do not contain formally assigned type localities, but the two hierarchically inferior subdivisions: "rudnyi tip" (ore type) and "rudnyi podtip" (ore subtype) occasionally do, e.g. Sumsar Pb–Zn type of Bogdanov and Kutyrev (1973); Kyzyl–Espeh Pb–Zn type of Magak'yan (1974); and Vitim Mo–W type of Shcheglov (1967). "Rudnyi tip" and "podtip" mostly serve the purpose to subdivide deposits included in a "rudnaya formatsiya" according to the age and stratigraphic affiliation of their host rocks.

Very recently, certain metallogenic divisions hierarchically superior to "rudnaya formatsiya" have been assigned type areas. Semenov et al. (1967) distinguished several subtypes of geochemical metallogenic provinces (e.g. Aldan, Ukraine, Baltic) based on the overall acidity/basicity of rocks. Shcheglov (1967) distinguished several varieties of activation zones with different metallogeny and named them after type areas (e.g. Rhodopean-type: polymetallic mineralization in median massifs; Czech—Bureya-type: Sn-W mineralization in median massifs; West African-type: Sn—W mineralization in shields; East African-type: rare-earth mineralization in shields). Several named types of metallogenic zones also appear in Magak'yan (1974, pp. 208—209).

In North America, several mineralization types have been widely used since the turn of this century, but in an unsystematic manner. Both characteristic ore or host-rock appearances (porphyry coppers; massive sulphides) and type localities (Algoma; Lake Superior; Clinton-types, all of iron; Mississippi Valley Zn—Pb-type; Ducktown Cu-type), have been common. Most of the ore type names originated in the mining industry, but spread rapidly into the economic geologic literature.

The past twenty years (a major revolutionary period in geological sciences) has brought about a sharp renewal of interest in mineralization types. Many existing types were extensively scrutinized, genetically modelled and accurate definitions attempted [e.g. Lowell and Guilbert (1970) for porphyry coppers; Snyder (1968) for the Mississippi Valley-type]. Numerous new types were introduced, some, initially, to fulfill the need of the mining industry to concisely identify modern exploration targets (e.g. McArthur River Zn—Pb; roll-type uranium), and others to differentiate within a broad category of mineral deposits of similar appearance of ore but different lithologic association and genesis (e.g. the Kuroko-, Cyprus- and Besshi-types within the group of massive sulphides; Hutchinson, 1973; Sawkins, 1972). The latter names have also been used to illustrate ore deposition in the various plate tectonic environments (Mitchell and Bell, 1973; Mitchell and Garson, 1976; Sawkins, 1976). The recent trend of application of mathematics and statistics to geology, and compilation of computerized data files, has also affected ore types (e.g. de Geoffroy and Wignall, 1972; Laznicka, 1973; Kimberley, 1978).

ORE TYPES VERSUS A COMPLETE CLASSIFICATION SYSTEM OF ORE DEPOSITS

A variety of mineral deposits classifications have been offered up to date (see Park and MacDiarmid, 1975, p. 211—219; Wolf in this volume). One classification devoted exclusively to stratiform and strata-bound deposits is provided in Gabelman (1976, Vol. 1 of this Handbook). None is perfect and each is controversial, which Graton (1968, p. 1706) attributes to the fact that ore deposits are so diverse and, therefore, so commonly devoid of consistent characterization even within single units — i.e. individual orebodies. Consequently, a perfect (even "relatively" perfect) classification as, for example, the biological taxonomy, could never be achieved.

The logical principles of classification require that: "(1) no two subsets have any element in common; and (2) all of the subsets together contain all of the members of the partitioned set — i.e. they are mutually exclusive and jointly exhaustive" (Körner, 1974).

A cursory examination of Table I of Gabelman (1976, pp. 81 and 83) shows that this is rarely so. For example, Arkansas as an "example-locality" of barite mineralization is representative of bedded barite constituting a whole-rock mineral concentration, of baritic shale, and of barite replacements in calcareous sedimentary rocks.

The above example is clearly one of the borderline cases – i.e. "objects that can with equal correctness be accepted or rejected as members of a class" (Körner, 1974) – that are so characteristic for classification of perceptual objects. Only a far-reaching generalization and oversimplification of categories, and the use of nonperceptual properties, can get rid of borderline cases in classification of ore deposits and provide a common denominator to make an overall classification possible. Genesis is an ideal property that is commonly non-perceptual for the majority of ore concentrations, hence the abundance of genetic classifications of ore deposits. This is not to say that there are no borderline cases among genetic classes of mineralizations. One that is particularly close to the subject of stratabound deposits comes immediately to mind: that of sedimentary–volcanic deposits grading into both purely volcanic ore concentrations on one side, and purely sedimentary deposits on the other (cf. Gilmour, 1976, ch. 4, Vol. 1 of this Handbook). The high degree of subjectivity in genetic classifications gives the author the power to draw a nonperceptual boundary between categories: as sharp as he chooses or dares.

The *Oxford English Dictionary* (1970 edition) devotes five densely printed columns and ten sets of definitions to the word "type". Type, defined as "the general form, structure or character, distinguishably a particular kind, group or class of beings or objects, a pattern or model after which something is made," is particularly fitting to ore type. There is no sharp difference between various classes within a complete classification of ore deposits, and ore type – the distinction is more or less a matter of feeling (Fig. 1). Some classes of a complete classification set are felt to be an ore type as well (for example, calcite-bearing silver veins; Schneiderhöhn, 1955, ore type *1* in Fig. 1), others are not [e.g. "ores originated in magmas by differentiation or in adjacent country rocks by injection" – modified Lindgren's classification in Ridge (1968)].

Classification misfits and cross-categories difficult to accommodate are particularly common among ore types. Massive sulphides are one of them. They are ores of distinct appearance and of characteristic form, but of variable genesis; thus in genetic classifications they become fragmented and are lost in different classes – for example, ore types *4* and *5* in Fig. 1. Porphyry coppers, red-bed deposits, sandstone uranium deposits, Mississippi Valley Zn–Pb-type, etc., are other examples.

As can be seen, ore types may be named after distinct appearance of ore (e.g., massive sulfide), after host-rock or lithologic association, after typical locality or region of distribution, and so on. The types named after localities are particularly characteristic, so they are considered by many as *the* ore type. A distinction should be made between an example locality and type locality. Example locality is representative of itself only. Type locality is more formal and is representative not only of itself, but represents all other localities assigned, with permissible deviations, to such type.

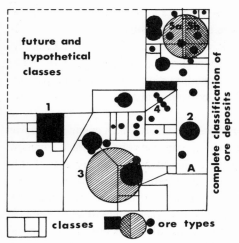

Fig. 1. Diagrammatic relationship among classes of overall (usually genetic) classification of ore deposits, and ore types.

Ore type *1* corresponds to an entire class. Ore type *2* is confined to, and is very characteristic of, the class *A*, but there are additional ore types in this class. Ore type *3* is common to several classes. *4* is a string of closely related ore types crossing a class boundary. Ore type *2* is independent. Ore type *5* is hierarchically subdivided into subtypes *5a* and *5b* which themselves contain ore types of even lower rank.

It appears that the main advantage of a mineralization type is that it can exist alone, without being an integral part of a complete classification. It can be introduced and withdrawn without destroying the whole system. Ore types can take good care of borderline cases, regardless of genesis. The three modes of distribution of barite in Arkansas in the above-mentioned example can all be included under, and considered minor varieties within, the Arkansas-type bedded barite deposits of Amstutz and Bubenicek (1967).

Another advantage of mineralization types, particularly locality types, is that their genesis can be ignored — a marked advantage in the case of controversial deposits of which the genetic interpretation changes every few years. An ore occurrence assigned to a locality ore type automatically changes its genetic interpretation with that of the type locality, and provided the selection of the type locality has been sensible and correct, such ore type is a longer-lasting category of reference than a genetic category. Hence the love affair of practical mining men with ore types. The genetic interpretation of many stratabound Zn–Pb deposits in carbonates, lacking conspicuously associated igneous rocks (Mississippi Valley-type) has changed in the past 30 years from hydrothermal (telethermal or apomagmatic) to sedimentogenic, and then back to hydrothermal in some cases. Despite the change in genetic ideas, the Mississippi Valley-type, closely related types and synonyma (Bleiberg, Upper Silesia, etc.) have remained unaffected.

Once a locality is assigned to an ore type, it becomes a member of a class rather than a concrete particular. This writer, however, after having visited some 1800 mineralized localities can state with confidence that not even two of them have been exactly alike, like two examples of *Picea glauca* (white spruce) or *Larus occidentalis* (western gull). It follows that to achieve an accuracy in assigning localities to ore types comparable to assigning a plant or a bird into the biologic taxonomy, as many ore types as individual mineralized localities would be needed.

The number of ore deposits of the world is not accurately known, because it essentially depends on where a boundary between an "ore deposit" (an economically mineable entity) and "ore occurrence" or "showing" (an entity with none, or with unknown, economical potential) is placed; a problem not only of the magnitude of geochemical concentration, but economic consideration as well. Moreover, the number of known ore deposits increases yearly with new discoveries. The minimum number is about 10,000 major deposits which, estimated on the basis of some quantitative data (MANIFILE — Laznicka, 1973), account for at least 90% of the past production and remaining reserves of all metals, on a worldwide basis.

A total of 10,000 ore types, however, is of course more than what can be reasonably handled by conventional methods. Any reduction of this number will, however, be necessarily compensated by lowered accuracy and this is exactly what experience, based on several tens of widely recognized and several hundred of locally valid ore types, indicates.

It could be concluded that:

(1) Ore types can exist alone, in a loose set full of gaps, and need not be part of a totally partitioned set. They may mutually overlap, merge, be confined to particular territories, subordinated to particular metals, tectonic regimes, etc. (Fig. 1).

(2) The entire system of ore deposits can theoretically be subdivided into ore types without gaps, but to do so immense quantity of data and sincere international cooperation is needed. Given the present political fragmentation of the world, an early success is unlikely.

(3) If ore types are ever integrated with genetic classifications (see discussions by Wolf, Chapter 1, this volume), the number of genetic categories will have to increase to several thousands due to the incredible variety of genetic combinations involved in the generation of polygenetic ore deposits. It now appears that polygenetic deposits are a commonplace, while simple ones are a rarity.

SCOPE AND RANK OF ORE TYPES

The ore types introduced by Schneiderhöhn (1955; see Wolf, Chapter 1, this volume) are subordinated to a very detailed system of all ore deposits. Within such a system, locality ore types occupy a relatively low rank: fifth to seventh order of partition.

The greater proportion of ore types in present use, however, is a result of partition of a much more restricted subject category, such as certain metal (e.g. iron ore types); limited geographic area (ore types of the Canadian Shield); a petrographic group of host-rocks (ore types in metalliferous conglomerates); a depositional environment or facies (ore types in activation zones) (see Table I).

The rank of ore types in all limited-scope sets of partition is variable. The over-whelming majority of ore types, particularly locality types, are of a relatively low rank — quite comparable with the 5th to 7th degree of partition in the Schneiderhöhn's genetic classification. Some have even a lower rank; these are variations within a single deposit, or even a single orebody. The Japanese Kurokō ore type is a well-known popular example. The Kurokō type is not named after a locality, but after the black, galena- and sphalerite-rich variety of volcanogenic massive sulphide that is characteristic for young ore deposits distributed throughout the "Green Tuff Province" of Japan. Ridge (1976, p. 248, Vol. 1 of this Handbook) says "Kurokō-type ores are composed of four ore types", and these are kurokō (black ore); okō (yellow ore); keikō (siliceous ore) and sekkokō (mineralized anhydrite and gypsum). Thus, kurokō is common to two ranks.

Most ore types of a higher rank have non-locality names (porphyry coppers; scheelite skarns; red-bed copper deposits; nickel laterites; massive sulphides), although the immensely popular Mississippi Valley-type is a notable exception. The relatively high rank of the Mississippi Valley-type results from its complexity, variety and controversial genesis. Because the type can form by both hydrothermal as well as purely sedimentogenic processes, it resists any unequivocal inclusion into a genetic system of ore deposits at the very beginning. But the main problem with this type is that it is named after a broad region that contains four major ore districts and several tens of minor districts and group-ings of occurrences. All have variable character of orebodies, hosts, mineralogy, age, etc. With the exception of a few locality types of lesser magnitude hierarchically subordinated to the Mississippi Valley-type (such as the Tri-State type, Hagni, 1976, Vol. 6 of this Handbook; Mine La Motte-type, Snyder and Gerdemann, 1968), no additional localities in the area have been introduced as ore types in the literature.

Authors trying to use Mississippi Valley-type as a standard with which to com-pare a certain ore deposit or an ore district elsewhere, as well as authors trying to define with reasonable accuracy characteristics of Mississippi Valley-type deposits, face problems of scale and hierarchy because single deposits or single districts are incomparable with entire mineralized areas. Valiant efforts to define Mississippi Valley-type deposits (e.g. Ohle, 1959; Snyder, 1968) are actually lists of common denominator-type properties of a great variety of mineralizations ranging from strictly stratiform mineralized beds to classic fissure veins, from deposits in carbonate rocks to deposits in shales, sandstones and even lamprophyre dykes, and from shallow, surficial deposits to those found at 1.5 km depths. Even the common denominator characteristics are based on the selection of the more common deposits present, with the atypical fringe ignored.

The time may be ripe to abandon Mississippi Valley-type in research writing and use

TABLE I

Scope and rank of ore types, with selected examples

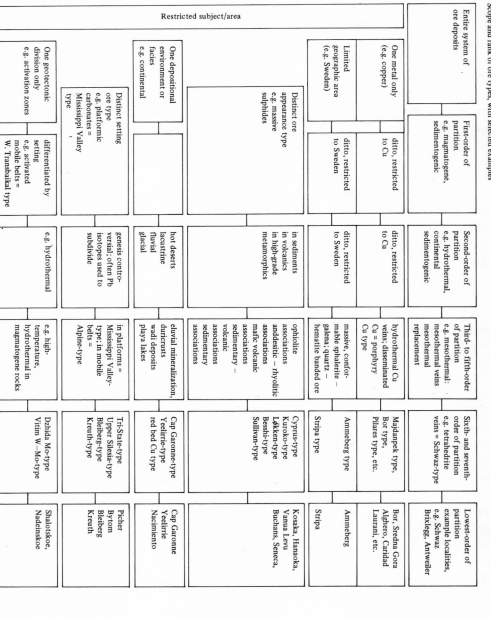

Entire system of ore deposits	First-order of partition e.g. magmatogene, sedimentogene	Second-order of partition e.g. hydrothermal, continental sedimentogene	Third- to fifth-order of partition e.g. mesothermal: mesothermal veins mesothermal replacement	Sixth- and seventh-order of partition e.g. tetrahedrite veins = Schwaz-type	Lowest-order of partition example localities, e.g. Schwaz Brixlegg, Antweiler
One metal only (e.g. copper)	ditto, restricted to Cu	ditto, restricted to Cu	hydrothermal Cu veins; disseminated Cu = porphyry Cu type	Majdanpek type, Bor type, Pilares type, etc.	Bor, Sredna Gora Alghero, Caridad Laurani, etc.
Limited geographic area (e.g. Sweden)	ditto, restricted to Sweden	ditto, restricted to Sweden	massive, conformable sphalerite – galena; quartz – hematite banded ore	Ammeberg type	Ammeberg
			Stripa type	Stripa	Stripa
One depositional environment or facies e.g. continental	Distinct ore appearance type e.g. massive sulphides	in sediments in volcanics in high-grade metamorphics	ophiolite associations andesitic – rhyolitic associations mafic volcanic associations sedimentary – volcanic associations sedimentary associations	Cyprus-type Kuroko-type Løkken-type Besshi-type Sullivan-type	Kosaka, Hanaoka, Vanua Levu Buchans, Seneca,
	Distinct setting ore type e.g. platformic carbonates = Mississippi Valley type	hot deserts lacustrine fluvial glacial	eluvial mineralization, duricrusts wadi deposits playa lakes	Cap Garonne-type Yeelirrie-type red bed Cu type	Cap Garonne Yeelirrie Nacimiento
		genesis controversial; often Pb isotopes used to subdivide	in platforms = Mississippi Valley-type; in mobile belts = Alpine-type	Tri-State-type Upper Silesia-type Bleiberg-type Kreuth-type	Picher Bytom Bleiberg Kreuth
One geotectonic division only e.g. activation zones	differentiated by setting e.g. activated mobile belts = W. Transbaikal type	e.g. hydrothermal	e.g. high-temperature, hydrothermal in magmatogene rocks	Dzhida Mo-type Vitim W-Mo-type	Shalotskoe, Nadeinskoe

Restricted subject/area

Examples based on Schneiderhöhn (1955), Borchert (1957), Shcheglov (1967), Magnusson (1970), Mitchell and Bell (1973), von Backström (1974), Rackley (1976) and Sawkins (1976).

it only for a rather inaccurate approximation especially in mining and exploration. However, it is realized that on the contrary, its popularity seems to be still rising particularly among authors whose distance from the type area prevents them from seeing its complexity.

THE REGIONAL VALIDITY, FREQUENCY, AND DEGREE OF FORMALITY OF ORE TYPES

The majority of ore types named after characteristic lithology of host association, distribution of ore substance, etc. (red-bed copper; Ni-laterites; podiform chromite) are used and understood worldwide. There are, naturally, local preferences over national territories reflecting variations in ore-type appearance, and these may have the strongest impression upon the observer. Porphyry coppers, for example, reflect in their name a conspicuous and persistent property of their igneous hosts in the minds of those who have investigated the classic area of the southwestern United States. The same property, however, is not so impressive in the equally classic area of a comparable type in the Lake Balkhash area, U.S.S.R. In the latter case, the single most conspicuous attribute of disseminated copper deposits is the profound wall-rock silicification which may cause many deposits, particularly the ones first discovered (Kounrad, Borly) to appear as erosion — resistant hills that dott the featureless Kazakh Steppe. Hence the synonym "disseminated copper deposits in secondary quartzites" (e.g. Rusakov, 1925; Nakovnik, 1968) is used for porphyry coppers in the U.S.S.R.

The matter is different when ore types named after localities are considered (Fig. 2). Only about 35 locality types out of 289 listed in the Appendix are used truly internationally — i.e. appear in both English and Russian language papers published in *Economic Geology, Mineralium Deposita* and *Geologiya Rudnykh Mestorozhdenii* (greater part of the latter is available in English translation in *International Geology Review* — an A.G.I. publication). Examples of truly international locality types are Witwatersrand-type gold; Kurokō-type massive sulphide; Mansfeld-type Cu; Sudbury-type Ni—Cu.

The "Western International" ore type category (Fig. 2) includes those types that are commonly understood by Anglo-Saxon, German, French, Scandinavian, Italian, Dutch, etc., geologists; but they do not appear in the Soviet literature. Commonly, this is only temporary until the information has been well digested.

The Soviet ore types appearing in the Soviet literature are not normally understood by the Western geological community despite the fact that these types may appear in English translations of Russian publications. The small number of Soviet locality types listed in the Appendix may be partly due to slight under-representation of Soviet literature in the sample of papers from which this list has been compiled; but it is also true that country-wide locality types are not widespread in the U.S.S.R., probably thanks to the relatively objective and very complete system of "ore formations". Examples of Soviet locality ore types are Dzhezkazgan-type Cu (Pb); Mirgalimsay-type Zn—Pb; Atasu-type Mn, Fe, barite, Zn, Pb.

458

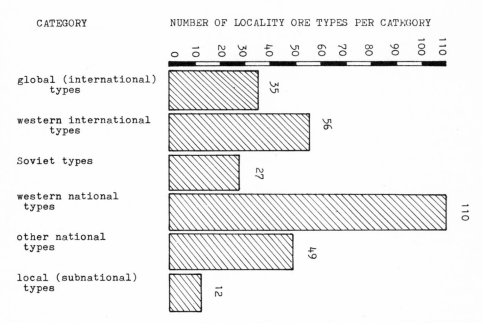

CATEGORY   NUMBER OF LOCALITY ORE TYPES PER CATEGORY

Fig. 2. Locality ore types as members of the six regional usage categories (based on the Appendix).

The "Western national" ore type, the strongest category represented in Fig. 1, includes locality ore types that are understood and/or used by geologists of separate western nations (e.g. U.S.A., Canada, Australia, F.R. of Germany, France, Sweden), but rarely understood outside their national boundaries (e.g. the Yxsjö—Hörken-type tungsten; Stripå-type iron; Åmmeberg-type Zn—Pb, all of Sweden; and Beaverlodge-type uranium; Kerr-Addison-type gold; Flin Flon-type Zn—Cu all of Canada. The Western national usage may include ore types located outside the national boundaries and introduced by influential national publications. For example, many non-German ore types of Schneiderhöhn (1955) are not used outside the sphere of influence of German literature (and very little has been translated into English, for example, which would have ascertained a more worldwide dissemination of the concepts).

There are few national ore types in the United States, because many (such as Mississippi Valley-type Zn—Pb; Clinton-type Fe) have entered the international category due to the leading role of the American economic geologic literature.

The category of "other national" ore types (Fig. 2) accounts for 49 examples in the Appendix and is certainly heavily under-represented because of language barrier and literature unavailability. Worst represented are some Asian, Arab and African countries and to a lesser extent South America. Most examples of this group in the Appendix (e.g. the Chinese "industrial ore types" — Ikonnikov, 1972) came from translations into European languages.

"Locally used ore types" are usually applied by a close-knit community of exploration and research geologists that operate in a specific area and because in many cases such types have not been published, their frequency is considerably under-represented in the Appendix. Example; Salmo-type Pb–Zn; Shuswap-type Pb–Zn; Sustut-type Cu; Anvil-type Zn–Pb; etc., popular with the Vancouver, British Columbia, geological and mining community.

The relative frequency with which locality ore types have appeared in the literature on ore deposits and metallogeny in the past 40 years, follows from Fig. 3. This figure is based on a sample of about 1000 international periodical papers and books, and although

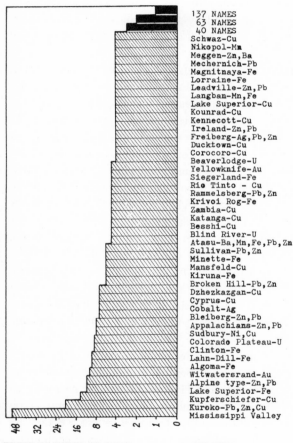

137 NAMES
63 NAMES
40 NAMES
Schwaz-Cu
Nikopol-Mn
Meggen-Zn,Ba
Mechernich-Pb
Magnitnaya-Fe
Lorraine-Fe
Leadville-Zn,Pb
Langban-Mn,Fe
Lake Superior-Cu
Kounrad-Cu
Kennecott-Cu
Ireland-Zn,Pb
Freiberg-Ag,Pb,Zn
Ducktown-Cu
Corocoro-Cu
Beaverlodge-U
Yellowknife-Au
Siegerland-Fe
Rio Tinto - Cu
Rammelsberg-Pb,Zn
Krivoi Rog-Fe
Zambia-Cu
Katanga-Cu
Besshi-Cu
Blind River-U
Atasu-Ba,Mn,Fe,Pb,Zn
Sullivan-Pb,Zn
Minette-Fe
Mansfeld-Cu
Kiruna-Fe
Broken Hill-Pb,Zn
Dzhezkazgan-Cu
Cyprus-Cu
Cobalt-Ag
Bleiberg-Zn,Pb
Appalachians-Zn,Pb
Sudbury-Ni,Cu
Colorado Plateau-U
Clinton-Fe
Lahn-Dill-Fe
Algoma-Fe
Witwatersrand-Au
Alpine type-Zn,Pb
Lake Superior-Fe
Kupferschiefer-Cu
Kuroko-Pb,Zn,Cu
Mississippi Valley

48 32 24 16 8 4 2 1 0

NUMBER OF PAPERS IN WHICH   ORE TYPE
ORE TYPE NAME APPEARED

Fig. 3. The frequency of locality ore type names in a random sample of about 1000 titles of international literature on ore deposits, 1954–1978. *Note:* A few listed names are not localities, but have locality implications (Kupferschiefer, Kuroko, Minette). Because type names are rated, an identical type may appear more than once under various synonyms.

every effort has been made to balance the sample by the language and country of origin, there is a strong bias towards the North American literature due to availability. The Mississippi Valley-type is the leader in popularity (mentioned in over 50 papers — about 5% of the papers consulted), followed by the Kurokō-type (20 papers) and Kupferschiefer (14 papers). Most locality types listed in the Appendix were recorded in a single publication only.

CORRELATIONS, MUTUAL TRANSITIONS AND SEQUENCES OF ORE TYPES

Correlation (i.e. a demonstration of equality of two or more geological objects, associations, units, etc.) is essentially accomplished by trial and error, and once established the correlation is usually considered usable until challenged and proven incorrect by subsequent work.

In geology, stratigraphic correlation is the most elaborate and best publicized one; the fundamental attribute that is correlated being geological time (age). Ore types are much more complex than ordinary lithological sequences and the majority of mineralizations are points on the map, rather than more or less continuous layers as most supracrustal rocks. Geological time is one of many attributes used to differentiate among ore types, but by far not the most important one. Porphyry coppers, for example, range from Archean to Quaternary, although Mesozoic and early Cenozoic porphyries are most common and more "typical". Mississippi Valley-type deposits have a range from Middle Proterozoic to Neogene. Only those ore types that correlate with the presence of higher vegetation (e.g. metal concentrations in coal; metals precipitated on plant debris such as a substantial proportion of the Colorado Plateau uranium-type, and redbed copper-type); ore types made possible only by a reducing atmosphere (Blind River—Elliot Lake uranium-type); ore types coincident with the transformation from reducing to oxidizing atmosphere (Superior-type iron) are controlled by evolutionary processes which in turn depend on geological time. Geological time — an increasingly better measurable property of rocks — is however of little help in *type*-correlating most ores, as only *stratigraphic* correlation may be possible.

Isotope geochemistry and fluid inclusions have become a powerful modern tool in interpretation of ore deposits; but so far, they have been less fortunate in providing an objective and accurate means of ore-type correlation: (1) establishment of genesis, even if with a high degree of confidence, does not particularly help ore-type correlation because most proposed ore types are the result of an effort to bypass genesis (i.e. they are descriptive); and (2) it doubles the controversy by splitting the geological community into "those who apply the incontrovertible laws of radiogenic decay with rigid conformity", and into "field geologists who attempt to interpret the complex facts of isotopic variation in the light of the infinite complexity of geological processes" (Brown, 1970, p. 109).

TABLE II

Examples of several increasingly more complex and uncertain tentative correlations of ore types (based on Appendix)

(1) Superior-Fe, or Animikie, or Labrador (international, N. America, Canada) ≅ Stripa (Sweden); Sydvaranger (Norway); Krivoi Rog (U.S.S.R.); Itabira (Brazil); Chingtiehshan (China)

(2) Cyprus-Cu (international) ≅ Ergani Maden (Turkey, Germany)

(3) Kupferschiefer-Cu, or Kupfermergel, or Mansfeld (international, Germany) ≅ Marl Slate (U.K.) < Copper Shale (international)

(4) Bilbao-Fe (Western) ≅ Eisenerz or Erzberg (Germany)

(5) Falun-Cu, Zn, Pb (Sweden) ≅ Flin Flon (Canada) ⪢ Rio Tinto (Spain)

(6) Kuroko (international, Japan) ⪡ west Tasmania or Rosebery (Australia) ≅ Altai (U.S.S.R.)

(7) Algoma (international) ∗ Kiruna (Sweden, international) ∗ Lahn – Dill (Germany, international)

(8) Kieslager-Cu (Germany, Western) > Besshi (Japan, international) ≅ Ducktown (U.S.A.) ∗ Röros (Norway) ∗ Kilzildere (U.S.S.R.) ∗ Miskin (Turkey)

(9) Broken Hill, N.S.W., Pb–Zn–Ag (Australia international) ≅ Åmmeberg (Sweden) ⪢ Shuswap (W. Canada)

(10) Porphyry Copper-Cu (Western) ≅ Cu in secondary quartzites (U.S.S.R.) < Chungtiao Shan (China); Kounrad (U.S.S.R.) Majdanpek (Yugoslavia, Germany); Bor (Yugoslavia, Germany); Pilares (Mexico, Germany); Copper Mountain (W. Canada)

(11) Copper Sandstones > Red Beds Copper > Dzhezkazgan (U.S.S.R.) ∗ Corocoro (Bolivia, Germany)
↗↗ Zambia, or Copperbelt, or Rhodesia (international)

(12) Copper Schists

Alpine (international, Germany) ∗ Bleiberg (Germany) or Mežica (Yugoslavia) > Kreuth (Germany)

(13) Mississippi Valley-Zn, Pb (international) < Appalachian (U.S.A.) ≅ Sumsar (U.S.S.R.)
Mississippi Valley (international, U.S.A.) ∗ Upper Silesia, or Oberschlesien, or Krakow–Silesia (Germany, Poland) Mirgalimsay (U.S.S.R.) ∗ Atasu (U.S.S.R.) ∗ Achisay (U.S.S.R.)

(14) Rammelsberg Zn, Pb, Cu (international, Germany) ∗ Meggen (Germany) ∗ Atasu (U.S.S.R.) ∗ Ireland (Western)

(15) Michigan-Cu, or Lake Superior Cu, or Keweenaw Cu (international, U.S.A.) ⪢ Nahe (Germany) < Tambillo (Bolivia) ∗ Azurita (Bolivia)

(16) Western States U (Western) < Colorado Plateau, or Sandstone Uranium, partly Red Beds Uranium (international, U.S.A.) > Wyoming (U.S.A.)

Explanations:

$A \cong B$ approximate equality.

$A < B$ A is of inferior rank and included in type B.

$A \ast B$ A close to B or by some authors equivalent.

$A \gg B$ B is believed to be geologically younger close equivalent of A (A usually Precambrian).

$A < \begin{array}{c} B \\ C \\ D \end{array}$ B, C, D of inferior rank included in A.

462

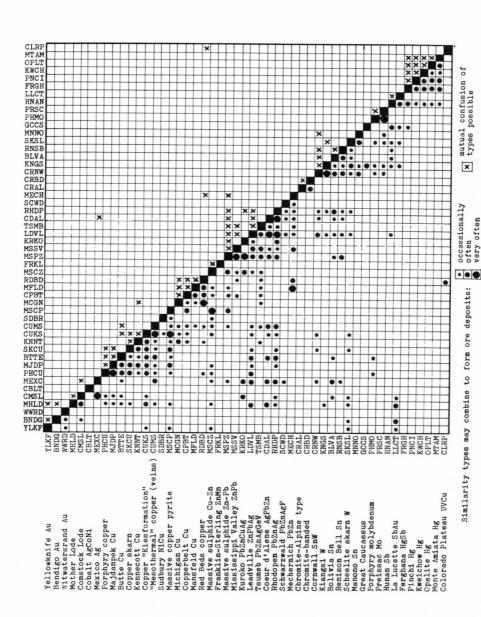

Fig. 4. Combinations, transitions and possible mutual confusion of "similarity types", used in MANIFILE. Redrafted after Laznicka (1973).

Brown (1970, p. 117), in conclusion to his paper, proposed to subdivide the Mississippi Valley-type into three categories, based on lead isotopes: (1) normal type (lead isotopes approximately indicate the age; e.g. the English Pennines); (2) B-type (lead isotopes older than deposition; e.g. Alpine deposits); and (3) J-type (lead isotopes suggest future age; e.g. central United States, Laisvall). These isotope characteristics have indeed been used for ore-type correlation, but with inconsistent terminology. B-type lead containing Mississippi Valley-type equivalents have been called Alpine-type (e.g. Morris et al., 1973; Heyl, 1967), whereas others understood Alpine-type (e.g. Wedow et al., 1973) or Appalachian-type (e.g. Gabelman, 1976; Vol. 1 of this Handbook) to be a Mississippi Valley-type equivalent located in mobile belts, regardless of isotopic lead composition. Even if the Mississippi Valley-type is restricted to localities with J-leads located in platformic setting, then several localities including the important Pine Point district, Canada, will have to be excluded on the basis of "wrong" lead isotopes (e.g. Cumming and Robertson, 1969).

The Mississippi Valley-type has now many synonyms and semi-synonyms (Table II) usually introduced to avoid or correct confusion, but these efforts have in the long run added considerably to confusion. Correlation in the absence of agreed-upon meaning is impossible.

Table II gives several examples of progressively more complex and uncertain tentative correlations of ore types, listed in the Appendix. Most uncomplicated and relatively reliable correlations are among the names given by various national schools to relatively distinct, repetitive and worldwide types of ore deposits, such as Precambrian banded iron formations, or native copper occurrences associated with altered amygdaloidal basalts.

Transitions among ore types are a commonplace. Figure 4 shows transitions among fifty "similarity types" listed in MANIFILE (Laznicka, 1973). Ore types, like many other geological objects, are rarely either black or white; consequently, it might be proper to conclude this paragraph with a quotation from the paper of Vokes and Gale (1976); "It would seem from the above cursory review that the Scandinavian Caledonian deposits can be shown to exhibit similarities to several of the 'type deposits' recently brought forward in the world literature, without really wholly corresponding to any of these."

ORE TYPES AND DATA FILES, GEOSTATISTICS AND GEOMATHEMATICS

Thirty years after humble beginnings, high-speed digital computers are a part of our lives, performing hundreds of transactions and retrievals. Twenty years ago, computers were thought to revolutionize geology: they did revolutionize the precision of the studies of the earth sciences dealing with measured and numerical data, but failed to do much with those parts of geology that are philosophical in scope: always inaccurate, always short of data, always split into several schools of thought. Ore types and ore genesis are for the greater part members of the second category.

In the 1960's and early 1970's numerous computerized files of geological data, including ore deposits, had been started (e.g. Hubaux, 1972). Few files however reached the degree of completion as originally intended, and only a very few have been published or otherwise made available to the public. Ore type tends to be included in most ore deposits inventory files, but in various forms. The file of ore deposits for the *Metallogenetic Map of Europe* (Laffitte and Permingeat, 1962) contains a data item termed "type genetique", which is essentially a category in the conventional genetic classification of mineralizations in the 1960's — e.g. "sedimentary" or "mesothermal" (ore deposits).

The *File of World Resources of Iron Ores,* prepared by nearly the same research group (École des Mines, P. Laffitte, written communication, 1967), lists similar "type genetique" narrowed in scope to suit iron deposits, as well as "Type Blondel", which is a mixture of petrographic, geometric and genetic types (e.g. "stratiform", "ferruginous sand", "laterites"), and locality types (Bilbao, Lake Superior, Magnitnaya, Taberg).

In many other ore deposit files, the ore type is not present as a separate data item. Its inclusion is considered unneccessary because most type-defining attributes such as ore mineralogy, host lithology, geometry of orebodies, environmental or facies information, geotectonic setting, etc., appear under specific data items. Therefore, many ore types can easily be retrieved by a complex request based on those characteristics, rather than on an inherently subjective and inaccurate ore-type name.

In MANIFILE — a computerized file of base and precious metal deposits of the world (Laznicka, 1973), porphyry coppers can be retrieved by a complex request that includes:

Data items	Entries in the file
CU-ASSURED (OR ESTIMATED) CONTENT	any tonnage figure
ASSOCIATED ROCKS	all or either, granodiorite, quartz monzonite, diorite, quartz porphyry
GENETIC TYPE	hydrothermal—plutonic, mesothermal
SHAPE OF OREBODIES	disseminations and veinlets in a stockwork
MINERALS	chalcopyrite, pyrite, molybdenite, chalcocite, quartz
ALTERATIONS	all or either quartz—sericite; potassic-feldspar, potassic-biotite; silicification; propylitic

Similar retrievals are objective to a certain degree, but in situations where equivocal interpretation of one or few attributes (which make the difference among two or more ore types) is involved, retrieval need not be satisfactory. The lack of data is an even greater obstacle. Compilation files based on internationally published data are especially rich in information gaps. If a reference to ore type is available at the data source, and more detailed information about the geology of the ore deposit is missing, the ore type becomes an important piece of data that certainly deserves to be stored. MANIFILE (Laznicka, 1973) contains a data item called "similarity type", that consists of about 50 locality and several attribute ore types (e.g. Mansfeld-Cu; Mother Lode-Au; see Fig. 4), and although quite subjective and often inaccurate these types make simple and rapid retrievals possible when a high degree of precision is not expected.

There have been several geostatistical studies done recently, involving ore types. The pioneering paper of Lowell and Guilbert (1970) presents an ideal or "most typical" image of porphyry copper deposits. It is based on a statistical evaluation of a control sample of 27 commercial porphyry coppers. The former control sample has been broadened to 58 commercial deposits in a subsequent paper by De Geoffroy and Wignall (1972).

Preoccupation with an "ideal type" of mineralization bears much similarity to limiting concepts in social sciences, such as the "typical bureaucrat". Körner (1974, p. 692) points out: "limiting concepts, which, though not exemplified in reality, serve nevertheless to explain the social behaviour of real people by concentrating on or even exaggerating certain features of people while ignoring others". Whatever the merit of exaggeration in social sciences, it is a distinct liability in geology. Statistical ranking of typicality, such as the one by De Geoffroy and Wignall (1972, table 4), is a welcome check on our subjective thinking regarding certain temporarily overemphasized properties of ore types, even if the control sample on which this study has been based is still very small and influenced by "elephants" — very large deposits which, some maintain, should be rejected like "ouragan" samples in mine assaying.

For example, before reading the De Geoffroy and Wignall's paper, this writer would list subjectively as the most typical characteristics of porphyry coppers: (1) quartz monzonite hosts to ore; (2) widespread zoned alteration, with dominant potassic alteration zone; and (3) andesite overlying or flanking an intrusive complex. De Geoffroy and Wignall's (1972) results, however, list; (1) phyllic alteration; (2) passive emplacement of intrusion; (3) regional fault control, as the highest ranking characteristics, with potassic (biotite) alteration coming fourth. Quartz monzonite host ranks 35th. The disproportion between subjective and statistical ratings is obvious.

Statistics, on the other hand, may be quite insensitive in differentiating among essential and accidental characteristics of ore deposits and ore types. The most typical porphyry copper-bearing complex, according to De Geoffroy and Wignall (1972, table 6), has a wall rock of Cretaceous basic volcanics altered to hornfels; NNW and/or NNE regional faults as the structural control of intrusion, etc. Here the basic volcanic wall rock and hornfelsing are clearly essential characteristics, while fault directions and to a considerable degree the geological age are clearly accidental. Most field oriented geologists would certainly hesitate to include accidental characteristics in the list of properties that determine ore type. Further discussion and refinement seems to be necessary.

DISCUSSION AND RECOMMENDATIONS

A critical review of the present status of ore (mineralization) type has been presented in the preceding paragraphs. From it follows that ore type is a widely used and very popular ingredient of economic geology and metallogeny, but one that is highly subjective, quite inaccurate and often misleading. Clearly, the present status of ore type resembles that of stratigraphic terminology prior to the publication of the *Code of Stratigraphic Nomenclature* (A.C.S.N., 1961). This code has brought organization and order into that field by international convention, and is recognized and formal at least in North America. It is desirable that the ore-type concept be brought before an international advisory body and subject to discussion that would recommend definitions, improve organization, and reduce synonymity. It is suggested that:

(1) The difference between an overall classification of ore deposits (mainly genetic) and ore types be recognized, and clearly defined.

(2) Ore types named after a characteristic feature (porphyry coppers, sandstone uranium), and ore types named after a type locality, be treated separately and not mixed. This does not exclude the possibility that an identical type could be a member of both sequences, but under various names (e.g. Kupferschiefer — feature type = Mansfeld — locality type).

(3) Existing and future ore types, particularly locality types, should be accurately defined or redefined and a comprehensive list or lexicon of ore type names published. The definition should include comprehensive description of the type locality complete with maps and sections, or alternatively reference should be made to a work in which such description has appeared earlier. The same locality name should not be applied to two or more different types, like (Lake) Superior-type Fe; Lake Superior-type Cu; and (Lake) Superior-type Zn, Cu, Ag, Au, Pb.

Also, a single binding name should be selected from among synonyms — a name that best represents the entity. For example Zambian Copperbelt-type (or NW Zambia-type) should be given preference to "Zambia-type" (there are many other mineralization types in that country), "Copperbelt-type" (there are other Copperbelts throughout the world), or "Rhodesia-type" (obsolete political status).

(4) It is recognized that any two deposits are not exactly alike, and consequently the correspondence of a compared locality to a type locality will not be absolute. The limits of deviation of type characteristics should be included in definition of ore types.

(5) Ore types need not be hierarchically subordinated, but when they are, it should also appear in the definition.

(6) One of the most serious confusions in the present use of locality ore types results from mixing of various size categories. A large mineralized territory obviously cannot serve as a type locality for a single ore deposit, and vice versa. It is recommended that the size category represented by a type locality always appears in the type name — for example, Mexican (Mexico) mineralized area (or metallogenic province) type; Tri-State

TABLE III

Proposed size categories of ore types

Category	Dimensions
(1) Metallogenic Province Type	$x00-x000$ km
(2) Metallogenic Subprovince Type	$x00$ km
(3) Ore District Type	$x0-$ca. 100 km
(4) Ore Subdistrict Type	$x0$ km
(5) Ore Deposit Type	ca. 1 km
(6) Ore Deposit Subtype (Orebody type)	<1 km

ore district type; Rammelsberg ore deposit type. The extended length of name is far outweighed by greatly increased accuracy. The proposed terminology of type localities that reflects their size is given in Table III.

The ore deposit types should ideally be named after an actual locality (e.g. Phoenix Mine), rather than after a better-known district town (e.g. Greenwood), county, mountain range, river valley, etc., because the broader locality can contain several ore types. The broader locality, however, can be used when only one ore type appears in its environs, or when one ore type located nearby has such a degree of publicity and/or importance that confusion with other little known or unimportant ore deposits and/or showings in the vicinity, is unlikely. Also in cases when a particular mine cannot be accurately defined because of large area continuous or discontinuous but uniformly similar, mineralization (e.g. as in the Picher Field, Oklahoma and Kansas).

(7) If there is a choice of type localities, preference should be given to those that have been studied in detail and are well described in universally available international publications. Localities that are of relatively "straight line" should also be given preference (e.g. a type locality of a vein deposit should not have gradation into replacement bodies, unless such gradation, rather than the vein, is what is intended to illustrate). Accessibility is of importance as well. In the jet age, political restrictions rather than distance are the main factor of accessibility. At present, it is easier for a Western European geologist to visit a locality in Australia, than in neighbouring Russia.

PRACTICAL TEST: AN ATTEMPT TO TYPIFY PREDOMINANTLY LEAD, ZINC, BARITE AND FLUORITE MINERALIZATION IN THE PLATFORMIC INTERIOR OF NORTH AMERICA

The Mississippi Valley-type is the most commonly mentioned ore type in the literature (Fig. 3). It is also one of the most confusing ones, and one most commonly criticized in this paper. Consequently, it is only natural that is has been selected as a guinea pig to test the feasibility of recommendations made above. The main emphasis here is methodologic. Although this writer has visited and sampled all the major and many minor mineralized districts and deposits in the area, it is admitted that the selection of type localities presented here can still be improved by local geologists in the future.

First- and second-order of magnitude divisions: metallogenic province and subprovince

Petrascheck (1965, p. 1622) defines metallogenic province as "the entity of mineral deposits that formed during a tectono-metallogenic epoch within a major tectonic unit and which are characterized by related mineral composition, form of the orebodies and intensity of mineralization". Its dimension should be at least "1000 km in one direction".

All the classic ore occurrences of the "Mississippi Valley-type" in the United States are located in Paleozoic, pre-Permian sediments (dominantly carbonates) that constitute a platform cover resting on the Precambrian crystalline basement of the North American craton (see Laznicka's Chapter 4, Vol. 8, of this Handbook). The cover is relatively lightly disturbed over the greater part of the area, but locally the disturbance is quite intense so that the platform may turn into an intracratonic mobile belt. The area discussed is certainly a major tectonic unit with a characteristic style of mineralization, although with uneven mineralization intensity, and is therefore a metallogenic province (or mega-province). Its western, eastern, southern and northern boundaries, both in outcrop and in subcrop, coincide with margins of Cordilleran, Appalachian, Ouachita and Innuitian marginal continental and partly intracratonic mobile belts (Fig. 5): (Laznicka, Chapter 4, Vol. 8 of this Handbook).

The continuity of Paleozoic sediments over the entire North American craton is, however, interrupted by the exposed Precambrian basement constituting the enormous Canadian Shield. Small Precambrian outcrops located close to the edge of the Shield (Sioux Uplift, Frontenac Axis), as well as at a great distance (Black Hills, Ozark Uplift, Llano Uplift), occur completely surrounded by the Paleozoic sediments.

In the Mississippi Embayment and over the greater part of the Western Interior Plains, the Paleozoic sequence is buried under the cover of Mesozoic and Cenozoic platform sediments that are overwhelmingly detrital (=clastic) in character (shales and sandstones). These younger sediments are sparsely mineralized, but when they do contain ore deposits (e.g., small bauxite occurrences in the Mississippi Embayment; manganese nodules in Cretaceous shales near Chamberlain, S. Dakota and Miami, Manitoba; uranium in the Black Hills area, western Dakotas and eastern Montana; etc.), they differ substantially in character compared with the mineralization in the Paleozoic rocks. The post-Paleozoic sediments belong clearly to another metallogenic province (Fig. 6).

Mainly copper mineralization in Permian red beds and evaporitic sequence in Oklahoma and Texas (Creta, Mangum, etc.) constitute still another metallogenic subdivision, sandwiched between the two provinces discussed above.

In the Canadian Arctic, in the Hudson Bay Basin, and under the Great Lakes, Paleozoic sediments are partly covered by water and by young, Pleistocene to Recent

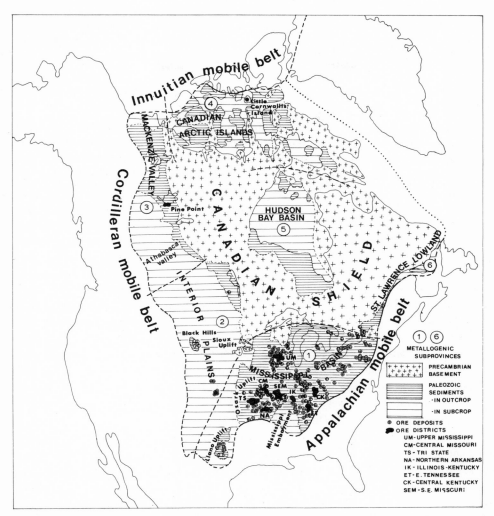

Fig. 5. Metallogenic provinces of the Paleozoic Platformic Cover of North American Craton, and its subdivisions
Metallogenic subprovinces: *1* = Mississippi Basin; *2* = Southwestern Interior Plains; *3* = Athabasca–Mackenzie Valleys area; *4* = Canadian Arctic Islands; *5* = Hudson Bay Basin; *6* = St. Lawrence Lowland.

glaciogenic sediments, that contain almost no metallic occurrences.

The relatively widespread mineralization in Paleozoic sediments in the U.S. mid-continent approximately south of the 40° parallel and west of the 95° meridian (e.g. Heyl, 1972), rapidly diminishes northward and westward, primarily due to deteriorating outcrop conditions. Scattered Pb–Zn occurrences have been recorded from drillholes

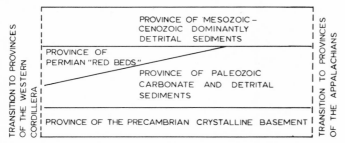

Fig. 6. Metallogenic provinces of the North American Craton as related to stratigraphic levels

The metallogenic megaprovince of the North American Craton (Fig. 5) is further subdivided into several stratigraphic levels each with distinct, and different from its neighbour, metallogeny so that each level corresponds to a separate metallogenic province. This creates the problem of showing these provinces on a map, because stratigraphically higher provinces (levels) are underlain by stratigraphically lower provinces (levels). Map plots of mineralizations in such cases should always be labelled as to their depth of occurrence.

in western Kansas (Dodge City area; D.L. Evans, 1962) and from the fringe of the Black Hills Uplift. The Paleozoic sediments that rim the western edge of the Canadian Shield in Manitoba, Saskatchewan and Alberta do not contain known mineralization, except for few occurrences of barite.

Devonian carbonates in the Mackenzie River valley in northwestern Canada, however, host the important lead—zinc district of Pine Point (e.g. Irvine et al., 1972), and Paleozoic carbonates on Cornwallis Island in the Canadian Arctic Archipelago contain similar mineralization, discovered recently (e.g. Sangster, 1974). No Pb—Zn occurrences have so far, been located in the Hudson Bay Basin area.

The remarkable similarity and continuity of major tectonic, stratigraphic, lithogenetic and metallogenic conditions throughout the entire outcrop and subcrop territory of the Paleozoic sediments of the North American Platform warrants its establishment as a type metallogenic area, designated by a tentative full name "Paleozoic Platform Cover of North American Craton, Metallogenic Province Type". Any shorter name (e.g. Mississippi—Mackenzie basins metallogenic province) would dilute the accuracy. Use of abbreviations (e.g. PPCNAC) is a suggested way to assure brevity in repeated use.

This enormous territory deserves further subdivision into smaller geographic units — metallogenic subprovinces. Six subprovinces have been recognized: Mississippi Basin; Southwestern Interior Plains; Athabasca—Mackenzie valleys area; Canadian Arctic Islands; Hudson Bay Basin; St. Lawrence Lowland (Table IV, Fig. 5). Their mutual boundaries are admittedly arbitrary.

The fundamental geological style of all subprovinces is comparable, although the intensity of known tectonic disturbance and the density of known mineralization is not. The Mississippi Basin subprovince is clearly anomalously rich in both latter attributes. It is a classical mineralized area, studied in considerable detail, so it is a natural choice for a type area, namely of the Mississippi Basin Metallogenic Subprovince type. The area occupied by this subprovince is described in considerable detail by Snyder (1968).

Third- and fourth-order subdivision: ore districts and subdistricts

A cluster of more or less genetically related ore deposits that appear distinct and separated from its surroundings as a consequence of conspicuous local increase in the intensity of mineralization, or due to different character and/or affiliation of its ore occurrences usually constitute an ore (or mineral) district. Ore district is usually a lower rank subdivision of either a metallogenic province, or a mineralized area, or an ore belt. Mining district, on the other hand, is an administrative unit that may, but need not, coincide with an ore district (e.g. Kalgoorlie ore district = geologic and geographic entity, located in the East Coolgardie Goldfield = administrative division). In most cases, however, there is a coincidence.

The size of ore districts vary; they tend to be smaller in densely and consistently mineralized metallogenic provinces, particularly in mobile belts (such as in the Colorado Mineral Belt) and larger in very irregularly mineralized vast territories (such as the North American Interior). The economically most important ore districts in the Mississippi Basin Metallogenic subprovince have a diameter or an elongation close to 100 km, and an area exceeding 1000 km^2. The Pine Point district has similar dimensions.

Some districts have a relatively uniform style and type of mineralization throughout (such as the Tri-State district), while others may show distinct variations. In the SE Missouri district Pb—Zn mineralization of different styles occurs, namely: (1) near the base of the local section of platform Paleozoic sediments (Lamotte and Bonneterre Formations); (2) near the top of the local Paleozoic section (Potosi and Eminence Formations); and (3) a few small Pb—Zn—Ag occurrences are even known in the Precambrian basement (e.g. Silver Mines or Edison, NW of Fredericktown). The bulk of mineralizations (1) and (2) is geographically separated, although overlap and local interference do occur. The former is named SE Missouri Lead Belt (more correctly lower stratigraphic level), the latter SE Missouri Barite—Lead Belt (upper stratigraphic level). Both were assigned a status of subdistrict in this test (Table IV). The mineralization (3) in the basement, presumably of Precambrian age, is part of another metallogenic province dominated by iron and some copper (Kisvarsanyi and Proctor, 1967).

The Mississippi Basin Metallogenic Subprovince contains about ten ore districts of uneven area and extremely variable magnitude of mineralization (Table V). This test is restricted to the four major districts of concentrated zinc, lead, barite and fluorite mineralization only: the Upper Mississippi district (in Wisconsin, Illinois and Iowa); the Tri-State district (in Kansas, Oklahoma and Missouri); the SE Missouri district; and the Illinois—Kentucky district. Less important districts (northern Arkansas; central Missouri, central Kentucky; etc.) and hundreds of scattered localities are not considered. It is quite obvious that each of the four ore districts is unique enough to constitute a type despite

TABLE IV

Subdivisions of the Paleozoic Platformic Cover of North American Craton Metallogenic Province, as proposed in this paper

(1) Metallogenic Province	(2) Metallogenic Subprovince	(3) Ore District	(4) Ore Subdistrict	(5) Ore Deposit
PALEOZOIC PLAT-FORMIC COVER OF NORTH AMERICAN CRATON	MISSISSIPPI BASIN	UPPER MISSISSIPPI VALLEY		MEEKERS GROVE OPEN-CUT MINE
				TREGO MINE, PLATEVILLE
				RODHAM MINE, SHULLSBURG
		TRI-STATE		PICHER-"SHEET GROUND"
				PICHER-"RUNS"
		SE MISSOURI	SE MISSOURI BARITE-LEAD	WASHINGTON COUNTY
				VIRGINIA-MERRIMAC MINES
			SE MISSOURI LEAD	MINE LA MOTTE
				HAYDEN CREEK MINE
				BONNE TERRE
				SCHULTZ BRECCIA

	ILLINOIS–KENTUCKY	MINERVA MINE, CAVE-IN-ROCK NANCY HANKS MINE ROSICLARE HUTSON MINE RIGGS MINE HICKS DOME
	other districts	
	scattered localities	
Southwestern Interior Plains	scattered localities	
Athabasca–Mackenzie valleys	Pine Point	
Canadian Arctic islands	Little Cornwallis Island	
Hudson Bay Basin	none known	
St. Lawrence Lowland	scattered localities	

Ore types appear in capital letters.

TABLE V

The magnitude of mineralization in the more important lead, zinc, barite and fluorite districts in the Mississippi Basin Metallogenic Subprovince

District	Approximate cumulative past production and remaining reserves (tons of contained metal)					
	Pb	Cu	Zn	Co	barite	fluorite
Upper Mississippi Valley	750,000	1,600,000	2,000		X,000	
Tri-State	2,650,000	11,000,000				
SE Missouri	35,050,000	2,120,000 *	300,000 *	70,000 *	11,500,000	
in it:						
Lead Belt	35,000,000	2,100,000 *	300,000 *	70,000 *		
Barite–lead Belt	50,000	20,000			11,500,000	
Illinois–Kentucky	70,000	250,000			50,000	15,000,000
N. Arkansas	X,000	X0,000			X0,000	
Central Kentucky	X,000	X0,000			X0,000	
Central Missouri	X0,000	X00,000			X00,000	
Others	X0,000	X0,000			X00,000	

* Subeconomic ore content, not actually recovered.

Based on data from Heyl et al. (1959); Cornwall and Vhay (1967); Heyl (1967); Brockie et al. (1968); Grogan and Bradbury (1968); Snyder and Gerdemann (1968); Brobst (1975); Pinckney (1976); Trace (1976); Davis (1977); and others.

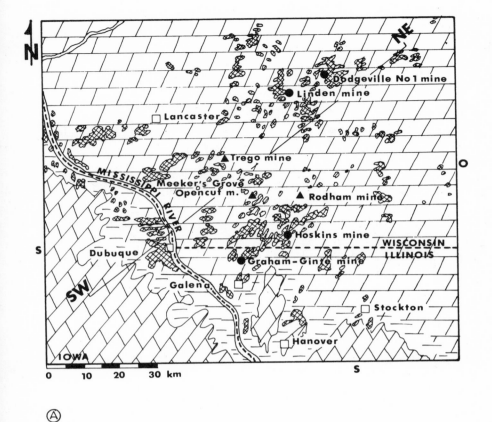

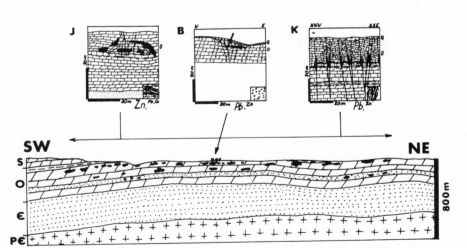

Fig. 7. The four type ore districts in the Mississippi Basin Metallogenic Subprovince: (A) Upper Mississippi Valley district; (B) Tri-State district; (C) Southeast Missouri district; and (D) Illinois–Kentucky district.
A compilation, based on data in Winslow (1894); Bain (1901); Buckley (1909); Currier (1944); Brown et al. (1954); Thurston and Hardin (1954); Heyl et al. (1959, 1965); Heyl (1967, 1972); Brockie et al.

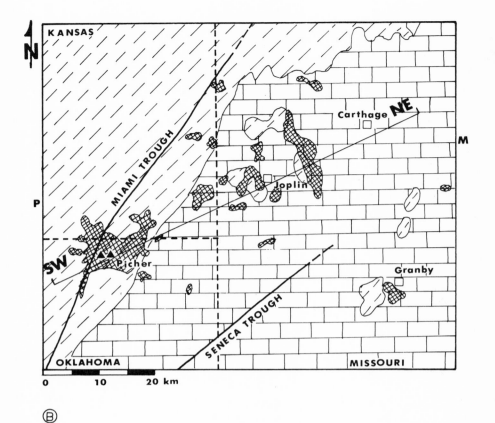

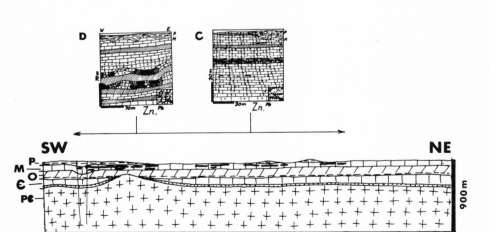

(1968); Brown (1968); McKnight and Fischer (1970); Hagni (1976); Pinckney (1976) Trace (1976) and Davis (1977).

See Fig. 9 for rock legend. Additional legend for locality plots: ▲ = ore deposit types from Fig. 9; and ● = non-type ore deposits listed in Fig. 8. Mineralization is shown by inclined and orthogonal mesh pattern in maps, and by solid black in sections. Orebodies in sections are generalized, greatly exaggerated and schematically projected on the line of section.

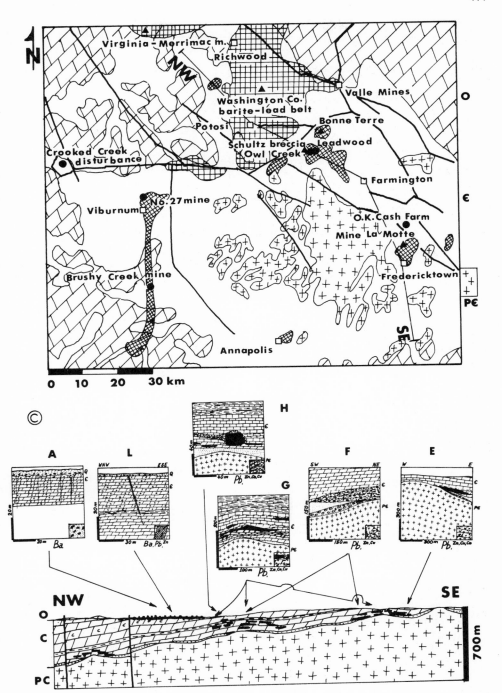

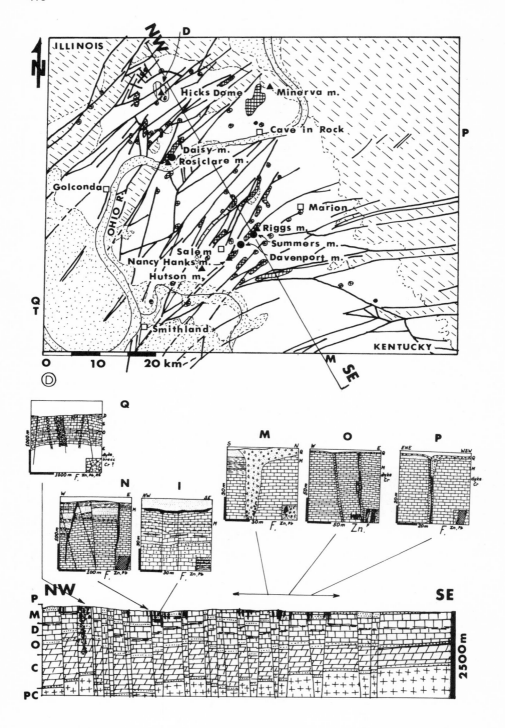

the numerous points of similarity they have in common, as will be demonstrated.

The geology and mineralization in these districts is shown diagrammatically in both map and section form in Fig. 7A–D. It has, however, not been considered essential to list their characteristics in this chapter, because excellent up-to-date descriptions of all four districts appear in the widely available book *Ore Deposits of the United States, 1933–1967*, edited by Ridge (Brockie et al., 1968; Heyl, 1968; Grogan and Bradbury, 1968; Snyder and Gerdemann, 1968).

The ore districts and subdistricts can usually be broken down into several variously named hierarchically inferior subdivisions (group of ore deposits, ore cluster, ore trend, local ore belt, mineralized faults and structures, etc.), which themselves contain several ore deposits. Ore deposits, however, may exist alone within an ore district, or even within a metallogenic province, and need not be part of any intermediate hierarchical category. The area occupied by ore deposits is very variable within a range of several tens of meters to several kilometers along their longest dimensions. In some cases, nearly interconnected mineralization, although of variable grade, underlies areas of tens to hundreds of square kilometers as in the Pitcher ore field, Tri-State district.

The term "ore deposit" almost requires the mineralization to be economically recoverable. In essentially non-economic studies such as this one, all other discrete occurrences of mineralization are added, regardless of their economic importance.

Ore deposits and occurrences may be further subdivided into discrete orebodies, ore shoots and zones (compare Table III). Mine and mining property are technologic and economic categories that in most cases are identical with ore deposits, but need not always be so. Large, continuous ore deposits may be exploited by several mines (e.g. in Broken Hill, N.S.W.), or a single mine may exploit several, often different, ore deposits (e.g. Mount Isa, Queensland). Ore deposit occurrence is no doubt the basic entity in metallogeny, and it is also the most frequent size category among ore types listed in the Appendix.

In this text, an attempt has been made to review all the ore deposits that occur within the limits of the four most important districts in the Mississippi Basin Metallogenic subprovince, and to arrange them into sets with sufficiently close internal similarity so that each set could be represented by a single type locality. This task, as originally conceived, would have involved a population of several thousands of ore deposits and occurrences (about 4000), and evaluation of several hundreds of characteristics. This would not only have been extremely time-consuming, but the results would not be substantially more accurate because available data are more scarce and more uncertain with increasing population when old, uneconomic, forgotten, and insufficiently described localities are drawn in. Instead, a sample population of 300 more important deposits and occurrences has been selected and evaluated, using a checklist of approximately 100 of the most common characteristics. Figure 8 shows diagrammatically a specimen set of 30 ore deposits to demonstrate the style of work and support the conclusions. Imposition of a standard set of questions to be answered greatly restricts the feelings and partiality of the evaluating

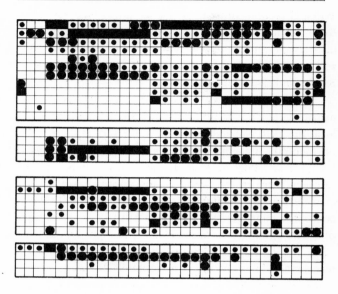

ore deposit number

ORE MINERALOGY
  sphalerite
  galena
  pyrite and marcasite
  chalcopyrite
  Ni-Co sulphides
  calcite
  dolomite
  quartz
  barite
  fluorite
  smithsonite,cerussite
  U,RE,Th minerals
HOST ROCK ALTERATION
  calcite leaching
  coarse recryst.calcite
  metasomatic dolomite
  silicification
DISTRIBUTION OF METALLIC
MINERALS IN ORE
  fine grained, even
  scattered crystals
  low concentration patches
  massive ore patches
  crystal lined vugs
  banded, ribboned ore
  ore in breccia
AVERAGE GRADE (% Pb+Zn)
  under 3%
  3-10%
  10-30%
  over 30%

Ⓐ

Fig. 8. A sample of thirty ore deposits out of three hundred located in the Mississippi Basin Metallogenetic Subprovince, showing attributes evaluated in selection of ore deposit types.

LOCALITY LIST (*A = an ore deposit type)

----

*A   (1) Washington County (SE missouri) residual barite (Winslow, 1894; Tarr; 1919; Brobst, 1975)

*B   (2) Meekers Grove Opencut Mine (Upper Mississippi Valley District), mineralized gravels (Heyl et al., 1959)

     (3) Linden, Wisconsin (Upper Mississippi Valley District) (Heyl et al., 1959)

*C   (4) Picher (Tri-State District), sheet ground ore deposits (Brockie et al. 1968; McKnight and Fischer, 1970)

*D   (5) Picher (Tri-State District), runs ore deposits (Brockie et al., 1968; McKnight and Fischer, 1970)

*E   (6) Mine La Motte (SE Missouri District) (Buckley, 1909; Snyder and Gerdemann, 1968)

*F   (7) Hayden Creek Mine (SE Missouri District) (Ohle, 1952; Snyder and Gerdemann, 1968)

*G   (8) Bonne Terre (SE Missouri District) area ore deposits (Buckley, 1909; Snyder and Gerdemann, 1968)

     (9) Brushy Creek Mine, Viburnum Trend (SE Missouri District) (D.L. Evans, 1977)

     (10) Viburnum No. 27 mine, Viburnum Trend (SE Missouri District) (Grundmann, 1977)

*H   (11) Schultz area, Leadwood (SE Missouri District), mineralized penecontemporaneous breccias (Snyder and Odell, 1958)

     (12) Owl Creek, Leadwood (SE Missouri District), mineralized breccias (Snyder and Odell, 1958)

     (13) Leadwood (SE Missouri District), mineralized breccias (Snyder and Odell, 1958)

*I   (14) Minerva Mine, Cave-in-Rock (Illinois–Kentucky District) (Currier, 1944; Anderson, 1953; Grogan and Bradbury, 1968)

ore deposit number

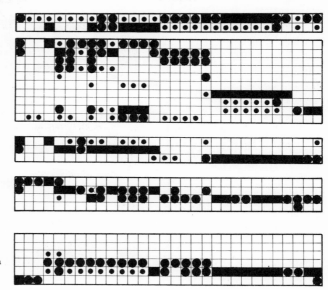

ORE-HOST RELATIONSHIP
  open-space filling
  replacement
SHAPE OF OREBODIES
  long lens approaching layer
  series of lenses
  wedges, "pitches"
  subhorizontal rings,strings
  subvertical columns
  stockworks
  fissure veins
  replacement veins
  mineralized breccias
  irregular bodies
RELATION OF ORE TO HOST DEPO-
SITIONAL FABRIC
  conformable
  semiconformable
  disconformable
DEGREE OF STRATIGRAPHIC CON-
FINEMENT OF MINERALIZATION
  within horizon X m thick
  within horizon XO m thick
  within horizon XOO m thick
  miner.within entire sedim.
PROBABLE TIMING OF MINERALI-
ZATION IN REGARD TO HOST
SEDIMENTOGENESIS
  syndepositional
  diagenetic,pre-lithification
  diagenetic,syn-lithification
  diagenetic,post-lithification
  epigenetic
  residual and clastic rework.

*J (15) Trego Mine, Platteville (Upper Mississippi Valley District) (Heyl et al., 1959)

    (16) Graham-Ginte Mine, Illinois (Upper Mississippi Valley District) (Heyl et al., 1959)

    (17) Hoskins Mine (Upper Mississippi Valley District) (Heyl et al., 1959)

    (18) Dodgeville No. 1 Mine (Upper Mississippi Valley District) (Heyl et al., 1959)

*K (19) Rodham Mine, Shullsburg (Upper Mississippi Valley District) (Heyl et al., 1959)

*L (20) Virginia-Merrimac Mines, Barite–lead Subdistrict (SE Missouri District) (Bain, 1901; Tarr, 1919)

*M (21) Nancy Hanks Mine, Kentucky (Illinois–Kentucky District) (Thurston and Hardin, 1954)

*N (22) Rosiclare Mine, Rosiclare, Illinois (Illinois–Kentucky District) (Weller et al., 1952; Baxter and Desborough, 1965; Grogan and Bradbury, 1968)

    (23) Daisy Mine, Rosiclare, Illinois (Illinois–Kentucky District) (Weller et al., 1952; Grogan and Bradbury, 1968)

    (24) Davenport Property, Moore Hill Fault System, Kentucky (Illinois–Kentucky District) (Thurston and Hardin, 1954)

    (25) Summers Mine, Moore Hill Fault System, Kentucky (Illinois–Kentucky District) (Thurston and Hardin, 1954)

*O (26) Hutson Mine, Kentucky (Illinois–Kentucky District) (Oesterling, 1952)

*P (27) Riggs Mine, Moore Hill Fault System, Kentucky (Illinois–Kentucky District) (Thurston and Hardin, 1954)

*Q (28) Hicks Dome, Illinois (Illinois–Kentucky District) (Brown et al. 1954; Baxter and Desborough, 1965; Baxter et al., 1967; Grogan and Bradbury, 1968)

    (29) O.K. Cash farm diatreme, Avon (SE Missouri District) (Kiilsgaard et al., 1963; Zartman et al., 1967)

    (30) Crooked Creek Disturbance, Missouri (near SE Missouri District) (Kiilsgaard et al., 1963)

482

HOST LITHOLOGY
  pure limestone
  pure dolomite
  impure limestone
  impure dolomite
  chert nodules present
  chert iterbeds
  shale
  sandstone
  gravel, sand
  granite boulder conglomerate
  lamprophyre, peridotite
"SPECIAL FACIES" IN THE AREA
  algal "reefs"
  facies variation inv."reefs"
     "   carbonates and shales
     "   carbonates and sands.
  intraform.slumps,breccias
  post-lithif.karsting,brecc.
AGE OF HOST ASSOCIATION
  Cambrian
  Ordovician
  Silurian
  Devonian
  Mississippian,Pennsylvanian
  Younger
PROXIMITY TO CRYSTALLINE
  BASEMENT
  X to XO meters
  XOO meters
  XOOO meters

©

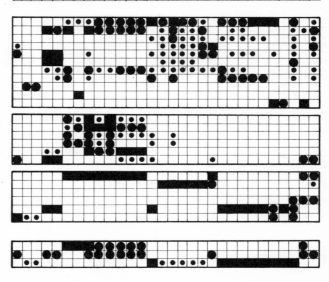

TECTONIC PREPARATION OF ORE
  HOST ASSOCIATION
  gentle folding
  light thrusts,bedding faults
  light fracturing,jointing
  intensive fracturing
  tectonic brecciation
  solution-collapse brecciation
RELATION OF MINERALIZATION TO
POST-BASEMENT MAGMATIC ROCKS
AND CRYPTOVOLCANIC FEATURES
  ore in,or at cont.of diatreme
  ore in intrusive dyke
  ore at contact of intr.dyke
  magm./cryptovolc.XOO m away
  magm./cryptovolc.XOOO m away

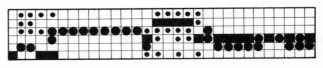

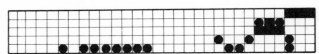

LEGEND:

most common, typical, important attribute
moderately common, typical, important attribute
least common, typical, important attribute
attribute not present

Ⓓ

operator, but some uncertainty and subjective interpretation still remains.

The set of sample deposits can be reasonably well represented by 17 ore deposit types, and these have been assembled in Fig. 9A—Q. The degree of similarity of members within certain groups is very close indeed (e.g. Trego Mine, Platteville type; compare the type locality — column *15,* and three members of this type — columns *16—*

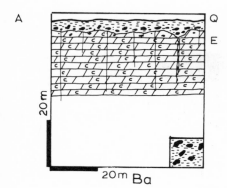

Fig. 9. A. Washington County (Missouri) Ore Deposit Type.

SE Missouri District. Location, Fig. 7C. Blocks and fragments of white crystalline barite with nodules and druses of chert and quartz occur near base of a red residual clay, blanketing weathered surface of Potosi and Eminence (Upper Cambrian) dolomite. Barite-bearing blankets sometimes grade into residuum-filled sinkholes. Residual, barite source in bedrock carbonates (Tarr, 1919; Brobst, 1975).

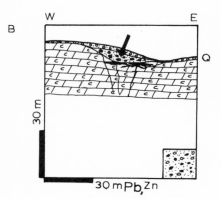

Fig. 9. B. Meekers Grove Opencut Mine, Gravels Ore Deposit Type.

Upper Mississippi Valley District. Location, Fig. 7A. Placer galena deposit. Scattered lumps of transported galena occur in stream gravel deposits located in depression and cracks in underlying platformic carbonate bedrock. The bedrock contains primary zinc–lead mineralization of the Trego Mine, Platteville type (Heyl et al., 1959; pp. 130, 131 and plate 3, p. 234).

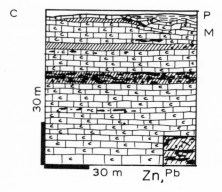

Fig. 9. C. Picher-Sheet Ground Ore Deposit Type.

Tri-State District. Location, Fig. 7B. Low grade but extensive blanket deposits of almost uniform thickness located in inconspicuously jointed Mississippian bedded chert and cherty limestone, containing scattered sphalerite crystals and blebs (Brockie et al., 1968, p. 422; McKnight and Fischer, 1970, p. 137).

Fig. 9. Diagrammatic sections and short descriptions of the seventeen ore deposit types located in Tri-State, Southeastern Missouri, Upper Mississippi Valley and Kentucky–Illinois ore districts

484

LEGEND

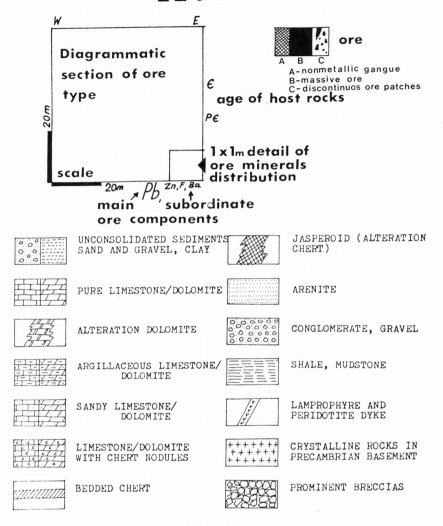

UNCONSOLIDATED SEDIMENTS SAND AND GRAVEL, CLAY

PURE LIMESTONE/DOLOMITE

ALTERATION DOLOMITE

ARGILLACEOUS LIMESTONE/ DOLOMITE

SANDY LIMESTONE/ DOLOMITE

LIMESTONE/DOLOMITE WITH CHERT NODULES

BEDDED CHERT

JASPEROID (ALTERATION CHERT)

ARENITE

CONGLOMERATE, GRAVEL

SHALE, MUDSTONE

LAMPROPHYRE AND PERIDOTITE DYKE

CRYSTALLINE ROCKS IN PRECAMBRIAN BASEMENT

PROMINENT BRECCIAS

AGE OF HOST ROCKS CODES:

Q   Quaternary
Cr Cretaceous
P   Pennsylvanian
M   Mississippian
D   Devonian
S   Silurian
O   Ordovician
Є   Cambrian
PЄ Precambrian (Middle Proterozoic)

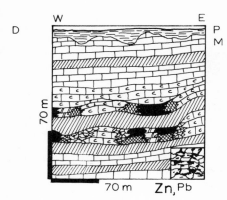

Fig. 9. D. Picher-"Runs" Ore Deposit Type.
Tri-State District. Location, Fig. 7B. Disseminated, vug-coating, breccia cementing and locally massive sphalerite > pyrite, marcasite > galena, occur at the boundary of crystalline dolomite alteration core and adjacent jasperoid in favourable horizons of Mississippian cherty limestone, as elongated to circular, channel to ring-like orebodies. Controlled by joint system and metasomatic zonality (Brockie et al., 1968; pp. 136–137, particularly sections A–B; C–D, plate 11; McKnight and Fischer, 1970).

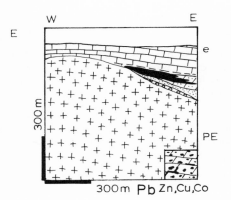

Fig. 9. E. Mine La Motte Ore Deposit Type ("pinchout type").
Southeastern Missouri District. Location, Fig. 7C. Disseminated and along bedding planes banded galena pyrite, sphalerite, chalcopyrite and siegenite form narrow linear and arcuate orebodies in bar-algal reef facies of Bonneterre (Cambrian) carbonates, located immediately above pinchouts of Lamotte (Cambrian) sandstone against buried knobs of Precambrian basement (Buckley, 1909; Snyder and Gerdemann, 1968 pp. 333, 334).

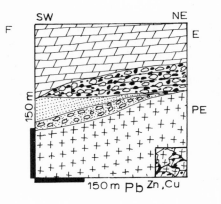

Fig. 9. F. Hayden Creek Mine (SE Missouri) Ore Deposit Type.
Southeastern Missouri District. Location, Fig. 7C. Galena crystals occur disseminated in strongly silicified matrix of Bonneterre (Cambrian) dolomite which cements granite boulder conglomerate adjacent to a Precambrian basement knob (Snyder and Gerdeman, 1968, p. 335 and fig. 4).

G

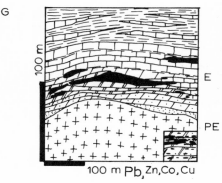

100 m Pb, Zn,Co,Cu

Fig. 9. G. Bonne Terre Ore Deposit Type.
Southeastern Missouri District. Location, Fig. 7C. Disseminated crystals of galena >> pyrite, sphalerite > chalcopyrite, siegenite, occur in large masses of metasomatically dolomitized Bonneterre (Cambrian) limestone, in an area of considerable facies variation particularly influenced by development of algal buildups ("reefs"); black shale zones: and dolarenite bars.

The intensity of mineralization is strongest along bedding planes, along fractures and in sedimentary structures. Orebodies are generally elongated and stratabound, but their extent, thickness and shape vary (Snyder and Gerdemann, 1968; pp. 338–343).

H

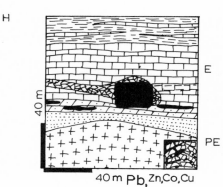

40 m Pb, Zn,Co,Cu

Fig. 9. H. Schultz Breccia (SE Missouri) Ore Deposit Type.
Southeastern Missouri District, Location, Fig. 7C. Galena and sphalerite occur as irregular veinlets, scattered blebs and massive replacements in dolomite–black shale matrix of penecontemporaneous breccia, that cements dolomite blocks. Orebodies, frequently multiple (several vertically stacked breccias) are always associated with underlying bedded Bonneterre-type ore, and cause considerable local increase in ore thickness (Snyder and Odell, 1958, pp. 921–924, and fig. 12) (Figure modified after Snyder and Odell, 1958.)

I

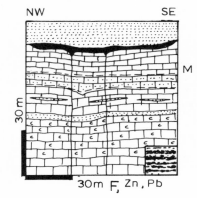

30m F, Zn , Pb

Fig. 9. I. Minerva Mine, Cave-In-Rock, Ore Deposit Type.
Illinois–Kentucky District. Location, Fig. 7D. Semiconformable, subhorizontal bodies of relatively coarse crystalline fluorite and calcite with some quartz, and variable content of sphalerite and galena, formed by hydrothermal replacement and partly by open space filling above fracture zones in a Mississippian limestone, topped by shale or sandstone (Currier, 1944; Anderson, 1953; Grogan and Bradbury, 1968).

J

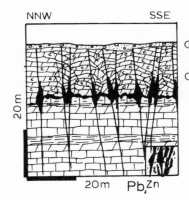

Fig. 9. J. Trego Mine, Platteville ("Pitch-and-Flat")
Ore Deposit Type.
Upper Mississippi Valley District. Location, Fig. 7A.
Reniform sphalerite >> pyrite, marcasite > galena,
with chert (jasperoid?) and recrystallized dolomite,
form banded veins, cavity coatings and replacements
in Ordovician dolomites along thrust and bedding
faults and fracture systems (Heyl et al., 1959, plate
11, pp. 115–119). (Figure modified after Heyl et al.,
1959.)

K

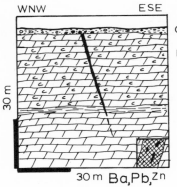

Fig. 9. K. Rodham Mine, Shullsburg ("Gash-Vein")
Ore Deposit Type.
Upper Mississippi Valley District. Location, Fig. 7A.
Galena >> sphalerite > chalcopyrite form short fracture filling and replacement veins along vertical joints
and horizontal bedding planes in lightly altered Ordovician dolomites (Heyl et al., 1959, pp. 130, 231).

L

Fig. 9. L. Virginia-Merrimac (SE Missouri) Mines Ore
Deposit Type.
Southeastern Missouri District, Barite–lead Subdis-
trict. Location, Fig. 7C. Crystalline barite with
scattered masses of galena and some sphalerite fills a
NNE-striking, steeply dipping fissure vein in cherty
dolomites of Potosi Formation (Cambrian), near
Palmer Fault zone (Bain, 1901; Tarr, 1919).

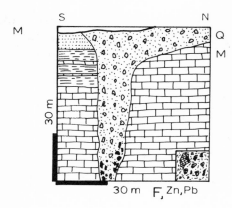

Fig. 9. M. Nancy Hanks Mine (Kentucky) Ore Deposit Type ("Gravel Spar").

Moore Hill area, Illinois—Kentucky District. Location, Fig. 7D. Residual fluorite fragments in matrix of clay, iron hydroxides and overburden fill a crevice formed by dissolution of a calcite—fluorite hydrothermal vein, and part of its limestone wallrock (Thurston and Hardin, 1954, p. 101 and plate 13). (Figure after Thurston and Hardin, 1954.)

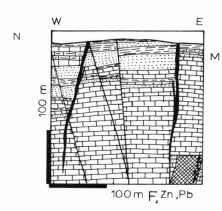

Fig. 9. N. Rosiclare Ore Deposit Type.

Illinois—Kentucky District. Location, Fig. 7D. Coarse crystalline, often banded, fluorite and calcite with variable content of scattered crystals and pockets of galena and sphalerite, form system of long, complex, NNE-striking steep hydrothermal fissure veins in lightly silicified Mississippian sediments: dominantly sandstones and limestones (Weller et al., 1952; Baxter and Desborough, 1965; Grogan and Bradbury, 1968).

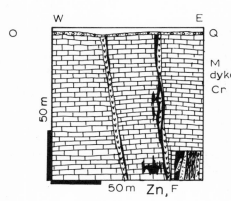

Fig. 9. O. Hutson Mine (Kentucky) Ore Deposit Type. Illinois—Kentucky District. Location, Fig. 7D. Hydrothermal fracture filling banded sphalerite—marcasite vein with some calcite and fluorite, occurs adjacent to, and within, an altered phlogopite peridotite dyke cutting Mississippian limestones. Replacement sphalerite forms local, discontinuous orebodies in limestones controlled by joints and proximity to dykes (Oesterling, 1952).

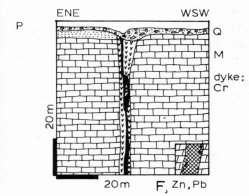

Fig. 9. P. Riggs Mine (Kentucky) Ore Deposit Type. Moore Hill area, Illinois–Kentucky District. Location, Fig. 7D. Hydrothermal fluorite–calcite >> sphalerite, galena vein healing fissure in a lamprophyre dyke, which intruded along a normal fault in Mississippian limestone (Thurston and Hardin, 1954, plate 16). (Figure after Thurston and Hardin, 1954.)

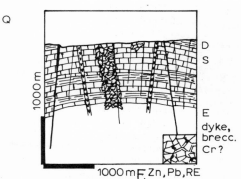

Fig. 9. Q. Hicks Dome (Illinois) Ore Deposit Type. Illinois–Kentucky District. Location, Fig. 7D. Explosive breccias (diatremes) of angular to subrounded fragments of sedimentary rocks and some nepheline syenite, carry sparsely and irregularly scattered crystal patches and veinlets of fluorite with some galena and sphalerite. Minor amounts of apatite, monazite, xenotime and other minerals with rare earths, thorium, uranium and beryllium are also present (Brown et al., 1954; Baxter and Desborough, 1965; Baxter et al., 1967; Grogan and Bradbury, 1968).

18, Fig. 8). Some types, however, show considerable variations, Hicks Dome type the greatest (compare column *28* – the type locality, with columns *29* and *30*, Fig. 8 – two members of this type).

There is a wide fluctuation among the number of examples represented by a certain ore deposit type that ranges from one (ore type that represents only the type locality itself – e.g. Hutson Mine type), to several thousand (e.g. Rodham Mine, Shullsburg, ore deposit type), as shown in Table VI. The size of members of types varies. Those types with the greatest number of representatives usually have small deposits.

Accurate naming of ore deposit types is an important part of typization. The requirement that the type name accurately identifies the particular deposit the author had in mind, makes it in most cases necessary to introduce long names. A typical name in this test consists of the mine name that actually represents the deposit because ore deposits themselves usually do not have names, plus a nearby town or state, followed by designation of the ore type category – e.g. Trego Mine, Platteville, ore deposit type. This is admittedly awkward, but the required accuracy cannot be achieved otherwise. In a few

TABLE VI

Number of deposits (occurrences) and approximate commodity tonnage represented by ore deposit types in the four major ore districts, Mississippi Basin ore subprovince

Ore deposit type	Very approximate, estimated number of localities	Very approximate estimated total tonnage (production plus reserves) of contained metal, CaF_2 and $BaSO_4$
Washington County	X00	11,000,000 t $BaSO_4$
Meekers Grove (Opencut Mine)	2	X00 t Pb
Picher-sheet ground	X0	2,000,000 t Zn
Picher-runs	~100	9,000,000 t Zn
Mine La Motte	~20	2,000,000 t Pb
		50,000 t Co
Hayden Creek Mine	1	X0,000 t Pb
Bonne Terre	~50	30,000,000 t Pb
Schultz Breccia	5	X00,000 t Pb
Minerva Mine, Cave-in-Rock	~10	5,000,000 t CaF_2
		100,000 t Zn
Trego Mine, Platteville	X00	1,500,000 t Zn
Rodham Mine, Shullsburg	X000	700,000 t Pb
Virginia-Merrimac mines	X0	500,000 t $BaSO_4$
		50,000 t Pb
Nancy Hanks Mine	X0	X00,000 t CaF_2
Rosiclare	~100	10,000,000 t CaF_2
		120,000 t Zn
		40,000 t Pb
Hutson Mine	1	40,000 t Zn
Riggs Mine	~3	10,000 t CaF_2
Hicks Dome	3	no production, no reserves

examples where a certain mine contained more than one type of ore deposit, it was necessary to add a short ore characteristic (e.g. Nancy Hanks Mine, Kentucky, gravel spar ore deposit type). If only one ore type occurs in a vicinity of a town, the town name is a sufficiently accurate type name (e.g. Rosiclare ore deposit type).

A collective (or group) type name has been used in areas of almost continuous mineralization, where individual ore deposits cannot be easily separated or identified (e.g. Picher-runs ore deposit type), and also in cases when a considerable variety of closely related and similar, but in detail different, ore deposits occur in one area (e.g. Bonne Terre ore deposit type).

Many mineralizations are transitional between two or more ore types; for example the ore deposit in the Viburnum No. 27 mine, SE Missouri district (Grundmann, 1977, particularly fig. 14), appears to be of the Bonne Terre-type, but transitional into the Mine La Motte-type due to its proximity to the Precambrian basement. Sometimes, it may be advantageous to give to a transitional locality a new ore type name. Otherwise, transitional mineralization can be listed with the type it most closely resembles, or defines quantitatively (e.g. locality X: 80% type A, 20% type B). The latter, however, can rarely be expressed that accurately.

In this test, a close correlation among ore deposit and ore district types has been found. For example, the Mine La Motte and Bonne Terre ore deposit types occur only in the SE Missouri district; the Picher-runs and Picher-sheet grounds ore deposit types occur only in the Tri-State disrict. Although there is much similarity in ore and host-rock appearance in hand-specimens (as seen on the waste dumps in the Picher area and, for example, in the Platteville area in the Upper Mississippi Valley district that mostly belongs to the Trigo Mine ore deposit type) both localities are different in structural setting, orebody geometry and the age of host rocks. Elsewhere, the apparent similarity of two or more ore types, as they appear on simplified sections, quickly fades away when one examines the actual rock and ore material from both ore deposits.

CONCLUSION

The test of practical typification of mineralizations in the North American Platform Interior demonstrates that the concept of ore types can be much upgraded and improved by organization, based on evaluation of a reasonably broad population of sample localities. It, however, also demonstrates that such upgrading is necessarily at the expense of brevity. Even after upgrading, however, the meaning of ore types is still not absolutely accurate, but at least approaches the accuracy of lithostratigraphic units in stratigraphy, namely Formations and Groups.

The existing ore types should be gradually redefined and their accuracy improved, and new ones added: best of all under the auspices of an internation, or multinational, organization. In the meantime, the existing ore types including many very inaccurate ones (Mississippi Valley-type) will no doubt continue their existence as a rapid means of approximate and subjective identification of geological objects and situations. They will fall into a similar category of accuracy as, for example graywacke, geosyncline, limonite, or granite (as a group name for most light-coloured intrusive rocks).

Even if there is a much upgraded and a more accurate list of ore types available in the future, the type comparison and correlation as well as the assignment of localities to ore types will always be equivocal to a certain degree. Very much will depend on wording. Locality X may: (1) be an ore type X, (2) correlate with ore type A, (3) strongly resemble ore type A, (4) resemble ore type A, (5) may have many features in commen with ore type A, (6) have some features in common with ore type A, (7) have a certain feature in common with ore type A, (8) have a certain feature similar to ore type A, etc. Careful wording is paramount.

APPENDIX

List of the more common ore (mineralization) types from the literature

Note: Each type name is followed by country of occurrence and country (or division) of use. References are to papers from which the examples listed have been taken, but these are not necessarily papers that *introduced* the types.

Achisay (Pb, Zn): U.S.S.R.; conformable as well as crosscutting massive pyrite > galena, sphalerite, mineralization in dolomitized carbonates (Smirnov, 1972)

Achisay—Mirgalimsay (Pb, Zn): U.S.S.R.; telethermal galena—sphalerite mineralization associated with Caledonian (= early Paleozoic) magmatism; in carbonates (Magak'yan, 1974)

Aillik (U): Canada; conformable, metamorphogenic disseminated uranium mineralization as in the Makkovik—Seal Lake area, Labrador (Ruzicka, 1971)

Aksoran (Pb, Zn): U.S.S.R.; veinlet and disseminated mainly galena mineralization in a skarn, developed on contact of Hercynian (= late Paleozoic) granodiorites in Kazakhstan (Magak'yan, 1974)

Akzhal (Pb, Zn): U.S.S.R.; galena and light-coloured sphalerite mineralization in a skarn, developed on contact of Hercynian granodiorites in Kazakhstan (Magak'yan, 1974)

Algoma (Fe): Canada, international; Precambrian, conformable, usually banded iron formation (hematite, magnetite or siderite) of submarine volcanic affiliation (Douglas, 1970; Eichler, 1976)

Almaden (Hg): Spain; cinnabar with pyrite and marcasite as impregnations in sandstone; strictly sedimentary association (Schneiderhöhn, 1955)

Alpine type chromite: international; chromite pods in deformed Alpine-type dunite or harzburgite bodies (Thayer, 1964)

Alpine type Pb—Zn: Europe, international; (1) conformable to semiconformable, even disconformable simple mineralization of sphalerite, galena ± pyrite in carbonate rocks of mobile belts; (2) equivalent of the Mississippi Valley-type with B-type lead isotopic composition (Morris et al. 1973)

Altai (Pb—Zn): U.S.S.R.; stratabound Pb—Zn pyritic massive sulphides associated with felsic submarine volcanism (Bogdanov and Kutyrev, 1973)

Altenberg (Sn—W): Erzgebirge; cassiterite-bearing greisen mineralization in cupola of granite porphyry (Chrt et al., 1966)

Ammeberg (Zn—Pb): Sweden; massive to disseminated conformable galena—sphalerite—pyrrhotite in gneisses and metacarbonates of the Precambrian "leptite series" (Magnusson, 1970)

Andrasman (Bi, Cu, Co, etc.): U.S.S.R.; hydrothermal complex veins with Bi, Cu, Co, Ni, Zn, Pb associated with granitic massifs in Kazakhstan (Ramović, 1968)

Animikie (Fe): Canada; synonym for (Lake) Superior type of banded Precambrian iron formations (Eichler, 1976)

Anshan (Fe): China; banded iron formation in pre-Sinian (early Precambrian) metamorphics (Ikonnikov, 1972)

Appalachian (Zn, Pb): U.S.A.; equivalent of the Mississippi Valley-type of platforms, occurring in mobile belts; Mascot—Jefferson district, Tennessee, is the most commonly considered example (Callahan, 1967)

Appalachian Manganese: U.S.A., U.S.S.R.; stratabound, "geosynclinal" Mn mineralization in "miogeosynclinal" limestones and dolomites as in Cambrian Roma Limestone (Varentsov, 1964)

Archean type (Cu, Zn, Pb, etc.); Canada, Western; "Archean" — a time period, here evidently implies massive sulphide mineralization in "greenstone belts" of Canadian Shield of Archean age; these carry volcanogenic sulphide deposits as in Noranda, but other types of mineralization as well; very inaccurate type name (Tuach and Kennedy, 1978) (Amstutz and Bubenicek, 1967)

Archean type (Fe): Europe; rare synonym for Algoma-type Fe-a Precambrian conformable iron formation of volcanogenic affiliation (Eichler, 1976)

APPENDIX (continued)

Arkansas (barite): U.S.A.; bedded, fine-grained, gray barite that forms conformable bodies in Mississippian gray to black shale limestone and chert; largest body is near Magnet, Arkansas (Amstutz and Bubenicek, 1967)

Atasu (Zn, Pb, Mn, barite): U.S.S.R.; rhythmic-bedded massive to disseminated stratiform galena—sphalerite—barite deposits in "flyschoid" association (Skripchenko, 1972)
stratabound, distal volcanogenic Pb—Zn mineralization in chemical sediments (Bogdanov and Kutyrev, 1973)

Azurita (Cu): Bolivia; native copper mineralization in sediments, interbedded with basalt flows; interpreted as exhalational (Ahlfeld and Schneider-Scherbina, 1964)

Banat (Cu, Zn, Pb; Fe): Rumania, Europe; pyrometasomatic (skarn) iron and base metal replacement in carbonates at contact with Laramide granodiorites (e.g. Dognacea) (Superceanu, 1971)

Barberton (Au): S. Africa; goldfields that occur in tectonically and metamorphically reconstituted ultramafics and mafics of Archean greenstone belts (Pretorius, 1976)

Bastnäs (Ce): Sweden; quartz-banded iron formation in leptite association, containing patches of actinolite "skarn" which carries cerite and other rare minerals (Magnusson, 1970)

Bathurst (Zn, Pb, Cu): Canada; conformable pyrite—sphalerite—galena—chalcopyrite, massive sulphide orebodies in volcanic—sedimentary association (Amstutz, 1959)

Bawdwin (Pb, Zn, Ag): Burma, Germany; disseminated to massive galena and sphalerite along shear zones, associated with felsic volcanics; believed mesothermal (Schneiderhöhn, 1955)

Bayan Obo (Fe): China; pod-like irregular replacement orebodies of hematite or magnetite with aegirine and arfvedsonite gangue in dolomitic limestones on contact with an alkaline complex (Ikonnikov, 1972)

Beaverlodge (U): Canada; simple pitchblende bearing veins in crystalline basement, located along major faults (Ruzicka, 1971; Tremblay, 1978)

Belgian Congo (Cu): obsolete, now Zaïre; very broad synonym for stratabound copper mineralization in Shaba (Katanga) (Schneiderhöhn, 1955)

Bendigo (Au): Australia; gold—quartz lodes in a strongly cleaved slate—graywacke association (Laznicka, 1973)

Benue (Pb): Nigeria, international; lenticular galena—sphalerite lodes mainly with siderite gangue in failed rift (aulacogen) geotectonic setting (Mitchell and Garson, 1976)

Besshi (Cu): Japan, international; bedded cupriferous pyrite deposits in pelitic metasediments contemporary with regionally developed submarine mafic flows and tuffs [Mitchell and Bell (1973) and Sawkins (1976) application of a type introduced by Kato (1937)]

Bilbao (Fe): Spain, Europe; hydrothermal — metasomatic siderite in carbonates (Walther and Zitzmann, 1972); strongly oxidized replacement carbonate (siderite) mineralization (Routhier, 1963)

Binnental type (Pb, Zn): Switzerland, Poland; a subdivision of Alpine- or Silesian—Cracovian-type of semiconformable galena, sphalerite and numerous rare minerals in metadolomites (Haranczyk and Galkiewicz, 1970)

Black Sea type of metallogenic environment — basins with reducing, anoxic environment and abundant degraded organic matter in which conformable sulphide deposits may presumably form (Ridge, 1976)

Blåfjell (Fe, Ti): Norway; massive to disseminated ilmenite and titanomagnetite forming veins and irregular orebodies in anorthosite (Holtedahl, 1960)

Bleiberg (Pb, Zn): Austria, Europe; apomagmatic, meso- to epithermal galena—sphalerite bodies in carbonates (Schneiderhöhn, 1955); a variety of "Alpine" Pb—Zn mineralization that consists of fine sphalerite and galena forming stratabound orebodies in beds of a quiet-water facies dolomite with geopetal structures (Höll and Maucher, 1976)

Bleiberg—Kreuth (Pb, Zn): Austria, Europe; synonym of Bleiberg type (Rösler et al., 1968a)

Blind River (U): Canada, international; detrital uranium minerals in the matrix of Proterozoic quartz conglomerate and arenite (Amstutz, 1959)

APPENDIX *(continued)*

Blind River—Elliot Lake (U): Canada, international; synonym of Blind River type (Wolf, 1976b)

Boleo (Mn, Cu): Mexico, Europe; stratabound Cu and Mn mineralization in sediments controlled by basement paleogeography and marine transgressions (Nicolini, 1970)

Bolivia (Sn, Ag): Europe, international; inaccurate designation most commonly of high-level (subvolcanic) hydrothermal vein to stockwork deposits that carry tin in both cassiterite and sulphides (teallite, cylindrite), as well as silver minerals as in Oruro or Potosi (Sierra et al. 1972; Laznicka, 1973)

Bor (Cu): Yugoslavia, Germany; meso- to epithermal copper-arsenic association, subvolcanic, impregnations and disseminations (Schneiderhöhn, 1955)

Broken Hill (Pb, Zn, Ag): Australia, international, massive galena, sphalerite, pyrrhotite, Mn-silicates, etc. in conformable bodies in high-grade metamorphics (Both and Rutland, 1976)

Burnier (Mn): Brasil, Germany; conformable manganese mineralization in micaschists, dolomitic marbles and quartz itabirites (Schneiderhöhn, 1955)

Butte (Cu): U.S.A., Europe; enargite — dominated Cu—As hydrothermal veins (Schneiderhöhn, 1955)

Cap Garonne (Pb, Cu): France, Germany, conformable disseminated Pb + Cu mineralization in arid-eluvial arenites (Schneiderhöhn, 1955)

Carguaicollo (Zn, Sn, Pb): Bolivia, Germany; subvolcanic epithermal würtzite—teallite veins (Schneiderhöhn, 1955)

Carpathian (Au, Ag): E. Europe, U.S.S.R. subvolcanic, high-level gold-bearing quartz veins in propylitized andesite—rhyolite association (Ramović, 1968)

Carthagena (Pb, Ag): Spain, Germany; subvolcanic impregnations and veins of Pb—Ag ores in and near propylitized continental volcanics (Schneiderhöhn, 1955)

Cerro Bolivar (Fe): Venezuela, Germany; Precambrian banded iron formation (Amstutz, 1959)

Changpu (Al): China; gibbsitic bauxite, as redeposited lateritic weathering crusts over basalts (Ikonnikov, 1972)

Chiaturi (Mn): U.S.S.R., Germany; stratiform concentrations of Mn oxides in shelf sedimentary association (Rösler et al. 1968a, b)

Chibougamau (Cu, Au): Canada, U.S.A.; lenses of massive pyrite—chalcopyrite along shears traversing anorthosites (Cox et al. 1974)

Chichiang (Fe): China; bedded oolithic hematite and siderite associated with Jurassic coal measures, in association with limnic sediments (Ikonnikov, 1972)

Chile (Cu): Europe; stratabound copper sulphides in upper parts of andesite lavas or rhyolite ignimbrites, related to calc-alkaline volcanism (Mitchell and Garson, 1976)

China (Sb): Germany; gold-free quartz—stibnite—pyrite hydrothermal veins (Schneiderhöhn, 1955)

Chingtiehshan (Fe): China; banded iron formation in Proterozoic (post-Sinian) metamorphics (Ikonnikov, 1972)

Chinkolobwe or Shinkolobwe (U): Zaïre, Germany; katathermal quartz—uranium veins (Schneiderhöhn, 1955)

Chungtiao Shan (Fe): China; veinlets, stockworks and disseminations of chalcopyrite—molybdenite in Proterozoic granodiorites intruding Archean? metamorphics (equivalent to porphyry coppers) (Ikonnikov, 1972)

Churchill (Cu): Canada; quartz—carbonate—chalcopyrite—pyrite fissure veins in Proterozoic sediments (mainly slates), intruded by numerous diabase dykes or sills (Kirkham, 1973)

Clinton (Fe): U.S.A., international; hematitic, typically oolithic or granular conformable bodies of ironstone in detrital sediments (as in the Clinton Formation of the Appalachians) (Bain, 1911)

Cobalt (Ag, Co, Ni): Canada, internat.; calcite, native silver, Co—Ni arsenides in numerous fissure veins, regionally associated with gabbro and diabase dykes and sills (Schneiderhöhn, 1955; Douglas, 1970; Laznicka, 1973)

Cobar (Pb, Zn, Cu): Australia; tabular, broadly stratabound but in detail discordant lenticular Pb, Zn, Cu pyritic massive sulphide bodies in highly cleaved detrital sediments (R.T. Russell and Lewis, 1965)

APPENDIX (continued)

Coeur d'Alene (Pb, Zn, Ag): U.S.A.; shear to fissure lodes of massive sulphides Pb—Zn and siderite—galena—sphalerite veins in a slate—schist belt (Laznicka, 1973)

Colorado Plateau (U, V): U.S.A., international; conformable, epigenetic deposits of fine uraninite and secondary oxides in sediments, mostly arenites (Amstutz, 1959; Gabelman, 1976)

Comstock Lode (Au, Ag): U.S.A.; high-level subvolcanic gold—silver-bearing veins, stockworks, etc., in propylitized association of andesite and rhyolite (Laznicka, 1973)

Copper Mountain (Cu): Canada; a variety of a porphyry copper formed in a silica deficient environment and located in comagmatic mafic volcanics and syenitic intrusions (Kirkham, 1973)

Copperbelt (Cu): Zambia, international; synonym to Zambian Copperbelt — stratabound copper orebodies in metasediments (Laznicka, 1973)

Corbach (Au, Se): Germany; plutonic-mesothermal gold, lead and selenium bearing veins and replacement deposits (Schneiderhöhn, 1955)

Cordilleran type (polymetallic): N. America; postmagmatic, high-level deposits in epizonal plutonic environments of young continental margin mountain chains; include many specific mineralization types (Sawkins, 1972)

Cornwall, Pennsylvania (Fe, Co): U.S.A.; metasomatic magnetite and cobalt-rich pyrite orebodies occur in limestones on contact with Triassic diabase (Brown, 1968)

Cornwall, England (Sn): Great Britain, Europe; high-temperature hydrothermal (or pneumatolytic) cassiterite bearing tin veins and stockworks, zonally arranged around granitic stocks (Laznicka, 1973)

Corocoro (Cu): Bolivia, Germany; Chalcocite and native copper bearing stratabound mantos, as well as fracture mineralization, in sediments of the red beds associations (Schneiderhöhn, 1955; Ahlfeld and Schneider-Scherbina, 1964)

Cripple Creek (Au, Ag): U.S.A., Germany; epithermal subvolcanic quartz—sericite—calcite—gold veins with Au—Ag tellurides (Schneiderhöhn, 1955)

Cyprus (Cu): Cyprus, international; pyrite and chalcopyrite conformable massive sulphide lenses are located in ophiolite association within tholeiitic or spilitic pillow lavas, or between pillow lavas and overlying pelagic sediments (Mitchell and Bell, 1973)

Dakota (U): U.S.A., Canada; infiltration uranium mineralization in lignites and coals (Ruzicka, 1971)

Dannemora (Fe, Mn): Sweden, Germany; bedded Mn—Fe silicates and magnetite in interbedded felsic metavolcanics and metacarbonates; polymetamorphic skarn (Magnusson, 1970; Walther and Zitzmann, 1972)

Dobschau, now Dobšiná (Cu, Co, Ni): Czechoslovakia, Germany; siderite—quartz—chalcopyrite—gersdorffite bearing fissure and shear veins (Schneiderhöhn, 1955)

Droškovac near Vareš (Pb, Zn): Yugoslavia; galena—sphalerite layers in sometimes brecciated siderite—hematite—limestone association, presumably submarine exhalative (Ramović, 1968)

Ducktown (Cu): U.S.A.; conformable, massive to disseminated pyrite—chalcopyrite in schists (Ross, 1935)

Dzhezkazgan (Cu, Pb): U.S.S.R., E. Europe; telethermal or sedimentary, stratabound to semistratabound copper mineralization in lagoonal—deltaic arenites, with superimposed fracture controlled galena (Bogdanov and Kutyrev, 1973; Magak'yan, 1974)

Dzhida (Mo): U.S.S.R.; molybdenite-bearing quartz veins and stockworks in fissure intrusions of granite porphyries in activated terranes (Shcheglov, 1967)

East Kwangtung (W): China; wolframite—scheelite fissure veins and stockworks in sandstones near granite contacts (Ikonnikov, 1972)

Ehrenfriedersdorf (Sn, W): Germany; combined mineralization of cassiterite in veins and greisen zones (Chrt et al., 1966; Rösler et al., 1968a, b)

Eisenerz (Fe): Austria, Europe; lightly oxidized stock-like hydrothermal metasomatic siderite masses (Routhier, 1963; Walther and Zitzmann, 1972)

Elliot Lake (U): Canada; detrital uranium minerals disseminated in matrix of quartz conglomerates and quartzites (Ruzicka, 1971)

496

APPENDIX (continued)

Ergani Maden (Cu): Turkey, Germany; stratabound Cu-pyrite in ophiolite association (Borchert, 1957)

Erzberg (Fe): Austria, Germany; synonym of Eisenerz; Paleozoic synsedimentary–diagenetic, submarine association of siderite with keratophyre; trace Mn > Mg; Ni/Co ratio = 10 (Höll and Maucher, 1976)

Exotica (Cu): Chile; copper oxides forming fracture and porosity infiltrations downstream from Chuquicamata porphyry copper, precipitated from groundwater; informal, Chilean mining community

Falun (Pb, Zn, Cu): Sweden; massive Cu–Zn sulphides in Precambrian felsic metavolcanics (Geijer, 1964; Magnusson, 1970)

Ferghana (Sb, Hg): U.S.S.R. Germany; stibnite–cinnabar impregnations and cavity fillings in limestone (Schneiderhöhn, 1955)

Flin Flon (Cu, Zn): Canada; massive Cu–Zn sulphides in metavolcanic association of Precambrian "greenstone" belts (Douglas, 1970)

Flöttum (Zn): Norway, Germany; conformable quartz–pyrite–sphalerite mineralization in schists, without a visible connection with intrusive rocks (Schneiderhöhn, 1955)

Fosen (Fe): Norway; banded magnetite ore in Precambrian metamorphics (Oftedahl, 1959)

Fouling (Fe): China; low-grade bedded oolithic hematite in Permocarboniferous sediments, beneath coal (Ikonnikov, 1972)

Franklin, or Franklin–Sterling (Zn, Mn): U.S.A.; conformable orebodies of Zn–Mn oxides and silicates (willemite, franklinite) in high-grade Precambrian metamorphics (Brown, 1968)

Freiberg (Ag, Pb, Zn): Germany; several traditional vein associations; hydrothermal fissure vein mineralization in gneisses in proximity to intrusive stocks (Schneiderhöhn, 1955)

Geyer (Sn); Germany; cassiterite-bearing stockworks of quartz veins in altered granite cupolas (Chrt et al., 1966)

Goldfield (Au): U.S.A., Germany; epithermal subvolcanic gold-bearing veins with abundant alunite (Schneiderhöhn, 1955)

Grängesberg (Fe): Sweden; apatite-bearing, magnetite conformable iron formation in Precambrian metamorphics (Magnusson, 1970)

Granite-Bimetallic (= Philipsburg district) (Pb, Zn, Cu, Ag): mesothermal quartz–pyrite–galena–tetrahedrite veins (Schneiderhöhn, 1955)

Great Bear Lake (U, Ag): Canada; pitchblende and native silver–Ni–Co-arsenides bearing fissure veins; used as synonym for Jáchymov type (Ruzicka, 1971)

Great Caucasus (W, Sb): U.S.S.R., E. Europe; ferberite and hübnerite–stibnite association in veins and conformable bodies in slate belts (Laznicka, 1973)

Grenville (U): Canada; uranium mineralization in pegmatitic alkalic granites and syenites, as in parts of the Precambrian Grenville Province of Canadian Shield (Ruzicka, 1971)

Harz (Fe): Germany; hydrothermal hematite fissure veins (Walther and Zitzmann, 1972)

Hauki (Fe): Sweden; Precambrian quartz–hematite ± magnetite banded ores in the Kiruna area (Parák, 1975)

Hawaii (sulphides): a hypothetical, not yet known mineralization type of massive sulphides, morphologically and compositionally similar to the Cyprus type, but developed within an intraoceanic, Hawaii-like volcanic association (Mitchell and Bell, 1973)

Hercynian type (Sb): France; meso- to epithermal stibnite ± pyrite, arsenopyrite vein mineralization, associated with microgranites; the granites are believed to have originated during the Hercynian (= late Paleozoic) orogeny; Brioude–Massiac district of the Massif Central is an example (Routhier, 1963)

Horní Luby (Hg): Czechoslovakia; cinnabar and pyrite, with quartz–calcite–barite gangue, forms impregnations and fracture fillings in zones of deformation (Chrt et al., 1966)

Hsiuwen (Al): China; thick, tectonically disrupted beds of diaspore redeposited bauxite, interbedded with shales (Ikonnikov, 1972)

APPENDIX (continued)

Hsuanlung (Fe): China; oolithic to granular sedimentary hematite in sandstone and shale (Ikonnikov, 1972)

Hunsrück (Fe, Mn): Germany; residual Fe + Mn oxides and hydroxides in karst and weathering profiles over schists and other silicate rocks (Schneiderhöhn, 1955)

Huta = Novoveska Huta (U): Czechoslovakia; effusive—sedimentary, conformable mineralization of uranium with some molybdenum in continental felsic volcaniclastics and sediments (Ruzicka, 1971)

Hüttenberg (Fe): Austria; synsedimentary diagenetic submarine siderite; trace Mn > Mg, Ni/Co ratio 3—5 (Höll and Maucher, 1976)

Ilimaussaq (U, Th, RE): Greenland; low-grade disseminated uranium and thorium minerals in nepheline syenites of an alkaline complex (von Backström, 1974)

Imini—Tasdremt (Mn): Morocco, U.S.S.R.; bedded polianite—psilomelane mineralization in limestone and dolomite associations (Varentsov, 1964)

Insbach (Cu): Germany; conformable bornite, chalcocite and native copper in red sandstone association (Schneiderhöhn, 1955)

Irish type, or Ireland (Pb, Zn): Ireland, Europe; stratabound replacements of marine limestones by fine-grained galena and sphalerite (M.J. Russell, 1976)

Iron Ranges (U): Canada; graphitic shales and schists bearing sedimentary—metamorphic uranium mineralization (Ruzicka, 1971)

Itabira (Fe): Brazil, Germany; banded Precambrian siliceous iron formation, hematitic, metamorphosed (Amstutz, 1959)

Jáchymov, formerly Joachimsthal (U, Ag, Ni, Co, Bi): Czechoslovakia; pitchblende-bearing fissure veins with association of native silver and Ni—Co arsenides (Ruzicka, 1971)

Jacobina (U): Brazil; detrital gold and minor uranium minerals in the matrix of Precambrian quartz pebble conglomerate, comparable to Witwatersrand (Wolf, 1976b)

Juneau (Au): U.S.A.; gold-bearing quartz lodes and stockworks of thin veinlets in sheared and cleaved slate—graywacke association (Ramović, 1968)

Kalgoorlie (Au, Te): Australia, Germany; gold-bearing impregnations and disseminations in silicate rocks (along shears); considered plutonic by Schneiderhöhn (1955)

Kantorp (Fe): Sweden; siliceous banded magnetite ore with abundant accessory feldspar, mica and anthophyllite, in leptite series; "pegmatite" iron ores (Magnusson, 1970)

Karadzhal (Mn): U.S.S.R.; conformable Mn mineralization in volcanic—sedimentary porphyry—siliceous-shaly association (Roy, 1976)

Karagaily—Kairakty (Ba, Pb): U.S.S.R.; barite—galena veins and masses along faults, genetically associated with late Hercynian granodiorites (Magak'yan, 1974)

Katanga (Cu): Africa; conformable copper mineralization in Proterozoic metasediments; synonym to Belgian Congo, Congo, Shaba (Amstutz, 1959)

Kavak (Cr): Turkey, Germany; schlieren and stocks of chromite ore occur in slightly serpentinized dunites (Borchert, 1957)

Kennecott (Cu): U.S.A., international; high-grade chalcocite veins, impregnations and mineralized breccias in limestones adjacent to mafic volcanics (Schneiderhöhn, 1955; Cox et al., 1974)

Kerr—Addison (Au): Canada; gold in quartz and carbonate veined, altered Archean ultramafics, adjacent metasediments and metavolcanics along a major lineament (Tihor and Crocket, 1977)

Keweenaw (Cu): U.S.A.; stratabound native copper in amygdaloidal basalts and interflow conglomerates, Proterozoic; synonym Michigan and Lake Superior Cu (Cox et al., 1974)

Khibiny (apatite): U.S.S.R., Germany; intrusive apatite—nepheline deposits in ijolites of alkaline complexes of the Kola Peninsula (Schneiderhöhn, 1955)

Khodzhertui—Shalot (Mo): U.S.S.R.; high-temperature quartz veins and muscovite greisens in alaskites, contain disseminated molybdenite and pyrite (Shcheglov, 1967)

Kiangsi (W): China, Europe; quartz—wolframite fissure veins in altered granite and hornfelsed country rocks (Ikonnikov, 1972)

APPENDIX (continued)

Kiruna (Fe): Sweden, Europe; liquidomagmatic—pneumatolytic, intrusive magnetite—apatite ore-bodies (Schneiderhöhn, 1955) possibly submarine—exhalative mineralization associated with spilite—keratophyre, metamorphosed (Amstutz, 1959)

Kizildere (Cu, Zn): U.S.S.R.; stratabound Cu—Zn pyritic massive sulphides located in "flyschoid", metasedimentary association of the Greater Caucasus (Skripchenko, 1972)

Konakry or Conakry (Fe): Guinea, France; ferruginous laterite (residual concentration) developed over ultramafic rocks (Routhier, 1963)

Kongsberg (Ag): Norway, Germany; native silver bearing calcite fissure veins in Proterozoic metamorphics, regionally associated with gabbro and diabase dykes (Schneiderhöhn, 1955)

Kounrad (Cu, Mo): U.S.S.R.; veinlet and disseminated chalcopyrite, molybdenite and pyrite in altered granodiorites in Kazakhstan; porphyry copper equivalent (Bilibin, 1955)

Kreuth (Pb, Zn): Austria, Germany; massive, conformable sphalerite mineralization with geopetal structures in mineralized reefs, (Höll and Maucher, 1976)

Krivoi Rog uranium; U.S.S.R., used in Canada; uranium mineralization formed through alkalic metasomatism of ferruginous rocks (Ruzicka, 1971)

Krivoi Rog (Fe): U.S.S.R., Europe; Precambrian siliceous banded iron formation (hematite, magnetite), metamorphosed, in the Ukrainian Shield (Walther and Zitzmann, 1972)

Kündikan (Cr): Turkey; chromite bodies and nodules that occur in intensively mylonitized and altered serpentinites (Borchert, 1957)

Kunghsien (Al): China; bedded, redeposited bauxite forming lenses in clay-filled depressions (Ikonnikov, 1972)

Kunming (Fe): China; low-grade bedded hematite in Triassic sandy shale (Ikonnikov, 1972)

Kupferschiefer (Cu): Germany, international; synonym of Mansfeld; stratiform chalcocite and bornite, dispersed and nodular in a thin bed of bituminous, carbonatic shale, as in the Harz Foreland of Germany; widespread use in the literature.

Kurokō (Pb, Zn, Cu); Japan, international; "Black Ore" variety of Miocene volcanogenic massive sulphide deposits of Japan, particularly of northeastern Honshu; the entire deposits containing the black ore, in addition to other ore varieties (Ishihara, 1974) subclass of massive sulphide deposits characterized by an intimate relationship to calc-alkaline, felsic pyroclastics erupted in submarine environment (Sawkins, 1976)

Kweichov (Hg): China, Europe: cinnabar veins, impregnations, disseminations in shales, sandstones and limestones (Laznicka, 1973)

Kyzyl-Espe (Pb, Zn): U.S.S.R.; galena-sphalerite bearing contact metasomatic deposits of Kazakhstan, associated with eraly Hercynian granitic rocks (Magak'yan, 1974)

Labrador (Fe): Canada; metamorphosed Proterozoic banded siliceous iron formation (Amstutz, 1959)

Lac Allard (Fe, Ti): Canada, France; magmatogene titanomagnetite mineralization that occurs in gabbros, norites and anorthosites (Routhier, 1963)

Lahn—Dill (Fe): Germany, international; submarine—exhalative and/or submarine hydrothermal, conformable hematite (red iron) mineralization in sedimentary—volcanic (keratophyre) association, as in the Devonian of the Rheinische Schiefergerbirge (Schneiderhöhn, 1955; Rösler et al; 1968a, b; Walther and Zitzmann, 1972; Quade, 1976)

Laisvall (Pb, Zn): Sweden; conformable disseminations and clusters of galena and subordinate sphalerite in the matrix of quartz arenites, near the base of the Scandinavian Caledonides autochton (Amstutz, 1959)

Lake Superior copper: U.S.A. international; see also Superior, synonym Keweenaw; conformable zeolite—chlorite—native copper association in Proterozoic amygdaloidal basalts and interflow conglomerates (Schneiderhöhn, 1955)

La Lucette (Sb, Au): France, quartz—stibnite—gold hydrothermal veins in metamorphics and granitic rocks (Laznicka, 1973)

Långban (Mn, Fe): Sweden, Europe; braunite, hausmannite and many rare Mn, Pb, As, etc., minerals

APPENDIX (continued)

occur in conformable "skarn" (tactite) orebodies in metadolomite, which is a part of the Precambrian leptite association (Schneiderhöhn, 1955; Magnusson, 1970)

L'Argentière (Pb, Zn): France; conformable orebodies of disseminated galena and sphalerite in Triassic arenite and dolomite, near the crystalline basement (Samama, 1976)

Leadville (Zn, Pb, Ag): U.S.A., Europe; massive to disseminated galena—sphalerite—pyrite mantos in metacarbonates near high-level intrusions of porphyry; kata- to mesothermal replacements after Schneiderhöhn (1955) (Laznicka, 1973)

Leksdal (pyrite): Norway; massive to disseminated pyrite as conformable orebodies in greenstone-sedimentary association (Schneiderhöhn, 1955)

Limaho (Cu, Ni): China; lenses, veins and layers of pyrite—pentlandite—chalcopyrite in peridotite and pyroxenite in small differentiated intrusions of Alaskan or Uralian type (Ikonnikov, 1972)

Lipetsk (Fe): U.S.S.R., Europe; conformable continental sedimentary siderite and limonite (Walther and Zitzmann, 1972)

Llallagua (Sn): Bolivia, Germany; subvolcanic hydrothermal veins and stockworks carrying cassiterite, in altered felsic porphyry stocks (Schneiderhöhn, 1955)

Lokris (Fe): Greece, Europe; iron hydroxide accumulation in residual weathering crusts over ultramafics (Walther and Zitzmann, 1972)

Lorraine (Fe): France, Europe; stratiform, low-grade, Jurassic marine oolithic limonite, siderite, chamosite and hematite; synonym Minette (Walther and Zitzmann, 1972)

Luchang (Cu): China; disseminated and nodular Cu sulphides occur in sandstones and shales of the red bed associations (Ikonnikov, 1972)

Luchiang—Chaoyuan (Cu): China; copper-bearing hydrothermal veins in acid and intermediate magmatic rocks intruding andesites (Ikonnikov, 1972)

Madhya Pradesh (Mn): India, U.S.S.R.; lenticular conformable bodies of Mn amphibole, rhodonite, braunite, hausmannite, spessartite, etc. in high-grade metamorphic Precambrian gondite (Varentsov, 1964)

Magnitnaya = correctly Gora Magnitnaya (Fe): U.S.S.R., Europe; contact metasomatic magnetite skarn in Paleozoic carbonates, adjacent to diorite intrusions (Routhier, 1963; Walter and Zitzmann, 1972)

Maidanpek (Cu): Yugoslavia, Europe; irregular veins, veinlets, impregnations, contact replacement bodies, etc., of chalcopyrite and pyrite in andesites, schists and granodiorite intrusions; corresponds to porphyry coppers (Schneiderhöhn, 1955)

Maikain (Ba, Pb—Zn, pyrite): U.S.S.R.; barite, polymetallic and pyrite mineralization associated with early Caledonian (= early Paleozoic) magmatism (Magak'yan, 1974)

Makkovik (U): Canada; pitchblende that occurs in granulite and other high-grade metamorphics, as in the Labrador (Ruzicka, 1971)

Manono (Sn): Zaïre; cassiterite and some columbite—tantalite bearing pegmatites (Laznicka, 1973)

Mansfeld (Cu): Germany, international; synonym to Kupferschiefer

Marl Slate (Cu): Great Britain; British equivalent (and extension) of the Permian Kupferschiefer (Brongersma-Sanders, 1967)

Matahambre (Cu): Cuba; massive chalcopyrite lenses and veins in slates, unrelated to volcanics or intrusive rocks (Cox et al., 1974)

Maubach—Mechernich (Pb, Zn): Germany; conformable disseminations and clusters of galena and minor sphalerite in Triassic arenites of a red beds association; synonym Mechernich (Amstutz, 1959)

McArthur (River) (Pb, Zn, Ag): Australia; fine-grained conformable galena, sphalerite and pyrite mineralization in tuffaceous mudstones and siltstones (Lambert, 1976)

Mechernich (Pb): Germany, Europe; conformable galena in the cement of sandstone, interpreted as possibly due to secondary hydrothermal enrichment (Schneiderhöhn, 1955)

Meggen (Zn, Pb, barite): Germany; bedded barite and massive pyrite containing fine-grained sphalerite

APPENDIX (continued)

and rare galena, in sediments proximal to submarine keratophyres (Amstutz, 1959; Ramović, 1968)

Meggen—Rammelsberg (Zn, Pb, barite): Germany; subvolcanic, submarine sedimentary bedded barite and massive sulphide Pb—Zn deposits (Rösler et al., 1968a, b)

Merensky Reef (Pt): S. Africa; layer of liquidomagmatic sulphides and platinum, palladium containing pyroxenite and diallag—norite occurs as a distinct member of the Bushveld Complex (Schneiderhöhn, 1955)

Mernik (Hg): Czechoslovakia; cinnabar in chalcedony—calcite veins in Neogene felsic volcanics, particularly in obsidian (Kuthan, 1968)

Metaline (Pb—Zn): U.S.A., Western Canada; conformable disseminated sphalerite and galena in unmetamorphosed carbonate rocks (Fyles, 1966)

Mexico (Ag): Germany, epithermal, subvolcanic, high-grade ("Bonanza") fissure silver veins (Schneiderhöhn, 1955)

Mežica (Zn, Pb): Yugoslavia; synonym to Bleiberg (Rösler et al., 1968a, b)

Midnite Mine (U): U.S.A.; uranium oxides occur within and near a fracture and shear zone, developed along contact of granite and schists (von Backström, 1974)

Mina Ragra (V): Peru, Germany; concordant lenses of vanadium sulphide occur in asphaltic anthracite, asphaltite, natural coke and sandy shale near an intrusive dyke (Amstutz, 1959)

Mine Lamotte (Pb, Co): U.S.A.; disseminated crystals of galena and minor pyrite and siegenite occur in narrow linear orebodies in Cambrian dolomite located above or near Cambrian basal sandstone pinchout against knobs of Precambrian basement (Snyder and Gerdemann, 1968)

Minette (Fe): France, Luxembourg, Europe; Jurassic bedded granular and oolithic sedimentary iron ore containing limonite, siderite, hematite and chamosite; synonym Lorraine (Schneiderhöhn, 1955)

Mirgalimsay (Pb, Zn): U.S.S.R.; disseminated galena and sphalerite in conformable bodies in dolomites, Karatau Range in Kazakhstan (Smirnov, 1972)

Miskin (Cu): Turkey; massive pyrite, chalcopyrite and minor sphalerite occur in calcareous slate, associated with mafic volcanics of the ophiolite association (Borchert, 1957)

Mission (Cu): U.S.A.; disseminated copper mineralization (porphyry copper) in altered metasediments (Davidson, 1965)

Mississippi Valley (Zn, Pb): U.S.A., international; very broad and controversial type. Applied to galena—sphalerite mineralization in predominantly carbonate association, unrelated to obvious intrusive rocks in the vicinity, and usually in undisturbed covers of stable platforms (Brown, 1970; Bush, 1970; Bogdanov and Kutyrev, 1973; Wedow et al., 1973; and many others) epigenetic, low-temperature, hydrothermal deposits of galena, sphalerite and pyrite, locally with barite, fluorite, chalcopyrite, etc.; lead is of the radiogenic Joplin type (Heyl 1972)

Moldava (F, Ba): Czechoslovakia; fluorite and barite fissure veins (Chrt et al., 1966)

Monserrat (near Poopo) (Sn, Zn): Bolivia, Germany; subvolcanic vein deposit of sulfostannates with würtzite (Schneiderhöhn, 1955)

Monte Amiata (Hg): Italy, Germany; cinnabar impregnations and fracture coatings in limestone and rarely trachyte, attributed to recent felsic volcanism in the area (Schneiderhöhn, 1955)

Morocco (Mn): U.S.S.R.; stratabound Mn ores in limestone—dolomite associations of platform regions (Varentsov, 1964)

Mother Lode (Au): U.S.A., Germany; plutonic—katathermal gold—quartz veins (Schneiderhöhn, 1955)

Mt. Keith (Ni): Australia, Canada; low-grade nickel in serpentinized ultramafics occurs as solid solution in silicate and oxide, and in minor sulphide (Duke, 1977)

Murgul (Cu): Turkey, Germany; chalcopyrite and pyrite veins, veinlets and impregnations in silicified volcanics and subvolcanic intrusive rocks (Schneiderhöhn, 1955)

Nahe (Cu): Germany; native copper associated with zeolites and epidote in mafic amygdaloidal flows (Schneiderhöhn, 1955)

APPENDIX (continued)

Nairne (pyrite): Australia; bedded pyrite in metasedimentary schist (Amstutz, 1959)

Nanshan (Fe): China; hydrothermal metasomatic veins and irregular masses of hematite and magnetite in metacarbonates, near contacts with diorite (Ikonnikov, 1972)

Nargeuchoum (Mn): Morocco, U.S.S.R.; disseminated Mn oxides (polianite) in red carbonate—terrigenous association (Varentsov, 1964)

Nikopol (Mn): U.S.S.R., Europe; stratiform bedded Mn oxides in quartz—glauconite—clay sandstones in littoral marine zone of the Russian Platform, south of the Ukrainian Shield basement outcrop (Varentsov, 1962; Roy, 1976)

Ninghsiang (Fe): China; bedded oolithic hematite in sandy shale or limestone (Ikonnikov, 1972)

Noda Tamagawa (Mn): Japan; manganese mineralization in volcanogenic greenstone—jasperoid association (Roy, 1976)

Oberharz or Clausthal (Pb, Zn, Ag): Germany; Meso- to epithermal quartz—calcite—galena—sphalerite fissue veins (Schneiderhöhn, 1955)

Obira (Au, Ag): Japan; high-level, subvolcanic, gold—silver bearing epithermal veins in propylitized volcanics (Miyahisa, 1960)

Oka (U, Nb): Canada; introduced to emphasize the by-product uranium content of pyrochlore-bearing fenites associated with a carbonatite intrusion (Ruzicka, 1971)

Olympic Peninsula (Mn): U.S.A.; conformable volcanic—sedimentary Mn mineralization in greenstone—siliceous association (Roy, 1976)

Opalite (Hg): U.S.A.; disseminated cinnabar in opal-replaced (silicified) young semiconsolidated volcaniclastics (Henderson, 1969; Laznicka, 1973)

Oriskany (Fe): U.S.A.; stratiform, bedded iron carbonate mineralization associated with Devonian Oriskany Sandstone of the Appalachians (Wright et al., 1968)

Orphan pipe (U): U.S.A., Canada; uranium, copper, lead, etc., minerals in an altered breccia pipe (Ruzicka, 1971)

Paiyinchang (Cu): China; conformable lenses of copper-bearing pyritic massive sulphide of submarine exhalative origin, in andesite—basalt—keratophyre association (Ikonnikov, 1972)

Pechtelsgrün (W): Germany; quartz—wolframite veins and stockworks inside or outside of granite stocks (Chrt et al., 1966; Rösler et al., 1968a, b)

Peine (Fe): Germany; detrital, redeposited debris of limonite in a nearshore marine sequence (Walther and Zitzmann, 1972)

Pennines (Pb, Zn): stratiform Pb—Zn mineralization with abundant barite and fluorite (Russell, 1976)

Pilares (Cu): Mexico, Germany; plutonic—subvolcanic impregnations of copper ores in eruptive breccias (Schneiderhöhn, 1955)

Pinchi Lake (Hg): Canada; cinnabar as fracture coatings, disseminations and veinlets in shattered carbonates, schists and ophiolites, along a suture zone (Laznicka, 1973)

Potosi (Ag, Sn): Bolivia, Germany; subvolcanic meso- to epithermal sulphidic silver—tin mineralization (Schneiderhöhn, 1955)

Pozanti (Cr): Turkey; chromite lenses in intensively serpentinized peridotites (Borchert, 1957)

Preissac (Mo): Canada; quartz and quartz—feldspar, molybdenite bearing veins, stockworks to disseminations, locally grading into pegmatites (Laznicka, 1973)

Preural (= Cisural) (Cu): U.S.S.R.; conformable copper mineralization in lacustrine—alluvial Permian sandstones (Bogdanov and Kutyrev, 1973)

Příbram (U): Czechoslovakia; pitchblende mineralization in hydrothermal fissure veins with simple (dolomite, calcite, pitchblende, pyrite) association (Ruzicka, 1971)

Přísečnice—Měděnec (Fe, Cu): Czechoslovakia; magnetite and minor chalcopyrite occur in pyroxene—garnet skarn in a metamorphic complex mantling a gneiss dome. (Chrt et al., 1966)

Pulacayo (Pb, Zn): Bolivia, Germany; subvolcanic meso- to epithermal lead—zinc bearing fissure veins (Schneiderhöhn, 1955)

Pyrenean type (= Pyrenées) (Cu, Fe, Mo, W): Europe; contact metasomatic mineralization in carbonates in contact with intrusions (Superceanu, 1971)

502

Queluz (Mn): Brazil and Germany; conformable, polymetamorphic rhodochrosite, spessartite, rhodonite, tephroite and Mn-oxides mineralization in schists (Schneiderhöhn, 1955)

Radium Hill (U): Australia, Canada; davidite-bearing vein mineralization (Ruzicka, 1971)

Rammelsberg (Zn, Pb, Cu): Germany, international; conformable zinc, lead and copper bearing massive sulphide, and conformable barite, in slate—graywacke association, with a distal volcanogenic influence (Amstutz, 1959; Laznicka, 1973)

Red Sea (Zn, Cu, Fe, Mn, Ag, etc.): international; unconsolidated metalliferous muds discovered in the
axial part of the Red Sea protooceanic gulf, precipitated from hot bottom brines (Lacy, 1974;
Ridge, 1976)

Redjang Lebong (Au, Se); Indonesia, Germany; epithermal subvolcanic quartz—sericite—calcite gold
and silver bearing veins with high selenium content (naumannite) (Schneiderhöhn, 1955)

Renison Bell (Sn): Australia; pyrrhotite—cassiterite replacement masses and veins in metacarbonates,
near contact with high-level granite porphyry stocks (Laznicka, 1973)

Rhein, or Rheinische Schiefergebirge (Pb, Zn): Germany; meso- to epithermal quartz—siderite—lead—
zinc fissure veins (Schneiderhöhn, 1955)

Rhodesia (Cu): obsolete, international; synonym with Zambian Copperbelt

Rhodopen (Pb, Zn, Ag): Bulgaria; subvolcanic hydrothermal fissure and replacement veins, stockworks
and irregular bodies in metamorphics and high-level granitic bodies, in a median massif setting
(Laznicka, 1973)

Richelsdorf (Co): Germany; short fracture veinlets in and under the stratiform Kupferschiefer, filled
by barite with Co-arsenides, and formed by mobilization by descending waters (Schneiderhöhn,
1955)

Rio Tinto (Cu, pyrite): Spain, European; conformable massive pyrite with minor chalcopyrite in sedimentary—volcanic association (Schneiderhöhn, 1955; Rösler et al. 1968a, b; Sierra et al., 1972)

Rödhammer (pyrite, Cu): Norway, Germany; pyrite, pyrrhotite and chalcopyrite orebodies developed
on contact of trondhjemite and volcanic—sedimentary association (Schneiderhöhn, 1955)

Röros (pyrite, Cu): Norway, Germany; conformable to semiconformable, massive to disseminated
pyrite and minor chalcopyrite, galena, sphalerite massive sulphides in micaschists and greenstones, near gabbro contacts (Schneiderhöhn, 1955)

Ross—Adams Mine (U): U.S.A.; uranium, thorium and rare-earth minerals disseminated in hornfelsed
rocks near an alkali granite stock, Alaska (von Backström 1974)

Rössing (U): Namibia; low-grade, disseminated uranium mineralization in alaskites; type example of
"porphyry uranium" mineralization (Armstrong, 1974)

Rožná (U): Czechoslovakia; sooty pitchblende, clay mineral and graphite gangue and locally calcite,
dolomite, fluorite and pyrite gangue with massive pitchblende occur along shear zones in metamorphics (Ruzicka, 1971)

Salmo (Pb, Zn): western Canada; conformable galena—sphalerite mineralization in folded, sheared and
greenschist facies metamorphosed carbonates (Fyles, 1966)

Salzgitter (Fe): Germany; limonite forming clasts in conglomerates (Schneiderhöhn, 1955); marine to
continental limonite debris and oolithic limonite (Walther and Zitzmann, 1972)

Sangasho (Cu) Japan; "Kieslager" — a conformable pyritic cupriferous massive sulphide related to
mafic metavolcanics (Muta, 1957)

Sangdong (W): Korea; scheelite-bearing skarn (contact metasomatic) mineralization (Zhelyaskova-
Panajotova et al., 1972)

Schemnitz = now Banská Štiavnica (Au, Ag, Pb, Zn, Cu): Czechoslovakia, Germany; high-level, subvolcanic gold, silver, lead, copper and zinc bearing hydrothermal veins, stockworks and
replacements in propylitized continental andesitic volcanics (Ramović, 1968)

Schmiedeberg (Co, Ni, U): Germany; calcite fissure veins carrying Co—Ni arsenides and pitchblende
(Schneiderhöhn, 1955)

Schneckenstein (Ba): Germany; fluorite free, red barite fissure veins (Chrt et al. 1966)

APPENDIX (continued)

Schneeberg (Ag, Bi, Co, Ni): Germany; silver, bismuth, Co—Ni arsenides bearing barite fissure veins (Schneiderhöhn, 1955)

Schönbrunn (F, Ba): Erzgebrige; fluorite—barite—carbonate fissure veins (Chrt et al., 1966)

Schwarzwald (Pb, Zn, F, Ba): Germany; meso- to epithermal fluorite—barite—galena—sphalerite fissure veins (Schneiderhöhn, 1955; Laznicka, 1973)

Schwaz (Cu, Sb, Ag, Hg): Austria, Germany; fracture veins and stockwork of veinlets in dolomite, mineralized by schwazite (Hg-tetrahedrite), along a major fault zone (Schneiderhöhn, 1955; Laznicka, 1973)

S.E. Hunnan (W): China; scheelite skarn mineralization in metacarbonates at granite contacts (Ikonnikov, 1972)

Shansi (Fe): China; Ordovician sedimentary bedded hematite deposits in sandstones and limestones (Ikonnikov, 1972)

Shihlu (Fe): China; hydrothermal metasomatic, irregular, high-grade hematite mineralization on contacts of strongly altered granite or diorite with metacarbonates (Ikonnikov, 1972)

Shuswap (Pb, Zn): western Canada; conformable galena—sphalerite—pyrrhotite massive sulphide in metamorphics of a mantling complex of a gneiss dome (Fyles, 1966)

Siebenbürgen = now Transylvania (Au, Ag): Germany; epithermal subvolcanic quartz—sericite—calcite gold and silver bearing veins, without tellurides (Schneiderhöhn, 1955)

Siegerland (Fe): Germany, Europe; hydrothermal siderite veins (Schneiderhöhn, 1955; Routhier, 1963; Walther and Zitzmann, 1972)

Silesian—Cracovian type (Zn, Pb): Poland; conformable disseminated to banded, low-temperature and often metacolloidal sphalerite, pyrite and galena in platformic carbonates as in the Bytom area (Haranczyk and Gałkiewicz, 1970)

Sredna Gora (Mn): Rumania, Bulgaria; stratabound, exhalative Mn deposits in volcanic—sedimentary association (Superceanu, 1971)

St. Andreasberg (Ag): Germany; complex calcite—zeolite silver bearing veins (Schneiderhöhn, 1955)

St. Joachimsthal (= now Jáchymov) (Ag, U, Ni, Co, Bi): Germany; quartz or dolomite, silver, pitchblende, bismuth and Ni—Co arsenides bearing fissure veins (Schneiderhöhn, 1955)

Steep Rock Lake (Fe): Canada; residual accumulation of limonitic iron ores, developed by alteration of Archean volcanic—sedimentary, low-grade carbonate and pyrite iron formation (Douglas, 1970)

Storgruvan (Fe): Sweden; magnetite irregularly distributed in andradite—pyroxene skarn, in metamorphics of the Precambrian leptite series (Magnusson, 1970)

Stripå (Fe): Sweden; Precambrian banded quartz—hematite ore (Magnusson, 1970)

Sudbury (Ni, Cu, Pt): Canada, international; massive to disseminated pyrrhotite—pentlandite—chalcopyrite developed along the contact of the basal part (gabbro, norite) of the differentiate Sudbury Complex of Precambrian age (Schneiderhöhn, 1955; Douglas, 1970)

Südharz (= southern Harz), or Lautenberg (Pb, Zn): Germany; meso- to epithermal quartz—barite Pb—Zn veins (Schneiderhöhn, 1955)

Sullivan Mine (in Kimberley, B.C.) (Pb, Zn, Ag): Canada, international; galena—sphalerite rich conformable and partly banded massive sulphide lens with a disconformable "feeder", located within a thick succession of terrestrial sediments of slate—subgraywacke association; Proterozoic (Fyles, 1966; Sawkins, 1976; Mitchell and Garson, 1976)

Sumsar (Pb, Zn): U.S.S.R.; stratabound, banded lead—zinc mineralization in coastal-marine carbonates (Bogdanov and Kutyrev, 1973)

Superior (Lake) (Fe) Canada, U.S.A., international; Precambrian banded siliceous iron formation without conspicuously associated volcanics; widely used (Douglas, 1970)

Superior (Lake) (Zn, Cu, Ag, Au, Pb): Canada; inaccurate type name that includes e.g. Noranda (far away from Lake Superior), which comprises mainly copper and zinc bearing pyritic massive sulphides in andesite—rhyolite complexes of Precambrian "greenstone belts" of the Canadian Shield (Lambert, 1976)

APPENDIX (continued)

Sydvaranger (Fe): Norway and Europe; metamorphosed quartz banded Precambrian hematite ores, of northeastern Lappland (Holtedahl, 1960)

Taberg (in Småland) (Fe, Ti): Sweden, Europe; titaniferous magnetite believed of liquidomagmatic origin in gabbros and norites (Routhier, 1963; Walther and Zitzmann, 1972)

Tambillo (Cu): Bolivia; veinlets of native copper and cuprite in amygdaloidal basalt flows in the eastern Altiplano (Ahlfeld and Schneider-Scherbina, 1964)

Tamiao (Fe): China; vanadium-bearing titanomagnetite veins and lenses in anorthosite or gabbro, or in a marble at a contact (Ikonnikov, 1972)

Tayeh (Fe): China; contact metasomatic iron deposits located at contacts of marble with acid to intermediate intrusive rocks (Ikonnikov, 1972)

Tellnes (Fe—Ti): Norway; Fine-grained ilmenite and magnetite mineralization in norite, with gradational boundary (Holtedahl, 1960)

Texas (U): U.S.A.; sedimentary uranium ores located in a near-shore, sandstone marine milieu (Wolf, 1976b)

Thackaringa (Pb, Zn, Ag): Australia; quartz, siderite, galena, sphalerite, etc., veins in gently dipping fractures in Precambrian metamorphics; introduced locally to contrast with the more important Broken Hill type conformable orebodies in the same area (Both and Rutland, 1976)

Topuk (Cr): Turkey; platy schlieren and layered chromite in almost fresh peridotites (Borchert, 1957)

Transvaal (Au): S. Africa; goldfields located in delta—open sea interface host environment associations (Pretorius, 1976)

Tsumeb (Pb, Zn, Cu, V, Ge): Namibia; replacement pipe, veins and disseminations of galena, sphalerite, tetrahedrite, tennantite and many rare minerals in carbonates (Schneiderhöhn, 1955; Laznicka, 1973)

Tungchuan (Fe): China; Precambrian, conformable, lightly metamorphosed iron formations in metasediments (Ikonnikov, 1972)

Tungchuan (Cu): China; layered as well as vein and stockwork copper sulphides in Precambrian metamorphics and gabbro intrusions (Ikonnikov, 1972)

Tungkuanshan (Cu): China; copper mineralization in garnet—diopside—magnetite skarns located on contact of carbonate and diorite stocks (Ikonnikov, 1972)

Upper Silesia (former Oberschlesien, now Górny Śłask) (Zn, Pb): Poland, Germany; stratabound, often banded and metacolloform sphalerite—galena—pyrite mineralization in platformic carbonates (Schneiderhöhn, 1955)

Usinsk (Mn): U.S.S.R.; stratabound rhodochrosite and manganocalcite mineralization in carbonates of the "eugeosynclinal" andesite—carbonate association of the Kuznetsk Alatau (Varentsov, 1964)

Vareš (Fe): Yugoslavia; volcanic—sedimentary (submarine—exhalative) stratabound siderite and hematite in Triassic limestone and siliceous shale association (Janković, 1962)

Vila Apacheta (Sn): Bolivia, Germany; colloidal cassiterite mineralization in sandstones (Schneiderhöhn, 1955)

Vitim (Mo, V): U.S.S.R.; molybdenite and wolframite in quartz—fluorite veins associated with porphyritic granites of the activated area of eastern Siberia (Shcheglov, 1967)

West Africa (Mn): conformable volcanic—sedimentary manganese mineralization in a greenstone—siliceous shale association (Roy, 1976)

Western States (U): U.S.A.; a comprehensive class of ore types of epigenetic uranium mineralization in sedimentary rocks of Wyoming, Montana, the Dakotas, Colorado, Utah and Arizona (Rackley, 1976)

Wittichen (F, Ba, Ag, Bi, Co): Germany; fluorite and barite fissure veins containing small quantities of silver and Ni—Co arsenides (Schneiderhöhn, 1955)

Witwatersrand (Au, U): S. Africa, international; conformable gold and uranium (in thucholite) mineralization in Proterozoic blanket conglomerates; widespread use
goldfields of S. Africa located in a fluvial fan—lacustrine interface environments associations (Pretorius, 1976)

APPENDIX (continued)

Wood River (Pb, Cu, Sb, Ag): U.S.A., Germany; mesothermal siderite–galena–tetrahedrite veins, Idaho (Schneiderhöhn, 1955).

Wyoming (U): U.S.A.; that portion of the "Western States type" of epigenetic uranium mineralization in sediments, that occurs without vanadium (Wolf, 1976b)

Yeelirrie (U): Australia; uranium oxide mineralization in calcrete derived from buried deposits in old drainage channels (von Backström, 1974; King, 1976)

Yellowknife (Au): Canada; large Archean gold–quartz lode systems, located along shears in metamorphosed volcanic–sedimentary ("greenstone") belts (Laznicka, 1973; Fyfe, 1978)

Yilgarn (Ni, Cu): Australia; stratabound Ni–Cu sulphides occur in or at contact of ultramafic lavas or shallow intrusives, and volcanogenic sediments, as at Kambalda (Lambert, 1976)

Yxsjö-Hörken (W): Sweden; disseminated scheelite, fluorite and accessory sulphides in a "skarn" (tactite), developed on contact of Precambrian metasediments and younger Precambrian palingenetic granites (Magnusson, 1970)

Zambia or Zambian Copperbelt (Cu): Africa, international; synonym for Rhodesia, Rhodesian Copbelt or Copperbelt; stratabound, disseminated copper mineralization in Proterozoic metasediments; widely used

REFERENCES

A.C.S.N. (American Commission on Stratigraphic Nomenclature) 1961. Code of stratigraphic nomenclature. *Am. Assoc. Pet. Geol. Bull.*, 45 (5): 645–660.

Agricola, Georgius, 1556. *De re metallica.* English translation and annotation by H.C. Hoover and L.H. Hoover. Dover Press, New Yrok, N.Y., 1950, 639 pp.

Ahlfeld, F. and Schneider-Scherbina, A., 1964. Los yacimientos minerales y de hidrocarburos de Bolivia. *Dep. Nac. Geol. Bol.*, No. 5, p. 388.

Amstutz, G.C., 1959. Syngenese und Epigenese in Petrographie und Lagerstättenkunde. *Schweiz. Mineral. Petrogr. Mitt.*, 39; 1–84.

Amstutz, G.C. and Bubenicek, L. 1967. Diagenesis in sedimentary mineral deposits. In. G. Larsen and G.V. Chillingar (Editors) *Diagenesis in Sediments,* Elsevier, Amsterdam, pp. 417–476.

Anderson, O.E., 1953. How Minerva Oil Company produces high-grade ceramic fluorspar. *Eng. Min. J.,* 154(3): 72–75.

Armstrong, F.C., 1974. Uranium resources of the future – "porphyry" uranium deposits. In: *Formation of Uranium Deposits.* Int. At. Energy Agency, Vienna, pp. 625–636.

Bain, H.F., 1901. Preliminary report on the lead and zinc deposits of the Ozark region. *U.S. Geol. Surv., 22nd Annu. Rep.*, pp. 23–277.

Bain, H.F. (Editor), 1911. *Types of Ore Deposits.* Publisher unknown, San Francisco, Calif., 378 p.

Bauchau, C., 1971. Essai de typologie quantitative des gisements de plomb et de zinc avec la répartition de l'argent. *Bull. Bur. Rech. Geol. Min.,* Sect. II, No. 3, pp. 1–72.

Baxter, J.W. and Desborough, G.A., 1965. Areal geology of the Illinois Fluorspar District, Part 2 – Karbors Ridge and Rosiclare Quadrangles. *Ill. State Geol. Surv., Circ.* 385, 40 pp.

Baxter, J.W., Desborough, G.A. and Shaw, C.W., 1967. Areal geology of the Illinois Fluorspar District, Part. 3 – Herod and Shetlerville Quadrangles. *Ill. State Geol. Surv., Circ.* 413, 41 pp.

Bilibin, Yu.A., 1955. Metallogenic provinces and metallogenic epochs. English translation by E. Alexandrov. *Geol. Bull., Dep. Geol., Queens College, Flushing, N.Y.,* Apr. 1968, 35 pp.

Blondel, F., 1951. La classification des gisements minéraux. *Chron. Mines Rech. Minière,* No. 175, pp. 1–6.

Blondel, F., 1955. Les types de gisements de fer. *Chron. Mines Colon.,* 23 (223): 226–244.

Bogdanov, Yu.V. and Kutyrev, E.I., 1973. Classification of stratified copper and lead–zinc deposits

506

and the regularities of their distribution. In: G.C. Amstutz and A.J. Bernard (Editors), *Ores in Sediments.* Springer, Berlin, pp. 59–63.

Borchert, H., 1957. Der initiale Magmatismus und die zugehörigen Lagerstätten. *Neues Jahrb. Mineral. Abh.* 91: 541–572.

Both, R.A. and Rutland, R.W.R., 1976. The problem of identifying and interpreting stratiform ore bodies in highly metamorphosed terrains: the Broken Hill example. In: K.H. Wolf (Editor), *Handbook of Strata-bound and stratiform Ore Deposits, 4.* Elsevier, Amsterdam, pp. 261–325.

Breithaupt, A., 1849. *Die Paragenesis der Mineralien.* Publisher unknown, Freiberg.

Brobst, D.A., 1975. Barium minerals. In: S.J. Lefond (Editor), *Industrial Minerals and Rocks,* Am. Inst. Min. Metall. Pet. Eng., New York, N.Y., pp. 427–442.

Brockie, D.C., Hare, Jr. E.H., and Dingess, P.R., 1968. The geology and ore deposits of the Tri-State District of Missouri, Kansas and Oklahoma. In J.D. Ridge (Editor), *Ore Deposits in the United States 1933–1967, Vol I.* Am. Inst. Min. Metall. Pet. Eng., New York, N.Y., pp. 400–430.

Brongersma-Sanders, M., 1967. Permian wind and the occurrence of fish and metals in the Kupferschiefer and Marl Slate. *Proc. 15th Int.-Univ. Geol. Congr.,* Univ. Leicester, Leicester, pp. 61–73.

Brooke, W.J.L., 1975. Cobar mining field. In: C.L. Knight (Editor), *Economic Geology of Australia and Papua New Guinea. Aust. Inst. Min. Metall. Monogr. Ser.,* 5: 683–693.

Brown, J.S., 1968. Ore deposits of the northeastern United States. In: J.D. Ridge (Editor), *Ore Deposits of the United States 1933–1967, Vol.* I. Am. Inst. Min. Metall. Pet. Eng., New York, N.Y., pp. 1–19.

Brown, J.S., 1970. Mississippi Valley type lead–zinc ores. *Miner. Deposita,* 5: 103–119.

Brown, J.S., Emery, J.A., Meyer, Jr., P.A., 1954. Explosion pipe in test well on Hicks Dome, Hardin County, Illinois. *Econ. Geol.,* 49(8): 891–902.

Buckley, E.R., 1909. Geology of the disseminated lead deposits of St. Francois and Washington Counties. *Mo. Bur. Geol. Mines,* Vol. 9, Part 1, 259 pp.

Bush, P.R., 1970. Chloride-rich brines from sabkha sediments and their possible role in ore formation. *Inst. Min. Metall. Trans. Sect. B,* 79: B137–B145.

Callahan, W.H., 1967. Paleophysiographic premises for prospecting for stratabound base metal mineral deposits in carbonate rocks. *Nev. Bur. Mines, Rep.* 13, Part. C, pp. 5–50.

Chrt, J., Bolduan, H., Bernstein, K.-H., Drbohlavová, J., Hora, Z., Hösel, G., Miňa, F., Pokamý, L., Sippel, H., 1966. Die postmagmatische Mineralisation des Westteils der Böhmischen Masse. *Sb. Geol. Věd, Ložisková Geol.,* 8: 113–192.

Cornwall, H.R. and Vhay, J.S., 1967. Cobalt and nickel. In: Mineral and Water resources of Missouri. U.S. Congr. 90th, 1st Sess., Senate Doc. 19. *Mo. Div. Geol. Surv. Water Resour. Rep.,* 2nd Ser., 43: 68–70.

Cox, D.P., Schmidt, R.G., Vine, J.D., Kirkemo, H., Tourtelot, E.B. and Fleischer, M., 1974. Copper. In: United States Mineral Resources. *U.S. Geol. Surv., Prof. Pap.,* 820: 163–190.

Cumming, G.L. and Robertson, D.K., 1969. Isotopic composition of lead from the Pine Point deposit. *Econ. Geol.,* 64: 731–732.

Currier, L.W., 1944. Geology of the Cave-in-Rock district. *U.S. Geol. Surv., Bull.* 942(Part 1): 1–72.

Davidson, C.F., 1965. A possible mode of origin of strata-bound copper ores. *Econ. Geol.,* 60: 942–954.

Davis, J.H., 1977. Genesis of the Southeast Missouri lead deposits. *Econ. Geol.,* 72: 442–450.

De Geoffroy, J. and Wignall, T.K., 1972. A statistical study of geological characteristics of porphyry copper-molybdenum deposits in the Cordilleran Belt – application to the rating of porphyry prospects. *Econ. Geol.,* 67: 656–668.

De Launay, L., 1900. Sur les types régionaux de gîtes métallifères. *C.R. Acad. Sci. Paris,* 130: 743–746.

De Launay, L., 1913. *Traité de métallogénie.* Béranger, Paris, 3 Vols.

Domarev, V.S., 1968. Ore types as historical–geological formations. Int. Geol. Rev, 12: 601–609.

Douglas, R.J.W. (Editor), 1970. Geology and economic minerals of Canada. *Geol. Surv. Can., Econon. Geol. Rep.* 1, 838 pp.

Duke, J.M., 1977. Mineralogy of serpentinized ultramafic rocks and associated nickel deposits. *Geol. Surv. Can. Pap.* 77-1A, p. 15.

Eichler, J., 1976. Origin of the Precambrian banded iron-formations. In: K.H. Wolf (Editor), *Handbook of Strata-bound and Stratiform Ore Deposits*, 7. Elsevier, Amsterdam, pp. 157–201.

Evans, D.L. 1962. Sphalerite mineralization in deep-lying dolomites of upper Arbuckle age, West Central Kansas. *Econ. Geol*, 57: 548–564.

Evans, D.L., 1977. Geology of the Brushy Creek Mine, Viburnum Trend, southeast Missouri. *Econ. Geol.*, 72: 381–390.

Fyfe, W.S., 1978. Crustal evolution and metamorphic petrology. *Geol. Surv. Can., Pap.* 78-10, pp. 1–3.

Fyles, J.T., 1966. Lead–zinc deposits in British Columbia. In: *Tectonic History and Mineral Deposits of the Western Cordillera. Can. Inst. Min. Metall. (Spec. Vol.)*, 8: 231-238.

Gabelman, J.W., 1976. Classification of strata-bound ore deposits. In: K.H. Wolf (Editor), *Handbook of Strata-bound and Stratiform Ore Deposits*, 1. Elsevier, Amsterdam, pp. 79–110.

Geijer, P., 1964. On the origin of the Falun-type of sulfide mineralization. *Geol. Fören. Stockholm, Förhandl.*, Part 1, 86 (516): pp. 3–27.

Gilmour, P., 1976. Some transitional types of mineral deposits in volcanic and sedimentary rocks. In: K.H. Wolf (Editor), *Handbook of Stratiform and Strata-bound Ore Deposits. 1,* Elsevier, Amsterdam, pp. 111–160.

Graton, L.C., 1968. Lindgren's ore classification after fifty years. In: J.D. Ridge (Editor), *Ore Deposits in the United States 1933–1967, Vol. II,* Am. Inst. Min. Metall. Pet. Eng., New York, N.Y., pp. 400–430.

Grogan, R.M. and Bradbury, J.C., 1968. Fluorite–zinc–lead deposits of the Illinois–Kentucky mining district. In: J.D. Ridge (Editor), *Ore Deposits of the United States 1933–1967, Vol. I,* Am. Inst. Min. Metall. Pet. Eng., New York, N.Y., pp. 370–399.

Grundmann, Jr., W.H. 1977. Geology of the Viburnum No. 27 mine, Viburnum Trend, southeast Missouri. *Econ. Geol,* 72: 349–364.

Hagni, R.D., 1976. Tri-State ore deposits: the character of their host rocks and their genesis. In: K.H. Wolf (Editor), *Handbook of Strata-bound and Stratiform Ore Deposits,* 6, Elsevier, Amsterdam, pp. 457–494.

Haranczyk, C. and Gałkiewicz, T., 1970. Consanguinity of the European zinc–lead ore deposits of Silesian–Cracovian type and their relation to alkaline–basic volcanites. In: Z. Pouba and M. Štemprok (Editors), *Problems of Hydrothermal Ore Deposition,* Schweizerbart, Stuttgart, pp. 84–95.

Henderson III, F.B., 1969. Hydrothermal alteration and ore deposition in serpentinite-type mercury deposits. *Econ. Geol.,* 64: 489–499.

Heyl, Jr., A.V., 1968. Some aspects of genesis of stratiform zinc–lead–barite–fluorite deposits in the United States. *Econ. Geol. Monogr.,* 3: 20–32.

Heyl, Jr., A.V., 1972. The 38th parallel lineament and its relationship to ore deposits. *Econ. Geol.,* 67: 879–894.

Heyl, Jr., A.V., Agnew, A.F., Lyons, E.J., Behre, Jr., C.H. and Flint, A.E., 1959. The geology of the zinc and lead deposits of the Upper Mississippi Valley district. *U.S. Geol. Surv. Prof. Pap.* 309, 310 pp.

Heyl, Jr., A.V., Agnew, A.F., Lyons, E.J., Behre, Jr., C.H. and Flint, A.E., 1965. Regional structure of the Southeast Missouri and Illinois–Kentucky mineral districts. *U.S. Geol. Surv., Bull.* 1202-B 20 pp.

Höll, R. and Maucher, A., 1976. The strata-bound ore deposits in the Eastern Alps. In: K.H. Wolf (Editor), *Handbook of Strata-bound and Stratiform Ore Deposits,* 5, Elsevier, Amsterdam, pp. 1–36.

Holtedahl, 0. (Editor), 1960. Geology of Norway. *Nor. Geol. Unders. No.* 208, 540 pp.

Hubaux, A., 1972. Geological data files – Survey of international activity. *CODATA Bull,* No. 8 (Nov. 1972), 30 pp.

Hutchinson, R.W., 1973. Volcanogenic sulphide deposits and their metallogenic significance. *Econ. Geol.,* 68: 1223–1246.

508

Ikonnikov, A.B., 1972. *Mineral resources of China.* (unpublished manuscript).

Irvine, W.T., Gondi, J. and Sullivan Mine Geological Staff, 1972 . Major lead–zinc deposits of Western Canada. *24th Int. Geol. Congr. Montreal, Que., Field Excursion* A24- C24 *Guideb.,* 36 pp.

Ishihara, S. (Editor), 1974. Geology of Kurokō deposits. *Min. Geol., Spec. Issue* No. 6, 435 pp.

Janković, S., 1962. Rudna ležista, metalogenetske epohe i metalogenetska prodrucja gvozdija u Jugos-slavii. *Rud. Glas.,* 4: 68–97.

Kato, T., 1937. *Geology of Ore Deposits.* Fuzambo, Tokyo (in Japanese).

Kiilsgaard, T.H., Heyl, A.V. and Brock, M.R., 1963. The Crooked Creek disturbance, southeast Missouri. *U.S. Geol. Surv. Prof. Pap.* 450-E: E14–E19.

Kimberley, M.M., 1978. Paleoenvironmental classification of iron formations. *Econ. Geol.,* 73: 215– 229.

King, H.F., 1976. Stratiform and strata-bound metal concentrations in Australia – summary. *Am. Assoc. Pet. Geol. Mem.,* 25: 426–429.

Kirkham, R.V., 1973. Tectonism, volcanism and copper deposits. In: I.F. Ermanovics (Editor), *Volcanism and Volcanic Rocks. Geol. Surv. Can., Open File Rep.,* 164: 129–151.

Kisvarsanyi, G. and Proctor, P.D., 1967. Trace element content of magnetites and hematites, Southeast Missouri iron metallogenic province, U.S.A. *Econ. Geol.,* 62: 449–471.

Konstantinov, R.M. and Sirotinskaya, S.V., 1974. Logiko-informatsionnye issledovaniya endogennykh rudnykh formatsii i variatsionnye ryady rudnykh mestorozhdenii. In: *Problemy endogennogo rudoobrazovaniya.* Nauka, Moscow, pp. 68–82.

Körner, S., 1974. Classification theory. In: *Encyclopedia Britannica, 4.* 1974 ed., W. Benton. Chicago, Ill., p. 691–694.

Kuthan, M., 1968. Young volcanic rocks of the Carpathians in Slovakia. In: M. Mahel' and T. Buday (Editors), *Regional Geology of Czechoslovakia, Part 2.* Academia, Prague, pp. 628–667.

Lacy, W.C., 1974. Review of the upper Paleozoic metallogenesis in the northern portion of the Tasman Orogenic zone. In: A.K. Denmead, G.W. Tweedale and A.F. Wilson (Editors), *The Tasman Geosyncline – A Symposium.* Geol. Soc. Aust. Brisbane, Qld., pp. 221–246.

Laffitte, P. and Permingeat, F., 1961. Maquette d'une carte métallogenique de France au 2.500.000^e. *Bull. Soc. Géol. Fr., Sér. 7,* 3: 481–485.

Lambert, I.B., 1976. The McArthur zinc–lead–silver deposit: features, metallogenesis and comparisons with some other stratiform ores. In: K.H. Wolf (Editor), *Handbook of Strata-bound and Stratiform Ore Deposits, 6,* Elsevier, Amsterdam, pp. 535–560.

Laznicka, P., 1973. *MANIFILE – the University of Manitoba File of Nonferrous Metal Deposits of the World.* Publ. No. 2, Cent. Precambrian Stud., Univ. Manitoba, Winnipeg, Manitoba, 3 Vols.

Lowell, J.D. and Guilbert, J.M., 1970. Lateral and vertical alteration–mineralization zoning in porphyry ore deposits. *Econ. Geol.,* 65: 373–408.

Magak'yan, I.G., 1974. *Metallogeniya.* Nedra, Moscow, 304 pp.

Magnusson, N.H., 1970. The origin of the iron ores in central Sweden and the history of their alterations. *Sver. Geol. Unders., Ser. C,* No. 643, Part. I, 127 pp.

McKnight, E.T. and Fischer, R.P., 1970. Geology and ore deposits of the Picher Field, Oklahoma and Kansas. *U.S. Geol. Surv., Prof. Pap.* 588.

Mitchell, A.H.G. and Bell, J.D., 1973. Island-arc evolution and related mineral deposits. *J. Geol.,* 81: 381–405.

Mitchell, A.H.G. and Garson, M.S., 1976. Mineralization at plate boundaries. *Miner. Sci. Eng.,* 8(2): 129–169.

Miyahisa, M., 1960. Geological studies on the ore deposits of Obira type in Kyushu. *Jpn. J. Geol. Geophys.,* 32: 39–54 (in Japanese).

Morris, H.T., Heyl, A.V. and Hall, R.B., 1973. Lead. In: *United States Mineral Resources, U.S. Geol. Surv., Prof. Pap.,* 820: 313–332.

Muta, K., 1957. On the genesis of the Sangasho-type "Kieslager" as related to diabases – A geochemical study. *Min. Geol.,* 7: 254–264 (in Japanese).

Nakovnik, N.I., 1968. *Vtorichnye kvartsity S.S.S.R. i svyazannye s nimi mestorozhdeniya poleznykh iskopaemykh.* Nedra, Moscow, 335 pp.

Nicolini, P., 1970. *Gîtologie des concentrations minérales stratiformes.* Gauthier-Villars, Paris, 792 pp.

Oesterling, W.A., 1952. Geologic and economic significance of the Hutson zinc mine, Salem, Kentucky. *Econ. Geol.,* 47: 316–338.

Oftedahl, C., 1959. A theory of exhalative–sedimentary ores. *Geol. Fören, Stockholm, Förh.,* 80(1): 1–19.

Ohle, E.L., 1952. Geology of the Hayden Creek lead mine, Southeast Missouri. *A.I.M.E. Trans.,* 193: 477–483.

Ohle, E.L., 1959. Some considerations in determining the origin of ore deposits of the Mississippi Valley type. *Econ. Geol.,* 54: 769–789.

Parák, T., 1975. Kiruna iron ores are not "intrusive-magmatic ores of the Kiruna type". *Econ. Geol.,* 70: 1242–1258.

Park, Jr., C.F. and MacDiarmid, R.A., 1975. *Ore Deposits.,* Freeman, San Francisco, Calif., 3rd ed., 529 pp.

Petrascheck, W.E., 1965. Typical features of metallogenic provinces. *Econ. Geol.,* 60: 1620–1634.

Pinckney, D.M., 1976. Mineral resources of the Illinois–Kentucky mining district. *U.S. Geol. Surv., Prof. Pap.* 970, 15 pp.

Pretorius, D.A., 1976. Gold in the Proterozoic sediments of South Africa: systems, paradigms, and models. In: K.H. Wolf (Editor), *Handbook of Strata-bound and Stratiform Ore Deposits, 7.* Elsevier, Amsterdam, pp. 1–28.

Quade, H., 1976. Genetic problems and environmental features of volcanosedimentary iron ore deposits of the Lahn–Dill type. In: K.H. Wolf (Editor), *Handbook of Strata-bound and Stratiform Ore Deposits, 7.* Elsevier, Amsterdam, pp. 255–294.

Rackley, R.I., 1976. Origin of Western-states type uranium mineralization. In: K.H. Wolf (Editor), *Handbook of Strata-bound and Stratiform Ore Deposits, 7.* Elsevier, Amsterdam, pp. 89–156.

Raguin, E., 1961. *Géologie des gîtes minéraux.* Masson, Paris, 686 pp.

Ramović, M., 1968. *Principles of Metallogeny.* Edited by Geogr. Inst., Nat. Sci. Fac, Univ. Sarajevo, Sarajevo, 265 pp.

Ridge, J.D., 1968. Changes and developments in concepts of ore genesis-1933 to 1967. In: J.D. Ridge (Editor), *Ore Deposits in the United States 1933–1967, Vol. II,* Am. Inst. Min. Metall. Pet. Eng., New York, N.Y., pp. 1713–1832.

Ridge, J.D., 1976. Origin, development and changes in concepts of syngenetic ore deposits as seen by North American geologists. In: K.H. Wolf (Editor), *Handbook of Strata-bound and Stratiform Ore Deposits, 1,* Elsevier, Amsterdam, pp. 183–297.

Rösler, H.J., Baumann, L., Lange, H., Fondrich, K. and Scheffler, H., 1968. Geosynklinalmagmatismus und submarin–hydrothermale Erzlagerstätten. *23th Int. Geol. Congr., Prague,* 7: 185–196.

Rösler H.J., Baumann, L. and Jung, W., 1968. Postmagmatic mineral deposits of the northern edge of the Bohemian Massif (Erzgebirge–Harz). *Guide to Excursions, 22 AC, 23th Int. Geol. Congr., East Berlin,* pp. 9–13.

Ross, C.S. 1935. Origin of the copper deposits of the Ducktown type in the southern Appalachian region. *U.S. Geol. Surv., Prof. Pap.,* 179, 165 pp.

Routhier, P., 1958. Sur la notion de "types" de gisements métallifères. *Bull. Soc. Géol. Fr., Sér.* 6, 8: 237–243.

Routhier, P., 1963. Les gisements métallifères. Masson, Paris, 2 Vols.

Roy, S., 1976. Ancient manganese deposits. In: K.H. Wolf (Editor), *Handbook of Strata-bound and Stratiform Ore deposits, 7.* Elsevier, Amsterdam, pp. 395–476.

Rusakov, M.P., 1925. Vtorichnye kvartsity i porphyry copper Kazakhskoi Stepi. *Vestn. Geol. Kom.,* No. 3, pp 10–18.

Russell, M.J., 1976. Incipient plate separation and possible related mineralization in lands bordering the North Atlantic. *Geol. Assoc. Can. Spec. Pap.* No. 14, pp. 339–349.

Russell, R.T. and Lewis, B.R., 1965. Gold and copper deposits of the Cobar district. In: J. McAndrew, (Editor), *Geology of Australian Ore Deposits 8th Commonw. Min. Metall. Congr., Melbourne, Vic.,* pp. 411–419.

Ruzicka, V., 1971. Geological comparison between east European and Canadian uranium deposits. *Geol. Surv. Can., Pap.* 70-48, 196 pp.

510

Samama, J.C., 1976. Comparative review of the genesis of the copper–lead sandstone type deposits. In: K.H. Wolf (Editor), *Handbook of Strata-bound and Stratiform Ore Deposits, 6.* Elsevier, Amsterdam, pp. 1–20.

Sangster, D.F., 1974. Geology of Canadian lead and zinc deposits. *Geol. Surv. Can., Pap.* 74-1(Part A): 141–142.

Sawkins, F.J., 1972. Sulfide ore deposits in relation to plate tectonics. *J. Geol.,* 80: 377–397.

Sawkins, F.J., 1976. Massive sulphide deposits in relation to geotectonics. *Geol. Assoc. Can., Spec. Pap.* No. 14, pp. 221–240.

Schneiderhöhn, H., 1955. *Erzlagerstätten.* Fischer, Jena, 3rd ed., 375 pp.

Semenov, A.I., Staritskiy, Yu.G. and Shatalov, Ye.T., 1967. Glavnye tipy metallogenicheskikh provintsii i strukturno–metallogenicheskikh (metallogenicheskikh) zon na territorii S.S.S.R. *Zakonomer. Razmeshcheniya Polezn. Iskop.,* 8: 55–78.

Shcheglov, A.D., 1967. Osnovnye cherty metallogenii zon avtonomnoi aktivizatsii. *Zakonomem. Razmeshcheniya Polezn. Iskop.,* 8: 95–138.

Sierra, J., Ortiz, A. and Burkhalter, J., 1972. The metallogenic map of Spain, 1 : 200,000. *Int. Geol. Congr. 24th, Montreal, Que., Sect. 4,* pp. 110–120.

Skripchenko, N.S., 1972. On lead–zinc ore formation in flyschoid formations. *Int. Geol. Congr., 24th, Montreal, Que., Sect. 4,* pp. 390–399.

Smirnov, V.I., 1972. The relations between syngenetic and epigenetic processes during the formation of stratiform ore deposits in the U.S.S.R. *Int. Geol. Congr. 24th, Montreal, Que., Sect. 4,* pp. 404–410.

Snyder, F.G., 1968. Geology and mineral deposits, midcontinent United States. In: J.D. Ridge (Editor), *Ore Deposits of the United States 1933–1967, Vol. I.* Am. Inst. Min. Metall. Pet. Eng., New York, N.Y., pp. 257–286.

Snyder, F.G. and Gerdemann, P.E., 1968. Geology of the Southeast Missouri district. In: J.D. Ridge (Editor), *Ore Deposits in the United States 1933–1967, Vol. I.* Am. Inst. Metall. Pet. Eng., New York, N.Y., p.p. 326–360.

Snyder, F.G. and Odell, J.W., 1958. Sedimentary breccias in the Southeast Missouri lead district. *Geol. Soc. Am. Bull.,* 69: 899–926.

Stelzner, A. and Bergeat, A., 1904–1906. *Die Erzlagerstätten.* Publisher unknown, Leipzig.

Superceanu, C.I., 1971. The eastern Mediterranean–Iranian Alpine copper–molybdenum belt. *Soc. Min. Geol. Jpn., Spec. Issue,* 3: 393–398.

Tarr, W.A., 1919. The barite deposits of Missouri. *Econ. Geol.,* 14: 46–67.

Thacker, J.L. and Anderson, K.H., 1977. The geologic setting of the Southeast Missouri lead district – Regional geologic history, structure and stratigraphy. *Econ. Geol.,* 72: 339–348.

Thayer, T.P., 1964. Principal features and origin of podiform chromite deposits, and some observations on the Guleman–Soridaǧ district, Turkey. *Econ. Geol.* 59: 1497–1534.

Thurston, W.R., and Hardin, Jr., G.C. 1954. Fluorspar deposits in Western Kentucky, Part 3: Moore Hill Fault System, Crittenden and Livingston Counties. *U.S. Geol. Surv., Bull.,* 1012E: 81–113.

Tihor, L.A. and Crocket, J.H., 1977. Gold distribution in the Kirkland Lake–Larder Lake area, with emphasis, on Kerr Addison-type ore deposits – a progress report. *Geol. Surv. Can. Pap.* 77-1A: 363–369.

Trace, R.D., 1976. Illinois–Kentucky fluorspar district. *U.S. Geol. Surv., Prof. Pap.,* 933: 63–74.

Tremblay, L.P., 1978. Uranium subprovinces and types of uranium deposits in the Precambrian rocks of Saskatchewan. *Geol. Surv. Can., Pap.,* 78-1A: 427–435.

Tuach, J. and Kennedy, M., 1978. The geologic setting of the Ming and other sulfide deposits, Consolidated Rambler Mines, northeast Newfoundland. *Econ. Geol.,* 73: 192–206.

Varentsov, I.M., 1964. *Sedimentary Manganese Ores.* Elsevier, Amsterdam, 119 pp.

Vokes, F.M. and Gale, G.H., 1976. Metallogeny relatable to global tectonics in southern Scandinavia. In: D.F. Strong (Editor), *Metallogeny and plate tectonics. Geol. Assoc. Can., Spec. Pap. No. 14,* pp. 413–441.

Von Bäckström, J.W., 1974. Other uranium deposits. In: *Formation of Uranium Diposits,* Int. At. Energy Agency, Vienna, pp. 605–624.

Von Groddeck, A., 1879. *Die Lehre von den Lagerstätten der Erze.* Publishers unknown, Leipzig.

Walther, H.W. and Zitzmann, A., 1972. On the metallogeny of iron in Europe – Presentation of the International Map of Iron Ore Deposits in Europe, 1 : 2,500,000. *Int. Geol. Congr. 24th, Montreal, Que., Sect. 4,* pp. 121–130.

Wedow, Jr., H., Kiilsgaard, T.H., Heyl, A.V. and Hall, R.B., 1973. Zinc. In: *United States Mineral Resources. U.S. Geol. Surv. Prof. Pap.,* 820: 697–711.

Weller, J.M., Grogan, R.M. and Tippie, F.E., 1952. Geology of the fluorspar deposits of Illinois. *Ill. Geol. Surv., Bull.* 76, 147 pp.

Winslow, A., 1894. Lead and zinc deposits. *Mo. Geol. Surv., Vol.* 7, 763 pp.

Wolf, K.H. (editor), 1976a. *Handbook of Strata-bound and Stratiform Ore Deposits.* Elsevier, Amsterdam, 7 Vols.

Wolf, K.H., 1976b. Conceptual models in geology. In: K.H. Wolf, (Editor), *Handbook of Strata-bound and Stratiform Ore Deposits, 1.* Elsevier, Amsterdam, pp. 11–78.

Wright, W.B., Guild, P.W., Fisch, Jr., G.E. and Sweeney, J.W., 1968. Iron and steel. In: *Mineral Resources of the Appalachian Region. U.S. Geol. Surv., Prof. Pap.,* 580: 396–416.

Zartman, R.E., Brock, M.R., Heyl, A.V. and Thomas, H.H., 1967. K–Ar and Rb–Sr ages of some alkalic intrusive rocks from central and eastern United States. *Am. J. Sci.,* 265(10): 848–870.

Zhelyaskova-Panajotova, M., Petrussenko, Sv. and Iliev, Z., 1972. Tungsten- and bismuth-bearing skarns of the Sangdong type from the region of the Seven Rila Lakes in Bulgaria. *24th Int. Geol. Congr. Montreal, Que., Sect. 4,* pp. 519–522.

Chapter 4

PLATFORM VERSUS MOBILE BELT-TYPE STRATABOUND/STRATIFORM ORE DEPOSITS: SUMMARY, COMPARISONS, MODELS

PETER LAZNICKA

INTRODUCTION

Metallogeny (de Launay, 1892, 1913) was born at a time when the progress in structural geology, stratigraphy, petrography and geological cartography, and the accumulated regional information, made it possible to produce compilation maps of large areas and plot onto them mineral deposits. This clarified the interrelationships and affiliations among ore deposits, as well as their genetic connections with host-rocks, not necessarily apparent during the phase of local field investigations. Out-of-scale symbols for ore deposits were introduced and they kept evolving to reach in many cases a high degree of complexity and sophistication. The oversized symbols proved to be a mixed blessing giving the reader an impression of geographic proximity and considerable density of mineral occurrences which is rarely factual, and this in turn has encouraged formulation of far-reaching hypotheses based on insufficient population of sample deposits. Many such exercises would not have been attempted had the deposit plots been executed on scale.

Metallogeny required a geologically justified framework for arrangement of mineral deposits, a framework that would have regional validity and could be shown on maps. A framework based on a conventional geologic map (stratigraphic—lithologic) soon appeared to be rather insensitive to the peculiarities of ore distribution; Pb—Zn veins, for example, occur in all geologic systems and in a variety of rocks. In the last century and the first half of the present century, the mainstay of ore deposits were hydrothermal veins. Many stratabound deposits were known and mined (Broken Hill, Mississippi Valley districts, Rammelsberg, Rio Tinto), but they were considered to be replacement deposits generated by endogenous hydrothermal fluids. Consequently, they were not thought to contradict the prevalent magmatogene epigenetic ideas.

Most ore deposits demonstrated, or were believed to demonstrate, genetic association with magmatic rocks. These, in turn, were controlled by tectonism. Coupled with the preoccupation with structural control of ore deposits in the 1930's and 1940's and the progress in geotectonics, it was not surprising that the tectonic map has become the universally used base for metallogenic maps; and, therefore, the regional distribution of mineral deposits started to be treated in a tectonic framework. Because a basic tectonic map shows the present structure, while most ore deposits formed in the past, an element of paleotectonic interpretation was necessary and this was provided by a sequence of progressively more complex and sophisticated geotectonic models offering historical synthesis of tectonism, magmatism, sedimentogenesis and metallogenesis. Since the beginning, however, the progress was plagued by numerous contradictions that are so characteristic of geology — a discipline at the boundary between the sciences and arts, and an uneasy symbiosis of evidence and imagination.

It appears that the most durable fundamental division of the dry land into variously named equivalents of mobile belts and platforms followed the widespread acceptance of the Dana's (1873) and Hall's (1883) concepts of geosynclines. The terms "folded belt" and "platform" are to a considerable degree geometric and observational. They are used interchangeably with semigenetic terms that include information on the behaviour of crustal units in the geologic history: "cratons" (Stille, 1936a, b) — a term that stresses the relative stability — and "orogenic belts" — inferring the opposite. Still another family of terms stresses the agent of, in most cases, deformation, e.g. "orogenic belts"; and in addition to it identifies the time period of final deformation: e.g. Hercynian orogen. The term "mobile belt", regardless whether the mobility is a result of depositional or deformational process, is usually attributed to Bucher (1933).

"Platform", in the contemporary international usage, usually means a large part of a continent covered by flat-lying sediments and underlain by a consolidated basement. If the basement age is Precambrian and if it outcrops over a large area, it is called "shield".

The original dual geotectonic division of the dry land into platforms and mobile (folded, orogenic) belts survived unchallenged until at least the early 1960's, and it appears in most currently available large-area tectonic compilation maps (e.g. of the United States, Australia, Canada, U.S.S.R., Europe, etc.; see P.B. King, 1969, pp.7—16). In the 1940's and 1950's, all mobile belts were automatically considered to be of the geosynclinal type. Terrains of intensive intracratonic deformation, sometimes complete with folding and with widespread development of intrusive and volcanic rocks, were added to geosynclinal mobile belts and platforms as the third fundamental geotectonic (and metallogenic) division of the Earth's crust. This was done by the Soviet school only in the 1960's (Nagibina, 1967; Shcheglov, 1967) under the name "zones of autonomous activation". The more recent synonym is "intracontinental (epiplatformic) orogenic belts" (Khain, 1971).

The oceanic domain, another fundamental division of the Earth's crust, was ignored or only cursorily mentioned in most publications discussing global geotectonic divisions until almost the onset of the "New Global Tectonics" in the late 1960's. The reason may be the practice of not showing geology of water covered areas on geologic maps. Following its addition, the oceanic domain has been further subdivided into oceanic platforms (thalassocratons) and intraoceanic mobile belts (Khain, 1971); see Table I.

The "New Global Tectonics" concept in 1968 and the following years rendered the geosyncline/orogenic belt hypothesis obsolete, particularly the genetic interpretation on which it relied. Many older concepts, however, were rehabilitated and rejuvenated; the Wegener's theory of continental drift in particular, as well as the observation of Suess (1883–1909) that there are two distinct types of continental margins, namely: (a) tectonically active, Pacific ones; and (b) tectonically inactive, Atlantic ones; a concept now universally accepted (e.g. Drake and Burk, 1974). Although the geosynclinal model is no longer in use, much can be saved and reinterpreted, as Dickinson (1971) has shown. The adjective "geosynclinal", formerly used for the mobile belts developed mainly along the ocean/continent interface (orthogeosynclines of M. Kay, 1951), was dropped or is being used in a very general sense only to stress a linear depositional trough accumulating greater thickness of supracrustal rocks, more mobile than its surroundings.

The traditional geotectonic subdivisions of the Earth's crust, developed in the days when the geosynclinal hypothesis dominated the thinking, however, have remained in use; but their modern definition is being increasingly more frequently based on the deep subsurface properties, rather than on their surficial expression applied originally. This concept is presently in transition, so that the framework of geotectonic domains used in this chapter may change in the near future.

There are now three second-order of magnitude crustal megadomains: the oceanic megadomain, the active ocean–continent transitional megadomain (mobile belt), and the continental megadomain which includes the continental platform and its inactive transition into the ocean (Table I, Fig.1). The majority of the oceanic and continental megadomains has the character of a platform (oceanic platform or thalassocraton, and continental platform), but both include one or more intervening mobile belts: intraoceanic and intracontinental mobile belts, respectively. Numerous fourth-order subdivisions shown in Fig.1 are transitional.

The active ocean–continent transitional domain requires an active interaction of continental and oceanic crusts, and this is satisfactorily fulfilled along the Pacific-type continental margins. Similar interaction, however, is difficult to demonstrate along the Atlantic-type margins, and the question into which domain the Atlantic-type margins possibly belong, is discussed in greater detail later.

The subdivision of the lithosphere into geotectonic domains (Fig.1, Table I) is not in competition with the plate tectonic hypothesis — it is complementary to it, and is more readily recognizable on a geologic or tectonic map (although with reconnaissance accuracy only), while plate-tectonic categories require interpretation that changes continually with development of our concepts. In this respect, the discussed categories are more permanent. Some of the above units (intraoceanic mobile belts and some intracontinental mobile belts that have developed into narrow oceans) coincide with lithospheric plate boundaries, while others do not. An average large lithospheric plate (e.g., the North American plate) contain all five geotectonic subdivisions. Belts resulting from crustal collisions (particularly continent–continent collisions) as well as transform fault systems,

TABLE I

Development of geotectonic divisions of the lithosphere – selected examples

Suess (1883–1909)	oceans		margins of the Pacific margins of the Atlantic			continents
Stille (1936a,b)	oceans		orthogeosynclines			cratons
Beloussov (1954)	oceans		geosynclines and orogenic belts		parageosynclines (semiplatforms)	platforms
von Bubnoff (1956)	oceans		geosynclines		shelves	blocks
Soviet school (1960's)	oceans		geosynclines and orogenic belts		regions of autonomous activation	platforms
P.B. King (1969)			foldbelts			platform areas
Demenitskaya (1967)	oceanic platforms	intraoceanic orogenic zones	orthogeosynclines		intracontinental mobile belts	continental platforms
Khain (1971)	oceanic platforms (thalassocratons)	intraoceanic orogenic belts	geosynclinal mobile belts; epigeosynclinal orogenic belts		intracontinental (epiplatformic) orogenic belts	continental platform (epeirocraton)
Douglas and Price (1972)	oceans		suboceanic pericratonic mobile belts		taphrogenic areas	continental cratons
Sloss and Speed (1974)	oceans		continental margins; miogeosynclines			cratons
Dickinson (1974) (crustal types)	oceanic crust	quasioceanic crust	orogenic domains	quasicontinental crust	transitional domains	continental crust
N.H. Fisher and Warren (1975)			active continental margin mobile belts			platform domains
This chapter	oceanic megadomain: oceanic platform	intraoceanic mobile belt			continental megadomain: intracontinental mobile belt	continental platform

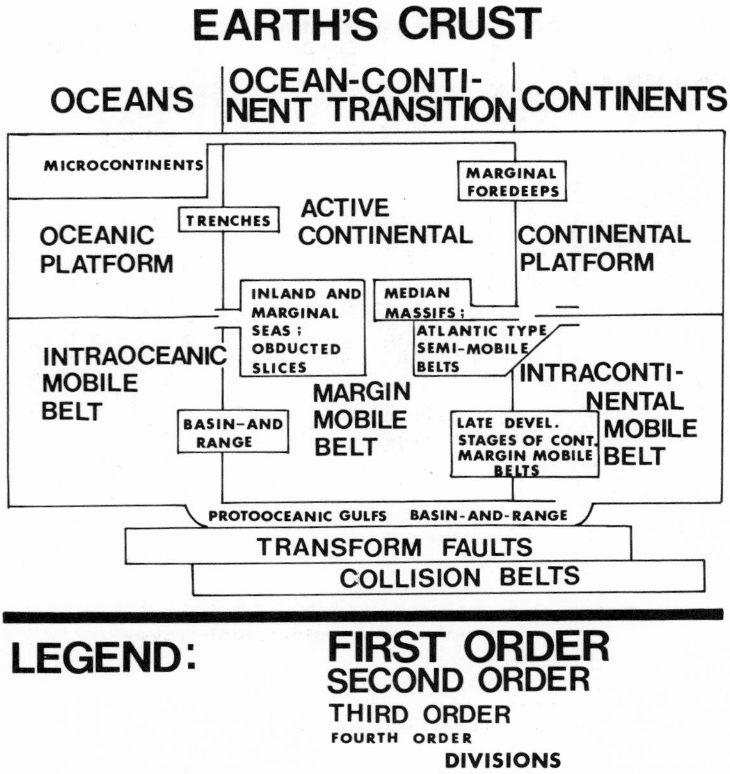

EARTH'S CRUST

| OCEANS | OCEAN-CONTI-NENT TRANSITION | CONTINENTS |

MICROCONTINENTS

OCEANIC PLATFORM

TRENCHES

ACTIVE CONTINENTAL

MARGINAL FOREDEEPS

CONTINENTAL PLATFORM

INTRAOCEANIC MOBILE BELT

INLAND AND MARGINAL SEAS ; OBDUCTED SLICES

MEDIAN MASSIFS ;

ATLANTIC TYPE SEMI-MOBILE BELTS

INTRACONTI-NENTAL MOBILE BELT

BASIN-AND RANGE

MARGIN MOBILE BELT

LATE DEVEL. STAGES OF CONT. MARGIN MOBILE BELTS

PROTOOCEANIC GULFS   BASIN-AND-RANGE

TRANSFORM FAULTS

COLLISION BELTS

LEGEND: **FIRST ORDER**
SECOND ORDER
THIRD ORDER
FOURTH ORDER
DIVISIONS

Fig. 1. The principal geotectonic units of the Earth's lithosphere, as used in this review.

do not readily fit into the above scheme and require a special approach.

The principal classic geotectonic division of the Earth's surface was soon applied in metallogeny (Table II). de Launay (1913), and numerous authors dealing with global or continental distribution of mineral deposits, make frequent reference to the contrasting style of mineralization (or the lack of it) in platforms, compared with mobile belts. The most rigid subdivisions of the global metallogeny into that of platforms (cratons) and mobile belts, presented as a contrast (either-or), appeared in the golden age of the geosynclinal hypothesis in the 1950's and 1960's (Bilibin, 1955; Staritskiy, 1958; Smirnov, 1962) and have had profound influence on our present thought. The awakening that not only the dry land, but the floor of water-covered basins could contain ore deposits, and therefore deserve to be considered in the global metallogenic schemes, has not come until after the appearance of Mero's (1965) book.

The arrival of the plate-tectonic paradigm generated a flow of papers in the early 1970's, relating metallogeny to the plate tectonic environments (Sillitoe, 1972; Mitchell and Bell, 1973; Sawkins, 1972; Strong 1976). Another recent trend of metallogenic

518

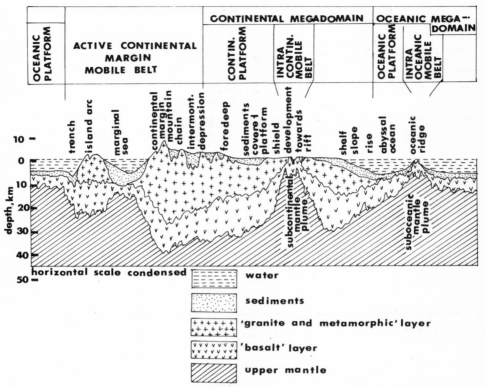

Fig. 2. The principal geotectonic units of the lithosphere, showing the usual interpretation of crustal layers. Plate-tectonic interpretation not shown.

studies follows the progress in oceanography and has, so far, contributed a wealth of information on submarine mineral deposits, many in the process of formation (e.g. Bonatti, 1975; Rona, 1978). Metallogeny, as related to the classic or modernized classic megadomains of the lithosphere, has not been recently very popular except, perhaps, in the Soviet literature (e.g. Smirnov and Kazanski, 1973).

The realization that the Pacific-type continental margins represent a major system of progradational conversion of oceanic crust into continental crust (e.g. Peyve et al., 1977), may cause the improved geotectonic domains of the lithosphere to become a meaningful framework for metallogenic study and description again. It can be demonstrated that many processes of ore deposition are related to the general pattern of progradational crustal development, within a wide range of scales, regardless of the location in the growing and often confusing family of the actualistic examples of plate environments. It

TABLE II

Development of the geotectonic framework for metallogeny – examples

Bilibin (1955)		geosynclinal mobile belts		platform regions
Abdullayev (1961)		geosynclines	platformal geosynclines; platform/geosyncline transitions	platformal zones
Laffitte and Rouveyrol (1965)	oceanic depths	orogenic domains	activation zones	covers
Shcheglov (1967)		geosynclinal mobile belts	zones of autonomous activation	platforms
Pereira and Dixon (1967)		geosynclines and orogenic belts	reworked basement; kratogenic volcanics	platform covers; Precambrian shields
N.H. Fisher and Warren (1975)		orogenic domain	transitional domain	platform domain
This chapter	oceanic megadomain (oceanic platform, intra-oceanic mobile belt)	active continental margin mobile belt	intracontinental mobile belt	continental megadomain (continental platform)

remains to be seen whether the hypothetical retrogradational [1] transitions of the felsic continental crust into the mafic oceanic crust (or its compositional equivalents) do exist and if so, what are their metallogenetic implications. If retrogradational metallogeny does take place, it does so most likely within the domain of intracratonic mobile belts.

Evolution is an important factor bearing on the distribution of geotectonic domains in the distant past (see Chapter 2 by Lambert and Groves, this volume). The geotectonic divisions discussed above are Neogean, actualistic ones, based on the present distribution of environments and on interpretation of ancient systems by analogy with the present systems. While this is very likely representative for the Phanerozoic time, different divisions may have existed in the past. The most recent hypotheses about the very early history of this planet, influenced to a considerable degree by the Moon landings, recognize an early Lunar or giant impacting stage (about 4.2—3.8 b.y.), followed by a nuclear or protocontinental stage of accelerated sialic differentiation (about 4.0—2.8 b.y.); see Pavlovskiy (1975) and Goodwin (1976) (Table III). The actualistic geotectonic division, therefore, may have been invalid in the very early stages and if mineralization formed demonstrably in the Lunar stage were ever identified, it would require treatment outside the framework as outlined in this review. The few examples of a very old crust believed to have formed in the Lunar stage (e.g., in the Aldan Shield, Siberia; Pavlovskiy, 1975), however, occur today incorporated into much younger mobile belts and the sparse mineralization they carry formed in the course of tectonomagmatic consolidation of the belt that encloses them.

Cratons and mobile belts, however, are recognized in the geological literature dealing with most eras, including a part of the Archaean Eon (e.g. Douglas and Price, 1972), although the ancient crust is believed to have been thinner and hotter (e.g. Fyfe, 1977; cf. Chapter 2 by Lambert and Groves, this volume). The deep erosional level and high degree of metamorphism and metasomatism of the oldest terranes, however, makes accurate delineation of both megadomains at present difficult, which leads to a mutual merger of cratons and mobile belts that is only partly due to the possibly different configuration of the Earth's crust in the distant past. Also, the opinion on the character of Archean mobile belts differs significantly and they have been compared with both the present-day Pacific-type margins (e.g. Goodwin, 1974) as well as with taphrogenic systems, possibly close to the present intracratonic mobile belts (e.g. Kröner et al., 1973).

---

[1] Progradational metallogeny — the one which is a part of a process leading towards the development of a more sialic, alkaline, lighter and more stabilized continental crust out of more mafic parents (e.g. Stille, 1936a; Auboin, 1965; Peyve et al., 1977).
Retrogradational metallogeny — the opposite; formation of mineral deposits in the process of continental crust destruction, for example by mafic metasomatism.

TABLE III

Geotectonic divisions of the lithosphere and evolution – probable progressive diversification

Stage of terrestrial development	Geotectonic domains of the lithosphere				
Protogean: Accretion and core generation ±4.6–4.2 b.y.	Universal lithosphere. At the beginning, probably non-differentiated Earth; accretion and segregation of the core, large-scale differentiation				
Lunar stage (giant impacting) ±4.4–3.8 b.y.	Universal lithosphere. At the beginning, probably developed crust (lithosphere), but undifferentiated laterally; meteorite impacting upon a comparatively thin and hot terrestrial crust influenced its differentiation and possibly established long-lived global convection systems				
Nuclear (protocontinental) stage ±4.0–2.8 b.y.	Sialic protocontinental crust produced by accelerated differentiation, operating deep thermal convection systems, first protocontinents (protocratons) formed; sedimentary basins and granitic domes formed. At the end: Protooceans			Protocontinents	
Archean ±3.5–2.4 b.y.	Protooceans		Protogeosynclines and mobile belts	Protoplatforms	
Neogean: Proterozoic to Recent	Oceanic platforms	Intra-oceanic mobile belts	Active continental margin mobile belts	Continental platforms	Intra-continental mobile belts

Based on data from Pavlovskiy (1975) and Goodwin (1976) for the Protogean, and this report for the Neogean.

OCEANIC PLATFORMS AND INTRAOCEANIC MOBILE BELTS

This is the most widespread subdivision of the Earth's crust. The absolute majority of oceanic platforms are covered by sea, most of it deep. Oceanic platforms are unique among the remaining geotectonic domains in having the highest ratio of supracrustal rock producing (mainly submerged) to non-producing (ancient, in the process of degradation) areas, which must be of the order of 10^4-10^5. This has been recently adequately explained by continuing subduction of oceanic lithosphere at consuming plate margins (e.g. Dewey and Bird, 1970), and this accounts for the relatively short lifespan of appearance of an oceanic-floor lithologic association in the Earth's history. Almost all the present oceans and seas floored by oceanic crust are younger than Jurassic, with the exception of the probably late Paleozoic (Burke et al., 1977) or Proterozoic (Volozh et al., 1975) floor under the northern Caspian Sea. Slices of ancient lithologic assemblages interpreted to represent former oceanic crust, and likely emplaced by the process of obduction, occur worldwide in many mobile belts. Rather sizeable portions of a very old oceanic crust, now modified beyond recognition by metamorphism and dismemberment, may be present in many Precambrian shields, e.g. in the Canadian Shield (Emslie, 1972).

The contemporary oceanic domain can be outlined on the basis of presence of a typical oceanic crust in its floor, and continuity. In marginal portions of oceans a transitional crust is developed and terrigenous sediments overlap into the oceanic domain, so the boundary of oceanic and continental crusts is gradational. The ancient stereotype about the flatness and monotony of the ocean floor has been shattered by the recent bathymetric maps and voluminous literature (e.g. Menard, 1964; Kort, 1970). If there is any "typical" part of oceans to be used as a model, it would correspond to what Menard (1964, fig.2.2) calls the "most normal regions", including mainly the abyssal plain.

Perhaps the most striking departure from "normality" within the oceanic megadomain is represented by intraoceanic mobile belts. They have a form of active, e.g. the Mid-Atlantic Ridge, as well as inactive, e.g. Nazca Ridge (Kort, 1970), rift mountain ranges along which spreading and crustal generation takes or used to take place. Most of them are submerged and grade into the more normal ocean floor by mixing of the indigenous oceanic sediments with the ridge-generated rocks. Locally, the ridges are exposed subaerially, most notably in Iceland. Numerous transcurrent faults cut across and usually offset the oceanic ridge and continue into the oceanic platform.

There is a growing literature on mineral deposition and occurrences in the oceans (e.g. Mero, 1965; Bonatti, 1975; Rona, 1978; see also Chapter 3, Vol. 9 of this Handbook). Most of the comprehensive publications, however, offer coverage of the oceans within their present geographic limits, i.e. include their margins underlain by continental or transitional crust; therefore utilizing parts of different geotectonic domains following the grouping used in this paper. At present, the limited commodities actually mined in the oceanic megadomain is recent guano (not yet a geological object but a future possible one), young phosphates developed by reaction between guano and the underlying biogenic limestones (e.g. Nauru Island), and occasional seawater-salt by evaporation. Some commodities will likely be mined in the future: ferromanganese nodules, particularly those with high Ni, Cu and Co contents (e.g., Horn et al., 1973, cf. Chapter 7, Vol. 7 of this Handbook). The majority of metalliferous occurrences described so far are at best showings with several orders of magnitude increased trace-metal content, interesting gene-tically, but far from being orebodies. (See also Chapter 3, Vol. 9.)

The peculiar nature of many oceanic mineral accumulations brings a new dimension to economic geology and ore petrology. Many terms in current use based on continental ore deposits will have to be modified. If ferromanganese nodules are mined, they will make most peculiar orebodies; 5 cm thick, several 10^8 cm long and wide, with gradational boundaries, and slowly replenishing themselves.

Oceanic volcanic associations

The stratigraphy of the igneous floor of ocean basins has been interpreted from seis-mic data (e.g. Ewing, 1969), drilling, coring and sample grabs along mid-ocean ridges (e.g. Bonatti, 1975), and from evidence of possibly correlative ancient sequences preserved on continents (e.g. Peterson et al., 1974).

An ultramafic upper mantle topped by the Mohorovicic's discontinuity (Moho) is overlain by a thick layer of mafic igneous rocks of tholeiitic affiliation with upward diminishing crystal size. The top of the igneous layer is represented by basalts (pillow lavas grading into hyaloclastites) and relatively minor volcaniclastic and volcanichemical sediments. Although mafic volcanic rocks are the major component of the oceanic crust that underlies the entire oceanic megadomain, their abundant outcrops, both subaerial and submarine, are limited to two environments: (1) spreading ridges; and (2) intraplate oceanic islands.

Spreading ridges (intraoceanic mobile belts). Recent studies on midoceanic ridges indicate that the mafic layer rather than being a regular, smooth plate carried away from the spreading centre, is intensely block-faulted, altered, metamorphosed, and interspersed with slices and rafts of serpentinite interpreted as a hydrated mantle. The intensity of metamorphism ranges from zero to amphibolite facies, and is very irregularly distributed (e.g. Aumento et al., 1971).

The mafic rocks, particularly the uppermost basalts in contact with seawater, un-dergo aging, while being carried away from the spreading centre. Cold-water alteration, comparable to diagenesis in sediments, is the major aging agent and it results in con-siderable local increase in alkalinity — particularly in potash content (Keen, 1975). The mafic rocks that become covered by an increasingly thicker pile of younger flows and sediments undergo burial metamorphism. In Iceland, the mineralogy of the amygdule fill in basaltic lavas is a very sensitive indicator of the metamorphic intensity (Walker, 1960).

With continuous aging of the mafic volcanic layer, felsic volcanic and intrusive rocks may locally appear in small quantity. They are probably most widespread and best studied in Iceland, where the majority of occurrences are of Neogene age, but many presently active volcanoes emit rhyolites as well (e.g. Walker, 1966).

The active submarine ridge crests are topped, or their volcanics interlayered, with sediments that are generated almost contemporaneously with the igneous emplacement. These are both chemical sediments (siliceous, calcareous, ferruginous) representing sedimented or accumulated mobile components leached by hydrothermal waters from the volcanics, or are detrital particles of usually strongly altered volcanics that disintegrated and fragmented by the hot magma/seawater interaction during effusion. Both types of components mix in all proportions.

The volcaniclastic products of spreading ridges that outcrop subaerally do not differ appreciably from those of plateau-basalts or Hawaiian-type volcanoes, except for the special class of subglacial emissions in Iceland (Walker and Blake, 1966; Jones, 1970), where hyaloclastites grading to palagonite tuffs with small proportions of volcaniclastic and chemical sediments form through alteration—dissipation and leaching of volcanics.

The association of ultramafic and mafic igneous rocks and related sediments at oceanic spreading ridges is lithologically close to ophiolites of ancient mobile belts now preserved within continents, and this has been frequently pointed out (e.g. Bonatti, 1975). Whether all reported ophiolites were former oceanic mobile belts, is not known. Because of the characteristic lithologic complexity involving more "continental" neighbours and rapid facies changes, many ophiolites have likely developed in "microoceans" within then active continental margin mobile belts, similar to the present Lau Basin (Hawkins, 1974). Regardless of their actual geotectonic site of generation, the ultramafics of ancient ophiolites contain both discordant and stratabound mineable deposits of chromite, platinum, nickel, copper, mercury and other metals.

Baturin and Rozanova (1975) described traces of Ni—Cu sulfide mineralization dredged from the Arabian—Indian Ridge of the Indian Ocean. This may have formed by relatively high-temperature reconstitution and migration of metals in ophiolites, involving ultramafics high in trace nickel.

Lower-temperature hydrothermal leaching and alteration of mafic rocks by the heated seawater and possibly also by "juvenile" fluids at active discharge zones on the ocean floor, causes far-reaching chemical changes. These are most conspicuous in basaltic hyaloclastites and fragmented lavas but occur as well in subsurface mafic and ultramafic rocks (Rona, 1978). Several transient metals in basalts (enriched as indicated by the following factors in relation to the clarke of the Earth's crust: $Fe = 1.5$; $Cu = 2—4$; $Mn = 1.5—2$; $Ti = 2—3$; $Au = 0.5—2$) are rendered mobile and precipitate under favourable conditions. Metal deposition taking place within a sub-floor igneous complex may result in the formation of mostly discordant mineralization, as exemplified by Fe-, Cu-, Ni- and Au-bearing disseminations, stockworks, veins and replacements (Bonatti et al., 1976).

Transient metals that migrate outside the igneous complex could accumulate in the immediately overlying or flanking sediments, or they could meet suitable traps at more distal sites within ocean-floor sediments. The Recent "metalliferous sediments" at and adjacent to the East Pacific Rise (e.g. Boström and Peterson, 1966; Kort, 1970; Dymond et al., 1973) are enriched in Fe, Mn, Cu, Ni, Co, Pb and Ba and are the best-studied examples of stratiform oceanic mineralization in, or close to, the process of formation (cf. also Chapter 7, Vol. 7 of this Handbook). These sediments, however, are no more than $\frac{1}{4}$ of their way to becoming ore deposits. The Cretaceous Cyprus umbers (Robertson and Hudson, 1973) could be a very similar ancient equivalent. Many other stratabound copper deposits (e.g. Betts Cove, Newfoundland; Upadhyay and Strong, 1973), manganese deposits (e.g. Bonatti, 1975), iron, and gold deposits in cherty sediments, spilites, altered volcaniclastics, have many although not all points of similarity with the East Pacific Rise.

Intraplate oceanic volcanic islands. Thousands of underwater, or partly subaerial, volcanoes rise from the ocean floor unrelated to the intraoceanic spreading ridges, but are frequently related to the transcurrent faults that disturb the ridges. They are particularly common in the Pacific. Characteristic features of these volcanoes include: (1) "non-orogenic" setting; (2) stratigraphic position above the spreading-generated mafic oceanic floor, and common piercing of the young ocean-floor sediments by the volcanics; and (3) increased alkalinity and petrographic diversification as compared with the volcanics of spreading ridges.

In places, however, there is no sharp dividing line among volcanics of the above type and volcanics of spreading ridges, and both at present obviously overlap in Iceland (Ward, 1971).

Volcanoes that reach above sea level generate subaerial lavas and pyroclastics strikingly different when compared with submarine lavas, mainly because of their lack of alteration. They considerably influence sedimentogenesis in their surroundings. Coarse volcaniclastic sediments accumulate on flanks of volcanoes and form well-sorted sand covers along beaches. Till, glacial drift and outwash sand plus gravel forms over glaciated islands, as in Iceland. In tropical zones, coral reefs are abundant, fringing the volcanic islands and forming atolls. The lagoonal sediments of atolls, geographically located in the middle of a vast area of basalts and pelagic sediments, represent a relatively rare but interesting microenvironment.

The present oceanic islands are almost devoid of metallic mineralization. Their basaltic lavas and scorias are normally fresh, grey to black or reddish when oxidized, with unfilled vugs. Fumaroles and solfataras along fissures are abundant in craters and on flanks of active volcanoes (e.g. on Kilauea, Hawaii) and they deposit sublimed sulfur on the surface and within argillized wallrocks.

Extensive solfatara fields that can be utilized for geothermal power generation, however, exist only in Iceland, where all the major fields are associated with manifestations of felsic volcanism (e.g. Tomasson, 1976). The volume of argillized volcanics that constitute steam reservoirs, hot geysers and mud volcanoes is considerable in some places. Sublimed sulfur is common, and fine impregnations of marcasite, pyrite, limonite and probably very rare cinnabar and realgar (Namafjall solfatara field, Mývatn area; own observation, 1978) can be seen impregnating thinly laminated gray mudstone formed by dehydration and compaction of mud discharged from mud volcanoes into shallow sedimentation pans. The Icelandic solfatara fields are very close in their surface and subsurface anatomy to those in Japan, New Zealand and elsewhere, from which interesting stratiform to semistratiform occurrences of Hg, Sb, Au and W mineralization have been described (e.g. Weissberg, 1969), although there are differences in geotectonic setting and deep-basement lithology. No metallic occurrences have so far been described from the Icelandic neovolcanics.

Copper, zinc, lead and molybdenum occurrences east of Höfn, southeastern Iceland, described by Janković (1972) as examples of mineralization on a mid-oceanic ridge, are located in the Neogene flood basalts, so that they are not confined to Iceland. Equivalent assemblages are scattered and mineralized through the Brito-Arctic province, for example in Scotland, Ireland and Greenland.

Subaerial pyroclastics exposed on the oceanic island of St. Helena in the Atlantic contain abundant manganese oxide mineralization — as both discordant infiltrations and reworked conformable accumulations (Mitchell-Thome, 1970).

The subaerial and near-shore sedimentogenesis on oceanic islands has all the features of corresponding sedimentogeneses within the same worldwide climatic zone regardless of the geotectonic setting. Pleistocene and older basalts in the tropical belt weather deeply, and aluminium-enriched portions of weathered basalts approaching ferruginous bauxites have been described from Hawaii (Patterson, 1971). Sorting of volcanogenic detritus in high-energy near-shore environments may generate beach sands enriched in magnetite and ilmenite, such as at Hilo, Hawaii and Papeete, Tahiti (Kort, 1970). Biogenic carbonates, particularly coral reefs fringing or capping volcanic islands in the tropical belt, are not known to contain metalliferous showings, although many restricted environments in atoll lagoons seem to be ideal sites to store and accumulate the sometimes strikingly high trace-metal content of certain organisms.

Oceanic sediments

The nature of oceanic sediments has been described in a voluminous literature (e.g. Bramlette, 1961; Berger, 1974). Fine-grained pelagic sediments contain variable proportions of the continent-derived terrigenous component and essentially consist of intrabasinally generated volcanogenic detritus in various stages of authigenic alteration, plus intrabasinal organogenic detritus. Cosmogenic detritus, although detectable, is subordinate. The sediment-lithology is controlled by a climatic zone, currents and bottom morphology.

In the "most normal" basins, where sedimentation rates are extremely slow the top layers of sediments are undergoing early diagenesis during the prolonged contact with seawater. The thickness of sediments varies with the ocean morphology, being thinner on tops and slopes of elevations (condensed sequences), and thicker in depressions. In many places, long-time intervals are represented by a few millimeters of sediments and correspond to diastems, as interpreted in ancient sequences.

Hemipelagic and terrigenous sediments dominate ocean margins and interfinger with truly oceanic sediments, but some turbidite deep-sea fans, submarine valley fills, rafted glacial erratics and wind-blow material may reach quite far into the centrally located "most normal" portions of oceanic basins. The composition of terrigenous components may vary according to the provenance, but as a rule they tend to be rich in quartz, feldspar and mica.

Much of the non-carbonate pelagic and hemipelagic sediments (or the insoluble residuum of carbonates) is slightly to strongly enriched in certain transient elements: Cu, Ba, V, Cr, Zn, Ni, Mo, U as well as pyrite (Cruickshank, 1974), but no striking metal concentrations have been reported so far. The ferromanganese nodules and encrustations dominate the literature (e.g. Menard, 1964; Mero, 1965; Glasby and Read, 1976, Chapter 7, Vol. 7 of this Handbook). The nodules form syndiagenetically by precipitation from seawater in areas of extremely slow sediment deposition, and the major ultimate source of metals lies no doubt mainly in the submarine basaltic volcanics. Several possible near-equivalents, described from ancient sediments (e.g. Bonatti et al., 1976), seem to be close genetically, but may have had more proximal setting to a continent.

Small-scale equivalents of oceanic domain

Numerous basins with a bottom-stratigraphy that corresponds or is close to the oceanic bottom-stratigraphy occur in many active continental margin mobile belts, separated from large oceans by island arcs, ridges with transitional or subcontinental crust, or orogenic belts. Such oceanic-crust-floored basins have been described from central parts of marginal seas (e.g. Lau Basin; Hawkins, 1974), some inland seas (e.g. Black Sea; Ross, 1974), or from basins formerly located along continental margins offset by strike-slip faults (e.g. the lower Tertiary basin west of Sierra Nevada, California; Nilsen and Clarke, 1975).

Small oceanic basins constitute a typical transitional geotectonic category that combines its small-scale oceanic characteristics and a large-scale affiliation with a Pacific- or Mediterranean-type continental margin mobile belt. The lithology and lithogenesis compares to those of oceans, but the proportion of terrigenous sediments along flanks and their diluting effect on bottom sediments of even the most central portions of basins is considerable. The depths are much shallower.

Little is known about the metallogeny of the Recent small oceanic basins, except for several occurrences of iron sulfides. The substantial role of taphrogenesis in their generation; the basinal restriction favourable for development of euxinic sediments such as in the Black Sea (Ross, 1974); and the proximity of island-arc volcanism, suggest considerable potential for stratiform base-metal mineralization possibly comparable with the Red Sea metalliferous muds and brines.

Deformed remnants of marginal and inland seas are probably common in orogenic belts and are believed to host widespread and important mineralization, particularly dominated by Pb, Zn and some Cu massive sulfides of various types (Laznicka, 1976).

Metallogeny of the post-depositional modification stage of oceanic rocks

As soon as the active deposition in an oceanic domain ceases, the rock association together with the contained mineralization may undergo reworking. The most impressive occurrence in the Neogene basalts of SE Iceland mentioned above, Ossura River (Svinhöler), carries scattered chalcopyrite in a rhyolite-bearing breccia pipe hosted by propylitized basalt, and is an example of reworking of the trace copper content of flood basalts. There is a striking similarity of this occurrence to several late Proterozoic showings in the Lake Superior area, for example the Tribag property near Batchawana, Ontario (Blecha, 1965).

Currently it is believed that most of the oceanic lithosphere ultimately disappears down subduction zones, where in the deep parts it undergoes partial melting and the products ascend to form a volcanic belt. It has been proposed that partially melted, metal-enriched pelagic sediments constitute the metal source of ore deposits emplaced in the near-surface level of the volcanic arc (Sillitoe, 1972).

Slices of oceanic crust added or preserved within a continent are subject to a wide range of processes of reworking. Hydration—dehydration during burial metamorphism of a tholeiitic lava pile may be one possible mechanism of generation of native copper deposits in amygdaloids, in a manner similar to the one described by Jolly (1974) from the Keweenaw peninsula, Michigan. Deposits of such type may range from stratabound to discordant. Shearing of basalt with concurrent metamorphism or alteration may rework earlier orebodies or generate new ones such as probably in the Triassic basalt—carbonate association at Kennecott, Alaska (Heiner et al., 1971). Batholithic "granitic" intrusions may have similar mobilizing effect — they commonly generate copper-bearing veins and "skarns".

CONTINENTAL PLATFORMS AND INTRACONTINENTAL MOBILE BELTS

Platform — "a part of a continent which is covered by flat-lying or gently tilted strata, mainly sedimentary, which are underlain at varying depth by a basement of rocks that were consolidated during early deformations" (A.G.I. *Glossary of Geology*, 1972 edition, p.547) — is used, particularly outside of North America, interchangeably with craton — a "part of the Earth's crust which has attained stability and which has been little deformed for prolonged periods" (above-cited *Glossary*, pp.163—164).

In metallogeny, platforms have served for a long time as reliable categories to be contrasted with mobile belts, but such contrast was usually inconsistent. Some authors included, others excluded metallogeny of shields. Some generalized that platforms are practically devoid of mineralization ignoring such giants as Bushveld and the Rand, while others dwelled on these examples at such length that the resulting impression was that these huge but unique and lonely giants are a fairly common and widespread feature of platform metallogeny. The controversy with regards to the metallogeny of platforms usually involved the failure to clearly differentiate between the pre-platformic and platformic stages of lithogenesis and metallogenesis. The first stage was in operation before the portion of the crust in question joined the platform (or became a platform), and the mineralization was inherited by the platform without any casual relation to the subsequent platformic development. Only the second, post-consolidation stage of mineralization has been an integral part of tectogenesis and lithogenesis of the crustal segment while a part of a platform, and bears relation to it.

Even if a clear distinction between the fundamental stages of mineralization had been made, the issue of platformic metallogeny has often been blurred by mixing the outcrop image of the "most normal" parts of a platform (the flat-lying sediments of interior basins) with the image of certain important (but anomalous) examples of mineralizations that do occur at the present erosional level within a craton, yet are genetically related to deep crustal or subcrustal processes (e.g. layered intrusions of the Bushveld-type). The "most normal" portions of platforms were generated by processes that were extremely inefficient mineralizers; hence ore deposits are very rare.

The relative scarcity of mineralization of the "most normal" portions of platforms is moreover magnified by their present geomorphologic expression: plains lacking outcrop. It can be said that until the onset of widespread drilling, most recorded mineralizations in platforms coincided with what is called "interrupted plains" in geomorphology — i.e. plains with locally developed considerable relief, such as in tablelands, plains with spaced hills and plains deeply dissected by river valleys.

In a broadest generalization, it appears that most of the platform mineralization tends to form at, or near, certain margins of geological contrasts of a variety of scales and settings. Many comparable contrasting margins also occur in mobile belts, and this considerably lessens the distinction formerly believed to exist between the metallogeny of platforms and of mobile belts.

Probably the most difficult and subjective exercise is to outline what constitutes the platform domain, at least for the sake of description in this paper. The "most normal" portions of platform, the vast exposure of flat-lying sediments of epicontinental seas now exposed subaerially are the archetypes of platforms. Also, the few presently active sea basins in comparable geotectonic setting that generate similar types of sediments, represent a modern depositional (productive) equivalent of platforms. They are, for example, the Baltic Sea, Hudson's Bay, and Sahul Shelf, among others. Drake and Burk (1974) call this type of depositional environments "interior shelves".

Coastal plains and shallow shelves over submerged portions of platforms at tectonically inactive (Atlantic-type) continental margins are also a natural extension of platforms, with sedimentary lithofacies of corresponding bathymetry much the same as in the epicontinental seas. Sloss and Speed (1974) have suggested inclusion of such shelves into cratons. This relationship was unnecessarily complicated in the geosynclinal/orogenic model, where shelves generally represented an actualistic example of miogeosynclines, e.g. the Atlantic shelf off eastern North America (Badgley, 1965); therefore a part of a geosynclinal mobile belt.

Several Recent depositional environments constitute close lithogenetic and possibly metallogenetic equivalents of platform environments at the upper (= present sea floor), level, but their relationship to large cratons is complicated. The Bahama Bank (a group of flat-topped shallow carbonate platforms surrounded by an ocean) may have an oceanic basement (Dietz et al., 1973), and consequently may actually be a part of the oceanic megadomain. "Quasi-oceanic" platforms in oceans, that are particularly abundant in the southwestern Pacific (e.g. the Campbell Plateau; Katz, 1974), are virtually impossible to fit into any category.

The Atlantic-type continental margin basinward from the stable shelf, is a zone of rapid subsidence and increased mobility. It may locally contain volcanics (Mayhew, 1974) and has a considerable lithologic contrast, particularly where it overlies extensively block-faulted margins, such as along the Atlantic coast of North America (e.g. Renard and Mascle, 1974; Sheridan, 1974). Such a zone then becomes a mixture of oceanic and quasi-oceanic elements (basaltic floor and basinward facies-transition into the oceanic megadomain; disappearance of the "granitic" crustal layer; remnants of intracratonic mobile belt-rifted basement under the sedimentary wedge), and platformic elements (continent-derived sediments; transition into the shelf; presence of common, more stable blocks having a distinct "microplatform" character). Placed into the classic geotectonic model, the continental slope and rise were lately considered as an actualistic example of eugeosynclines (Dietz, 1972). This "eugeosyncline" is, however, substantially different compared with the one along a Pacific-type margin.

It follows that the outer (oceanward) portions of Atlantic-type continental margins display a degree of interaction of the oceanic and continental lithospheres, most of it inherited from the stage of development when the margin site was part of a rift system. It has some features similar to the Pacific-type continental margins, but other features are different. Named "Atlantic-type semi-mobile belts" in this report, it is a transitional category shown in Fig.1.

The boundary of a platform against a Pacific-type continental margin is even more complicated. At most Pacific-type margins, the craton edge formed a basement over which parts of the supracrustal rocks of the future mobile belt were deposited. As a rule the craton edge was subsequently involved in an orogeny that deformed the flanking sediments, and became part of the ensialic portion of the resulting mobile belt. There has never been an agreement regarding the position of the craton—mobile belt boundary at the onset of basin filling, and moreover the boundary kept shifting with geologic time — most dramatically during the orogeny. The boundary of platforms and mobile belts should consequently be considered as a broad three-dimensional transition zone, instead of a sharp line. Foredeep basins, that frequently accumulated a great thickness of sediments, are developed along the outermost (continentward) edge of orogenic mountains, but rest on the depressed craton.

Precambrian shields and consolidated deeply eroded mobile belts are usually considered to be part of a platform (craton) in its present outline. Vast areas of deeply eroded Phanerozoic orogenic systems continuously covered by flat-lying younger platformic sediments, are also mostly considered as platforms (young platforms in the Russian usage — epiplatforms and plates, e.g. the West Siberian Plate). There is, however, a controversy whether deeply eroded former mobile belts, that once in their history became a portion of a stable craton (were "cratonized") but were subsequently (usually very recently) rejuvenated by an uplift and exhumed, are part of platforms or not. Logically, they should be. The central and western European deeply eroded massifs cratonized in the Paleozoic (such as the Massif Central or the Bohemian Massif with their intervening basins filled by younger flat-lying sediments, exemplified by the Paris Basin or the South German Basin) all have features equivalent to Precambrian shields and their platformic cover, except for the age and usually smaller areal extent. The common European practice, however, is that if such terranes are contrasted with the younger, Alpine mobile systems, they are referred to as platforms (or cratons); for example, as in the "Germanic" (platformic) and "Alpine" ("geosynclinal") lithofacies of Mesozoic sediments. If contrasted with the Baltic Shield, the Paleozoic massifs are presented as segments of the Hercynian mobile belt.

We may now be ready to accept examples such as those cited above into the platform family, but the same criteria should apply to once cratonized and subsequently exhumed portions of the crust that show striking elongation and continuity, as for instance the Urals or the Appalachians. They also contain incorporated Precambrian blocks, like the discussed European massifs, although not necessarily centrally located (e.g. blocks of the Precambrian Grenville Province along the western edge of the Appalachians), and young cover-sediments generated at broad coastal plains and epicontinental embayments (e.g. the Gulf of Mexico coastal plain and the Mississippi Embayment).

Both belts mentioned above, however, are prototypes of mobile belts; indeed, the Appalachians were the first "geosyncline" interpreted as such by Dana and Hall, and later used by M. Kay (1951) as type area of "geosynclinal mobile belts", that has achieved worldwide recognition. This concept is still too much alive to ignore the Appalachians as a mobile belt and lump them together with platforms.

Clearly, we are dealing with confusing semantics which is the result of mixing of categories as they are at present (e.g. the Appalachians as part of the present enlarged North American craton) with categories as they were in the geologic past during the actual progressive generation (e.g. the Appalachians during the Paleozoic — a mobile belt system marginal to the then smaller North American craton). The classification of mobile belts in a historical sense, usually by assigning them to an orogeny and/or a geological time of ultimate consolidation, is now a basic technique of construction of modern tectonic maps (see e.g. Chapter 3, Vol. 6 of this Handbook) — but it was not so 20—30 years ago. Some tectonic maps (e.g. the *Tectonic Map of the U.S.S.R.* — Spizharsky, 1966) make even a distinction between "geosynclinal systems" or active mobile belts (= mobile belts still in the process of generation and deformation) and "folded systems" (= past mobile belts in which the process of deformation has been completed). As expected, there is no sharp boundary between the period of activity and inactivity of mobile belts, but despite this, it is imperative that categorization of geotectonic domains and metallogeny be considered in the historical sense. If it is not, or if such consideration is inconsistent, examples and categories become meaningless.

The argument regarding the constituents of platforms so far has considered lithologic domains or environments generated in the course of progressive development; i.e. leading towards the more stable and more felsic unit, the craton. There is, however, also a retrogressive development when portions of a craton are destabilized, initially by vertical tectonics. A trend of increasing tectonic intensity becomes established that ultimately results in the formation of distinctly mobile belts developed within cratons: intracratonic or intracontinental mobile belts. Boundaries of platforms and intracontinental belts are even more gradational than any other boundary, and are discussed in more detail later.

Metallogeny of undisturbed sedimentary sequences of platforms, with minimum of endogenic influence

Stage of active sedimentation. Epicontinental basins of deposition and shelves were until recently considered "atectonic", generated by epeirogenesis — i.e., gentle sagging. It appears that it might not have been so in all cases. Many basinal edges are block-faulted, but the fault morphology may be masked and smoothed by initial continental sediments so that the marine transgression may enter a "prepared" basin. If the basin is really atectonic, however, it is an important restraint on endogenetic interference: volcanism, intrusions, fracture-conducted hydrothermal fluids would be rare to absent.

Basins receive detrital sediments brought in from the surrounding land under erosion, as well as biochemical and chemical sediments produced intrabasinally from colloidal or ionic solutions or from leaching the terrigenous detritus. The ultimate provenance of the dissolved matter is, nevertheless, in the surrounding continent as well. Newly supplied detritus mixes with relic sediment, the latter being a remnant from an earlier environment. Material for lithogenesis unrelated to the above sources is rare, for example: ash-fall from distant volcanic explosions, pumice float, and extraterrestrial matter.

The lithogenic components (rock fragments, minerals, solutions) reflect the composition of the source area and the magnitude of exogenetic reworking superimposed on it. The intensity of this reworking, or departure from the original composition of the detrital source, is influenced by climate, morphology, length of transportation, number of cycles imposed on a particle and may range from almost none (proximal glaciogenic debris) to absolute breakdown (tropical soils).

Advanced exogenetic reworking causes a considerable degree of convergence — i.e. lithological contrasts in the source area, influencing the newly formed sediment, are gradually being obliterated and the resulting sediment becomes almost uniform, under conditions of the same climate and bathymetry.

Following the deposition of the initial basal sediment (often continental), which on crystalline basement is usually of the first cycle, all the following stratigraphically higher sediments receive progressively higher percentage of second and n-cycle detritus recycled from the underlying and laterally outcropping sediments. Additionally, the proportion of intrabasinally generated biogenic and chemogenic components increases. The recycling takes place mainly during erosional intervals in parts of the basins. These are more common than previously thought. Ronov et al. (1970) argued that during the 600 m.y.-long history of the Russian platform at the average three quarters of the total area has been occupied by land, which underwent erosion.

The repeated process of sediment recycling generates thick rock sequences of considerable textural and mineralogical maturity and uniformity, distributed over large areas within the internal parts of basins. The uniform distribution of the main chemical elements is paralleled by a sub-clarke, almost uniform distribution of trace transient metals. The surplus transient metals originally brought in from the continent obviously reside in seawater. In basins with good circulation, they are lost to the world ocean; in basins with restricted circulation they continue to build up in the water and can under certain conditions be withdrawn by chemically contrasting media (e.g. strongly reducing or sulphur-rich). Such media are, however, very rare in the "most normal" portions of basins.

As a result of the extreme lithologic uniformity and "normality" of the interior and non-basal portions of platform basins, ore concentrations are almost entirely absent because every deposit represents an anomaly — a departure from normality. It follows that the "least normal" portions of basins will be more favourable for ore deposition, and this has been confirmed repeatedly by observation. Proximity to a source of the first-cycle detritus, particularly if this detritus is enriched in transient metals or if it comes from terranes with pre-existing ore deposits, and circulation—restriction under suitable climatic conditions, are among the most important factors of metal accumulation.

The sites proximal to the source of first-cycle detritus, little depleted in its content of transient metals, are to be found along basinal margins and, because in most basins lithofacies are diachronous reflecting the transgressive advance as the basin grew, a considerable width of the former basinal margin will be preserved at the base of the sequence. The only possibility that the first-cycle detritus will appear elsewhere, is through a basement uplift, or another sudden and profound disruption of the established depositional pattern, such as continental glaciation. For example, in the northern Bering Sea near Nome (Alaska), a close example of a Recent platform depositional domain, low-grade but

extensive beach and off-shore placer (clastic) gold deposits occur in relict gravels overlying glacial drift and far removed from the bedrock source of gold (see Chapter 5, Vol. 3
of this Handbook). They are the result of episodic advance of Quaternary glaciers beyond
the present shoreline of Seward peninsula (Nelson and Hopkins, 1972).

The circulation—restriction is not confined to basinal margins, although it is there
most common. The result of circulation—restriction in tropical seas is the generation of
carbonate, sulfate and chloride evaporites. Although evaporite salts have normally an
extremely low transient-metal content, they represent a great repository of sulfur which
may become a factor of ore deposition later during diagenesis or metamorphism.

Pustovalov (1965) distinguished four categories of paleoenvironments and/or conditions involved in the lithogenesis and metallogenesis of the Russian Platform in the
Ukraine: (1) shallow waters of the normal, open, epicontinental sea; (2) restricted portions of such seas in lagoons and bays, enriched in evaporites; (3) weathering crusts on
emerged portions of platforms and continental, e.g. river, deposition; and (4) partly subaerial deltas and fans developed within and at the edge of the platform. Paleoenvironments (3) and (4) are transitional into those of shields and intracratonic belts.

The ore deposition in platform sediments and elsewhere depends on numerous factors, some of which are discussed below. Prior to this discussion, however, it has to be
stressed that due to economic considerations, ores are metal accumulations with widely
variable concentration factors. The concentration factor of a 30% iron deposit is about 6,
the factor of an 0.5% copper deposit is about 150, and of a 5% lead deposit about 3000.
The grades used above are recent world averages. It follows that a mildly anomalous process is capable of generating an economic iron deposit; hence iron deposits of depositional
(= syngenetic) — diagenetic origin occur in "almost normal" platform sequences (e.g. the
"minette" of Lorraine and Luxemburg), but an extremely efficient concentration process
is needed to generate lead deposits of mineable size. Because the probability that such a
process will be related to the normal conditions of basinal deposition is very slim, an outside, accidental, "exotic" factor added has been sought by generations of ore geologists
(cf. Chapter 2 by Kimberley, Vol. 9 in this Handbook, on oolitic iron ores).

Factors influencing mineralization in platform basins under ordinary depositional
conditions, are as follows:

(1) Metal availability in erosional source area and area preparation (e.g. suitable stage
of weathering).

(2) Efficient agent of transportation.

(3) A site within the depositional basin with conditions capable of selectively or
preferentially removing and retaining one or more metals being brought into the basin,
thus preventing their dispersal.

These factors are directly proportional. The best chance for ore deposition is when
these factors operate at maximum efficiency. Additional enhancing factors are:

(4) Long-lasting steady-flow condition of the mineralization process.

(5) Repeated multi-cycle progression of events moving in the "right direction" —

towards metal concentration. For example, a gradual shift of the depositional system that results in placing a sediment slightly enriched by a metal during the sedimentation episode 1, into the position of a source in the episode 2. This may cause the product of the episode 2 to become metal-enriched, and so forth.

This mechanism has been treated several times in the literature, for example by Gruner (1956) as a multiple migration—accretion hypothesis for sedimentary uranium; by Killeen and Richardson (1978) as an idea of generation of higher uranium enrichment at intersections of belts of lower uranium enrichment (see models by Wolf, 1976a; and fig.3b, Chapter 2, Vol. 1 of this Handbook), and by Pretorius (1976, Chapters 1 and 2, Vol. 7 of this Handbook) for generation of the Witwatersrand gold. The latter may represent a case of maximum efficiency of interplay of all the factors listed above, rarely achieved in the generation of the remaining platform mineralizations known.

If one fundamental factor is "strong" and others "normal", ore concentration rarely occurs. This seems to be commonplace: a major source of disappointment in exploration and a major source of disillusionment with the literature on ore genesis when concentrating on a single geological factor ignoring the others, with the consequence that few proposed models find application in actual fieldwork.

It appears that factor (3) received the greatest attention in the recent literature dealing with stratabound deposits probably for one reason: because this parameter can be reasonably well reconstructed both in the Recent sedimentary environments as well as during studies of ancient stratigraphic piles.

The following are selected examples of economic mineralizations in typical platform sediments, where the metal is believed to have come from the same source area as the detritus and dissolved matter that form the major basinal sediments, and at approximately the same time as the major rock components:

(1) Unconsolidated heavy-mineral concentrations in river estuaries, beach placers and offshore, such as ilmenite, zircon, rutile, monazite, etc., of Georgia and Florida coasts, Nile delta, etc. (e.g. Neiheisel, 1962; Overstreet et al., 1968; Adams and Staatz, 1973); diamonds of the Namibia coast (Murray et al., 1970); gold off Nome, Alaska (Nelson and Hopkins, 1972).

(2) Consolidated ("fossil") placers in sediments formed mainly in deltaic and beach environments. Ilmenite, rutile, monazite and zircon in Cretaceous and Tertiary sandstones and conglomerates, Wyoming (Adams and Staatz, 1973); detrital cassiterite in metasediments, Gierczyn, Poland (Jaskołski, 1960); gold in Tertiary and Cretaceous conglomerates, Wyoming (Antweiler and Love, 1967), and in Proterozoic conglomerates, Witwatersrand (Pretorius, 1976, Chapters 1 and 2, Vol. 7 of this Handbook); detrital uraninite, brannerite and other minerals in Proterozoic conglomerates, Elliot Lake, Ontario (Roscoe, 1969). The principal controlling factors here were (1) to (5), with factor (1) pre-eminent.

(3) Euxinic muds in Recent epicontinental seas and on stable shelves, enriched in metals but not economic at present, such as certain Norwegian fiords with up to 600 ppm U in sediment (Nicolini, 1970, p.564); Pb, Zn, Co and Ni enrichment on the Atlantic shelf

off Walvis Bay, Namibia (Calvert and Price, 1970).

(4) Possible consolidated equivalents of (3): phosphate- and uranium-bearing Devonian—Mississippian Chattanooga Shale of central U.S.A. (Conant and Swanson, 1961); uraniferous Cretaceous Mancos Shale of Colorado Plateau (Shawe, 1976); Cambro-Ordovician Alum Shale, Sweden (Svenke, 1956); Cu-, Zn-, Pb-, etc., bearing "Kupfer-schiefer" of the European Zechstein [e.g. Deans, 1950 (see also Chapter 7, Vol. 7 of this Handbook)].

(5) Minor occurrences (nodules, lenses) of Mn-carbonates on the present Baltic Sea floor (Manheim, 1964) or in Cretaceous Pierre Shale near Chamberlain, South Dakota (Bodenlos and Thayer, 1973).

(6) Accumulations of trace metals in coal ash, such as germanium in the Pennsylvanian Lower Kittanning Bed in eastern Ohio (Stadnichenko et al., 1953); lead and uranium in Carboniferous coal (Radvaňovice and Bečkov, NE Bohemia); accumulation of iron in pelosiderite concretions and seams ("blackband"), e.g. in the Appalachian Plateaus Province (Wright et al., 1968). The pre-eminent controlling factor has been factor (3), i.e. a reducing environment.

(7) Large bedded deposits of diagenetically fixed iron and manganese oxides sometimes grading into carbonates, developed along basinal edges. Jurassic "minette" iron ores of Lorraine and Luxembourg; Carboniferous iron ores of Shati Valley, Libya (Goudarzi, 1970); Tertiary Mn-ores of Nikopol, Ukraine and Chiaturi, Soviet Georgia (Varentsov and Rakhmanov, 1977), and the Kalahari Basin, South Africa (de Villiers, 1970), are important examples. The majority of bedded iron deposits of the world also show affiliation with this category, although the depositional conditions of some of them, in particular the siliceous banded Proterozoic ores commonly known as of the Lake Superior type, had likely been different when compared with present conditions. Moreover, many types of bedded iron associations show variable degrees of indirect to direct volcanogenic influence. For example, the common type of bedded hematite, often oolithic and usually known as Clinton type (cf. Chapter 2, by Kimberly, Vol. 9 of this Handbook, on oolitic iron ores), may display a wide range of lithofacies and subtypes from sedimentogenic to submarine exhalative, frequently within a single period and single basin (e.g. in the Ordovician of the Barrandian Basin, Bohemia). It appears that factor (1) (= metal source) was pre-eminent among the rest.

(8) Non-residual (redeposited) bedded bauxite, forming lenses and beds in lacustrine and marine sediments (e.g. in the Ukraine; Belevtsev, 1974), or the Gulf Plain of Georgia and Alabama (e.g. Andersonville; Zapp, 1965).

(9) Phosphate accumulations occur in both recent and ancient sediments of: (a) stable shelves with warm currents, located particularly along eastern coasts (Florida and Georgia, U.S.A.; Agulhas Bank near the Indian Ocean coast off southern Africa; Siesser et al., 1974); (b) on stable shelves or in epicontinental seas (e.g. in the Devonian Oriskany Sandstone, eastern United States); and (c) at shelf edges in warm climates, marked by upwelling of nutrient-rich deeper waters (Permian Phosphoria Formation of western

U.S.A., and Cambrian of northern Queensland) (Cathcart and Gulbrandsen, 1973; see Chapter 11, Vol. 7 of this Handbook). Most phosphates contain increased trace-metal content, often recoverable as a by-product, particularly uranium. Factor (3) has been pre-eminent, factors (4) and (5) very important.

Stage of early diagenetic reconstitution

Early diagenesis (syndiagenesis) starts as soon as the sedimentary particles come to rest while still in contact with seawater. The "most normal" diagenesis, however, starts when the sediment becomes separated from circulating seawater by a thickness of overlying sediments and undergoes compaction and lithification (for a recent summary, see Wolf, 1976a).

The system under diagenesis includes sedimentary particles and pore water. The pH—Eh conditions change compared with the period of sedimentation, but the five factors discussed above continue operating. Metals dissolved in seawater may become insoluble under diagenetic milieus and precipitate. The "in situ" precipitation from pore water, however, has a negligible effect on ore deposition so it is important to have channelways through which the solution can move [factor (2)], in a long-lasting steady flow [factor (4)], over favourable sites [factor (3)], where the metals become trapped to form a mineral concentration. The pore fluids are not the only possible source of metals, not even the most important one. Other sources are those sediments enriched in metals (e.g. volcanic ash) already at the time of sedimentation and early diagenesis; detrital minerals and rocks, and possibly some skeletal fragments as well, that suffer diagenetic breakdown associated with the release of metals; metals carried in solutions in groundwater from the drainage basin; and, finally, metals present in soft organic bodies.

The most profound diagenetic changes can be expected in the lithologically most varied and least mature first-cycle sediments that accumulate near the sediment—basement interface, so that the general favourability for metal concentration includes the same sites favourable in the sediment-accumulative stage. The diagenetic alteration of the "most normal" monotonous sediments in the central portions of basins is of little metallogenic significance. Examples of early diagenetic mineralizations, or anomalous metal enrichments in platform sequences are:

(1) The uranium—vanadium deposits in carbonaceous strata of Jurassic Morrison Formation in the Slick Rock district, Colorado Plateau, presumably precipitated from connate water of the Cretaceous Mancos Shale, expelled during compaction (Shawe, 1976; see Wolf, 1976a).

(2) Portion of the peneconcordant U—V mineralization of the "Western States" type that occurs in thin beds of marine sandstone and limestone, for example along the edge of the western stable interior of the United States (Finch, 1967). Most of the "Western States" type of mineralization, however, is in continental sediments in foreland belts and tectonically bound basins, and the period of emplacement of the epigenetic

uranium peaks during the late diagenetic stage (see Chapter 3, Vol. 7 of this Handbook).

(3) Part of the stratabound mineralization of the "copper-sandstone" type, located within gray (reduced) layers interbedded with oxidized layers in the "red-bed" lithologic association, e.g. Dzhezkazgan in Kazakhstan, Samonov and Pozharisky (1977) (see also Chapter 7, Vol. 6 of this Handbook, describing the Kupferschiefer).

Stage of sedimentogenic alteration or reworking of previously consolidated sediments — very-late-diagenesis (epigenesis and katagenesis) and hypergenesis

Late diagenetic and hypergenetic (hypergenesis — a term introduced by Fersman, 1934) changes are due to meteoric waters, in contrast to formational waters available during early diagenesis, that retrogressively alter consolidated sediments. The late diagenetic waters tend to be reducing, whereas the hypergenetic waters are oxidizing, but both may be members of the same contemporary system representing different depth levels.

In contrast to depositional (= syngenetic) and early diagenetic systems, where the process of sedimentation and lithification is rather uniformly distributed, the late diagenesis processes affect only limited portions of the sedimentary pile so that they are strongly dependent on localization: channelways, fractures and zones of porosity. The late-diagenetic waters typically leach relatively mobile components from basinal rocks along their passage and redeposit them later elsewhere when decreased solubility causes crystallization. The most omnipresent late-diagenetic minerals are non-metallic: quartz, calcite, dolomite, gypsum, etc. — accumulated as fracture veinlets, breccia cement or intergranular cement. They usually closely reflect the lithology of the source rock; for example quartz veinlets are most common in quartzites, calcite veinlets in limestones, thus indicating a very proximal source relationship. Less frequent late-diagenetic precipitates are phosphates (e.g. wavellite, variscite; phosphorus leached mainly from fossils); fluorite, barite, celestite, strontianite, etc. Metallic minerals, other than the omnipresent pyrite, occur relatively frequently outside of economic ore deposits, but in very small quantities. Among the more common are chalcopyrite, sphalerite, galena, millerite, linnaeite, bravoite and siegenite. Their abundance as a rule increases in ferruginous and manganiferous sediments (e.g. sphalerite and galena in calcite-filled short fractures in oolithic iron ores of Wabana, Newfoundland; millerite in Ordovician shales, Barrandian Basin of Bohemia; Laznicka, 1965); in reducing lithologies (e.g. in bituminous or coal-bearing sequences), or in both. The crystallized minerals found in septarian cavities of argillaceous siderite concretions in underclays of many coalfields are well known to collectors. All the above occurrences obviously reflect intrasedimentary metal-sources and sedimentogenic agents of extraction and redeposition.

If a sequence, originally enriched in trace metals during sedimentogenesis and early diagenesis, undergoes superimposed late-diagenetic or hypergenetic reworking, the chance of generation of an economic deposit increases. Karst-plumbing systems are among the best known hypergenetic (upper levels) and late diagenetic (lower levels) systems and indeed many economic deposits mainly of zinc and lead in carbonates, attributed to karsting (see Chapter 4, Vol. 3 of this Handbook), also contain regionally associated fine detrital sediments anomalously enriched in Zn and Pb in their pre-late-diagenetic history. An example are the Pennsylvanian shales with scattered galena and sphalerite crystals and nodules in the Tri State district (McKnight and Fischer, 1970).

Late-diagenetic systems, including those that are karst-forming, generally leach metals from the upper levels and deposit them at the lower levels. This is exactly opposite to most endogenetic mineralizations, but because the endogenetic concept had been a leading theory of ore deposition since the 1900's, the reverse cause-and-effect relationship has gained acceptance only very slowly. Many "Mississippi Valley"-type districts in platform carbonates contain regionally distributed shales positioned stratigraphically above the ore-bearing level, that may have been the lead and zinc sources (see Wolf, 1976a, for discussions). Unfortunately, the shales are now eroded at most localities, so confirmation of this hypothesis is rarely possible.

The recent uranium mineralization in calcrete at Yeelirrie, west Australia (H.F. King, 1976) is one of the few examples of exogenetic mineralization where, owing to the hydrogeological conditions of semi-arid regions, the leaching of metals took place underneath the site of deposition (i.e., the uranium was precipitated by one variety of low-T and low-P, non-igneous, hypogene solution).

Subaerial weathering and leaching superimposed on platform sequences which are mildly enriched in metals not only contributed to the possible concentration of the mobile (leachable) metal component, but caused a residual "in situ" enrichment of various immobile metals as well. Examples are the bauxite and laterite deposits developed on Tertiary sediments around Weipa, Queensland (Evans, 1975); the "circle deposits" of residual barite in collapse and sink structures in the SE Missouri mineral district (Winslow, 1894), and the "land-pebble" phosphate deposits of Florida, developed over the primary Miocene low-grade phosphate-bearing carbonates (Cathcart, 1966).

Metallogenesis in platformic sediments due to "exotic" influence

In the above paragraphs it has been assumed that, since their deposition, the basinal sediments behaved as a closed system: nothing has been brought in from the outside. This may have been close to the truth in the "most normal" portions of the basins. Most mineralizations characterized by a high factor of metal concentration occur close to the basinal edge and along or over zones of increased mobility, where the evidence of tectonomagmatic activity is often abundant.

Tectonism, particularly extensional faulting typical for platforms, generates steep vertical gradients for contrasting sedimentation; improves communication of metal-bearing solutions; and generates steeper thermal gradients. These alone can enormously improve the efficiency of purely sedimentogenic metal concentration processes discussed above, but they also open the avenues of access to subcrustal magmas and their "avant-garde" hydrothermal fluids. These can interfere and combine in all proportions with the sedimentogenic fluids, and add an endogenic component to almost every mineral-depositing process. Locations of many important districts bearing conformable mineralization (for example many districts of the "Mississippi Valley" zinc–lead type) coincide with the presence of important structural lineaments in the subsurface (e.g. in the Tri State district, McKnight and Fischer, 1970; in the Pine Point district, Campbell, 1967).

This makes differentiation of "exogenetic" and "endogenetic" metal sources difficult and uncertain (see Wolf's Chapter 1, this volume, for terminological discussions), despite the recent progress in lead-isotope geochemistry (cf. various chapters in several volumes of this Handbook). It appears that many high-factor of concentration mineralizations located within platforms have a mixed (endogenic + exogenic) source of either metals, or ore-forming anions such as sulfur, or both. They are hardly a simple product of essentially equilibrating processes of the "most normal" platform lithogenesis and are transitional into mineralizations related to intracratonic mobile belts.

Metallogenetic consequences of load-metamorphism superimposed on platform sequences

Metallogenetic conditions of load ("anorogenic") or "burial" metamorphism super-imposed on platform sequences have, so far, received less attention than they deserve (see, however, summary by Wolf, 1976a, on influence of compaction to metamorphic processes). This seems to be due to two main reasons: (1) load-metamorphosed platforms are commonly confused with mobile belt assemblages, and (2) the metamorphic overprint is ignored and the rocks are described in terms of pre-metamorphic equivalents. Many large and important districts of Precambrian conformable mineralizations in platforms and intracratonic mobile belts belong here; for example, the Shaba–Zambia Copperbelt (cf. Chapter 6, Vol. 6 of this Handbook); the Kodar–Udokan copper province of Siberia; the copper occurrences in the Belt Supergroup, Montana and Idaho; and the Aitik copper deposit in the Swedish Lappland (Zweifel, 1972).

Most of the mineralizations mentioned above are probable depositional–diagenetic accumulations, isochemically metamorphosed and with many primary fabrics preserved. Alternatively, metamorphism may cause lesser or greater relocation of the ore substance and alter or obliterate the original ore–host-rock relationship. Schistosity-conformable orebodies can be generated from previously non-conformable ones, and vice versa. In the Aitik pit (Sweden), the originally conformable disseminated chalcopyrite is locally remobilized into schlieren, nests and veinlets within and along bodies of anatectic pegmatite, observable at a variety of scales.

In conclusion, continuous metamorphism and metasomatism leads to a large-scale convergence during which contrasts between mineralizations originally conformable and disconformable, originally exogenic and endogenic, originally in sediments, volcanics or intrusive rocks, and originally in platforms or mobile belts, become completely or to a considerable degree either modified or even totally obliterated.

Metallogeny of crystalline shields and massifs, exposed within platforms (cratons)

Most crystalline basement outcrops in cratons are conspicuously well-mineralized, if contrasted with the surrounding, fringing or overlying platform sediments. Most of the basement-mineralization, however, formed in the early development history, when present shields or massifs were a portion of a mobile belt. Such mineralization was simply inherited when the former mobile belt became part of a craton, and subsequent long-lasting erosion levelled and/or removed most of the high-level mineralization. The metallogenic contrast between old shields and young mobile belts is believed by some to be more a function of deeper erosional level, rather than due to a process of crustal diversification that parallels evolution. The inherited, precratonic metallogeny of shields terminates at the shield—platform cover interface and, with rare exceptions, does not overprint into platform sediments.

The pre-cratonic mineralization of shields consists of both stratabound and discordant deposits, the ratio of both being slightly to considerably higher in old shields than in young mobile belts. This is the consequence of the majority of important shield deposits, such as the volcanogenic massive sulfides of the Canadian Shield or the "iron formations", being products of essentially subsiding depositional basins favourable for generation of conformable deposits, which during the same time have a high preservation potential (Laznicka, 1973).

The "cratonic" (or post-orogenic) metallogeny of shields developed after the shields have become part of a stable craton prior, during, and after the deposition of platform cover-sediments. This type of mineralization can, and commonly does, overprint into the overlying platform sediments and the same mineralization process often generated ore-bodies of striking contrast, depending on the ore host-rock. The majority of the cratonic stage mineralizations in shields are discordant veins (e.g., Cobalt, Ontario; Great Bear Lake, Canada; both Ag—Co—Ni veins; Lake Athabasca uranium area, Saskatchewan U veins).

A special transitional class of ore deposits are those located at, or along, the basement/cover unconformity or on the surface of shields with a cover-rock sequence not yet developed. Their systematics is ambiguous because the mineralization may be considered as either a product of the "cratonic" metallogeny of shields, or as an initial member of the platform sedimentation cycle. Suitable examples are residual ore deposits forming by weathering of the exposed basement surface (for example laterites) or accumulations resulting from redeposition of weathered residuum (such as placers).

A brief selection exemplifying the cratonic metallogeny of shields and cratonic massifs would include:

(1) Nickel laterites, sometimes also containing residual chromite, developed over

ultramafics; e.g. in the Voronezh Massif and the Bug area, Ukraine (Belevtsev, 1974).

(2) Zinc (Zn-silicate and smithsonite-bearing) laterites developed over Zn-rich metamorphics, Vazante, Brazil (Carvalho et al., 1962).

(3) Residual manganese oxide deposits developed over manganiferous metamorphics of the Guinea Shield; e.g. in Tambao, Upper Volta, and Nsuta, Ghana (Bodenlos and Thayer, 1973).

(4) Residual bauxites with high Ti, Zr, rare earths and Nb trace contents developed over nepheline syenites as in Saline County, Arkansas (Gordon and Murata, 1952).

(5) Soft, high-grade residual "direct-shipping" iron oxide ores developed over primary siliceous iron formations in most Precambrian shields of the world (e.g. Mesabi Range, Minnesota; Krivoi Rog, Ukraine).

Uranium mineralization is particularly common at or near unconformities separating deeply eroded cratons from overlying detrital sediments (e.g. Tremblay, 1978), and has a variety of forms depending on the nature of the host-rock.

Placer gold as well as platinum, tin and diamonds have been won in most districts that contain the primary mineralization of the respective metals in the basement, except for those primary districts recently scoured by Quaternary glaciers, such as in the Canadian Shield.

Intracratonic mobile belts

In the culmination period of the geosynclinal/mobile belt hypothesis, all mobile belts were automatically considered to have resulted from reversal of a geosyncline during a multiphase geotectonic cycle. Although numerous faults, grabens, rifts, volcanic fields, even folded terrains located within cratons were well-documented in the geological literature, they were considered as local anomalies that depart little from the overall stability of cratons. It was not until the 1960's when the extent and importance of the variety of intracratonic disturbances for metallogenesis was recognized and summarily named "regions of autonomous activation" in the U.S.S.R. (e.g. Shcheglov, 1967).

The original meaning of zones of autonomous activation was that they were large fault systems and downwarps usually filled by volcanogenic and terrigenous, often coal-bearing deposits, superimposed on stabilized cratons. Their history often terminated with folding, and there was commonly a widespread granitic magmatism. The upper and lower limits of zones of autonomous activation as a class were poorly defined, and their contrast with "geosynclinal" mobile belts overemphasized. Any portion of a "geosynclinal" mobile belt involving reactivated cratonic basement would generate similar effects.

In this regard it is very instructive to compare maps of distribution of fluorite deposits in the United States (e.g. fig.12 in Shawe, 1976a) with a map showing the outline of the North American craton at the end of the Precambrian (P.B. King, 1974, fig.29E). A close correlation between both maps is obvious.

Fluorite occurs throughout the entire former cratonic area, regardless of whether its portions were later incorporated into marginal "orthogeosynclinal" belts or not. The deep crustal structure, as well as the presence of extensional tectonics and locally higher heat flow, apparently controls the fluorite as well as some other mineralizations, rather than the presence or absence of folding, so stressed in earlier hypotheses.

Despite its transitional nature and lack of complete uniqueness, tectonomagmatic reactivation of cratons represents a useful concept in metallogeny because it controls, directly or indirectly, all the "exotic", or depth-derived, mineralizations in platforms. The accurate separation of metallogenies of undisturbed and activated platforms is, however, impossible because the activation is a superimposed feature gradually increasing in intensity from zero, and not always accompanied by the development of laterally distinct sedimentary lithofacies.

Intracratonic mobile belts are controlled by deep-reaching faults, their mobility being predominantly vertical, and marked by progressive thinning of the continental crust ("granitic layer") and thickening of the underlying oceanic crust ("basaltic layer"). An anomalous crust develops occasionally. Any single diabase dyke in a continental crust represents an initial stage of crustal extension that may progressively grow to reach an end-product of the category: a fully developed proto-oceanic gulf or narrow ocean, along which lithospheric blocks have been completely separated (Dickinson, 1974).

Compared with the majority of continental margin mobile belts that have passed through most of the successive phases of the classic geotectonic cycle, the majority of the preserved intracratonic belts have been left in various stages of completion and preservation. Those that were successfully completed constitute now the continental basement of Atlantic-type continental margins. Alternatively, if the process of crustal separation was soon reversed, both margins on opposite sides of a former rift may have been forcibly reunited along a collisional suture, and modified beyond recognition.

Recently a sequence of events causing "rifting" and their consequences, have been reviewed in the literature (e.g. Degens and Ross, 1976; and Chapter 4, Vol. 4 of this Handbook). The survey that follows concentrates on selected subdivisions of intracratonic mobile belts and their mineralization.

Domes and arches in the cratonic basement. The activation of continental crust starts with an uplift (doming), followed by incipient rifting. If the process of activation stops following the arching stage, or if the subsequent rifting shifts laterally, the genetic connection between both may not be obvious. It is by no means certain where to draw a line between basement arching as an initiation of a continental breakup, and "normal" arching that is commonplace in both platforms and mobile belts.

Activation-doming is associated with increased heat flow. Rhyolite and rhyodacite in the form of flows, tuffs and plugs plus central volcanoes, as well as basalts, are surface manifestations of activation, as in Yellowknife, Wyoming (e.g. Christiansen and Blank, 1972). Ring-complexes, rounded stocks and dykes or peralkaline rhyolites, quartz porphyries, high-level syenites, aegerine and riebeckite granites, albitites, etc., appear at subvolcanic levels, e.g. in the Aïr Massif, Niger; Jos Plateau, Nigeria; Brandberg and Erongo, Namibia; Oslo Graben, etc. Peralkaline granites, syenites and pegmatites may represent deep levels.

Associated mineralization is abundant, but in most cases it is discordant: veins, stockworks, disseminations and replacements of fluorite, cassiterite and wolframite, columbite—tantalite and other, mainly lithophile metals. Stratiform mineralization of "primary" nature forms rarely: for example the conformable fluorite at the Hammer deposit, Colorado, formed in Tertiary coluvium mineralized by hot springs (Tweto et al., 1970), or fluorite and barite accumulated in lacustrine tuffs at Castel Giuliano near Rome, Italy (Spada, 1969); "Secondary" stratiform mineralization forms by surficial reworking of primary mineralizations of all types, as weathering crusts, or as placers (e.g. the Sn, W, Ta and Nb placers of Nigeria; Williams et al., 1956).

Intracratonic folded (mountain) belts. Intracratonic folded belts are initially related to the pre-rifting-doming in activated zones of platforms. Ensialic portions of continental margin ("orthogeosynclinal") mobile belts may, however, undergo similar development, so that both types of belts may merge and could be difficult to distinguish. Many intracratonic tensional folded belts are furthermore confused with collisional suture belts, but since many of these are a result of cyclical ocean opening and closing during a "Wilson cycle" (Burke et al., 1977), the difference may not be sufficiently significant, and products of extensional lithogenesis and metallogenesis may also be included in belts that terminate by continental collision.

Prominent isolated intracratonic folded belts are relatively uncommon. One of them coincides with the Wyoming and Colorado Rocky Mountains and their southern continuation, located between the main mass of the North American craton and the relatively stable block of the Colorado Plateau. In the early Paleozoic, the region received thin platform sediments, and was repeatedly reactivated during later Paleozoic, Mesozoic and Cenozoic times. Reactivation caused basement uplift into elongated ridges subdivided by troughs. The troughs received thick clastic deposits that were subsequently folded together with the earlier cratonic cover and the basement, and penetrated by many intrusions (P.B. King, 1969). One of the densest mineralized belts of North America, the Colorado Mineral Belt, resulted. The bulk of related mineralization there is associated with high-level intrusive rocks and is non-conformable. Gold-bearing placers have been the most valuable stratiform mineral concentration to date (Parker, 1974).

The Transbaikalia mobile belt of Siberia; the Pampean Ranges of Argentina; the Cordillera Real of Bolivia; and the Great Bear Lake area, Canada (Hoffman and Cecile, 1974), have many features in common with the Colorado Rockies.

Many high-grade metamorphic belts, lacking both a conspicuous "eugeosynclinal" sequence preceding the orogenic deformation and any evidence of collisional interaction of distinct crustal blocks, may be deeply eroded equivalents of intracratonic foldbelts; for example the Proterozoic Grenville Province of the Canadian Shield (Wynne-Edwards, 1972).

Terranes dominated by extensional faults (taphrogenes). Extensional faulting ("rifting") generates contrasting gradients and steep-walled depositional troughs that increase the metallogenic productivity even if nothing more than sedimentogenic differentiation is involved. The system of extensional faults, however, serves frequently as a plumbing system for subcrustal magmas. These form dominantly plateau-type basalt (flood basalt, trap) at the surface level, and dolerite or gabbro dykes, sills and differentiated layered complexes at the subvolcanic and deep levels. The basalt flows often cover immense areas of undisturbed, "normal" platforms; thus in this sense they are an integral part of the platform sequence. Due to their deep origin, however, they are not directly related to the sediments with which they are interbedded. A closer correlation and interrelation of flood basalts and intervening sediments exists, if both formed and interacted in a structurally restricted area, for example a graben.

Plateau basalts are not confined to platforms and occur in young mobile belt systems as well (e.g. Columbia River and Snake River Plateau basalts of the western Cordillera, U.S.A.), but the plateau basalt complexes of platforms are better known and cover larger areas (e.g. Karoo of southern Africa, Cox, 1972; Paraná Basin of Brazil; Deccan Traps of India).

The metallogeny of flood basalts is a paradox. The fresh (gray) basalts are among the most barren host-rocks of ore known on Earth. The altered (green) basalts, particularly if interbedded with sediments, contain locally important copper mineralization of a unique type (such as in northern Michigan), where stratabound to semistratabound orebodies with native copper and chalcocite prefer the interface of quartz—epidote and prehnite—pumpellyite metamorphic assemblages. Many closely equivalent mineralizations are known worldwide, although of much more modest economic importance, not only within shields and platforms (e.g. Coppermine area, Arctic Canada; Kindle, 1972), but within mobile belts as well (e.g. Buena Esperanza Cu deposit near Tocopilla, Chile; Losert, 1973; White River occurrence, Yukon; Cairnes, 1915).

Continental, lagoonal or shallow-marine sediments are often regionally associated with plateau basalts filling grabens. They are dominantly the product of arid environments and the "red-bed" facies is typical. This is almost certainly the result of the high preservation potential of immature sediments and associated volcanics in arid climates. Stratabound copper dominates the metallogeny. In many districts, two distinct facies of sediments and copper mineralization can be distinguished:

(1) A relatively older, more proximal, and coarse-grained facies (conglomerate, sandstone) is often interbedded or coeval with plateau basalts. Native copper, chalcocite, bornite or chalcopyrite usually occur in the cement of conglomerates and sandstones, and are epigenetic in origin. Example: the Hecla—Calumet mineralized conglomerate of Keweenaw peninsula, Michigan (White, 1968).

(2) A younger, more distal, usually marine and fine-grained facies, influenced by chemical sedimentation (shale, bituminous shale, siliceous shale, dolomitic shale), usually younger than the basalts. Low-sulfur copper minerals are believed to be syngenetic or

early diagenetic, and occur dispersed, in thin laminae, or as nodules. Example: None-such Shale of Michigan (Ensign et al., 1968; see Chapter 1 by Brown, Vol. 9 of this Handbook).

Layered differentiated mafic to ultramafic complexes have been in most moderately eroded areas interpreted as subvolcanic to deep-seated equivalents of the plateau basalt volcanism. They filled chambers generated by extensional tectonism and underwent magmatic differentiation that closely approximated the classic model of Bowen. The Bushveld Complex is one of the best-known prototypes and its stratiform magmatogenic mineralization of chromite, platinum metals, copper—nickel sulfides and vanadium-bearing titanomagnetite (Willemse, 1969) is repeated in many similar complexes worldwide; e.g. in the Stillwater Complex, Montana; Burakovo—Aganozero Intrusion, northwest U.S.S.R. (Garbar et al., 1977).

Other complexes show incomplete range of rock differentiates and contain a limited selection of mineralization types, such as Cu—Ni sulfides in basal gabbros and sublayer hybrid breccias (Sudbury, Ontario; Hawley, 1962; Norilsk, Siberia; Glazkovsky et al., 1977); titanomagnetite in anorthosites (Adirondack Mts., New York, eastern Quebec); or layered chromite (Kemi, Finland; Campo Formoso, Brazil; Fiskenaesset, Greenland; Thayer, 1973).

Alkaline magmatic rocks associated with tensional faults. Alkaline magmatic rocks are characteristically developed in platforms, but occur also in some ensialic portions of continental margin mobile belts. In platforms, they are controlled by deep-reaching extensional faults that tend to have a surface expression of horsts and grabens in their youthful stage (e.g. along the East African Rift), but these may be inconspicuous in deeply eroded areas. Small outcrops of many kimberlites and carbonatites are scattered over large territories and their fault-control may be obscure. Numerous classifications of alkaline rock complexes have been presented in the literature (e.g. Sørensen, 1974; Currie, 1976), but only a simplified subdivision is used for the purpose of this review: (a) "alkaline ultramafics" — mostly kimberlites; (b) carbonatites; (c) differentiated complexes of alkaline syenites to gabbros (nepheline syenite dominates); (d) association of olivine to feldspathoid basalts, phonolite, trachyte; and (e) gneissic alkaline rocks — e.g. nepheline syenite gneiss. Association (a) has no surface equivalents. Association (b) has rare equivalents (e.g. the volcano Oldoinyo Lengai, Tanzania; J.B. Dawson, 1966), and association (c) is often in part the deep-seated equivalent of (d).

Alkaline rocks carry distinct mineralization. Association (a) carries diamonds, pyrop, zircon, and sometimes fluorite. Association (b) contains widespread niobium mainly in pyrochlore, and sometimes thorium, uranium, zirconium and rare earths. Chalcopyrite with titanomagnetite and some Ni, U and rare earths are recovered from the Palabora (South Africa) carbonatite.

Association (c) may carry Zr, Ti, U, Th, and rare earths in eudialyte, lujavrite and other rare minerals associated with apatite, and some orebodies are broadly conformable with igneous layering (Khibiny and Lovozero complexes, Kola peninsula). Association (d) is rarely mineralized, except for uneconomic Pb—Zn veins (Roztoky, Bohemia) and more frequent fluorite veins or replacements along contacts. Association (e) contains molybdenite at Mt. Copeland near Revelstoke, British Columbia (Fyles, 1970).

Almost all the mineralization mentioned above is discordant, although some "layered" or "dyke" carbonatites may be fairly conformable with the surrounding rocks and some controversy remains whether such bodies are intrusive or metasomatic (e.g. in Kaiserstuhl, F.R. of Germany; Wimmenauer, 1966; and Yenisei Ridge, Siberia; Zabrodin and Malyshev, 1977).

Stratiform accumulations of metals genetically associated with alkaline complexes may presumably form: (1) contemporaneously with the magmatic activity when the igneous rock itself, or possibly a hydrothermal system that it generates and triggers, reaches the surface; and (2) subsequently through weathering, erosion and sedimentogenic reworking.

Category (1) has few documented examples. Certain soda lakes along the East African Rift may contain material from reworked lava and ash of carbonatite volcanoes (e.g. Deans, 1966; see Chapter 4, Vol. 9 of this Handbook), and common enrichment in fluorite and trace metals. Indications of stratiform uranium mineralization in lake beds coeval with the activity of the Khanneshin (Afghanistan) Quaternary carbonatite volcano have been described by Alkhazov et al. (1977). The lacustrine evaporites, mainly carbonates, of the Eocene Green River Group of Wyoming contain rare Nb, Zr, Ti, rare earths and uranium minerals (Milton and Fahey, 1960). This raises an interesting speculation of carbonatite volcanism as a possible contributor to the Green River sedimentogenesis or alternatively provides a possible point of evidence for the idea of generation of alkaline magmatism through assimilation of evaporites and brines by basaltic magma, suggested by Ayrton (1974).

Category (2) contains more examples of stratiform mineralization, as could be expected: the South African and Siberian diamond placers; Brazilian baddeleyite and zircon placers; Bohemian pyrope and zircon placers. In situ weathering of alkaline magmatic rocks generated Neogene high-titania bentonitic clays in the Ohře Graben, Bohemia; bauxite, overlying the Magnet Cove complex in Arkansas; and other occurrences.

Proto-oceanic gulfs and notes on some additional categories of taphrogenes. The recent literature dealing with metallogeny of intracratonic mobile belts does so usually under various headings some of which are summary headings (e.g. "rifts"), while others imply a distinct position and function in the plate-tectonic geometry.

"Rifts" are used in the broadest sense, including both single rift valleys and proto-oceanic gulfs of Dickinson (1974, p.13), as well as entire rift systems such as the East African "Rift", complete with pre-rift arches.

Proto-oceanic gulfs develop by flooding of rift valleys and adjacent terrane by sea, as in the case of the Red Sea. The development of a rift-valley lake, such as the Dead Sea

(Bender, 1975), appears to be an incipient stage of the same process. Metallogeny of the Red Sea (an archetype of proto-oceanic gulfs) has received sufficient consideration by Degens and Ross in Chapter 4, Vol. 4 of this Handbook, hence there is no need for further treatment. The model of the recent mineralization in the axial zone in the Red Sea has become an enormously attractive tool for genetic interpretation of many stratiform ore deposits worldwide, e.g. in McArthur River and Mt. Isa areas, Queensland (Plumb and Derrick, 1975), and elsewhere.

Aulacogen, a term originally applied by Shatsky to the Donetsk Basin in the U.S.S.R., has recently been reactivated in the Western literature as a term indicating a failed arm of a triple junction of lithospheric plates, that occurs as a transverse trough extending from an "orthogeosyncline" into the adjacent platform (Hoffman et al., 1974). Although aulacogens may have a distinct position in the plate-tectonic geometry, their lithology and metallogeny is fully comparable with any of the various extensional systems discussed above. They contain common plateau basalt flows and their deep-seated equivalents, as well as continental to shallow-marine sediments, and stratabound copper deposits often tend to be abundant and characteristic (e.g. in the Adelaide "Geosyncline", Rutland, 1973; or in the Donetsk Basin). A distinct feature of aulacogens is that they gradually fade away onto the craton and, consequently, may have a pronounced dip towards the ocean marked by the thickening of sedimentary fill. This may ultimately bury the aulacogen and convert it into an ordinary cratonic edge-basin, such as in the Benue— Abakaliki Trough (Hogue, 1977) or in the Donetsk Basin. Proterozoic "sinking reentrants" along the western edge of the North American craton, particularly in the Belt Basin (Harrison et al., 1974), are of similar nature.

Many aulacogens have a long-lasting polyphase development, often terminating with folding, as in the Donetsk Basin (Nagornyi and Nagornyi, 1976) and this suggests transition into intracratonic fold-belts. Many more worldwide-occurring structures have been recently interpreted as aulacogens and the Pb—Zn orebodies seem to be particularly abundant. They are of both discordant vein and replacement types (e.g. the Abakaliki—Zurak belt in the Benue Trough, Nigeria; Mitchell and Garson, 1976), as well as conformable orebodies (Mt. Isa, Queensland; Burke et al., 1977; and Sullivan, British Columbia; Sawkins, 1976).

Basin-and-range terranes

The Basin-and-Range Province of the western United States and northern Mexico is a highly mineralized and classical area of metallogeny. It has been interpreted recently as a result of multiple subparallel deep-seated crustal extension of the western margin of North America, initiated as a consequence of this continent's overriding the East Pacific oceanic ridge (e.g. McKee, 1971). It is a controversial area for several reasons. One of them is the fact that the Basin-and-Range Province appears to be part of the Cordilleran mobile belt, and thus it might seem that any discussion in the context of platform metallogeny may be out of order. A considerable part of its basement had been, however, an integral part of the North American platform, at least till the late Paleozoic (P.B. King, 1969), as the remarkable continuity of Paleozoic sedimentary facies indicates. The basin-and-range tectonic regime, however, is relatively recent, so that only the Neogene mineralization is representative of the basin-and-range metallogeny.

The Neogene mineralization is mostly related to the activity of hot springs, recent shallow hydrothermal systems, and pore brines of interior salt (playa) lakes, that mutually interfere. Many salt brines have a high content and large resources of recoverable trace metals, such as tungsten (Searles Lake, California); lithium (Silver Peak, Nevada), and a major boron content (Boron and Searles Lake, California; Hobbs and Elliott, 1973, and Norton, 1973). These brines are metal carriers — not actual orebodies in the established sense — and the geological features of the possibly resulting orebodies (or an anomalous metal-enrichment) will depend on the nature of the brine trap. Stratabound orebodies form when a metal-bearing brine is discharged into an ephemeral lake in which it crystallizes as evaporite (e.g. Boron, California), or in which the trace metal precipitates by adsorption as in the case of the Golconda, Nevada, stratabound exotic deposit that contains up to 2.78% WO_3 in a manganese oxide layer underlying Quaternary travertine (Kerr, 1940).

ACTIVE CONTINENTAL MARGIN MOBILE BELTS

A typical mobile belt of today should be located along a continental margin, embracing both a portion of the continental edge, and a portion of the oceanic edge. It should be reasonably active — i.e. demonstrate seismic, volcanic and sedimentogenic activity that can be interpreted in terms of continuing interaction of both neighbouring megadomains. The mainstay of active continental margins, i.e. the Pacific-type margin, should be situated along and above a Benioff seismic zone for at least a reasonable length within its total extent, and should receive and process a reasonable proportion of newly (originally from the mantle) added materials.

It should have an uninterrupted linear continuation of thousands of kilometers, rugged relief, and ideally be flanked by Mesozoic and Cenozoic folded (mountain) systems in approximately parallel orientation. It should, finally, contain relatively young mineralization that can still be related to the present or nearly past regimen. Such a belt would stand close to the former eugeosyncline. In the "New Global Tectonics" model it has been broken down into several fourth-order of magnitude subenvironments that are extensively listed and described in the recent literature (e.g. Dickinson, 1971; R.L. Fisher, 1974; Seely et al., 1974).

The mobile belts of the Pacific type are sometimes subdivided into a part undergoing predominantly supracrustal deposition ("depositional domain"), and a portion undergoing predominantly erosion ("erosional domain"). Volcanism is common to both.

The first part is mainly marine, the second mainly continental. Despite the dominance of either regime, exceptions exist in both divisions. The most notable exception in the predominantly depositional domain are the subaerial tops of island arcs undergoing erosion, whereas intermontane basins represent the most obvious depositional environment within the predominantly erosional domain.

In the classic geotectonic model, these differences in style were treated sequentially. The "depositional domain" was usually assigned to the "geosynclinal stage", the "erosional domain" to the "orogenic" and "late" stages. When this was done in a static manner and considered representative for an entire contemporary "geosynclinal" belt, a telescoping of events took place and the correct time—site perspective was distorted. On the other hand, it is true that no matter how inaccurate the sequential interpretation in the classic model may have been, it recorded at least in a general way the common and repeated interactions of earlier-formed supracrustal rocks with subsequently emplaced depth-generated rocks, that no doubt served as a key to a sensible metallogenic interpretation. These interactions were to a considerable degree bypassed in the early stages of the development of the "New Global Tectonics", and more research is clearly needed.

It is not recommended, for reasons already discussed in the chapter on platforms, to add the aseismic or Atlantic-type continental margins to typical mobile belts, because of the lack of interaction between them and the underlying lithosphere, which they merely overlie and over which they prograde. What seems obvious in the present environments, however, is not always clear when ancient sequences are interpreted. Far-reaching convergence affects fourth-order of magnitude lithogenetic environments and corresponding facies distribution, and many local basins will generate an equal lithology essentially influenced by bathymetry and climatic zone regardless of whether such a basin had been a part of a Pacific- or an Atlantic-type margin. Particularly hard to interpret are those local portions of Pacific-type margins that do not face an active Benioff zone; for example shelves and slopes adjacent to a transform fault margin, or shelves and slopes that face marginal seas ("marginal coasts" of Inman and Nordstrom, 1971). Also, shelves and slopes of a former Pacific-type margin, that due to the shift in position of an oceanic spreading ridge have been placed into a mid-plate position, have a newly acquired character of an Atlantic-type margin (e.g. the Pacific side of Baja California).

Many Atlantic-type continental margins, on the other hand, ultimately merge with Pacific-type margins when the configuration of lithospheric plates, and especially the location and dip of the Benioff zones, changes.

Numerous additional points of view proposed to differentiate among mobile belts include the useful concept of ensimatic belts generated within or over simatic (oceanic) crust, and ensialic belts generated within a previously formed continental lithosphere. Most subenvironments that result in the origin of Pacific-type mobile belts are, however, somewhere between these two extremes and the overall igneous rock basicity of broad mobile belts decreases towards the continent, paralleling the direction of dip of a Benioff zone. The overall basicity/acidity of a mobile belt or its portion, e.g. an island arc, may be taken as a measure of "maturity" of the unit (i.e. more mature — approaching the average composition of continental crust; Laznicka, 1973). Semenov et al. (1967) distinguished several types of "geochemical metallogenic zones" in the U.S.S.R. on the basis of overall basicity/alkalinity of mobile belts, which in turn influence the predominant types of ore deposits — particularly the different metal associations.

Long inactive mobile belts are no more related to forces and processes that built them and either undergo peneplanation and gradually become parts of older platforms, or undergo reworking: in part or in full, immediately after completion or at any time later, within a superimposed mobile belt. Fossil mobile belts that were once cratonized during their history and were subsequently exhumed without an appreciable tectonomagmatic activation are technically no more mobile than the remaining portions of platforms despite the fact that some of the best studied examples (the Appalachians, European Hercynides, The Ural, Tien Shan, Tasman "Geosyncline", and others) belong to this category.

Lithogenesis and metallogenesis in mobile belts

Active and recently active continental margin mobile belts outcropping on dry land, are without doubt the most heavily mineralized segments of the Earth's crust. The abundance of *known* mineral deposits is to a considerable degree influenced by the mostly superior outcrop situation in mountains.

Mobile belts are dynamic and display a rapid change in facies; rapid variation over small areas of rock groupings of various ages, various derivations, various levels of emplacement, various grades of metamorphism, various structural styles and various metallogenesis. Only part of this variety is due to basinal conditions at the time of the original sedimentogenesis and volcanism. In belts that have passed through the "orogenic stage", the effect of structural dismemberment, intrusions and metamorphism at deeper levels intervenes — often to the point of obliteration of the depositional configuration. Both depositional and orogenic regimes are mutually dependent: the depositional conditions are influenced by the initial movements that later develop into a full-scale orogeny. The effects of the "orogenic stage", particularly in the case of the igneous rocks, are at least at the near-surface levels considerably influenced by country rocks plus the physico-chemical milieus they intruded. Most metallic deposits are the result of interaction of predominantly laterally differentiated facies resulting from supracrustal deposition, and vertically differentiated facies in "orogenic" systems, governed by the downward increase of temperature, pressure and decrease of available water. Any metallogenetic study considering only one of the two fundamental crust-forming systems (for example metallogenesis as related to surficial depositional environments only) is partial and incomplete in investigating mobile belts.

Most progressive (i.e. leading towards a more "sialic" and alkaline, particularly potassic, crust) crust-forming processes and sequences show an essentially similar geochemical balance in the course of their development in both major and minor elements; and metallogeny is clearly tied to this progression.

Whether there is a retrogressive (i.e. leading towards a more mafic crust) crust-forming process other than a mechanical one (fracturing and distention), is debatable. The concept of essentially metasomatic basification (oceanization) that occasionally surfaces (e.g. Beloussov, 1969) has reached its lowest status in the time of appearance of the plate-tectonic paradigm, although it appears that it is again returning as a possible alternative (e.g. Katz, 1974). This author believes that partial basic metasomatism controls certain contrasting (mafic/felsic) lithologic associations and is an alternative hypothesis to the deep-Earth's interior degassing in explaining the enrichment in the "granitophile" suite of elements, such as Pb, Sn, F, Be, Li and U, in terranes marked by distention.

The progressive crust-generating process is marked by the increase in silica, alkalies and alumina and decrease in magnesium, iron and calcium. Among the minor elements, chromium, nickel and cobalt sharply decrease from their maximum at the ultramafic end of the sequence; copper reaches its maximum in basalts and gabbros and decreases both upwards and downwards; lead and tin increase upwards. The progressive crust-forming process can be accomplished by several mechanisms, including magmatic differentiation from an initial melt; partial melting of an initial solid or semi-solid; metamorphism and metasomatism; and sedimentogenic differentiation, and there is a remarkable convergence of these processes and their combination leading towards a common end-product.

The sequence of substantial minor-metal accumulations, in both economic ore deposits and anomalous trace-metal enrichments, essentially follows the general progressive trend in such a way that chromite and nickel deposits are to be found within or close to ultramafics; copper deposits in genetic affiliation (not necessarily direct) with mafics, and tin in genetic affiliation with felsics.

But petrogenesis of even a compositionally equivalent rock (such as granite) has no doubt a bearing on metallogeny influencing not only the mineralization type, but also whether or not a sufficient trace-metal accumulation capable to generate an ore deposit is possible in the given petrogenetic system. For example, an ultramafic rock generated as a residuum after partial melting of a basaltic solid will not possibly accumulate a sufficient amount of chromium to form chromite deposits, in contrast to an ultramafic rocks formed by magmatic differentiation of an undepleted, primitive (pyrolite?, lherzolite?) melt in a large reservoir. Depositional lead accumulation can hardly result from magmatic differentiation or partial melting of a primitive ultramafic or mafic magma, because the initial trace-lead content is insufficient to satisfy even the normal trace-lead requirement of the more felsic end-products. It, however, could likely occur during a partial or full granitization of a suitable, lead-enriched argillaceous sediment or arkose, where significant amounts of trace-lead in excess of the trace-lead requirement of the resulting granite, is set free.

Some transient metals are highly selective with respect to the petrogenesis of their mineralization parents (e.g. Nb, Cr), while others, like gold, appear less selective. This is reflected in the variety of mineralization types and a number of cycles (Laznicka, 1970) in which the accumulation of a given metal may take place within one geotectonic cycle.

The sequence of elemental and trace-metal variations, as well as ore deposits of specific metals and repetitive ore types, is well expressed across the strike of most continental margin mobile belts. The progression is generally from the mafic (oceanic) to felsic (continental) assemblage in the direction from the ocean edge towards the continental interior, but there are common local reversals essentially due to agglomeration of numerous former "micromargins" or mafic crust invasions.

The usual progression of zones of metallic enrichment and characteristic mineralization in continental margin mobile belts is: Fe; Fe, Cu—Au, Au; Cu; Cu—Zn; Cu—Mo; Zn, Zn—Pb, W (scheelite); Pb, Pb—Sn, Sn—W (wolframite), Mo; Sn. This is representative for at least the "typical portions" of the Cordilleran and Andean active belts, and Appalachian, Tasman, Uralian, etc., cratonized but recently exhumed former mobile belts (Laznicka, 1970; Gabelman, 1976, Chapter 3, Vol. 4 of this Handbook). Chromium and nickel do not seem to have a fixed position in the common sequence, despite their distinct petrogenetic affiliation. This is due to the failure of their ultramafic associates to form a broad, outcropping petrogenetic belt of consistent setting within a continental margin. In an average mobile belt, ultramafics occur along deep-reaching lineaments that in detail do not correlate with progression of the more orderly supracrustal rocks lithofacies.

The emergence of the "New Global Tectonics" focussed interest on a definable environment, and the final event of mineralization in the plate-tectonic interpretation of metallogeny of mobile belts. Several excellent papers have been written along these lines (e.g. Sawkins, 1972; Mitchell and Bell, 1973; Mitchell and Garson, 1976), but it appears that at the present state of knowledge only a minority of ore deposits can be safely assigned to the plate-tectonic environment in which they originated. Although the degree of uncertainty is high, an arrangement stressing the present-day lithologic associations hosting ore deposits have been selected as a framework for the following short review of stratabound metallogeny of mobile belts.

Mobile belts with prominent ultrabasics. It is the prevalent opinion that the ultimate origin of ultramafics is in the upper mantle, so their presence in mobile belts implies a thin lithosphere, deep-reaching faults, or obduction mechanism. The present outcrops of ultramafics in the continental crust are exposed at different levels of erosion, so compositionally identical complexes may have either an intrusive or an extrusive expression. Also, the deep-reaching fault control makes it impossible to accurately separate ultramafics of cratons and of mobile belts, because many occur along boundary faults between both. In this sense, the category treated here is transitional to the one already discussed within intracratonic mobile belts.

The following varieties and/or associations involving ultramafics are most distinct metallogenically:

(1) Lavas, marked by spinifex textures.

(2) Bodies of mainly serpentinized peridotite associated with submarine basalts, gabbros, diabases, products of sodic metasomatism (spilites, keratophyres, carbonatized rocks) and cherts.

The lithologies listed in (2) correspond to an ophiolite assemblage (Dewey and Bird, 1970) and are part of the classic "Steinmann's Trinity".

(3) Serpentinites of the "Alpine type", injected in solid or semi-solid state along faults.

(4) Small differentiated and often layered complexes of peridotite, pyroxenite and gabbro that are frequently folded or tilted. Those in Phanerozoic orogenic belts tend to be approximately circular and designated as of the Alaska (Taylor, 1967) or Ural-types. Lithologically similar complexes in Precambrian orogenic belts often have the character of sills, e.g. the Bird River Sill in Manitoba (Davies et al., 1962).

Massive to disseminated cumulus chromite conformable with layering is frequently present in type (4) (Bird River Sill; Kempirsai Massif, U.S.S.R.; Pavlov and Grigor'eva, 1977). Disseminated stratiform as well as non-stratiform, platinum and rare osmium—iridium occur in clinopyroxenite—dunite association, corresponding to type (4), in central Ural (Razin, 1977), Goodnews Bay, Alaska (Mertie, 1976), and elsewhere.

All four types of ultramafic associations are enriched in trace nickel as expected, but stratabound economic nickel mineralization has various forms. In the simplest form, nickel occurs as an elevated trace content present in solid solution in silicate and oxide minerals, e.g. in the Dumont ultrabasic body, Quebec. This nickel is, however, uneconomic to recover if under 1% Ni (Duke, 1977). Comparable low-grade disseminated mineralization but in the form of nickel sulfide (pentlandite), could be economic and is represented by the "sill-type" orebodies in Western Australia, e.g. Mt. Keith (Eckstrand, 1974, cf. Chapter 6 by Groves and Hudson, Vol. 9 of this Handbook). The majority of pyrrhotite—pentlandite—chalcopyrite orebodies associated with ultramafics occur along margins of peridotite and pyroxenite (e.g. the majority of orebodies in the Yilgarn Nickel Province of Western Australia; Chapter 3 in Knight, 1975, and Chapter 6 by Groves and Hudson, Vol. 9 of this Handbook; and Pechenga district, U.S.S.R.; Glazkovsky et al., 1977), or even outside the ultramafic body itself, usually in high-grade metamorphics, such as in the Thompson belt, Manitoba (Zurbrigg, 1963).

A variety of genetic explanations on the origin of hypogene nickel mineralization in ultrabasics have been proposed, ranging from late crystallization (or deuteric stage of the ultramafic magma) to metamorphogenic remobilization and metasomatic sulfurization (e.g. Naldrett and Cabri, 1976), but all authors agree that the ultramafics were the source of nickel.

Nickel hydrosilicate-bearing laterites are products of tropical weathering (cf. Chapter 3, Vol. 3 of this Handbook) of ultramafics located in all types of geotectonic settings, that includes mobile belts (Cuba, New Caledonia, Colombian Andes, etc.). Cobalt-bearing (asbolite) manganese laterites of New Caledonia have a similar origin.

Placers derived from ultramafics yield platinum metals (Nizhny Tagil, Nevyansk, Syssertsk in the Urals; Goodnews Bay, Alaska), chromite (Oregon coast; Griggs, 1945), and magnetite.

Conformable disseminations of bornite and chalcopyrite locally occur in the gabbro phase of type (4) complexes, such as at Volkovo deposit in the Urals (Samonov and Pozharisky, 1977).

The primitive mantle and komatiitic association is known to be enriched in gold (Saager, 1973), and this may be an important source of the very extensive gold mineralization in the volcanic-sedimentary ("greenstone") belts of Precambrian shields. Many of the gold deposits in "greenstone belts" show close association with altered ultramafics controlled by fault lineaments (e.g. Timmins; Pyke, 1975, Kirkland Lake—Larder Lake, Longlac and other belts, Ontario; Barberton Mountain Land, South Africa; Viljoen et al., 1970). Both conformable and discordant gold-bearing orebodies occur; the Kerr Addison mine in the Larder Lake Belt exemplifying the conformable type (Ridler, 1970).

Belts with prominent basalts and associated marine sediments. Lithologic association of basalt flows, hyaloclastites and volcaniclastics (mostly "graywacke"), argillite, and chert ± limestone constitute the archetype of the former "Eugeosynclinal Furrow" (Auboin, 1965) fill. It sometimes partly correlates with, or grades into, ophiolites. In inactive belts, this association is commonly metamorphosed, the "most normal" being the greenschist facies. The basalts and basaltic andesites are usually oceanic to island-arc tholeiites, but Hawaiian-type basalts may be present. Intensive pervasive alteration (albitization, silicification, carbonatization) that often causes virtual dissipation and retexturing of volcanics to form metasomatic pseudovolcanics (certain keratophyres) or sediments (limestones and ankeritites, layered albitites, cherts, jaspers, iron formations and bedded sulfides), is very characteristic, and this alteration is an event that necessarily mobilizes and sets free the original trace-metal content of the basalts to possibly form depositional concentrations (e.g. Oelsner, 1960; Smirnov, 1968a).

Although many submarine-volcanic alteration systems are generated at the time of hot-lava emplacement and its reaction with seawater, similar alteration can be the result of diagenetic (post-cooling) interaction with pore water, or greenschist facies burial-metamorphism. This has an important metallogenic implication, because it vastly increases the number of possible events and sites of ore deposition. In addition to a metal-supply resulting from submarine autometasomatism of volcanics, "exhalations" are also believed to provide metal for submarine metallogenesis.

The question as to why some submarine volcanics are profoundly altered and others are not, and why some are richly mineralized whereas others are completely barren, has so far not been satisfactorily explained.

Metallic mineralization occurs within both the submarine volcanics and associated sediments. The mineralization in mafic volcanics is dominated by pyrite and copper sulfides, and ranges from stratiform (usually lens-like) orebodies located in pillow lavas or hyaloclastites, or on top of pillow lavas (e.g. Troodos Complex, Cyprus; Hutchinson, 1973; Løkken, Norway; Vokes and Gale, 1976) to loosely stratabound or distinctly discordant disseminations, veins and stringers (e.g. Springdale peninsula, Newfoundland; Smitheringale and Peters, 1974). Presently similar mineralizations are usually interpreted as "syngenetic" (= syndepositional) in origin. Chalcopyrite, however, is almost always paragenetically younger than pyrite and while pyrite masses tend to conform with the depositional fabric, the advancing chalcopyrite shows in many cases conformity with the superimposed deformation fabric, therefore suggesting that some orebodies may have had a multistage genesis.

Stratabound mineralization in sediments adjacent to submarine volcanics includes siliceous and carbonatic iron formations (e.g. the Lahn—Dill type of Germany; Schneiderhöhn, 1955 and Chapter 6, Vol. 7 of this Handbook) and bedded manganese carbonates or oxides in cherts and jaspers (Olympic peninsula, Washington; Park, 1946; and Čevljanovići, Bosnia).

Submarine keratophyres and fine-grained solid volcaniclastics associated with limestones, black shales, and cherts, contain stratabound lenses of pyrite and barite with galena and sphalerite in Meggen, F.R. of Germany (Ehrenberg et al., 1954; and Chapter 9 by Krebs, Vol. 9 of this Handbook), as well as in Mojkovac, Montenegro (Brskovo mine); Horní Benešov, Moravia; and elsewhere.

Fe, Mn and Zn mineralization with abundant barite of the Atasu type at Zhairem and elsewhere, Kazakhstan (Smirnov and Gorzhevsky, 1977), contain both conformable and discordant orebodies. Several metamorphosed stratabound deposits in the Leptite Series of central Sweden, such as Åmmeberg Zn—Pb, Långban Fe—Mn, and others (Magnusson, 1970), may have originally formed in a similar setting.

The mineralized associations dominated by relatively light-colored sodic submarine volcanics and associated sediments depart significantly from the simpler, basalt-only dominated belts, and represent a transitional group. They appear, however, distinct from the differentiated andesite—dacite—rhyolite association that follows.

Belts or centres with differentiated (basalt)—andesite—dacite—rhyolite, plus related sediments, association. This is an association abundantly present in contemporary, as well as ancient, island arcs that developed on continental or transitional crusts, or that contain continental crustal blocks (as in Japan). The volcanic rocks occur as lavas and a variety of volcaniclastics and are both subaerial and submarine in origin. Most of the preserved sediments are, however, marine. The association is extensively mineralized. The best-known stratabound to semistratabound sulfide deposits are almost always genetically associated with the waning stage of the felsic (rhyolite) phase, and is dominated by pyrite with variable amounts of zinc, lead and copper sulfides. The best-known example of mineralization is the Kuroko type (e.g. Ishihara, 1974; see also Lambert's Chapter 12, Vol. 6 of this Handbook), which consists of several varieties. In the Kuroko deposits, lead and zinc content tends to be much higher than the copper content.

In massive sulfides of comparable affiliation that occur in "greenstone belts" of Precambrian shields, lead is often missing or its content is reduced (e.g. Hutchinson, 1973).

The geology and metallogeny of this association has been described in considerable detail in the literature (e.g. Sangster, 1972; Mitchell and Bell, 1973, including this Handbook series: Sangster and Scott's Chapter 5, Vol. 6; Solomon's Chapter 2, Vol. 6).

Belts dominated by fine-grained detrital marine sediments and metasediments with subordinate (meta-) volcanics (many "slate" and "schist" belts). "Slate" and "schist" belts are among the most monotonous lithologic associations of mobile belts and one of the most difficult ones to interpret. The unfortunate incompatibility of petrographic classifications where igneous rocks are named mainly according to mineralogical composition, detrital sedimentary rocks according to the grain size and metamorphic rocks according to the metamorphic grade, adds to the confusion and obscures the provenance and affiliation of "slates" and "schists". Commonly, two contrasting associations among "slates", etc., are treated in the literature: (1) terrigenous, generally quartz-rich epiclastic association believed to have been derived mostly from the continent; (2) volcanogenic association derived from oceanic, intrabasinal, island-arc or continental-margin volcanism.

Association (1) may be present along both Atlantic- and Pacific-type margins. Association (2) is characteristic for Pacific-type margins, although it may exist also in the most distal parts of Atlantic-type margins. Virtually identical rocks may also form in the many types of depositional basins associated with intracratonic mobile belts.

Any of the three predominantly volcanogenic associations of active mobile belts discussed above grade, with increasing distance from the source of detritus, into "slate" belts. Slate belts are also transitional into carbonate-dominated associations on one side, and coarse clastic (e.g. "flysch") associations on the other. Gradation of slate belts into coarser and more proximal equivalents aids interpretation, but the most reliable method of environmental interpretation is the distribution of regional megafacies. This, unfortunately, has been complicated in most ancient belts because former connections may have been disturbed or obliterated.

Slate and schist belts, often interbedded with "graywacke", that are associated with or grade into elongated lineament-controlled strings of strongly deformed ultramafics and "greenstones", commonly host "gold belts" (e.g. in and near the contact of Archean Temiscaming Group of the Abitibi "greenstone belt", Ontario and Quebec; Goodwin et al., 1972; Jurassic Mariposa Formation, Sierra Nevada, California; Duffield and Sharp, 1975; Cambro-Ordovician slates west of the Heathcote Axis in the Victoria Goldfields, Australia; Bowen and Whiting, 1975). The most characteristic single rock type here is a dark lustrous slate that often carries abundant arsenopyrite. Chert and a lean banded "iron formation" is frequently present, and so are melanges (e.g. in the Mother Lode belt, California; Duffield and Sharp, 1975). The slates of gold belts display tight folding, penetrative deformation and intensive cleavage, and most economically recoverable gold occur in quartz veins controlled by the deformational fabric (e.g. saddle reefs, shear lodes, fissure veins). Regional control of mineralization by pre-metamorphic lithology, i.e. stratiform depositional gold enrichment, has recently become apparent in several fields (e.g. Barberton, South Africa; Viljoen et al., 1970). It is probable that many gold-bearing slate belts are filled oceanic trenches and/or transform-fault grabens.

The "fåhlbands", conformable schist bands enriched in copper, cobalt and arsenic in the Modum (Skutterud) district of southern Norway (Gammon, 1966), also show spatial association with regionally developed ultramafics, but are not gold-bearing.

Slate belts that grade into ophiolites and basalt-dominated associations often carry stratiform lenses of massive to disseminated pyrite—chalcopyrite to pyrite—chalcopyrite—sphalerite (e.g. occurrences listed as of "Besshi-type" by Sawkins, 1976; see Chapter 12, Vol. 6 of this Handbook). Many mineralizations in similar settings contain pyrite only (e.g. Nairn in Kanmantoo Trough, South Australia), or pyrite and manganese carbonates (Chvaletice, Bohemia).

Mineralized belts with a spilite—keratophyre association such as the Rheinische Schiefergebirge and Harz in Germany, or the Vareš area in Bosnia (Ramović, 1968), have particularly well-developed stratiform Pb—Zn ± Cu and barite mineralization in more distal sediments. The Meggen deposit (see Chapter 9 by Krebs, Vol. 9 of this Handbook) is in sediments but close to keratophyres. The Rammelsberg ore (Kraume, 1955; Chapter 10 by Hannak, Vol. 9 of this Handbook) is entirely in sediments although with minor volcaniclastic interbeds; and so are the stratiform Pb—Zn—barite showings in the Vareš area, for example Rupica Hill near Borovica.

Several important stratiform lead—zinc ± barite mineralizations as in the Macmillan Pass district, Howard's Pass and Anvil district (e.g. K.M. Dawson, 1977; Tempelman-Kluit, 1972), have been recently found and explored in the otherwise sparsely mineralized Paleozoic slate belt of the eastern Cordillera in the Yukon, Canada. The first two occurrences, of which Howard's Pass (Summit Lake) is of first magnitude with a possible content of several tens of millions tons of lead and zinc, occur in very monotonous dark-gray slates. The nearest, possibly coeval, occurrences of mafic metavolcanics are known several tens of kilometers away.

The eastern Canadian Cordillera slate belt appears to be a relatively distal portion of an Atlantic-type continental margin, situated along the western edge of the North American craton, that shows an excellent sequence of progressively more proximal to continent lithofacies in the easterly direction (e.g. Wheeler and Gabrielse, 1972; cf. also Chapter 2 by Thompson and Panteleyev, Vol. 5 of this Handbook). The geotectonic setting of the Rheinische Schiefergebirge and the Harz is more complicated (e.g. Krebs and Wachendorf, 1974).

Slate belts also carry stratiform orebodies of scheelite (e.g. Lanersbach in Tux Valley, Australia; Höll and Maucher, 1976, Chapter 1, Vol. 5, and Maucher, Chapter 10, Vol. 10 of this Handbook); stibnite (Stadt Schlaining, Austria, Maucher and Höll, 1968; Pezinok, Slovakia; Cambel and Böhmer, 1955, Murchison Range, South Africa; Anhaeusser and Button, 1976, Chapter 7, Vol. 5 of this Handbook); cinnabar (Horní Luby,

Bohemia); vanadium and molybdenum (Balasanskandyk in the Karatau Range, Kazakhstan; Kholodov, 1973); barite (Magnet Cove, Arkansas); molybdenum; arsenic; uranium; rare earths; thallium; siderite; and magnesite. The geotectonic position of most of the above mineralizations is uncertain and some coincide with a marginal to a platform setting, already discussed earlier.

"Flysch" belts. "Flysch" is a lithofacies composed of alternating, characteristically graded-bedded arenites, and argillites believed to have been deposited by turbidity currents. As such, it may occur in numerous environments and at various stages of geotectonic development. The composition of detritus is naturally variable, being dependent on the source area. Two most common flysch assemblages are "terrigenous flysch" mainly of granitic, metamorphic and recycled sedimentary derivation, and "volcanogenic flysch", typically derived from terranes of intermediate to felsic explosive volcanism that flank island arcs. The process of flysch formation, particularly in the type area of the Alps, does not favour syndepositional ore concentration, and consequently, one would not expect stratiform deposits to be common in typical flysch sequences.

Thin beds of impure siderite (Bezkydy Mts.) and impure Fe–Mn carbonates (Kyšovce–Švabovce, Slovakia; Fusán, 1963) occur in the Cretaceous and Paleogene flysch of the western Carpathians and could be interpreted as an example of the "most normal" metallogeny of flysch.

The literature survey indicates that since the revival of interest in flysch following its modern genetic interpretation in the late 1940's, several mineralized associations have been interpreted as "flysch", but in most cases rather atypical — rich in distal, fine detrital and chemical sediments. Such a "flysch" is gradational into "slate belts", described earlier.

A selection of conformable deposits described as in "flysch"–host association includes the bedded barite in Stanley Shale of Arkansas (Morris, 1974), Sullivan Pb–Zn orebody in Proterozoic Aldridge Formation, British Columbia (Freeze, 1966; see Index in Vols. 4 and 7 of the Handbook for references to this deposit), and the gold lodes of Victoria Goldfieds, Australia (Bowen and Whiting, 1975).

Lithogenesis and metallogenesis in the period of orogenic culmination

In the classic geotectonic cycle (e.g. Bilibin, 1955) and classic magmatic-hydrothermal approach to ore genesis (e.g. Lindgren, 1933; Bateman, 1951; Schneiderhöhn, 1955), the orogenic stage during which transformation of depositional basins into mountain belts, folding, metamorphism and batholitic intrusion took place, was of fundamental importance because it generated the bulk of hydrothermal deposits.

Intrusive rocks, which dominate the "orogenic stage", form in cores of volcanic island arcs, often as deeper-level equivalents of their comagmatic effusives. In ensimatic (immature) island arcs, these intrusives are Na- and Mg-rich primitive rocks (diorite dominates) and have associated Fe, Au and Cu mineralization. In the classic geotectonic model, these were usually placed into the "geosynclinal" stage, or into the transitional period between the former and the "orogenic" stage.

Comparable activity that was taking place in island arcs, with more continental crust than the above assemblage (as in Japan), produced a more "mature" magmatic suite with a greater variety of associated ores (Cu, Au, Ag, Pb, Zn, Mo, Sn, W). In the classic model, similar intrusions were usually placed into the main "orogenic stage".

Rocks intrusive into active continental margins (Andean-type mobile belts), given the same maturity of the intruded crust, could be lithologically and metallogenically closely equivalent with those of island arcs, and both belts could be synchronous. In the classical model, Andean and similar belts were usually placed into the "late" orogenic stage. The products of subaerial volcanism in Andean-type belts, termed "subsequent volcanics" (Stille, 1936b; Schneiderhöhn, 1955) in the classic model, are substantially different compared with the bulk of mainly water-laid, autometasomatically and diagenetically altered volcaniclastics of island arcs.

All the above-listed categories of magmatic rocks carry in their high levels widespread mineralization, but it is all epigenetic and mostly discordant; in the form of veins, disseminations and replacements.

Strata-related mineralizations are rare in intrusive-dominated ancient mobile belts. However, many manto-type orebodies in carbonates, adjacent to intrusions, show a considerable degree of conformity (e.g. Ophir Cu, Ag, Zn, Pb, and Tintic Pb, Ag, Zn districts, Utah; Maria Christina Zn—Pb manto near Copiapo, Chile; Renison Bell and Mt. Bischoff tin deposits, Tasmania; Knight, 1975a), but they formed by hydrothermal replacement of "favourable" (reactive) horizons in supracrustal rocks at or near the contacts with intrusive bodies. Some manto-shaped orebodies in intrusive exocontacts (aureoles) (for terminology, see Smirnov, 1968b) could be modified former stratabound deposits. Many iron- and manganese-bearing "skarns" probably belong to this category. The magnetite and iron silicate ("skarn") iron ores near Babbitt, Minnesota, have demonstrably developed at the contact of sedimentary Biwabik (Proterozoic) iron Formation and Duluth Gabbro, and can be traced into nonmetamorphosed sedimentary iron ore equivalents of the Mesabi Range (Bonnichsen, 1975).

Conformable mantos of disseminated chalcopyrite, bornite, chalcocite or tetrahedrite in subaerial or nearshore intermediate volcaniclastics commonly occur in island-arc associations intruded by granodiorites; associations that frequently contain porphyry coppers as the principal economic type. Examples: Sustut and Sam Goosly copper deposits in British Columbia (Ney et al., 1972; Harper, 1977); Mala district, coastal Peru (Ripley and Ohmoto, 1977); Lo Aguirre mine near Santiago airport, Chile (staff of Compañía Minera Pudahuel, oral communication, 1977).

As in many other settings, supergene alteration of pre-existing discordant ore deposits or metal-enriched rocks causes redistribution of ore substances that can under favourable circumstances generate secondary, commonly stratiform deposits. The laterally or downward migrating metals or minerals may chemically precipitate from solutions, such as the copper oxide and silicate mantos adjacent to several porphyry coppers in arid regions (e.g. Exotica near Chuquicamata, Chile; Roethe, 1975; Mineral Creek orebody in Ray district, Arizona; Phillips et al., 1971). Alternatively, they may form clastic accumulations, such as the tin placers of Malaya and Indonesia; euxenite and columbite placers in the Bear Valley, Idaho (Mackin and Schmidt, 1956), or gold placers (e.g. Sierra Nevada, California).

In the old geotectonic model, widespread intrusive activity during the "orogenic" stage used to be listed together with large-scale crustal shortening and deformation, particularly in thrust and nappe belts, as principal agents influencing mineralization. Recent work, however, indicates much looser, even antipathethic relationship between both. The far-reaching crustal shortenings were recently attributed to collisions of crustal blocks of various sizes: arc—arc, arc—continent, continent—continent (Mitchell and Reading, 1969; Dewey and Bird, 1970).

Epidermal tectogenesis, however, has generally had a negative metallogenetic influence, unless accompanied by volcanism or intrusive activity. It destroyed and dismembered much more pre-existing deposits, than it has created. Terranes with spectacularly developed thrusts and nappes (such as the Swiss Alps, Himalayas, the Scottish Highlands, the Rocky Mountains of southern Canada) are, despite excellent outcrops, almost devoid of mineralization.

Collisions of large lithosphere blocks that caused juxtaposition and doubling of crust thicknesses with resulting crustal melting (Burke et al., 1977), however, may have generated neomagmas and batholiths. In areas where both Benioff-zone-related and collision-related batholiths possibly coexist, the former tend to be conspicuously well mineralized, the latter are nearly barren. There seems to be a reason. The Benioff-zone-controlled magmatism received at least a portion of the first-cycle undepleted, mantle- or lower-crust-generated, partial melts in addition to the recycled crustal sources, and there probably was an abundant supply of sulfur. The system was furthermore reasonably open, long-lasting, capable of developing good zonality, and with a plumbing system. The "magmatic" water was augmented by pore and surface waters to feed and keep in motion hydrothermal plumbing systems. The igneous petrogenesis was probably to a degree influenced by reaction with surrounding rocks, but the essential petrographic variations are likely due to true magmatic differentiation. This accepted concept of trace-metal separation, movement and accumulation has been maintained as valid since the days of Emmons.

The collision-related intrusions, on the other hand, depended on the local (crustal) sources for their generation and reflect the pre-collisional local petrographic variations. They behaved more as a closed system, with a restricted internal mobility. The system was water-poor, and mobilized components that have escaped into the roof rarely formed mineralization.

Lithogenesis and metallogenesis in the period of predominantly non-marine orogenic adjustments

This rather awkwardly worded subheading has been chosen in order to avoid a reference to temporal relationships within an entire mobile belt. In the classic geotectonic cycle, the erosion of folded or uplifted mountains and trough-filling, was the backbone of the "late stage". If, however, this process is operative over the emerged portions of the various examples of magmatic belts discussed in the previous paragraphs, this relationship will not apply in all cases.

The nature of weathering and erosion in mobile belts as well as the nature of their products, do not differ qualitatively from weathering and erosion on platforms, particularly so in intraplatform mobile belts. Only the rate is faster, because due to the continuous isostatic uplift much more material is processed. Tensional tectonics, graben and horst formation, and block-faulting initiate the sedimentogenesis and control the magmatic activity. The erosion-wasted material, usually coarse-grained and poorly sorted, accumulates in three principal types of basins: (a) foredeeps (between mountain belt and craton; setting transitional into platforms); (b) intermontane deeps; and (c) back-deeps (between mountain belt and ocean).

The resulting lithologic associations are commonly described in the literature as "molasse". The molasse in the type area — the Outer Alpine Foredeep — has been derived from the most external Alpine nappes, has sedimentary clasts with much limestone, no associated volcanism, and no ore deposits. Many sedimentary associations in comparable setting have been reported: for example the Cretaceous to Paleogene sediments in the Cordilleran foredeep in Canada (e.g. Eisbacher et al., 1974) as well as the Pennsylvanian sediments in the Appalachian foredeep in the United States. They are virtually unmineralized, save for the increased trace content of certain elements in coal (e.g. Ge, Ga, In, U) in the occasionally present argillaceous siderite beds, and in concretions also in close association with coal.

The presence of important uranium deposits in the Wyoming portion of the Cordilleran foredeep (Powder River Basin — Fig.3) may contradict this thesis, but the similarity between the Wyoming and southern Canadian (Alberta) Cordilleran foredeeps is not complete. The Powder River basin is adjacent to a much more external belt in what is summarily, on physiographic grounds, called the North American Cordillera. The external belt — the Wyoming and Colorado Rocky Mountains — is marked by distinct uplift and doming of the Precambrian basement, with which is associated magmatic activity coeval with the lithogenesis in the foredeep. It is an example of an intracratonic foldbelt already discussed in this paper. The crystalline basement weathering products and shoshonitic volcanism, have probably supplied the uranium in Powder River basin, as many authors believe (for review, see Rackley, 1976, Chapter 3, Vol. 7 of this Handbook).

Comparably favourable metal-source—depositional-basin relationships did not exist in the foredeep of the non-activated thrust belts (composed of imbricate sheets of sedi-

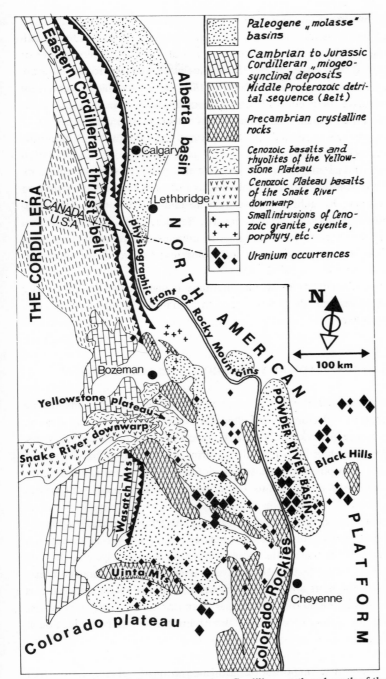

Fig. 3. Eastern front of the North American Cordillera north and south of the U.S.A.–Canada boundary, to illustrate the uneven distribution of uranium mineralization in Tertiary sediments of foredeep basins.

ments) adjacent to the Canadian Rocky Mountains.

The extensive copper mineralization in the upper Proterozoic and Cambrian sediments in marginal troughs flanking the Baikal mobile belt along the southern border of the Siberian Platform (e.g. Narkelyun et al., 1975), has also been controlled by availability of metal (copper) in the source area of clastics. The copper had likely come from mafic extrusive and intrusive rocks controlled by taphrogenic tectonism; thus basement activation adjacent to foredeep troughs has again been of considerable importance.

Intermontane troughs are presented in the literature (e.g. Bilibin, 1955; Smirnov, 1962; Narkelyun et al., 1975), as the most frequently used type-area for mineralization generated in late and terminal stages of the geotectonic cycle. The most commonly quoted examples are the Carboniferous stratabound copper mineralization in red-bed associations of Dzhezkazgan, Kazakhstan; the chalcocite and native copper mineralization in Tertiary arenites in Corocoro, Bolivia and San Bartolo, Chile; and the uranium oxides associated with carbonized plant remnants in arenites and conglomerates (e.g. Challis area, Idaho; Fuki—Donen, southern British Columbia; Ningyo—Toge and other localities, Japan; e.g. Katayama et al., 1974).

Most conformable mineralizations in intermontane troughs are not substantially different in both accumulated metals and ore type from graben-like settings that are a part of intracratonic mobile belts.

DISCUSSION AND CONCLUSION: PLATFORM VERSUS MOBILE BELT METALLOGENY

The neat stereotype of the 1950's dividing the geology and metallogeny of the Earth's surface into that of "platforms" and "mobile belts" plus the usually ignored remainder (i.e. oceans), has been weakened by the geological progress in the past 30 years. We can still recognize examples of both typical platforms and typical mobile belts, but the division of the entire lithosphere into those two domains plus oceans is not an easy task because of the ambiguity of numerous transitional situations, many still poorly understood.

There is still considerable contrast when comparing the "most normal" regions of both platforms and mobile belts, but boundaries between them are rarely simple and sharp, in both the lateral and vertical spatial sense as well as in both relative and absolute geologic time. It may not be premature to state that most differences and contrasts between platforms and mobile belts are quantitative rather than qualitative, and that it is an oversimplification to use an either-or approach, as in the past. This, of course, makes an elegant pigeonholing and listing of contrasts almost impossible, and simple models without important exceptions do not exist.

There is a conflict of scale. A second-order of magnitude (continental size) platform or a mobile belt is equivalent to the platform or the mobile belt, respectively, for all purposes of discussion. Virtually the same entity of smaller dimensions ($x00$ km), existing alone or separated from a large continuous platform or mobile belt by an intervening mobile belt or a rigid block, tends to be described as part of the predominant surrounding entity. For example, a rigid and stable median massif tends to be considered as a segment of a mobile belt, despite the many internal characteristics it has in common with large-scale continuous cratons. The crystalline basement of median massifs may be the same as that of shields and large platforms. The young sedimentary cover of median massifs may be identical to the sedimentary cover of young platforms, and so may be the metallogeny of both. The density of mineralization of median massifs, however, is almost always much greater compared with the metallogeny of large platforms. This seems to be primarily a consequence of a "margin effect". It has been shown that in platforms most mineralizations are located at contrasting "margins", rather than in the monotonous "centres". The ratio of "margins" to "centres" in median massifs is naturally much greater than the same ratio calculated for the area of an entire platform. Also, the effects of tectonomagmatic activation are most abundant in median massifs and contribute to mineralization (e.g. Shcheglov, 1971).

If median massifs are included as part of a mobile belt, as they almost always are, and if such composite mobile belts are then contrasted with large platforms, the difference in metallogeny is diluted, because distribution and generation of ore deposits is normally controlled by relatively small-scale conditions that could have been operative in a local setting, regardless of the dimension of the host global geotectonic domain. For example, a kilometer-long ultrabasic body could carry nickel mineralization regardless of whether it is surrounded by an oceanic, mobile belt, or continental domain.

The conflict in scale discussed above cannot be avoided by introducing a numerical size-limit that would, for example, consider all median massifs over 100 km in size as platforms, and the smaller ones as parts of mobile belts. The most striking contrast between both (the relative rigidity and relative mobility) continues down to the dimension of a sand-size particle.

The status of marginal and inland seas, that form a fourth-order of magnitude subdomain within the second-order of magnitude mobile belt megadomain is quite comparable to that of median massifs. Marginal seas are outliers of the oceanic domain, so that they can be expected to possess the oceanic metallogeny, modified by the high ratio of the "marginal"-to-"central" facies.

Continental margin mobile belts ($M_{2nd\ order}$[1]) can therefore be viewed as a mixture of minor relatively immobile ($P_{3,4th\ order}$), and major relatively mobile ($M_{3,4th\ order}$) components within a predominantly mobile second-order framework, or $M_2 = \Sigma M_{3,4} > \Sigma P_{3,4}$.

Platforms ($P_{1st,\ 2nd\ order}$), both continental and oceanic, are on the other hand a

---

[1] M = "mobile belt" component.
[2] P = "platform" component.

mixture of relatively mobile ($M_{3,4\text{th order}}$) and major relatively immobile ($P_{3,4\text{th order}}$) components within a predominantly immobile second-order framework, or $P_{1-2} = \Sigma P_{3,4} > \Sigma M_{3,4}$.

The interference of mobile and immobile components and processes, as well as lateral and vertical tectonic motions, is overwhelming and is additionally complicated by the dimension of geological time.

There is a conflict of semantics: what exactly constitutes a platform, and what a mobile belt? Our terminology is to a considerable degree a matter of heritage and is strongly influenced by an author's bias, the "degree of thoroughness" of a scientific investigation, and the data available. In the past 100 years, when most geologic concepts currently in use have been formulated, little was known about the early crust, about the deep crustal structure, and about the interplay of subcrustal and supracrustal forces.

The geotectonic domains, initially platforms and mobile belts, were established mostly on the basis of surficial observations, and a variety of treatments have appeared in the literature. The generally older approach, mainly of a reconnaissance nature as widely used in textbooks and classifications, has been the *general* mode of treatment. This treatment tends to divide the world into "platforms" and "mobile belts" on the basis of a kind of cumulative impression unconfined, or only very broadly confined, to a particular time. In this respect, most of the present physiographic plains of North America were included under platforms, whereas the two most conspicuous linear belts of mountains and highlands fringing the plains along both the Pacific and Atlantic coasts were labeled "mobile belts" (Fig.4A). This type of treatment has rarely been consistent, and areas posing interpretative problems or localities of a transitional nature were generally left unconsidered. In the general approach, shields were usually included as part of a platform.

The *historical* (or modern tectonic-map style) approach to the study of geotectonic domains has concentrated on the presence or absence and timing of the latest (youngest) orogenic deformation. As a rule, a more extensive sequence of time orogens and a less-extensive sequence of timed platform cover-rocks has been shown. The "latest-effect" feature (orogeny), however, cancelled all previous effects, so that portions of different former domains incorporated into a mobile belt cannot be distinguished. Compared with the general approach, the historical treatment results in a greater proportion of mobile belts (Fig.4B) because of inclusion of the greater part of shields among "orogens". Additional variations of the historical approach have been introduced by either inclusion or exclusion of water-covered continental margins and oceanic basins (Fig.4C, D).

In the historical mode of investigation of the geotectonic domains, both lateral and vertical boundaries of mobile belts and platform cover regions are of importance, but there has never been a universal agreement as to their location. The lateral boundary has been briefly discussed in various places in this paper. The boundary between the cratonized former mobile belt and overlying undisturbed platform sediments, however, deserves a short discussion below.

In a textbook example of a shield–platform cover-rock boundary, the unmeta-morphosed Paleozoic or Mesozoic, shallow-marine sediments rest on a deeply eroded, almost completely levelled Archean crystalline basement, separated by a sharp noncon-formity. This occurs under the central portion of the North American Interior Plains, where for example in Manitoba the nonconformity represents a hiatus about 2 b.y. long.

Elsewhere, the situation is more complex. In the "Kaapvaal Craton" of South Africa (Haughton, 1969), the "true" deformed and metamorphosed basement (Swaziland System) is overlain by several "old" platform sequences: the Witwatersrand Triad, Trans-vaal System, Waterberg System and Nama System, that are mostly undeformed and mostly unmetamorphosed, but local exceptions of deformed and metamorphosed rocks do occur. They themselves are unconformably overlain by a "young" platform sequence, i.e. the late Paleozoic–early Mesozoic Karoo System. The latter departs to some degree from the conventional stereotype of platform cover sequences, consisting essentially of what has previously been termed the stratigraphic/lithologic "marginal" or "basal" por-tion of a platform, considerably influenced by intracratonic mobility.

The vertical contrast between the cratonized mobile-belt basement and stratified cover sequence in the above examples is more gradational, and in places disputable. Along the northern edge of the Lake Superior Basin in Ontario, the folded and metamorphosed Archean basement rocks are unconformably overlain by Proterozoic Huronian and Animikie Supergroups that in the north closely correspond to a platform sequence, but thicken and become deformed southward. The boundary between the "platform" and "mobile belt" portions is, however, inconsistently placed, with the result that the Elliot Lake uraniferous conglomerates – one of the pillars of platform metallogeny – are situ-ated in the Penokean Fold (mobile) belt of Card et al. (1972).

The basement–cover relationship is especially difficult to interpret in cases where a prolonged hiatus between the formation of the folded basement and deposition of the platform cover was not present. Examples: in central and western Europe where the lower Carboniferous is still distinctly "orogenic" and therefore part of a mobile belt; the upper Carboniferous and Permian being transitional; and the Triassic "platform" (e.g. von Bubnoff, 1956).

Metallogeny has adopted geotectonics as a framework for organizing the regional distribution of ore deposits, and the variety of choice of geotectonic frameworks, coupled with the variety and changing nature of genetic hypotheses, has stretched the number of possible interpretations considerably.

Ore deposits are natural anomalies that have been given an out-of-proportion degree of publicity, like criminals in the human society, and both are most difficult to relate to the prevalent "normal" pattern in both geology and sociology. Even worse, many ore deposits have the "deplorable" characteristic to concentrate along boundaries in man-made classificatory categories. As a consequence, the global-scale metallogenic studies have so far enjoyed an unprecedented freedom of interpretation to such an extent that about two-thirds of the world's ore deposits can now be placed either into platforms or

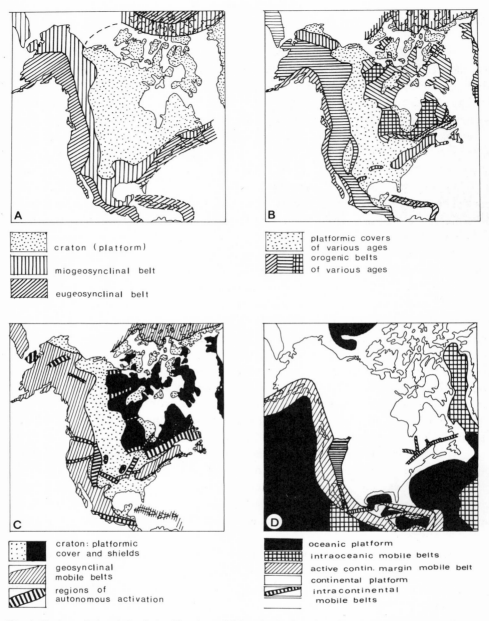

Fig. 4. Various alternatives of classification of lithospheric domains, shown on the example of North America.

A. The "general approach" to geotectonic domains, used as a popular framework for metallogenic classifications in the 1940's and 1950's. Modified after M. Kay (1951).

B. The "historic" (or tectonic-map style) approach to geotectonic domains shows several orogens (orogenic belts) and platformic domains. Maps like this have been used as a base of metallogenic maps in the 1960's and 1970's. This example is simplified after the *Tectonic Map of North America* 1:2,500,000 (P.B. King, 1969).

C. A modified "general approach" to geotectonic domains, showing tri-fold division of domains (geosynclinal mobile belts; regions of autonomous activation; platforms). Commonly used in the 1960's and 1970's.

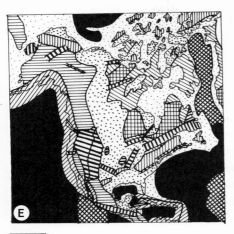

oceanic platform and marginal seas
intraoceanic mobile belts
recent and ancient contin. margin mobile belts
continental platform – subaerial and submarine portions
recent and ancient intracontinental mobile belts

LITHOSPHERIC PLATES

 hanging wall of Benioff zones

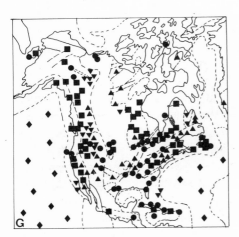

CONFORMABLE DEPOSITS GENERATED IN

♦ oceanic megadomain
■ continental margin mobile belt
▼ transition mobile belt – platform
● continental platform and shelf
▲ intracratonic mobile belts

D. Actualistic approach, showing five geotectonic domains, presently active. Water-covered areas are also considered. The "platform" (or craton, stable block) includes inactive mobile belts, and undergoes contemporary platformic mineralization.

E. Combined approach, closest to the "historic" approach, (B). Presently active domains are shown, and the inactive domains are classified according to past domains that contributed most to their lithogenesis and metallogenesis.

F. Distributions of Recent lithospheric plates. It is apparent that only limited correlation with geotectonic domains exists.

G. A plot of the major conformable ore (and phosphorite, barite, fluorite) deposits of North America.

into mobile belts by geoextremists, if required. The limited, although not necessarily always accurate, consistency of metallogenic interpretation during the past 40 years is probably the consequence of the leading effect of benchmark papers (such as Bilibin, 1955).

There is a widespread inconsistency in metallogenetic classifications regarding the sites (domains) of generation and sites (domains) of present occurrence of lithologic associations, including ores. Rocks and ores generated under equivalent conditions, but with a different post-depositional history, tend to be separated, listed and classified under different categories. For example, the association of subaerial flood-basalts with continental "red-bed" sediments (e.g. Keweenawan Supergroup, Karoo System), universally placed into the platform megadomain, is closely equivalent to many "initial assemblages" of "geosynclinal" mobile belts, such as the Verrucano or Werfener Schichten of the Apennines, Alps and Carpathians.

Most former mobile belts now eroded and dissected (so that contrasting depth levels occur close to the present erosional surface) contain collections of mineral deposits developed in different geologic settings during different stages of development. The present-day Appalachians contain: (1) ore deposits that were generated within a platform, before a subsequent "orogeny" incorporated them into the mobile belt (e.g. Mascot—Jefferson Zn, Tennessee; northwestern Newfoundland, Zn–Pb); (2) ore deposits that formed by supracrustal and subvolcanic processes in a mobile depositional system (geosyncline), located at the site of the present range (e.g. Bathurst, New Brunswick, massive Zn–Pb); (3) ore deposits formed primarily by endogenous processes in a mobile orogenic system (e.g. gold—lode deposits; porphyry copper and skarn, Murdochville, Quebec); (4) endogenous and exogenous deposits formed during the period of levelling of the Appalachian orogenic belt, or after the levelling has been practically completed and cratonic conditions established [e.g. Gays River, Nova Scotia, "Mississippi Valley-type" Zn–Pb in carbonates; Silvermines (Talisman, Salmon River) deposit of disseminated galena in arenites]; and (5) endogenous and exogenous mineralizations that resulted from Triassic activation processes superimposed on the levelled, cratonized Appalachians along extensional structures of the Newark Trough and elsewhere (e.g. Cornwall, Pennsylvania, metasomatic magnetite sheets in carbonates along contact with gabbro sills; Triassic copper-bearing sediments of northwestern Nova Scotia, Connecticut Valley and southeastern Pennsylvania; Tourtelot and Vine, 1976).

Out of the five groups distinguished, only three [(2)–(4)] are directly related to the mobile belt development; while the remaining two [(1) and (5)] are only accidentally related, having been incorporated into the belt, (1), or superimposed over its former remnants, (5). The accidentally related deposits often have "siblings" that escaped incorporation, so they occur in different geotectonic domains. The polygenetic and polyphase nature of mineralization of most former mobile belts cannot be separated on ordinary tectonic maps because of overlaps, but could be distinguished on a _series_ of paleotectonic maps (cf. Chapter 3, Vol. 6 of this Handbook).

The multistage genesis of many, if not most, mineralizations has been recognized and applied to interpretation of ore genesis only very recently (e.g. Gruner, 1956, Pretorius, 1976 in Chapter 1, Vol. 7; see also conceptual models by Wolf, 1976a, b, Chapter 2, Vol. 1 of this Handbook). It is a strong factor that makes any tabulation and categorization of ore deposits according to geotectonic units in time of ore generation only partially true and still more imperfect. Nickel laterites are an especially instructive example.

The nickel laterite in Greenvale, Australia (Fletcher and Couper, 1975) is a mined deposit with 1.57 wt.% Ni, formed during the Tertiary when the area was part of the Australian craton. The mineable accumulation of nickel in laterite, however, was only possible because of the ultrabasic bedrock containing 0.2 wt.% Ni, so the nickel concentration factor of lateritization was only about 8. The ultramafic was emplaced in the Devonian, when the area was a mobile belt and the 0.2 wt.% Ni in the peridotite, contrasted with the about 80 ppm Ni crustal clarke value, gives a factor of concentrations of 25. The event of peridotite emplacement was a much stronger factor than lateritization, bearing on the formation of the Greenvale orebody, so that it should be incorporated into any data tabulation. Unfortunately, it is not; in Table IV a note is made, but the table still cannot be considered to be fully consistent as a whole, because most multistage-ore geneses remain unknown or disputed. The ultimate origin of the Greenvale peridotite and its nickel-enrichment had been in the mantle, so the peridotite emplacement into a mobile belt has been only a transient event. The mantle, however, is not part of the lithosphere, and therefore not a lithospheric geotectonic unit. As a consequence, the lithospheric geotectonic units incorporate mantle-generated mineralization only by proxy.

There is often an insensitive averaging of quantitative data related to metallogeny of geotectonic domains that may create wrong impressions and obscure the actual controls and style of mineralization. The majority of ore deposits are linked to a very small order of magnitude (x to $x0$ km) structural and lithogical anomalies, which in many cases are, but need not be casually linked to the prevalent style of large-size ($x00-x000$ km) total area of geotectonic domains. Large territories of platforms are completely devoid of mineralization, but small segments may be richly mineralized. These mineralized segments are in most cases tectonically disturbed areas; i.e. sites where the predominantly lateral lithogenesis and rare low-concentration metallogenesis at the surface, constituting the bulk of the platform, was interferred by another, predominantly vertical, subsurface lithogenesis and metallogenesis, thus considerably enhancing the probability of ore formation. If the sites of vertical disturbance were of sufficiently concentrated nature and part of an obvious regional lineament with demonstrable mobility, they were commonly excluded from the undisturbed platform domain and treated under the "intracratonic mobile belt" heading. But because of the high degree of mobility of ore-bearing solutions, a minor disturbance may have introduced, and precipitated a large quantity of ore and still remain undetected. This part would be considered part of the undisturbed domain in metallogenetic classification, and constitute a source of major inconsistency.

The question arises: should the mineralization in anomalous portions of platforms (in particular the "Mississippi Valley" Pb—Zn type deposits) be added, and the cumulative tonnage divided by the area of the platform to demonstrate that, for example, an average platform carries 50 kg of economic Pb + Zn per square kilometer, in a similar sense in which Canada contains two inhabitants per square kilometer? Certainly not. The more sensible conclusion is that the bulk (80—90%) of platforms are almost completely devoid of a high factor of concentration mineralizations, but limited areas along platform margins (including internal margins and basement uplifts) and terranes affected by the incipient stage of processes that later result in the formation of intracratonic mobile belts, do contain mineralization the density of which is comparable to many portions of mobile belts. The ratio of almost nonmineralized to reasonably mineralized terranes in platforms is about 8—9 : 2—1.

Oceanic platforms seem to show comparable relationships. Ferromanganese nodules, however, are the most notable exception. Their distribution is unique; the nodules are quite evenly distributed so that the calculation of a mean content of nodules per square kilometer over large portions of the ocean floor is quite proper and justified (e.g. Mero, 1977; see also Chapter 7, Vol. 7 of this Handbook).

Mobile belts, in contrast to platforms, are conspicuous by their denser and much more even distribution of ore occurrences. Some terranes conspicuously devoid of mineralization, however, also occur. The more notable ones at the present level of erosion are: "typical" flysch belts (as in the outer Carpathians); broad alluvial valleys and coastal plains (as in the Fraser River delta, British Columbia); median massifs with thick platformic cover (e.g. the Pannonian Basin, Hungary); young plateau-basalt fields (e.g. Columbia and Snake River Plateaus); detritus-filled foredeeps (e.g. the outer Alpine Molasse); deeply eroded crystalline massifs (e.g. the Swiss Alps); upper levels of collisional belts (the Himalayas and Tibetian Plateau); ignimbrite-covered plateaus (e.g. the western edge of Altiplano, Bolivia); thrust-belts of sedimentary rocks (e.g. Front Ranges of the Rocky Mountains, Canada); and deeply eroded batholiths (e.g. the Coast Ranges Batholith, British Columbia).

The ratio of almost-barren to mineralized terranes in mobile belts varies between 1 : 3 and 1 : 1. In young mobile belts undergoing erosion, the ratio may first decrease (shallow covers being removed), but with continuing erosion the ratio increases steadily (removal of the more densely mineralized 0—5 km deep surface layer and exposure of barren roots).

A conspicuous variation in density of mineralization along an approximately evenly eroded uniform structural and lithofacies belt with alternating densely and sparsely mineralized segments, is attributed to a mineralizing agent unrelated to the prevalent lithogenesis within the belt.

Table IV lists the usual geotectonic setting of conformable deposits as it was (or

TABLE IV

Conformable ore deposits and their usual geotectonic setting (in time of origin)

Type of mineralization, example locality	Geotectonic domains of the lithosphere				
	oceanic platform including oceanic islands	intra-oceanic mobile belt	active continental margin mobile belt	continental platform including Atlantic margin	intra-cratonic mobile belt
ESSENTIALLY NONVOLCANIC					
Continental					
Residual Deposits					
Nickel laterite (Nicaro, Moa–Cuba)		a	P; XX	XX	P; X
Manganese laterite (gossan) – Nsuta		a	P; X	XX	P; X
Residual limonite (Guinea)	X	a	P; X	XX	P; X
Residual bauxite (Al-laterite); Weipa	X	a	"X"	XX	X
Placers – Unconsolidated					
Gold placers, mainly valleys			XX	XX	X
Ilmenite, zircon, monazite, etc., placers – NE Florida			X	XXX	X
Cassiterite, wolframite, columbite placers – Malaya			X	X	XXX
Magnetite, chromite placers (New Zealand, Oregon)	X	a	XX	X	X
Consolidated (ancient) Placers					
Gold, minor uranium (Witwatersrand)				X	XXX
Uranium, almost no gold (Elliot Lake)				X	XX
Ilmenite, zircon, monazite, etc. – Goodrich Quartzite				X	X
Marine/continental					
Chemically precipitated metallic minerals					
Copper in arenites (mainly red bed associations)			X	X	XXX
Lead–zinc in arenites (Mechernich)			X	X	XX
Peneconcordant uranium in arenites ("Western States")			X	"XX"	X

(continued)

TABLE IV (continued)

Type of mineralization, example locality	Geotectonic domains of the lithosphere				
	oceanic platform including oceanic islands	intra-oceanic mobile belt	active continental margin mobile belt	continental platform including Atlantic margin	intra-cratonic mobile belt
Peneconcordant silver in arenites (Silver Reef)			a	X	X
Peneconcordant cinnabar in arenites, carbonates (Idrija)			"X"	"X"	X
U, Ge, Ga, In, etc., recoverable from coal ash (Radvaňavice)			"X"	a	a
Vanadium in asphaltites (Mina Ragra)			"X"	a	a
Lead–zinc sulfides in carbonates (Mississippi Valley)			X	X	XXX
Fluorite in carbonates (Avallon–France)			X	X	XXX
Barite and witherite (bedded) in carbonate association			X	X	XX
Bedded willemite, hemimorphite, smithsonite (Beltana)			a	"X"	X
Willemite–franklinite, etc. in metasediments (Franklin)			a	a	X
Copper sulfides in shale, argillite, slate (Mansfeld)			X	"X"	XX
Copper sulfides in schist (Aitik, Zambia)			X	"X"	XXX
Lead–zinc sulfides in shale, argillite, slate (McArthur, R.)			X	"X"	XX
Marine					
Nearshore iron ores (Lorraine, Clinton)			X	XXX	X
Nearshore manganese ores (Nikopol)			X	XXX	X
Nearshore (transported) bauxites (Andersonville)			X	XX	X
Phosphorites in carbonate association, trace U (Florida)			X	XX	a
Phosphorites in chert–shale association, trace U, V (Phosphoria)			X	XX	a
Bedded barite in chert–shale association (Magnet, Ark.)			X	XX	X
Uranium-bearing shales (Kolm, Chattanooga)			a	XX	X
Vanadium-bearing shales (Pariatambo Formation)			X	a	a
TRANSITIONAL VOLCANOGENIC/NONVOLCANIC, MOSTLY MARINE					
Mn–Fe oxide nodules and crusts (Pacific Ocean)	XXX	XX	X	X	X
Mn–Fe carbonate nodules (Baltic Sea)	X	X	X	X	X
Low-grade metalliferous sediments (Fe, Mn, Ba, Cu)	X	XXX	X		XX
High-grade metalliferous muds and brines (Red Sea)	a	a	a	a	XXX

Deposit type			
Conformable cinnabar in phyllite association (Horni Luby)			XX
Co–Ni arsenides in schist (Modum)			XX
Disseminated and banded arsenopyrite in slate (Meguma Gr.)			XX
Cerite and rare earth minerals in metamorphics (Bastnäs)	X		X
VOLCANIC–SEDIMENTARY			
Bertrandite and fluorite impregnated tuffs (Spor Mt.)	X		"XX"
Mn oxides impregnated tuffs (Charco Redondo)	XX		X
U–Mo impregnations in acid volcaniclastics (Novoves, Huta)	X		X
Banded iron formation in exhalites (Algoma)	XXX		X
Banded Mn-carbonates in exhalites (Olympic peninsula)	XXX		
Gold in exhalites (Lead, S. Dak.; Timmins)	XXX		"X"
Chalcopyrite, thin banded in exhalite (Boston Creek, Ont.)	XX		X
Zn–Cu massive sulfides in exhalites (Canadian Shield)	XXX		X
Pb–Zn–Ag (Cu) in exhalites and felsic volcaniclastics (Kuroko)	XXX		"X"
Massive pyrite–Cu in mafic exhalites and ophiolites (Cyprus)	X	"XX"	"X"
Massive Pb–Zn (Cu) + barite in sedimentary keratophyre association (Meggen)	X	"XX"	"X"
Lithium-, tungsten-, boron-, etc., bearing brines (Searless Lake)	X		
Disseminated Cu-sulfide mantos in intermediate volcaniclastics (Sustut)	XXX		XXX
VOLCANOGENIC			
Magnetite lava flows (El Laco)	X		
Disseminated native Cu-chalcocite in amygdaloid basalts (Portage L., Keweenaw)	X	"X"	XX
Disseminated gold in komatiitic flows (protore only)	X		X
INTRUSIVE			
Porphyry coppers (non-conformable)	XXX		XXX
Layered chromite ± Pt, Pd in pyroxenite, etc. (Bushveld)	X		XXX
Titanomagnetite in gabbro, anorthosite (Grenville Province, Quebec)			X
Layered chalcopyrite–bornite in gabbro (Volkovo)	X		
Disseminated Ni-silicates in peridotite sills (Amos, Quebec)	X		a
Disseminated Ni-sulfide in peridotite sills (Mt. Keith, W. Australia)	X		a
CONTACT METASOMATIC			
All hydrothermal replacement mantos (Fe, Cu, Pb, Zn, etc.)	XXX		X

X = rare; XX = common; XXX = very common; a = possible; P = "preconcentration" in previous stage; "X" = transitional, atypical, etc. "XX" =

TABLE V

Comparative lithogenesis and metallogenesis of the five lithosphere domains

	Oceanic platform including oceanic islands (OP)	Intraoceanic mobile belts (IOB)
General characteristics bearing on conformable mineralization	the most extensive domain; appears to be lightly mineralized but this is due partly to the thin sampling and prevalence of un-consolidated rocks; it is a domain with the most limited development; practically only syndeposi-tional metallogenesis is involved, diagenetic lithification and post-lithification stages are missing	very dynamic domain but only spotty, mainly surface sampling done; like OP, IOB is a domain of limited development (deposi-tional metallogeny only), althoug more advanced mineralization can be observed in probable ancient equivalents added to CMB
Subaerial exposure influencing surficial mineralization	very limited (less than 1/100 of the area)	very limited (less than 1/20 of the area)
Ores generated by reworking of pre-existing orebodies in subaerial conditions (gossans, placers)	none	none
Ores generated by concentration of trace metals in subaerial con-ditions (laterites, placers)	rare	none recent; common in ancient IOB, incorporated into CMB
Mineralization in the process of syngenetic redeposition of weathering derived metals in nearshore areas	very rare on fringe of volcanic islands	not known
Mineralization in the process of diagenesis and katagenesis of sedimentary sequences	far-reaching early (pre-lithifica-tion) diagenesis generates Mn nodules, lithification and later diagenesis beyond reach in recent environments, ancient rocks eliminated by subduction	proximal end of gradation into OP
Volcanism and conformable mineralization	widespread and essential; domi-nantly basalts — "quiet out-pourings", some hyaloclastites form oceanic floor, but mostly covered by sediments; oceanic islands — mainly basaltic central volcanoes; no significant recent mineralization	mainly basalts, emplaced at spreading ridges; widespread autometasomatism at contact with seawater, generation of exhalites and metalliferous sediments; recent mineraliza-tions no economic grade, but deposits occur in ancient IOB now incorporated into CMB

...ve continental ...in mobile belt ...B)	Continental platform (CP)	Intracratonic mobile belts (ICB)
dynamic and very complete ...ain; a classic in which most ...logenic studies done; prod- ...of all stages of the geotec- ...cycle are represented, dis- ...ng increasing complexity ...elative interaction in time; ...ormable deposits dominate ...the early and late stages ...ed by widespread sedimento- ...sis, while non-conformable ...sits dominate the stage of ...rmation, magmatism and ...morphism	extensive, but least dynamic domain; long lasting sedimentation and slow but steady vertical movements generated very regular but monotonous sedimentary sequences on a grand scale, but very few ore deposits of low factor of concentration; most recorded mineralizations are due to reworking of former mineralizations and accessory minerals in crystalline basement under platformic conditions, or due to interference from ICB	a dynamic domain, but restricted in scope; its unique system of deep plumbing and steep-walled, restricted depositional basins favours movement of metal leaching/metal depositing fluids and provides efficient trapping mechanism; the recycled crustal metals involved in ore deposition are likely supplemented by mantle derived metals that enter the crust for the first time; growing number of large stratiform deposits is being attributed to taphrogenic setting following the recent example in the Red Sea
...sequal to submarine; varia- ...due to development stages	subaerial slightly exceeding subaqueous	subaerial slightly exceeding subaqueous
...common	common	common
...ion	very common	common
...on, but mostly small-size, ...le type deposits	very common; many large bodies of low factor concentration metals; very regular controls, orderly progression of facies	conditions transitional to CP; more proximal mineralization, increased proportion of high-factor of concentration metals
...ges of diagenesis took place ...eir effects were to a con- ...ble degree obliterated by ...imposed metamorphism ...arge area; consequently ...etic mineralization rarely ...ished	most stages of diagenesis took place, effects well preserved; most deposits are diagenetically fixed syngenetic accumulations; rare deposits with diagenetic metal supply	conditions transitional to CP; diagenetically fixed syngenetic deposits as well as common and important ones of diagenetically supplied metal
...subaerial and submarine ...ism very significant, explo- ...desite—dacite—rhyolite ...ism most typical; volcani- ...s more common than lava, ...e of sedimentary detritus; ...rmable sulfide orebodies ...y result of reaction of sub- ...us volcanics with seawater ...ttom sediments	"most normal" CP do not contain volcanics; distal ash falls may become significant source of uranium in some marginal portions of platforms; plateau basalts overlap into CP from ICB	significant volcanism; "contrast differentiated" basalt—rhyolite association, basalt most common; alkaline volcanism; most volcanics subaerial, so syn-emplacement conformable deposits rare; most conformable deposits form by reworking

TABLE V (continued)

	Oceanic platform including oceanic islands (OP)	Intraoceanic mobile belts (IOB)
Intrusive magmatism and conformable mineralization	Not exposed in Recent oceans	peridotites, gabbros and diorites occur at spreading ridges; non-conformable Ni—Cu sulfides reported from a Recent ridge; abundant in ancient ophiolite associations added to CMB
Metamorphism	None exposed in Recent oceans	hydrothermal metasomatism and high-pressure metamorphism in deeper levels of Recent spreading systems and in many ophiolite belts added to CMB
Horizontal (compressional) tectonism	None in Recent oceans	none in Recent IOB
Vertical (tensional) tectonism	Probably widespread, masked to considerable degree by unconsolidated sediments	fundamental at spreading ridges; conduit for magmatism

believed to have been) in the time of ore generation. The table has to be considered as a very rough approximation only, because of the formidable problems, inaccuracies and inconsistencies in placing mineralizations into the framework of geotectonic domains, discussed throughout the paper. Many exceptions, transitions and uncertainties have had to be disregarded. Only conformable mineralizations are listed although "porphyry coppers" were added as a standard; thus the table is not representative for all types of ore deposits.

e continental n mobile belt)	Continental platform (CP)	Intracratonic mobile belts (ICB)
widespread and characteristic lkaline intrusive rocks range ultramafics to alaskites, but diorite most common; abun- on-conformable mineraliza- conformable ores mostly dary, formed by reworking vious mineralization	absent in "most normal" CP	abundant and important in deeply eroded ICB; peralkaline granite— syenite association carries mostly reworked conformable deposits; layered mafic intrusions of Bushveld type carry important and characteristic magmatogene stratiform Cr, Pt, Fe—Ti—V and Ni—Cu mineralization; alkaline complexes, carbonatite, kimber- lites — "secondary" mineralization
lant, mainly-low pressure, emperature metamorphism; rinted over pre-existing bound orebodies causes y textural changes tamorphosed orebodies); norphogenic orebodies less on, many disputed	absent in "most normal" CP; load metamorphism in deeply buried sedimentary wedges of Atlantic continental margins	abundant and highly variable metamorphism in deeply eroded ICB; reaching granulite facies in probable equivalents of intra- cratonic foldbelts (Grenville, Namaqualand); anorthosites with conformable Fe—Ti, metamor- phosed older deposits
on and widespread folding, tion of oceanic lithosphere ontinent, thrust faulting ds craton; mineralization on only if magmatic activity ated, most orebodies non- rmable	absent in "most normal" CP	none
read throughout the geotectonic cycle, but apparent in late stages; es conduits for magmatism drothermal fluids	present, intensity decreases from ICB; apparent mostly in crystalline basement	widespread and fundamental; generates vertical gradients, framing basins, conducts magmas; responsible for thermal gradients that may convert surficial waters into heated metal leaching and depositional brines

From Table IV it is apparent that certain mineralization types occur in most domains, while others are confined to a single domain. Ferromanganese nodules and crusts,
for example, are one of the few mineralization types that have formed in all domains,
provided that such domains contained a waterfilled basin. Porphyry coppers, on the other
hand, are consistently present in a single domain only — in mobile belts along the Pacific-
type continental margins, specifically in segments presumably situated above Benioff
zones.

This is clearly due to the degree of complexity of ore deposition, which is rather low in the first case (manganese precipitated from the water), but complex and unique in the second case [copper accumulated as a product of crust and mantle(?) melting above the Benioff zone, within a distinct depth interval (Mitchell and Garson, 1972)].

Table IV is from data based on studies of ore deposits, and there may be some conflict between "actually present" and "theoretically possible" mineralizations. For example, bauxite has formed over basalts in Hawaii (Patterson, 1971), which is part of the oceanic platform domain, but none has been reported from the intraoceanic mobile belt domain because no sufficiently large Tertiary islands within this domain occur in the tropical belt. This irregularity in the present distribution of islands in oceans — one of the many shifting conditions through geological time — should not therefore be used as a generally valid rule that oceanic islands do (and intra-oceanic ridge islands do not) contain bauxite in the geologic record. A similar complication has almost disappeared in regard to the mineralization types with the large populations of examples utilized.

Table V compares the various types of lithogeneses and metallogeneses in the five geotectonic domains of the lithosphere, and many differences are readily apparent. The oceanic megadomain contrasts with all the other domains in its complete lack of consolidated sediments, so that it is devoid of all possible mineralizations that are linked genetically to any of the stages of sedimentogenic development later than the early (pre-lithification) diagenesis.

The presence of continental (dry land-generated) ore deposits, including supergene deposits is primarily a function of the abundance of subaerial outcrop, proper climatic zone and suitable relief, and only secondarily a function of the geotectonic domain that controls the presence of the weathered source rock. In the case of common metals and low-factor-of-concentration deposits (such as iron), suitable source rocks are thus independent of the geotectonic domain.

The majority of conformable deposits are water-laid, and orginated within a sedimentary sequence with both weathering — recycled (epiclastic), volcanogenic, and still poorly-defined miscellaneous endogenous and transitional sources. The natural consequence is that the ratio of conformable/nonconformable ore deposits in any geotectonic domain will roughly approximate the ratio of water-laid sediments to other rocks. This ratio is highest in the oceanic platform, and indeed all the known mineralizations and metal enrichments in recent oceanic sediments are stratiform. The second-highest ratio have continental platforms.

In undisturbed platforms (i.e. both oceanic and continental), the abundance of conformable mineralization is a function of the progress and degree of diversification of supracrustal lithogenesis and metallogenesis. In disturbed geotectonic domains (most typically in active continental margin mobile belts), the abundance of conformable deposits is proportional to the progress and degree of diversification of supracrustal lithogenesis, minus those deposits generated in the former period removed or converted during orogeny. In portions of mobile belts where layered supracrustal rocks have been largely removed, such as in large batholithic belts (e.g. the Coast Batholith of British Columbia), conformable deposits are entirely missing, although they may have been present originally. The former presence of such deposits in the area is indicated by several large pre-batholithic stratabound orebodies, such as Tulsequah, Grandux, Anyoc and Britannia (e.g. Thompson and Panteleyev 1976, Chapter 2, Vol. 5 of this Handbook), but occur along the immediate "granite" contact in supracrustal rocks, or even in large metasedimentary/metavolcanic rafts and xenoliths completely surrounded by "granite".

The destructive development in regard to intrusive lithogenesis and metallogenesis, on the other hand, generates supracrustal rocks and conformable ore deposits in the closing (adjustment and levelling) stages of a mobile-belt history.

Consequently, conformable deposits occur most commonly in those mobile belts and their portions in which large tracts of supracrustal rocks are preserved; for example in the Scandinavian Caledonides, the Ural Mts., or the eastern belt of Canadian Cordillera.

The interplay of deep-reaching, relatively open conduits that span a wide range of thermal gradients in the Earth's crust (which makes them suitable for mobilizing and moving metal-extracting—metal-depositing solutions), with restricted-circulation depositional basins (suitable for trapping and the conformable accommodation of introduced metals), has been most efficient in intracratonic mobile belts. Such belts, therefore, contain the highest proportion of hybrid conformable deposits composed of depth-derived metals which are accommodated by supracrustal processes.

As already mentioned in the introductory part, there is an approximate correlation between the surficial expression of lithospheric domains and the deep crustal structure (Fig.1). The deep crustal structure is without doubt of fundamental importance for metallogeny, particularly as far as the source of ore metals is concerned (e.g. Smirnov, 1968a; Krauskopf, 1971; Noble, 1974). There is, obviously, an entire series of metal (as well as the ore-forming anions, particularly sulfur) derivations in which the continental crust on one side, and the mantle (or even the asthenosphere) on the other, are but two contrasting end-members.

Metallogenesis is a salient feature of the Earth's crust transformation, and is most intensively developed in transitional phases of that process. The prevalent contemporary opinion based on the plate-tectonic model, as well as the more conservative but upgraded traditional thinking in terms of the static model (e.g. most Soviet writing between 1968 and 1974), assumes progressive conversion of the oceanic lithosphere into the continental lithosphere, which is best documented along Pacific-type continental margins.

A new approach in the compilation of regional tectonic maps (and possibly also metallogenic maps in the near future) based on the above assumption, has recently been announced in the U.S.S.R. (Peyve et al., 1977). The principal criterion of this methodology is the time of development of the consolidated continental crust and/or the time of formation of the granite-metamorphic crustal layer. Practical results of these studies, demonstrated by the map of development of the continental crust in Eurasia (Peyve et al., 1977, fig.1) indicate that there can hardly exist a sharp boundary between a "craton" on one side and a "mobile belt" on the other. It appears that such a boundary is gradational and it would be more accurate to speak about a degree of "continentization".

If "continentization" could be reliably calculated using numerical data and using a geologically sensible approach, based on the ratio of unconverted (oceans) to converted (continents) crusts, it appears that such a ratio at oceanic spreading ridges would be close to 100 : 0; in immature (ensimatic) island arcs such as the Marianas, about 80—90 : 20—10; in island arcs that contain parts of older continental crust such as in Japan, about 70—50 : 30—50; and in the continental-edge mobile belts such as the Andes, 50—30 : 50—70. The repeatedly erosion-levelled cores of the oldest cratons could approach the ratio of 0 : 100.

The postulated retrogressive conversion (continental crust into oceanic) could also be graded, using inverse ratios (continents : oceans). An abortive rift, for example, could have a value of 70 : 30 whereas a fully separated rift would become a proto-oceanic gulf and ultimately an ocean with a ratio close to 0 : 100, and the circle of development would close.

A preliminary research done so far by the author indicates that the peaks in concentration and accumulation of the various metals, as well as the peaks of appearance of certain mineralization types, parallel in the general sense the development discussed above. This would, however, appear clearly only as a result of a careful statistical processing of most global ore deposits (population of over 10,000), the work on which is presently in progress. This trend, however, becomes lost if few handpicked examples of ore deposits or mineralized areas are treated as in the case of most recent studies. The reason for a considerable interference is the high degree of lithologic, structural and stratigraphic heterogeneity and complexity of the individual increments of the developing crust.

REFERENCES

Abdullayev, Kh.M., 1961. Printsipy vydeleniya strukturno-geologischeskikh zon, rudno-petrografiches-kikh provintsii i rudnykh rayonov. *Uzb. Geol. Zh.*, No. 4, pp.46—58.
Adams, J.W. and Staatz, M.H., 1973. Rare-earth elements. *U.S. Geol. Surv., Prof. Pap.*, 820: 547—556.
Alkhazov, V.Yu., Atakishiyev, Z.M. and Azimi, N.A., 1977. Geology and mineral resources of the early Quaternary Khanneshin carbonatite volcano (southern Afghanistan). *Int. Geol. Rev.*, 20(3): 281—285.
Anhaeusser, C.R. and Button, A., 1976. A review of southern African stratiform ore deposits — their position in time and space. In: K.H. Wolf (Editor), *Handbook of Strata-bound and Stratiform Ore Deposits, 5.* Elsevier, Amsterdam, pp.257—319.
Antweiler, J.C. and Love, J.D., 1967. Gold-bearing sedimentary rocks in northwest Wyoming — a preliminary report. *U.S. Geol. Surv., Circ.*, 541, 12 pp.
Auboin, J., 1965. *Geosynclines.* Elsevier, Amsterdam, 335 pp.
Aumento, F., Loncarevic, B.D. and Ross, D.J., 1971. Hudson geotraverse: geology of the Mid-Atlantic Ridge at 45°N. *Philos. Trans. R. Soc. London, Ser. A,* 268: 623—650.
Ayrton, S., 1974. Rifts, evaporites and the origin of certain alkaline rocks. *Geol. Rundsch.,* 63: 430—450.
Badgley, P.C., 1965. *Structural and Tectonic Principles.* Harper and Row, New York, N.Y., 521 pp.
Bateman, A.M., 1951. *Economic Mineral Deposits.* Wiley, New York, N.Y., 916 pp.

Baturin, G.N. and Rozanova, T.V., 1975. Ore mineralization in the rift zone of the Indian Ocean. In: A.P. Vinogradov and G.B. Udintsev, (Editors), *Rift Zones of the World Ocean*. Halsted, Jerusalem, pp.431–441.

Belevtsev, Ya.N. (Editors), 1974. *Metallogeniya Ukrainy i Moldavii*. Nauk Dumka, Kiev, 510 pp.

Bellido, E.B. and de Montreuil, L.D., 1972. Aspectos generales de la metalogenia del Peru. *Peru, Serv. Geol. Min., Geol. Econ.*, No. 1, 150 pp.

Beloussov, V.V., 1954. *Basic Problems in Geotectonics*. McGraw-Hill, New York, N.Y., 1962, 816 pp. (English translation).

Beloussov. V.V., 1969. Interrelations between the Earth's crust and upper mantle. In: P.J. Hart, (Editor), *The Earth's Crust and Upper Mantle*. Am. Geophys. Union, Geophys. Monogr., 13: 698–712.

Beloussov, V.V. and Shantser, E.V., 1970. Primary features of the structures of the Earth's crust. In: *Great Soviet Encyclopedia, 9*. Collier–Macmillan, New York, N.Y., 3rd ed., pp.58–59 (English translation).

Bender, F., 1975. Geology of the Arabian Peninsula – Jordan. *U.S. Geol. Surv., Prof. Pap.*, 560-I.

Berger, W.H., 1974. Deep-sea sedimentation. In: C.A. Burk and C.L. Drake, (Editors), *The Geology of Continental Margins*. Springer, Berlin, pp.213–241.

Bilibin, Yu, A., 1955. Metallogenic provinces and metallogenic epochs. *Geol. Bull., Dep. Geol., Queens Coll., Flushing, N.Y.*, 1968, 35 pp. (English translation).

Blecha, M., 1965. Geology of the Tribag mine. *Can. Min. Metall. Bull.*, 58: 1077–1082.

Bodenlos, A.J. and Thayer, T.P., 1973. Magnesian refractories. *U.S. Geol. Surv., Prof. Pap.*, 820: 379–399.

Bonatti, E., 1975. Metallogenesis at oceanic spreading centers. *Annu. Rev. Earth Planet. Sci.*, 3: 401–431.

Bonatti, E., Guerstein-Honnorez, B.M. and Honnorez, J., 1976. Copper–iron sulfide mineralization from the equatorial Mid-Atlantic Ridge. *Econ. Geol.*, 71: 1515–1525.

Bonnichsen, B., 1975. Geology of the Biwabik Iron Formation, Dunka River area, Minnesota. *Econ. Geol.*, 70: 319–340.

Boström, K. and Peterson, M.N.A., 1966. Precipitates from hydrothermal exhalations on the East Pacific Rise. *Econ. Geol.*, 61: 1258–1265.

Bowen, K.G. and Whiting, R.G., 1975. Gold in the Tasman Geosyncline, Victoria. In: C.L. Knight, (Editor), *Economic Geology of Australia and Papua New Guinea, 1. Metals. Aust. Inst. Min. Metall. Monogr. Ser. 5*: 647–658.

Bowers, S.D., 1966. Geology of the Arabian Peninsula, Yemen. *U.S. Geol. Surv., Prof. Pap.*, 560-B.

Bramlette, M.N., 1961. Pelagic sediments. In: M. Sears (Editor), *Oceanography. Am. Assoc. Adv. Sci. Publ.*, 67: 345–366.

Bucher, W.H., 1933. *The deformation of the Earth's Crust*. Princeton University Press, Princeton, N.J., 518 pp.

Buddington, A.F. and Chapin, T., 1929. Geology and mineral deposits of southeastern Alaska. *U.S. Geol. Surv., Bull.*, 800, 398 pp.

Burke, K., Dewey, J.F. and Kidd, W.S.F., 1977. World distribution of sutures – the sites of former oceans. *Tectonophysics*, 40: 69–99.

Cairnes, D.D., 1915. Upper White River district, Yukon. *Geol. Surv. Can., Mem.*, 50.

Calvert, S.E. and Price, N.B., 1970. Minor metal contents of Recent organic-rich sediments off West Africa. *Nature (London)*, 227:593–595.

Cambel, B. and Böhmer, M., 1955. Cajlanské antimonové a pyritové ložiská a chemismus malokarpatských rud. *Geol. Práce, Geotech.*, 8: 1–54.

Campbell, N., 1967. Tectonics, reefs and stratiform lead–zinc deposits of the Pine Point area, Canada. *Econ. Geol. Monogr.*, 3: 59–70.

Card, K.D., Church, W.R., Franklin, J.M., Frarey, M.J., Robertson, J.A., West, G.F. and Young, G.M., 1972. The Southern Province. In: R.A. Price and R.J.W. Douglas (Editors), *Variations in Tectonic Styles in Canada, Geol. Assoc. Can. Spec. Pap.*, No. 11, pp.335–380.

Carvalho, P., Guimaraes, D. and Dequech, D., 1962. Jazida plumbo–zincifera do Municipio de Vazante, Minas Gerais (Brazil). *Braz. Dep. Nac. Prod. Miner., Div. Fomento, Bull.*, 110, 119 pp.

584

Cathcart, J.B., 1966. Economic geology of the Fort Meade Quadrangle, Polk and Hardee Counties, Florida. *U.S. Geol. Surv., Bull.,* 1207.

Cathcart, J.B. and Gulbrandsen, R.A., 1973. Phosphate deposits. *U.S. Geol. Surv., Prof. Pap.,* 820: 515–525.

Christiansen, R.L. and Blank, Jr., H.R., 1972. Volcanic stratigraphy of the Quaternary rhyolite plateau in Yellowstone National Park. *U.S. Geol. Surv. Prof. Pap.,* 729-B.

Conant, L.C. and Swanson, E.E., 1961. Chattanooga Shale and related rocks of central Tennessee and nearby areas. *U.S. Geol. Surv., Prof. Pap.,* 357.

Cox, K.G., 1972. The Karoo volcanic cycle. *J. Geol. Soc. London,* 128: 311–336.

Cruickshank, M.J., 1974. Mineral resources potential of continental margins. In: C.A. Burk and C.L. Drake (Editors), *The Geology of Continental Margins,* Springer, Berlin, pp.965–1000.

Currie, K.L., 1976. The alkaline rocks of Canada. *Geol. Surv. Can., Bull.,* 239.

Dana, J.D., 1873. On some results of the earth's contraction from cooling including a discussion of the origin of mountains and the nature of the earth's interior. *Am. J. Sci.,* 5: 423–443; 6: 6–14, 104–115, 161–171.

Davies, J.F., Bannatyne, B.B., Barry, G.S. and McCabe, H.R., 1962. Geology and mineral resources of Manitoba. *Manit. Dep. Mines Nat. Resour., Mines Branch,* 190 pp.

Dawson, J.B., 1966. Oldoinyo Lengai — an active volcano with sodium carbonatite lavaflows. In: O.F. Tuttle and J. Gittins (Editors), *Carbonatites.* Interscience, New York, N.Y., pp. 155–168.

Dawson, K.M., 1977. Regional metallogeny of the northern Cordillera. *Geol. Surv. Can., Pap.,* 77-1A: 1–4.

Deans, T., 1950. The Kupferschiefer and the associated lead–zinc mineralization in the Permian of Silesia, Germany and England. *18th Int. Geol. Congr., London, Part. 7,* pp.340–352.

Deans, T., 1966. Economic geology of African carbonatites. In: O.F. Tuttle and J. Gittins (Editors), *Carbonatites.* Wiley, New York, N.Y., pp.388–413.

Degens, E.T. and Ross, D.A., 1976. Strata-bound metalliferous deposits found in or near active rifts. In: K.H. Wolf (Editor), *Handbook of Stratabound and Stratiform Ore Deposits, 4.* Elsevier, Amsterdam, pp.165–202.

de Launay, L., 1892. *Formation des gîtes métallifères ou métallogenie.* Béranger, Paris.

de Launay, L., 1913. *Traité de métallogenie.* Béranger, Paris, 3 Vols.

Demenitskaya, R.M., 1967. *Kora i mantiya zemli.* Nedra, Moscow.

de Villiers, P.R., 1970. The geology and mineralogy of the Kalahari manganese field north of Sishen, Cape Province. *S. Afr. Geol. Surv., Dep. Mines, Mem.,* 59, 84 pp.

Dewey, J.F. and Bird, J.M., 1970. Mountain belts and the new global tectonics. *J. Geophys. Res.,* 75: 2625–2647.

Dickinson, W.R., 1971. Plate tectonic models of geosynclines. *Earth Planet. Sci. Lett.,* 10: 165–174.

Dickinson, W.R., 1974. Plate tectonics and sedimentation. *Soc. Econ. Paleontol. Mineral., Spec. Publ.,* 22: 1–27.

Dietz, R.S., 1972. Geosynclines, mountains, and continent-building. *Sci. Am.,* 226(3): 124–133.

Dietz, R.S., Holden, J.C. and Sprou, W.P., 1973. Geotectonic evolution and subsidence of Bahama Platform. *Geol. Soc. Am. Bull.,* 81: 1915–1928.

Douglas, R.J.W. and Price, R.A., 1972. Nature and significance of variations in tectonic styles in Canada. In: R.A. Price and R.J.W. Douglas (Editors), *Variations in Tectonic Styles in Canada.* *Geol. Assoc. Can., Spec. Pap.,* 11: 626–688.

Drake, C.L. and Burk, C.A., 1974. Geological significance of continental margins. In: C.A. Burk and C.L. Drake (Editors), *The Geology of Continental Margins,* Springer, Berlin, pp.3–10.

Duffield, W.A. and Sharp, R.V., 1975. Geology of the Sierra foothills melange and adjacent areas, Amador County, California, *U.S. Geol. Surv., Prof. Pap.,* 827.

Duke, J.M., 1977. Mineralogy of serpentinized ultramafic rocks and associated nickel deposits. *Geol. Surv. Can., Pap.,* 77-1A: 15.

Dymond, J., Corliss, J.B., Heath, G.R., Field, C.W., Dasch, E.J. and Veeh, H.H., 1973. Origin of metalliferous sediments from the Pacific Ocean. *Geol. Soc. Am., Bull.,* 84: 3355–3372.

Eckstrand, O.R., 1974. Significance of some Australian and African occurrences for Canadian Archean nickel deposits. *Geol. Surv. Can., Pap.,* 74-1 (*Part A*): 133–134.

Ehrenberg, H., Pilger, A. and Schröder, F., 1954. Das Schwefelkies – "Zinkblende" – Schwerspat-lager von Meggen (Westfalen). *Beih. Geol. Jahrb.*, 12, 352 pp.

Eisbacher, G.H., Carrigy, M.A. and Campbell, R.B., 1974. Paleodrainage pattern and late-orogenic basins of the Canadian Cordillera. In: W.R. Dickinson (Editor), *Tectonics and Sedimentation, Soc. Econ. Paleontol. Mineral. Spec. Publ.*, 22: 143–166.

Emslie, R.F., 1972. Oceanic crust and the identification of ancient oceanic crust on the continents: a summary. In: *The Ancient Oceanic Lithosphere, Can. Contrib. No. 11 to the Geodynamic Project*, pp.153–156.

Ensign, Jr., C.O., White, W.S., Wright, J.C., Patrick, J.L., Leone, R.J., Hathaway, D.J., Trammell, J.W., Fritts, J.J. and Weight, T.L., 1968. Copper deposits in the Nonesuch Shale, White Pine, Michigan. In: J.D. Ridge (Editor), *Ore Deposits of the United States 1933–1967*. Am Inst. Min. Metall. Pet. Eng., New York, N.Y., pp.460–488.

Evans, H.J., 1975. Weipa bauxite deposit, Queensland. In: C.L. Knight (Editor), *Economic Geology of Australia and Papua New Guinea, 1. Metals. Aus. Inst. Min. Metall., Monogr. Ser.*, 5: 959–963.

Ewing, J., 1969. Seismic model of the Atlantic Ocean. In: P. Hart (Editor), *The Earth's Crust and Upper Mantle. Am. Geophys. Union, Upper Mantle Sci. Rep.*, 21: 220–225.

Fersman, A.Ye., 1934. *Geokhimiya*. Reprinted in: A.Ye. Fersman-Izbrannye Trudy, 1955, *Tr. Akad. Nauk S.S.S.R.*, 3: 798.

Finch, W.I., 1967. Geology of epigenetic uranium deposits in sandstone in the United States. *U.S. Geol. Surv., Prof. Pap.* 538, 121 pp.

Fisher, N.H. and Warren, R.G., 1975. Outline of the geological and tectonic evolution of Australia and Papua New Guinea. In: C.L. Knight (Editor), *Economic Geology of Australia and Papua New Guinea Metals, 1 Aust. Inst. Min. Metall., Monogr. Ser.*, 5: 27–42.

Fisher, R.L., 1974. Pacific-type continental margins. In: C.A. Burk and C.L. Drake (Editors), *The Geology of Continental Margins*, pp.25–41.

Fletcher, K. and Couper, J., 1975. Greenvale nickel laterite, North Queensland. In: C.L. Knight (Editor), *Economic Geology of Australia and Papua New Guinea, 1. Metals, Aust. Inst. Min. Metall., Monogr. Ser.*, 5: 995–1000.

Freeze, A.C., 1966. On the origin of the Sullivan orebody, Kimberley, B.C. In: H.C. Gunning and W.H. White (Editors), *Tectonic History and Mineral Deposits of the Western Cordillera*, Can. Inst. Min. Metall., pp.263–294.

Fusán, O. (Editor), 1963. *Vysvetlivky k prehladnej geologickej mape Č.S.S.R. 1 : 200,000, M-34-XXVII*. Vysoké Tatry, Bratislava, 216 pp.

Fyfe, W.S., 1977. Crustal evolution and metamorphic petrology. *Geol. Surv. Can. Pap.*, 78-10: 1–3.

Fyles, J.T., 1970. Jordan River area. *B.C. Dep. Mines Pet. Resour. Bull.* 57, 64 pp.

Gabelman, J.W., 1976. Strata-bound ore deposits and metallotectonics. In: K.H. Wolf (Editor), *Handbook of Strata-bound and Stratiform Ore Deposits*, 4. Elsevier, Amsterdam, pp.75–164.

Gammon, J.B., 1966. Fåhlbands in the Precambrian of southern Norway. *Econ. Geol.*, 61: 174–188.

Garbar, D.I., Sakhnovskaya, T.P. and Chechel, E.K., 1977. Geologic structure and ore occurrences of the Burakovo–Aganozero massif. *Int. Geol. Rev.*, 20: 637–647.

Glasby, G.P. and Read, A.J., 1976. Deep-sea manganese nodules. In: K.H. Wolf (Editor), *Handbook of Strata-bound and Stratiform Ore Deposits*, 7. Elsevier, Amsterdam, pp.295–340.

Glazkovsky, A.A., Gorbunov, G.I. and Sysoev, F.A., 1977. Deposits of nickel. In: V.I. Smirnov (Editor), *Ore Deposits of the U.S.S.R.*, 2. Pitman, London, pp.3–79.

Goodwin, A.M., 1962. Structure, stratigraphy and origin of iron formation, Michipicoten area, Algoma District, Ontario, Canada. *Geol. Soc. Am. Bull.*, 73: 561–586.

Goodwin, A.M., 1974. The most ancient continental margins. In: C.A. Burk and C.L. Drake (Editors), *The Geology of Continental Margins*, Springer, Berlin, pp. 767–780.

Goodwin, A.M., 1976. Giant impacting and the development of continental crust. In: B.F. Windley (Editor), *The Early History of the Earth*. Wiley, New York, N.Y., pp.77–95.

Goodwin, A.M., Ridler, R.H. and Annels, R.N., 1972. Precambrian volcanism of the Noranda–Kirkland Lake–Timmins–Michipicoten, and Mamainse Point areas, Quebec and Ontario. *24th Int. Geol. Congr., Montreal, Que., Excursion A 40-C 40, Guideb.*, 93 pp.

586

Gordon, Jr. M. and Murata, K.J., 1952. Minor elements in Arkansas bauxite. *Econ. Geol.*, 47(2): 169–179.

Goudarzi, G.H., 1970. Geology and mineral resources of Libya – a reconnaisance. *U.S. Geol. Surv., Prof. Pap.*, 660.

Green, A.R., 1977. The evolution of the Earth's crust and sedimentary basin development. In: J.G. Heacock (Editor), *The Earth's Crust. Am. Geophys. Union Monogr.*, 20: 1–18.

Griggs, A.B., 1945. Chromite-bearing sands of the southern part of the coast of Oregon. *U.S. Geol. Surv. Bull.*, 945-E: 113–150.

Gruner, J.W., 1956. Concentration of uranium in sediments by multiple migration–accretion. *Econ. Geol.*, 51: 495–520.

Hall, J., 1883. Contribution to the geological history of the American continent. *Proc. Am. Assoc. Adv. Sci.*, 31: 29–69.

Harper, G., 1977. Geology of the Sustut copper deposit in British Columbia. *Can. Inst. Min. Metall. Bull.*, Jan. 1977, pp.97–105.

Harrison, J.E., Griggs, A.B. and Wells, J.D., 1974. Tectonic features of the Precambrian Belt Basin and their influence on post-Belt structures. *U.S. Geol. Surv., Prof. Pap.* 866, 15 pp.

Haughton, S.H., 1969. Geological history of southern Africa. *Geol. Soc. S. Afr.*, Cape Town, 535 pp.

Hawkins, Jr., J.W., 1974. Geology of the Lau Basin, a marginal sea behind the Tonga Arc. In: C.A. Burk and C.L. Drake (Editors), *The Geology of Continental Margins*. Springer, Berlin, pp.505–520.

Hawley, J.E., 1962. The Sudbury ores, their mineralogy and origin. *Can. Mineral.*, 7 (Part 1): 1–207.

Heiner, L.E., Wolff, E.N. and Grybeck, D.G., 1971. Copper mineral occurrences in the Wrangell Mountains–Prince William Sound area, Alaska. *Miner. Ind. Res. Lab., Fairbanks, Alaska, Rep.* No. 27.

Hobbs, S.W. and Elliott, J.E., 1973. Tungsten. *U.S. Geol. Surv., Prof. Pap.*, 820: 667–678.

Hoffman, P.F. and Cecile, M.P., 1974. Volcanism and plutonism, Sloan River Map-area (86 K), Great Bear Lake, District of Mackenzie. *Geol. Surv. Can. Pap. 74-1 (Part A)*: 173–176.

Hoffman, P.F., Dewey, J.F. and Burke, K., 1974. Aulacogens and their genetic relation to geosynclines, with a Proterozoic example from Great Slave Lake, Canada. In: R.H. Dott, Jr. (Editor), *Modern and Ancient Geosynclinal Sedimentation. Soc. Econ. Paleontol. Mineral., Spec. Publ.*, 19: 38–55.

Hogue, M., 1977. Petrographic differentiation of tectonically controlled Cretaceous sedimentary cycles, southeastern Nigeria. *Sediment. Geol.*, 17: 235–245.

Höll, R. and Maucher, A., 1976. The strata-bound ore deposits in the Eastern Alps. In: K.H. Wolf (Editor), *Handbook of Strata-bound and Stratiform Ore Deposits*, 5. Elsevier, Amsterdam, pp.1–36.

Horn, D.R. Delach, M.N. and Horn, B.M., 1973. Metal content of ferromanganese deposits of the oceans. *Tech. Rep. Off. Int. Decade Ocean Explor.*, 3, 57 pp.

Hutchinson, R.W., 1973. Volcanogenic sulfide deposits and their metallogenic significance. *Econ. Geol.*, 68: 1223–1246.

Inman, D.L. and Nordstrom, C.E., 1971. On the tectonic and morphologic classification of coasts. *J. Geol.*, 74: 1–21.

Ishihara, S. (Editor), 1974. Geology of Kuroko deposits. *Min. Geol. Tokyo, Spec. Issue* No. 6, 435 pp.

Janković, S., 1972. The origin of base-metal mineralization on the Mid-Atlantic ridge. *24th Int. Geol. Congr., Sect. 4*, pp.326–334.

Jaskolski, S., 1960. Beitrag zur Kenntnis über die Herkunft der Zinnlagerstätten von Gierczyn (Giehren) im Iser-Gebirge, Niederschlesien. *Neues Jahrb. Mineral., Abh.*, 94: 181–190.

Jolly, W.T., 1974. Behaviour of Cu, Zn and Ni during prehnite–pumpellyite rank metamorphism of the Keweenawan basalts, northern Michigan. *Econ. Geol.*, 69: 1118–1125.

Jones, J.G., 1970. Intraglacial volcanoes of the Laugarvatn region, southwest Iceland, II. *J. Geol.*, 78: 127–140.

Katayama, N., Kubo, K. and Hirono, S., 1974. Genesis of uranium deposits of the Tono mine, Japan. In: *Formation of Uranium Deposits*, Int. At. Energy Agency, Vienna, pp.437–452.

Katz, H.R., 1974. Margins of the southwest Pacific. In: C.A. Burk and C.L. Drake (Editors), *The Geology of Continental Margins*, Springer, Berlin, pp.549–565.

Kay, M., 1951. North American geosynclines. Geol. Soc. Am. Mem. 48, 143 pp.

Kay, R., Hubbard, N.J. and Gast, P.W., 1970. Chemical characteristics and origin of oceanic ridge volcanic rocks. *J. Geophys. Res.*, 75: 1585–1613.

Keen, M.J., 1975. The oceanic crust. *Geosci. Can.*, 2: 36–43.

Kerr, P.F., 1940. Tungsten-bearing manganese deposit at Golconda, Nevada. *Geol. Soc. Am. Bull.*, 51: 1359–1390.

Khain, V.Ye., 1971. Regional'naya geotektonika. Nedra, Moscow, 548 pp.

Kholodov, V.N., 1973. *Osadochnoi rudogenez i metallogeniya vanadia.* Nauka, Moscow, 262 pp.

Killeen, P.G. and Richardson, K.A., 1978. The relationship of uranium deposits to metamorphism and belts of radioelement enrichment. *Geol. Surv. Can. Pap.*, 78-1B: 163–168.

Kindle, E.E., 1972. Classification and description of copper deposits, Coppermine River area, District of Mackenzie, *Geol. Surv. Can. Bull.* 214, 109 pp.

King, H.F., 1976. Stratiform and strata-bound metal concentrations in Australia – summary. *Am. Assoc. Pet. Geol. Mem.*, 25: 426–429.

King, P.B., 1969. The tectonics of North America – a discussion to accompany the Tectonic Map of North America, scale 1:5,000,000. *U.S. Geol. Surv., Prof. Pap.* 628, 95 pp.

King, P.B., 1974. Precambrian geology of the United States – An explanatory text to accompany the geologic map of the United States. *U.S. Geol. Surv. Prof. Pap.*, 902, 85 pp.

King, P.B. and Beikman, H.M., 1974. Explanatory text to accompany the geologic map of the United States. *U.S. Geol. Surv., Prof. Pap.* 901, 40 pp.

Knight, C.L. (Editor), 1975a. *Economic Geology of Australia and Papua New Guinea, 1. Metals. Aust. Inst. Min. Metall. Monogr. Ser.,* 5.

Knight, C.L., 1975b. Mount Bischoff tin orebody. In: C.L. Knight (Editor), *Economic Geology of Australia and Papua New Guinea, 1. Metals. Aust. Inst. Min. Metall., Monogr. Ser.,* 5: 591–592.

Kort, V.G. (Editor), 1970. *Tikhii okean – osadkoobrazovaniye v Tikhom okeane, 1.* Nauka, Moscow, 427 pp.

Kraume, E., 1955. Die Erzlager des Rammelsberges bei Goslar. *Beih. Geol. Jahrb.*, 18, 394 pp.

Krauskopf, K.B., 1971. The source of ore metals. *Geochim. Cosmochim. Acta,* 36: 643–659.

Krebs, W. and Wachendorf, H., 1974. Faltungskerne im mitteleuropäischen Grundgebirge – Abbilder eines orogenen Diapirismus. *Neues Jahrb. Geol. Paleont. Abh.*, 147: 30–60.

Kröner, A., Anhaeusser, C.R. and Vajner, V., 1973. Neue Ergebnisse zur Evolution der präkambrischen Kruste im südlichen Afrika. *Geol. Rundsch.*, 62: 281–309.

Laffitte, P. and Rouveyrol, P., 1965. *Carte minière du globe sur fond tectonique au 1:20,000,000.* Bur. Rech. Géol. Min., Paris.

Laznicka, P., 1965. Millerite occurrences in the Bohdalec Formation in Prague. *Časopis Mineral. Geol.*, 10: 281–288.

Laznicka, P., 1970. *Quantitative Aspects in the Distribution of Base and Precious Metal Deposits of the World.* Ph.D., Thesis, Univ. of Manitoba, Winnipeg, Man. (unpublished).

Laznicka, P., 1973. Development of nonferrous metal deposits in geological time. *Can. J. Earth Sci.*, 10: 18–25.

Laznicka, P., 1976. Lead deposits in the global plate tectonic model. In: D.F. Strong (Editor), *Metallogeny and Plate Tectonics, Geol. Assoc. Can., Spec. Pap.* No. 14, pp.243–270.

Lindgren, W., 1933. *Mineral Deposits.* McGraw-Hill, New York, N.Y., 3rd ed. 909 pp.

Losert, J., 1973. Genesis of copper mineralizations and associated alterations in the Jurassic volcanic rocks of the Buena Esperanza mining area (Antofagasta Prov., N. Chile). *Dep. Geol., Univ. Chile, Santiago, Publ.* No. 40.

Mackin, J.H. and Schmidt, D.L., 1956. Uranium- and thorium-bearing minerals in placer deposits in Idaho. *U.S. Geol. Surv., Prof. Pap.* 300: 375–380.

Magnusson, N.H., 1970. The origin of the iron ores in central Sweden and the history of their alterations. *Sver. Geol. Unders. Ser. C,* No. 643, Part I, 127 pp.

Manheim, F.T., 1964. Recent manganese deposits in the Baltic Sea. In: *Program, 1964.* Annu. Meet. Geol. Soc. Am., Miami Beach, Fla., p.127.

Maucher, A. and Höll, R., 1968. Die Bedeutung geochemisch-stratigraphischer Bezugshorizonte für die Altersstellung der Antimonitlagerstätte von Schlaining im Burgenland, Österreich. *Miner-Deposita*, 3: 272–285.

Mayhew, M.A., 1974. Geophysics of Atlantic North America. In: C.A. Burk and C.L. Drake (Editors), *Geology of Continental Margins*, Springer, Berlin, 409–427.

McKee, E.H., 1971. Tertiary igneous chronology of the Great Basin of western United States – implications for tectonic models. *Geol. Soc. Am. Bull.* 82: 3497–3502.

McKnight, E.T. and Fischer, R.P., 1970. Geology and ore deposits of the Picher Field, Oklahoma and Kansas. *U.S. Geol. Surv., Prof. Pap.*, 588, 165 pp.

Menard, H.W., 1964. *Marine Geology of the Pacific*. McGraw-Hill, New York, N.Y., 271 pp.

Mero, J.L., 1965. *The Mineral Resources of the Sea*. Elsevier, Amsterdam, 312 pp.

Mero, J.L., 1977. Economic aspects of nodule mining. In: G.P. Glasby (Editor), *Marine Manganese Deposits*, Elsevier, Amsterdam, pp. 327–355.

Mertie, Jr., J.B., 1976. Platinum deposits of the Goodnews Bay District, Alaska, *U.S. Geol. Surv., Prof. Pap.*, 938.

Milton, Ch. and Fahey, J.J., 1960. Classification and association of the carbonate minerals of the Green River Formation. *Am. J. Sci.*, 258-A: 242–246.

Mitchell, A.H.G. and Bell, J.D., 1973. Island-arc evolution and related mineral deposits. *J. Geol.*, 81: 381–405.

Mitchell, A.H.G. and Garson, M.S., 1972. Relationship of porphyry copper and circum-Pacific tin deposits to palaeo-Benioff zones. *Trans. Inst. Min. Metall. Sect. B*, 81: B 10–B 25.

Mitchell, A.H.G. and Garson, M.S., 1976. Mineralization at plate boundaries. Miner. Sci. Eng., 8(2): 129–169.

Mitchell, A.H.G. and Reading, H.G., 1969. Continental margins, geosynclines and ocean floor spreading. *J. Geol.*, 77: 629–646.

Mitchell-Thome, R.C., 1970. *Geology of the South Atlantic Islands*. Bornträger, Heidelberg, 376 pp.

Morris, R.C., 1974. Sedimentary and tectonic history of the Ouachita Mountains. In: W.R. Dickinson (Editor), *Tectonics and Sedimentation, Econ. Paleontol. Mineral., Spec. Publ.*, 22: 120–142.

Murray, L.G., Joynt, R.H., O'Shea, D.O., Foster, R.W. and Kleinjan, L., 1970. The geological environment of some diamond deposits off the coast of South-West Africa. *Inst. Geol. Sci. London Rep.*, 70/13: 119–141.

Nagibina, M.S., 1967. Tectonic structures related to activation and revivation. *Geotectonics*, 4: 213–218.

Nagornyi, Yu, N. and Nagornyi, V.N., 1976. Geological evolution of the Donetsk Basin. *Geotectonics*, 10: 45–52.

Naldrett, A.J. and Cabri, L.J., 1976. Ultramafic and related mafic rocks: their classification and genesis with special reference to the concentration of nickel sulfides and platinum-group elements. *Econ. Geol.*, 71: 1131–1158.

Narkelyun, L.F., Bezrodnykh, Yu.P. and Trubachev, A.I., 1975. Tectonic position of cupriferous sandstones and shales. *Geotectonics*, 9: 90–96.

Neiheisel, J., 1962. Heavy-mineral investigation of Recent and Pleistocene sands of Lower Coastal Plain of Georgia. *Geol. Soc. Am. Bull.*, 73: 365–374.

Nelson, C.H. and Hopkins, D.M., 1972. Sedimentary processes and distribution of particulate gold in the northern Bering Sea. *U.S. Geol. Surv., Prof. Pap.* 689.

Ney, C.S., Anderson, J.M. and Panteleyev, A., 1972. Discovery, geologic setting and style of mineralization, Sam Goosly deposit, B.C. *Can. Inst. Min. Metall. Bull.*, 65: 53–64.

Nicolini, P., 1970. *Gîtologie des concentrations minérales stratiformes*. Gauthier-Villars, Paris.

Nilsen, T.H. and Clarke, Jr., H., 1975. Sedimentation and tectonics in the early Tertiary continental borderland of central California. *U.S. Geol. Surv., Prof. Pap.*, 925, 64 pp.

Noble, J.A., 1974. Metal provinces and metal finding in the western United States. *Miner. Deposita*, 9: 1–25.

Norton, J.J., 1973. Lithium, cesium, and rubidium – the rare alkali metals. *U.S. Geol. Surv., Prof. Pap.*, 820: 365–378.

Oelsner, O., 1960. Bemerkungen zur Bedeutung von Assimilationsvorgängen bei der Intrusion initialer

Magmen zur Genese oxidischer Geosynklinallagerstätten. *21st Int. Geol. Congr. Norden, Copenhagen*, Part 16, pp.29—42.

Overstreet, W.C., White, A.M., Whitlow, J.W., Theobalt, Jr., P.K., Cuppels, N.P. and Caldwell, D.W., 1968. Fluvial monazite deposits in the south-eastern United States. *U.S. Geol. Surv. Prof. Pap.*, 568.

Park, C.F., 1946. Spilite and manganese problems of the Olympic Peninsula, Washington. *Am. J. Sci.* 244: 305—323.

Parker, Jr., B.H. 1974. Gold placers of Colorado. *Q. Colo. Sch. Mines, Golden, Colo.*, Vol. 69, Nos. 3 and 4.

Patterson, S.H., 1971. Investigations of ferruginous bauxite and other mineral resources of Kauai and a reconnaisance of ferruginous bauxite deposits on Maui, Hawaii. *U.S. Geol. Surv., Prof. Pap.*, 656, 74 pp.

Pavlov, N.V. and Grigor'eva, I.I., 1977. Deposits of chromium. In: V.I. Smirnov (Editor), *Ore Deposits of the U.S.S.R.*, Pitman, London, pp.179—236.

Pavlovskiy, Ye.V., 1975. Origin and evolution of the continental crust. *Geotectonics*, 9: 333—340.

Pereira, J. and Dixon, C.J., 1967. A statistical investigation of mineral occurrences in western Europe. *Proc. 15th Int. Univ. Geol. Congr. 1967, Univ. Leicester, Leicester*, pp.251—270.

Peterson, J.J., Fox, P.J. and Schreiber, E., 1974. Newfoundland ophiolites and the geology of oceanic layer. *Nature (London)*, 247: 194—196.

Peyve, A.V., Yanshin, A.I., Zonenshayn, L.P., Knipper, A.L., Markov, M.S., Mossakovskiy, A.A., Perfil'yev, A.S., Pusharovskiy, Yu.M., Shlezinger, A.Ye. and Shtreys, N.A., 1977. Development of continental crust in northern Eurasia. *Geotectonics*, 11: 309—318.

Phillips, C.H., Cornwall, H.R. and Rubin, M., 1971. A Holocene ore body of copper oxides and carbonates at Ray, Arizona. *Econ. Geol.*, 66: 495—498.

Plumb, K.A. and Derrick, G.M., 1975. Geology of the Proterozoic rocks of the Kimberley to Mount Isa region. In: C.L. Knight (Editor), *Economic Geology of Australia and Papua New Guinea*, 1. Metals. *Aust. Inst. Min. Metall., Monogr. Ser.* 5, pp.217—252.

Pretorius, D.A., 1976. Gold in Proterozoic sediments of South Africa: systems, paradigms, and models. In: K.H. Wolf (Editor), *Handbook of Strata-bound and Stratiform Ore Deposits*, 7. Elsevier, Amsterdam, pp.1—28.

Pustovalov, L.V., 1965. Izuchennost' i geologicheskie perspektivy rudonosnosti osadochnogo chekhla Russkoi platformy. In: *Rudonosnost' Russkoi platformy*, Nauka, Moscow, pp.3—28.

Pyke, D.R., 1975. On the relationship of gold mineralization and ultramafic volcanic rocks in the Timmins area. *Ont. Div. Mines, Misc. Pap.*, 62, 23 pp.

Rackley, R.I., 1976. Origin of western-States type uranium mineralization. In: K.H. Wolf (Editor), *Handbook of Strata-bound and Stratiform Ore Deposits*, 7. Elsevier, Amsterdam, pp.89—157.

Ramović, M., 1968. Principles of metallogeny. Edited at University of Sarajevo, Sarajevo, 265 pp.

Razin, L.V., 1977. Deposits of platinum metals. In: V.I. Smirnov (Editor), *Ore Deposits of the U.S.S.R., III.* Pitman, London, pp.100—124.

Renard, V. and Mascle, J., 1974. Eastern Atlantic continental margins: various structural and morphological types. In: C.A. Burk and C.L. Drake (Editors), *Geology of Continental Margins.* Springer, Berlin, pp.285—291.

Ridler, R.H., 1970. Relationship of mineralization to volcanic stratigraphy in the Kirkland—Larder Lakes area, Ontario. *Geol. Assoc. Can. Spec. Pap.*, 21: 33—42.

Ripley, E.M. and Ohmoto, H., 1977. Mineralogic, sulfur isotope and fluid inclusion studies of the stratabound copper deposits of the Raul Mine, Peru. *Econ. Geol.*, 72: 1017—1041.

Robertson, A.H.F. and Hudson, J.D., 1973. Cyprus umbers: chemical precipitates on a Tethyan ocean ridge. *Earth Planet. Sci. Lett.*, 18: 93—101.

Roethe, G., 1975. Silikatische Kupferlagerstätten in Nord-Chile. *Geol. Rundsch.*, 64: 421—456.

Rona, P.A., 1978. Criteria for recognition of hydrothermal mineral deposits in oceanic crust. *Econ. Geol.*, 73: 135—160.

Ronov, A.B., Migdisov, A.A. and Barskaya, N.V., 1970. Tectonic cycles and regularities in the development of sedimentary rocks and paleogeographic environments of sedimentation of the Russian Platform (an approach to a quantitative study). *Sedimentology*, 16: 137—185.

Roscoe, S.M., 1969. Huronian rocks and uraniferous conglomerates in the Canadian Shield. *Geol. Surv. Can. Pap.,* 68-40, 205 pp.

Ross, D.A., 1974. The Black Sea. In: C.A. Burk and C.L. Drake (Editors), *The Geology of Continental Margins,* Springer, Berlin, pp.669–682.

Rutland, R.W.R., 1973. On the interpretation of Cordilleran orogenic belts. *Am. J. Sci.* 273: 811–849.

Saager, R., 1973. Metallogenese präkambrischer Goldvorkommen in den volkano-sedimentären Gesteinskomplexen (greenstone belts) der Swaziland-Sequenz in Südafrica. *Geol. Rundsch.* 62: 888–901.

Samonov, I.Z. and Pozharisky, I.F., 1977. Deposits of copper. In: V.I. Smirnov (Editor), *Ore Deposits of the U.S.S.R., II.* Pitman, London, pp.106–181.

Sangster, D.F., 1972. Precambrian volcanogenic massive sulphide deposits in Canada: a review. *Geol. Surv. Can. Pap.,* 72-22, 39 pp.

Sawkins, F.J., 1972. Sulfide ore deposits in relation to plate tectonics. *J. Geol.,* 80: 377–397.

Sawkins, F.J., 1976. Massive sulfide deposits in relation to geotectonics. In: D.F. Strong (Editor), *Metallogeny and Global Tectonics.* Geol. Assoc. Can., Spec. Pap., 14: 221–240.

Schneiderhöhn, H., 1955. *Erzlagerstätten.* Fischer, Jena, 3rd ed., 375 pp.

Seely, D.R., Vail, P.R. and Walton, G.G., 1974. Trench slope model. In: C.A. Burk and C.L. Drake (Editors), *Geology of Continental Margins,* Springer, Berlin, pp.249–260.

Semenov, A.I., Staritskii, Yu.G. and Shatalov, Ye.T., 1967. Glavnye tipy metallogenicheskikh provintsii i strukturno-metallogenicheskikh zon na territorii S.S.S.R. *Zakonomern. Razmeshcheniya Polezn. Iskop.,* 8: 55–78.

Shawe, D.R. (Editor), 1976a. Geology and resources of fluorine in the United States. *U.S. Geol. Surv., Prof. Pap.,* 933, 99 pp.

Shawe, D.R., 1976b. Geologic history of the Slick Rock district and vicinity, San Miguel and Dolores Counties, Colorado. *U.S. Geol. Surv., Prof. Pap.,* 576-E, 19 pp.

Shcheglov, A.D., 1967. Osnovnye cherty metallogenii zon avtonomnoi aktivizatsii. *Zakonomern. Razmeshcheniya Polezn. Iskop.,* 8: 95–138.

Shcheglov, A.D., 1971. *Metallogeniya sredinnykh massivov.* Nedra, Leningrad, 148 pp.

Sheridan, R.E., 1974. Atlantic continental margin of North America. In: C.A. Burk and C.L. Drake (Editors), *The Geology of Continental Margins.* Springer, Berlin, pp.391–407.

Siesser, W.G., Scrutton, R.A. and Simpson, E.S.W., 1974. Atlantic and Indian ocean margins of southern Africa. In: C.A. Burk and C.L. Drake (Editors), *The Geology of Continental Margins.* Springer, Berlin, pp.641–654.

Sillitoe, R.H., 1972. Relation of metal provinces in western America to subduction of oceanic lithosphere. *Geol. Soc. Am. Bull.,* 83: 813–818.

Sloss, L.L. and Speed, R.C., 1974. Relationships of cratonic and continental margin tectonic episodes. In: W.R. Dickinson (Editor), *Tectonics and Sedimentation.* Soc. Econ. Paleontol. Mineral., Spec. Publ., 22: 98–119.

Smirnov, V.I., 1962. Metallogeniya geosinklinalei. *Zakonomern. Razmeshcheniya Polezn. Iskop.* 5: 17–81.

Smirnov, V.I., 1968a. The sources of ore-forming material. *Econ. Geol.,* 63: 380–389.

Smirnov, V.I., 1968b. Pyritic deposits. *Int. Geol. Rev.,* 12: 881–908; 1039–1058.

Smirnov, V.I., 1976. Geology of mineral deposits. MIR, Moscow, 520 pp. (in English).

Smirnov, V.I. and Gorzhevsky, D.I., 1977. Deposits of lead and zinc. In: Smirnov (Editor), *Ore Deposits of the U.S.S.R., II.* Pitman, London, pp.182–256.

Smirnov, V.I. and Kazanski, V.I., 1973. Ore-bearing tectonic structures of geosynclines and activized platforms in the territory of the U.S.S.R. *Z. Dtsch. Geol. Ges.,* 124: 1–17.

Smitheringale, W.G. and Peters, H.R., 1974. Volcanogenic copper deposits in probable ophiolitic rocks, Springdale Peninsula. In: D.F. Strong (Editor), *Plate Tectonic Setting of Newfoundland Mineral Occurrences.* Guideb. NATO Adv. Study Inst. Met. Metallog. Plate Tectonics, St.John's, Nfld., pp.95–108.

Solomon, M., 1976. "Volcanic" massive sulphide deposits and their host rocks – a review and an explanation. In: K.H. Wolf (Editor), *Geology of Strata-bound and Stratiform Ore Deposits, 6.* Elsevier, Amsterdam, pp.21–54.

Sørensen, H. (Editor), 1974. *The Alkaline Rocks.* Wiley, New York, N.Y., 622 pp.

Spada, A., 1969. Il giceimento di fluorite e baritina esalative-sedimentarie (facies) lacustre, in tes calate nei sedimenti piroclastici delle zone de Castel Giuliano. In: *Provincei di Roma. Ind. Min.,* 20, 37 pp.

Spizharsky, T.N. (Editor), 1966. *Tectonic Map of the U.S.S.R.* VSEGEI, Moscow, Scale 1:2,500,000, 16 sheets.

Stadnichenko, T. et al., 1953. Concentration of germanium in the ash of American coals – a progress report. *U.S. Geol. Surv., Circ.,* 272, 34 pp.

Staritskiy, Yu.G., 1958. Some features of magmatism and metallogeny of platform regions. *Int. Geol. Rev.,* 5: 402–430.

Stewart, J.H., 1971. Basin-and-Range structure: A system of horsts and grabens produced by deep-seated extension. *Geol. Soc. Am. Bull.,* 82: 1019–1044.

Stille, H., 1936a. Wege und Ergebnisse der geologisch-tektonischen Forschung. *25 Jahr. Kaiser Wilhelm Ges.,* 2: 77–97.

Stille, H., 1936b. Die Entwicklung des amerikanischen Kordillerensystems in Zeit und Raum. *Preuss. Akad. Wiss. Phys.-Math. Kl., Sitzungsber.,* 15: 134–155.

Stille, H., 1940. Einführung in den Bau Amerikas. In: H. Stille, 1964, *Izbrannyie Trudy.* Mir. Moscow, pp.202–274 (translation from the Russian).

Stille, H., 1950. Der "subsequente" Magmatismus. In: H. Stille, 1964, *Izbrannye Trudy.* Mir. Moscow, pp.686–702 (translation from the Russian).

Strong, D.F. (Editor), 1976. Metallogeny and plate tectonics. *Geol. Assoc. Can. Spec. Pap.,* 14, 660 pp.

Suess, E., 1883–1909. *Das Antlitz der Erde,* 1–3. Freytag, Leipzig.

Svenke, E., 1956. The occurrence of uranium and thorium in Sweden. In: *United Nations, Geology of Uranium and Thorium.* Proc. Int. Congr. Peaceful Uses At. Energy, Geneva, 6: 198–199.

Taylor, Jr., H.P., 1967. The zoned ultramafic complexes of southeastern Alaska. In: P.J. Wyllie (Editor), *Ultramafic and Related Rocks.* Wiley, New York, N.Y., pp.97–121.

Tempelman-Kluit, D.J., 1972. Geology and origin of the Faro, Vangorda and Swim concordant zinc–lead deposits, central Yukon Territory. *Geol. Surv. Can. Bull.,* 208, 73 pp.

Thayer, T.P., 1973. Chromium. *U.S. Geol. Surv. Prof. Pap.,* 820: 111–121.

Thompson, R.I. and Panteleyev, A., 1976. Stratabound mineral deposits of the Canadian Cordillera. In: K.H. Wolf (Editor), *Handbook of Strata-bound and Stratiform Ore Deposits,* 5. Elsevier, Amsterdam, pp.37–108.

Tomasson, J., 1976. Hydrothermal alteration in Icelandic geothermal fields. *Visindafélag Isl.,* 5.

Tourtelot, E.B. and Vine, J.D., 1976. Copper deposits in sedimentary and volcanogenic rocks. *U.S. Geol. Surv., Prof. Pap.,* 907-C, 34 pp.

Tremblay, L.P., 1978. Uranium subprovinces and types of uranium deposits in the Precambrian rocks of Saskatchewan. *Geol. Surv. Can. Pap. 78-1A:* 427–435.

Tweto, O., Bryant, B. and Williams, F.E., 1970. Mineral resources of the Gore Range–Eagles Nest Primitive area and vicinity, Summit and Eagle Counties, Colorado. *U.S. Geol. Surv., Bull.,* 1319-C, 127 pp.

Upadhyay, H.D. and Strong, D.F., 1973. Geologic setting of the Betts Cove copper deposits, Newfoundland: an example of ophiolite suite mineralization. *Econ. Geol.,* 68: 162–167.

Varentsov, I.M. and Rakhmanov, V.P., 1977. Deposits of manganese. In: V.I. Smirnov (Editor), *Ore Deposits of the U.S.S.R.,* 1. Pitman, London, pp.114–178.

Viljoen, R.P., Saager, R. and Viljoen, M.J., 1970. Some thoughts on the origin and processes responsible for the concentration of gold in the Early Precambrian of Southern Africa. *Miner. Deposita,* 5: 164–180.

Vokes, F.M. and Gale, G.H., 1976. Metallogeny relatable to global tectonics in southern Scandinavia. In: D.F. Strong (Editor), *Metallogeny and Plate Tectonics. Geol. Assoc. Can. Spec. Pap.,* 14: 413–441.

Volozh, Yu.A., Sapozhnikov, R.B. and Tsimmer, V.A. 1975. Structure of the crust in the Caspian Depression. *Int. Geol. Rev.,* 19: 25–33.

von Bubnoff, S., 1956. *Einführung in die Erdgeschichte.* Akademie-Verlag, Berlin, 3rd ed., 808 pp.

592

Walker, G.P.L., 1960. Zeolite zones and dike distribution in relation to the structure of the basalts of Eastern Iceland. *J. Geol.*, 68: 515–528.

Walker, G.P.L., 1966. Acid volcanic rocks in Iceland. *Bull. Volcanol.*, 29: 375–402.

Walker, G.P.L. and Blake, D.H., 1966. The formation of a palagonite breccia mass beneath a valley glacier in Iceland. *Q. J. Geol. Soc. London*, 122: 45–61.

Ward, P.L., 1971. New interpretation of the geology of Iceland. *Geol. Soc. Am. Bull.*, 82: 2991–3012.

Weissberg, B.G., 1969. Gold–silver ore grade precipitates from New Zealand thermal waters. *Econ. Geol.* 64: 95–108.

Wheeler, J.O. and Gabrielse, H., 1972. The Cordilleran structural province. In: R.A. Price and R.J.W. Douglas (Editors), *Variations in Tectonic Styles in Canada. Geol. Assoc. Can., Spec. Pap.*, 11: 1–81.

White, W.S., 1968. The native copper deposits of northern Michigan. In: J.D. Ridge (Editor), *Ore Deposits in the United States 1933–1967, 1*: 303–326.

Willemse, J., 1969. The geology of the Bushveld igneous complex, the largest repository of magmatic ore deposits in the world. *Econ. Geol., Monogr.*, 4: 1–22.

Williams, F.A. et al., 1956. Economic geology of the decomposed columbium-bearing granites, Jos Plateau, Nigeria. *Econ. Geol.*, 51: 303–332.

Wimmenauer, W., 1966. The eruptive rocks and carbonatites of the Kaiserstuhl. Germany. In: O.F. Tuttle and J. Gittins (Editors), *Carbonatites*, Interscience, New York, N.Y., pp.183–204.

Winslow, A., 1894. Lead and zinc deposits. *Mo. Geol. Surv.*, Rep., 6, 763 pp.

Wolf, K.H., 1976a. Ore genesis influenced by compaction. In: G.V. Chilingar and K.H. Wolf (Editors), *Compaction of Coarse-Grained Sediments, II*, Elsevier, Amsterdam, Ch. 5, pp.475–676.

Wolf, K.H., 1976b. Conceptual models in geology. In: K.H. Wolf (Editor), *Handbook of Strata-bound and Stratiform Ore Deposits, 1*. Elsevier, Amsterdam, Ch. 2, pp.79–110.

Wright, W.B., Guild, P.W., Fish, Jr., G.E. and Sweeney, J.W., 1968. Iron and steel. In: Mineral Resources of the Appalachian Region. *U.S. Geol. Surv. Prof. Pap.*, 580: 396–416.

Wynne-Edwards, H.R., 1972. The Grenville Province. In: R.A. Price and R.J.W. Douglas (Editors), *Variations in Tectonic Styles in Canada. Geol. Assoc. Can., Spec. Pap.*, 11: 263–334.

Zabrodin, V.Yu. and Malyshev, A.A., 1977. A new association of basic-alkaline rocks and carbonatites on the Yenisey Ridge. *Int. Geol. Rev.*, 20: 517–536.

Zapp, A.D., 1965. Bauxite deposits of the Andersonville district, Georgia. *U.S. Geol. Surv., Bull.* 1199-G, 37 pp.

Zurbrigg, H.F., 1963. Thompson mine geology. *Trans. Can. Inst. Min. Metall.*, pp.227–236.

Zweifel, H., 1972. Geology of the Aitik copper deposit. *24th Int. Geol. Congr., Montreal, Que.*, Sect. 4, pp.463–473.